Good Science 10

Second Edition

Authors

Cristy Herron (series editor)

Rebecca Cashmere

Amy Constant

Aleryk Fricker

Peter Saffin

Contributing authors

Emma Craven

Aaron Elias

Investigations reviewer

Carli Hinck

Good Science 10
Second edition

Some content in Good Science 10, Second edition has been sourced and adapted from material contained in titles previously published by Matilda Education: Good Science NSW Stage 4 and Good Science NSW Stage 5 (first published in 2019); Good Science Victoria Year 7, Good Science Victoria Year 8, Good Science Victoria Year 9 and Good Science Victoria Year 10 (first published in 2020); Good Science NSW Stage 4, 2e and Good Science NSW Stage 5, 2e (first published in 2024 and 2025, respectively). The authors of these titles are: Rebecca Cashmere, Amy Constant, Thomas Cosgrove, Emma Craven, Sarah Edwards, Aaron Elias, Aleryk Fricker, Shirley Gilbert, Haris Harbas, Cristy Herron, Kiran Jagpal, Nirupma Kumar, John Ley, Vicki Maggs, Rachael Riley, Peter Saffin, Natalie Stinson.

Publisher: Melinda Schumann

Production manager: Hannah Tatton

Copy editor: Robyn Flemming

Project editor: Siobhan Thomas

Proofreader: Gabrielle Brady

Cover and text design: Paul Ryan

Typesetters: Nikki M Group, Jo-Anne Ridgway

Illustrator: QBS Learning

Indexer: Master Indexing

Warning: It is recommended that Aboriginal and Torres Strait Islander Peoples exercise caution when viewing this publication as it may contain images of deceased persons.

Acknowledgement of Country

We acknowledge the Aboriginal and Torres Strait Islander Peoples of this nation. We acknowledge the Traditional Custodians on whose unceded lands we have created this resource. We pay our respects to ancestors and Elders past and present.

Note about language

Please note that we use the terms 'Aboriginal and Torres Strait Islander' and 'First Nations' interchangeably. We acknowledge that there is no one term that is universally accepted and we use these terms with respect.

First published in 2026 by **Matilda Education Australia**, an imprint of Meanwhile Education Pty Ltd

Melbourne, Australia

Telephone: 1300 277 235

Email: customersupport@matildaed.com.au

Web: www.matildaeducation.com.au

Publication data

Title: Good Science 10, Second edition

ISBN: 9780655093879

A catalogue record for this book is available from the National Library of Australia

Printed in Malaysia by Vivar Printing

November 2025

Contents

Learning Ladder for Victoria Year 10

Steps in progression	Biological science: Genetics	Biological science: Evolution	Chemical science: The periodic table	Chemical science: Chemical reactions	Earth and space science: Climate change	Earth and space science: The universe	Physical science: Motion
5	I can evaluate the implications of technologies that examine or manipulate genetic inheritance	I can evaluate evidence for how the diversity of life has changed over time	I can evaluate how the periodic table has been developed and used over time	I can evaluate the impact our understanding of reactions has in various industries	I can discuss how humans impact Earth's systems and evaluate strategies to mitigate climate change	I can evaluate models and techniques used to explain and explore the origin of the universe	I can evaluate the impact of Newton's laws on our development of technologies and policies
4	I can analyse the outcomes of cell division and inheritance	I can analyse a range of evidence for evolution	I can analyse the properties of atoms based on trends in the periodic table	I can analyse the effect of changing reactant and reaction conditions on chemical reactions	I can analyse interactions between greenhouse gas emissions and energy exchanges within and between Earth's systems	I can discuss the advantages, disadvantages and impacts of space exploration	I can apply Newton's laws to analyse and predict the motion of objects in a system
3	I can explain the processes that underpin heredity and genetic diversity	I can explain the theory of evolution using examples	I can explain relationships between the organisation of the periodic table and atomic structure	I can explain how different factors affect reactions	I can explain patterns of global climate with reference to causal factors	I can describe the scientific evidence for the big bang theory, and methods used in space exploration	I can explain Newton's laws of motion using examples
2	I can describe a range of features related to genetic inheritance	I can describe different types of evidence that support the theory of evolution	I can describe patterns and trends in the periodic table	I can describe and represent changes that occur in a range of chemical reactions	I can describe trends and patterns of global climate	I can sequence key events in the origin and evolution of the universe and space exploration	I can describe the motion of objects using Newton's laws and a range of representations
1	I can identify the features and functions of cells required for cell division and inheritance	I can identify processes involved in natural selection	I can identify characteristics of elements based on their position in the periodic table	I can identify different types of chemical reactions	I can identify and differentiate between the features of Earth's systems	I can identify features of the universe and technologies used for space exploration	I can identify features, properties and laws of motion
Vic V2.0 Codes →	VC2S10U04	VC2S10U05	VC2S10U07	VC2S10U09	VC2S10U11	VC2S10U12 VC2S10U13	VC2S10U17 VC2S10U15

Science understanding

Steps in progression	Nature and development of science	Use and influence of science	Questioning and predicting	Planning and conducting	Processing, modelling and analysing	Evaluating	Communicating
5	I can analyse how advances in technologies enable advances in science	I can analyse the key factors that contribute to scientific knowledge being adopted more broadly by society	I can develop explanatory models when investigating scientific questions, problems and claims	I can evaluate scientific methods regarding safety, ethical and procedural considerations	I can analyse the quality of data using descriptive statistics	I can evaluate the validity and reproducibility of investigation methods	I can justify scientific ideas, findings, arguments and proposals for diverse audiences
4	I can discuss how scientific knowledge is validated, including the role of publication and peer review	I can discuss how scientific information and misinformation may inform personal and social decision-making	I can discuss what is needed for a question to be investigable or a prediction to be reasoned	I can design and conduct reproducible investigations that consider safety, ethical and procedural factors	I can discuss relationships and anomalies that emerge in processed data	I can construct evidence-based arguments to justify conclusions, address ethical issues or assess claims	I can communicate scientific findings and arguments effectively for specific purposes to specific audiences
3	I can explain how science has contributed to developments in technologies and engineering	I can explain how the values and needs of society influence the focus of scientific research	I can develop a hypothesis that predicts the relationship between investigation variables	I can use equipment to generate and record data with precision, to obtain replicable data, using digital tools as appropriate	I can identify and explain trends and/or patterns in a range of dataset representations	I can discuss ways to improve the quality of data and validity of conclusions and claims	I can use digital technologies and/or scientific representations to communicate data and information
2	I can describe how scientific knowledge is refined over time	I can describe scientific knowledge that may be interpreted in different ways	I can formulate questions to investigate scientific problems	I can develop and follow risk assessments that consider safety and ethical issues	I can process data by using mathematical relationships and/or constructing graphs	I can describe the impact of assumptions and errors, and propose ways to reduce them	I can prepare a variety of presentation formats to communicate ideas and findings
1	I can identify technologies that have enabled advances in science	I can identify scientific knowledge that can address socio-scientific issues	I can make simple predictions based on what I know and observe	I can select appropriate equipment to collect precise data for scientific investigations	I can organise data and information using tables, keys and/or models	I can identify assumptions and types of errors in an investigation	I can select appropriate formats, content and vocabulary to communicate scientific ideas and findings
	VC2S10H01 VC2S10H02	VC2S10H03 VC2S10H04	VC2S10I01	VC2S10I02 VC2S10I03	VC2S10I04 VC2S10I05	VC2S10I06 VC2S10I07	VC2S10I08
	Science as a human endeavour		Science inquiry				

Year 10 curriculum correlation grid

CHAPTERS ⟶	1.0 Genetics
Science understanding	
Genetics **VC2S10U04** genetic inheritance involves the function of DNA, chromosomes, genes and alleles, and the roles of mitosis and meiosis in passing on genetic information to the next generation; the principles of Mendelian inheritance can be used to predict ratios of genotypes and phenotypes in monohybrid crosses involving dominant and recessive traits	✓
Evolution **VC2S10U05** the theory of evolution by natural selection includes the processes of variation, isolation and adaptation and is supported by evidence including the fossil record, biogeography and comparative embryology; the theory explains past and present biodiversity and demonstrates how all organisms have some degree of relatedness to each other	
The periodic table **VC2S10U07** the organisation of the elements in the periodic table is related to the structure and properties of atoms; patterns and trends include the significance of rows and periods, metallic and non-metallic properties, atomic size and reactivity	
Chemical reactions **VC2S10U09** chemical reactions include synthesis, decomposition and displacement reactions and can be classified as exothermic or endothermic; reaction rates are affected by factors including temperature, concentration, surface area of solid reactants, and catalysts	
Climate change **VC2S10U11** the dynamics of global climate change can be modelled and explained by examining the interactions between greenhouse gas emissions and energy exchanges within and between Earth's systems; mitigating human-induced climate change requires addressing various activities including power generation, deforestation, manufacturing, transportation, food production and resource consumption	
The universe **VC2S10U12** space exploration seeks to expand knowledge of the origins and structure of the universe and to resolve the challenges of humans travelling and living away from Earth's surface **VC2S10U13** the universe contains features including galaxies, stars, solar systems and black holes; the big bang theory models the origin and evolution of the universe and is supported by evidence	
Motion **VC2S10U15** the law of conservation of energy can be analysed in systems, including Earth systems, by assessing the efficiency of energy inputs, outputs, transfers and transformations **VC2S10U17** Newton's laws of motion can be used to quantitatively analyse the relationship between force, mass and acceleration of objects	

2.0 Evolution	3.0 The periodic table	4.0 Chemical reactions	5.0 Climate change	6.0 The universe	7.0 Motion
✓					
	✓				
		✓			
			✓		
				✓	
					✓

CHAPTERS →	1.0 Genetics
Science as a human endeavour	
Nature and development of science **VC2S10H01** scientific knowledge is contestable and is validated and refined over time through expanding scientific methods, replication, publication, peer review and consensus	✓
VC2S10H02 advances in technologies have enabled advances in science, while science has contributed to developments in technologies and engineering	✓
Use and influence of science **VC2S10H03** the use of scientific knowledge to address socio-scientific issues and shape a more sustainable future for humans and the environment may have diverse projected outcomes that affect the extent to which scientific knowledge and practices are adopted more broadly by society	✓
VC2S10H04 scientific knowledge may be interpreted in different ways by individuals and groups in society; the values and needs of society can influence the focus of scientific research	✓

Science inquiry	
Questioning and predicting **VC2S10I01** investigable questions, reasoned predictions and hypotheses can be used in guiding investigations to test and develop explanatory models and relationships	
Planning and conducting **VC2S10I02** valid, reproducible investigations to answer questions and test hypotheses can be planned and conducted, including identifying and controlling for possible sources of error and bias in sampling or in making observations; safe, ethical investigations include undertaking risk assessments and following protocols when accessing cultural sites and artefacts on Country and Place	
VC2S10I03 equipment can be selected and used to generate and record data sets that show precision, including consideration of sample size and using digital tools as appropriate	
Processing, modelling and analysing **VC2S10I04** data and information can be organised, processed and summarised by selecting and constructing representations including tables, graphs, descriptive statistics, models, symbols, formulas and mathematical relationships	✓
VC2S10I05 information and processed data can be analysed and compared to identify and explain qualitative and quantitative patterns, trends, relationships and anomalies	✓
Evaluating **VC2S10I06** the validity and reproducibility of investigation methods and the validity of conclusions and claims can be evaluated, including by identifying assumptions, conflicting evidence, biases that may influence observations and conclusions, sources of error and areas of uncertainty	
VC2S10I07 arguments based on a variety of evidence can be constructed to support conclusions or evaluate claims, including consideration of any ethical issues and cultural protocols associated with accessing, using or citing secondary data or information	
Communicating **VC2S10I08** communicating and justifying scientific ideas, findings and arguments for diverse audiences involves the selection of appropriate presentation formats, content, scientific vocabulary, conventions, models and other representations, and may include the use of digital tools	✓

2.0 Evolution	3.0 The periodic table	4.0 Chemical reactions	5.0 Climate change	6.0 The universe	7.0 Motion
✓ ✓	✓ ✓	✓ ✓	✓ ✓	✓ ✓	✓ ✓
✓ ✓	✓ ✓	✓ ✓	✓ ✓	✓ ✓	✓ ✓
✓		✓		✓	
		✓ ✓			✓ ✓
	✓ ✓				✓ ✓
✓ ✓	✓ ✓		✓ ✓		
			✓	✓	

1.0 Genetics

All living things on Earth, at a cellular level, are controlled by their genetics. An organism's genetic code, or instructions, is contained within its DNA, which it inherits from its parents. Our DNA determines our physical characteristics, or traits, and even whether we have certain inherited diseases. Almost every individual has a unique genetic code. Scientists use specific methods to sequence and record these codes to help us understand how our bodies work and to track how traits and inherited diseases are passed from person to person.

The Learning Ladder for each chapter maps the Science Understanding, Science as a Human Endeavour and Science Inquiry strands that will be covered. Each ladder has five levels of progression, called steps. To climb the ladders, you need to develop fluency at each step. This will help you develop the ability to complete tasks that are more complex.

Steps in progression	Science understanding: Biological science: Genetics	Science as a human endeavour: Nature and development of science	Science as a human endeavour: Use and influence of science
5	I can evaluate the implications of technologies that examine or manipulate genetic inheritance	I can analyse how advances in technologies enable advances in science	I can analyse the key factors that contribute to scientific knowledge being adopted more broadly by society
4	I can analyse the outcomes of cell division and inheritance	I can discuss how scientific knowledge is validated, including the role of publication and peer review	I can discuss how scientific information and misinformation may inform personal and social decision-making
3	I can explain the processes that underpin heredity and genetic diversity	I can explain how science has contributed to developments in technologies and engineering	I can explain how the values and needs of society influence the focus of scientific research
2	I can describe a range of features related to genetic inheritance	I can describe how scientific knowledge is refined over time	I can describe scientific knowledge that may be interpreted in different ways
1	I can identify the features and functions of cells required for cell division and inheritance	I can identify technologies that have enabled advances in science	I can identify scientific knowledge that can address socio-scientific issues

Figure 1.1: Almost every individual has a unique genetic code. In 2022, two baby Galapagos tortoises were born in Switzerland. One baby was black like its parents, while the other was albino. Albinism is a rare genetic condition where the body produces little or no melanin, the pigment that gives hair, skin and eyes their colour. Genetic differences between individuals can result in unique and beautiful physical characteristics, such as this albino Galapagos tortoise – the only one ever born in captivity.

Science inquiry

Steps in progression	Processing, modelling and analysing	Communicating
5	I can analyse the quality of data using descriptive statistics	I can justify scientific ideas, findings, arguments and proposals for diverse audiences
4	I can discuss relationships and anomalies that emerge in processed data	I can communicate scientific findings and arguments effectively for specific purposes to specific audiences
3	I can identify and explain trends and/or patterns in a range of dataset representations	I can use digital technologies and/or scientific representations to communicate data and information
2	I can process data by using mathematical relationships and/or constructing graphs	I can prepare a variety of presentation formats to communicate ideas and findings
1	I can organise data and information using tables, keys and/or models	I can select appropriate formats, content and vocabulary to communicate scientific ideas and findings

1·1 ▸ Arrangement of genetic information

Learning intention

At the end of this lesson, I will be able to:

- identify that all organisms contain a genetic code
- compare DNA, chromosomes and genes, in relation to their roles in genetics.

Key terms

allele: a variation of a gene

chromosome: a tightly coiled strand of DNA

deoxyribonucleic acid (DNA): the carrier of genetic information; located in the nucleus of a cell

double helix: the structure of a DNA molecule; a double-stranded spiral

gene: a segment of DNA, the basic functional unit of heredity

genome: an organism's entire sequence of DNA

genotype: the genetic code for an organism

heredity: the passing on of traits from parents to their offspring

nucleotide: the building block of DNA, consisting of a sugar, a phosphate and a nitrogenous base (adenine, guanine, thymine or cytosine)

phenotype: how the genotype is physically expressed

plasmid: a small circular DNA molecule in bacteria and some other microscopic organisms

Investigation 1.1

Extracting DNA from strawberries, p. 310

Key idea: Form and function

Your genome is a blueprint unique to you. It contains all the instructions for your genetic traits, passed down from your mother and father. This blueprint is arranged in three distinct levels: genes, which are sections of DNA, which is stored as chromosomes.

DNA is made up of nucleotides

Our **deoxyribonucleic acid (DNA)** contains all our biological instructions. Located in the nucleus of nearly every cell, DNA is in the shape of a twisted ladder called a **double helix**. The sides of the 'ladder' are the backbone of DNA and are made up of units called **nucleotides**. Each nucleotide consists of a phosphate group, a five-carbon sugar and a nitrogen-containing base. In DNA, there are four nitrogenous bases – adenine, thymine, guanine and cytosine. They form pairs to make the 'rungs' of the ladder. The bases only partner in specific pairs: adenine with thymine, and guanine with cytosine.

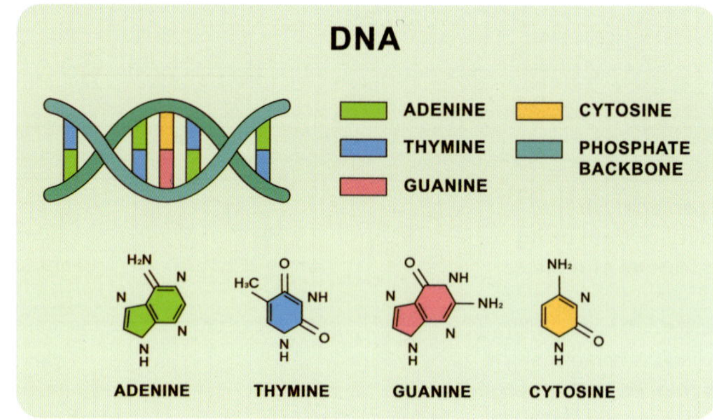

Figure 1.2: DNA is made up of nucleotide bases. Collections of these bases make up genes that code for proteins.

Genes are segments of DNA

One section of DNA is called a **gene**. A gene is the basic functional unit of **heredity**. Genes are how traits are passed down from a mother and a father to their offspring. Every human has two copies of each gene – one from their mother and one from their father.

Some genes code for proteins, which are the basic building blocks for our bodies. They:

- are used to repair and make new tissue
- make up much of our skin, hair, muscle and bone
- make up the chemical messenger system of hormones, blood and antibodies.

Other genes code for a variety of molecules with different functions. Some genes are common (such as those for brown eyes) and some are rare (such as those for blue eyes). Variations of genes that control the same trait are called **alleles**.

Genes can vary in size – from a few hundred base pairs to millions of base pairs in length. Your genes make up your **genotype** – the genetic code for a specific gene or genes – and determine your unique genetic identity (your **genome**). The way the genes are expressed is known as your **phenotype**, which can be visible traits such as height, hair colour or eye colour, or characteristics such as blood type. Differences in phenotype are a result of the alleles people inherit.

Condensed chromosomes contain tightly coiled DNA

In cells, DNA often exists as disorganised long pieces. Under a microscope, DNA can look like a bunch of squiggles. At various stages (especially during cell division), it becomes tightly coiled and packed. This packaged DNA is known as a **chromosome**.

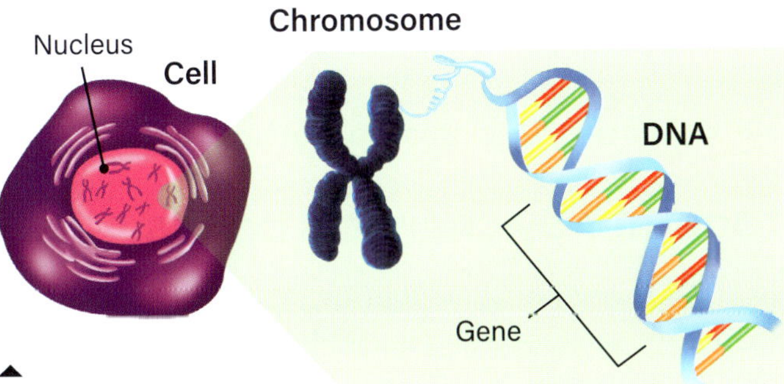

Figure 1.3: Cells contain chromosomes (made up of DNA) that contain our genes. The totality of all our genes is known as our genome.

Humans normally have 23 pairs of chromosomes (46 in total). Each chromosome is made up of two identical chromatids. The two chromatids are joined so that they have short and long arms. Each chromatid is made up of one long DNA molecule that has been tightly coiled and packed into the small structure. To ensure the DNA is effectively coiled and packed, it is wrapped around small protein balls called histones.

All organisms have a genetic code

All living things, whatever their size and shape, have a genetic code, containing specific genes that code for particular traits. These genes are linked as sections of DNA, which are stored in chromosomes.

In complex eukaryotic organisms, such as humans, condensed chromosomes typically form an 'X' shape, whereas in simple prokaryotic organisms, the chromosomes are circular and are called **plasmids**.

Learning Ladder

Genetics

1. Identify the main components in the structure of DNA.
2. Describe the relationship between chromosomes, DNA, genes and alleles.
3. Explain how the structure of genetic materials is related to the traits expressed.

Nature and development of science

1. Identify a technology that has helped scientists to view the structure of DNA.
2. Describe an example of how our knowledge of DNA changed when researchers were able to study the DNA of prokaryotic organisms. (Prokaryotic DNA is 'free floating' and is arranged in circles within the organism.)
3. Explain how knowledge of DNA, genes and chromosomes may contribute to scientific advances in treatments for diseases such as those that affect our physical characteristics.

Communicating p. 262

1. Identify an appropriate way to communicate information about the structure of genes and chromosomes.
2. Select one type of genetic material from this section and construct a labelled three-dimensional representation of it.
3. Construct a digital diagram to demonstrate the relationship between DNA and the phenotype, including as many steps in between as possible.
4. Based on the information above, write a TEEL essay explaining the importance of DNA in living things. For assistance with this, refer to 'Writing evidence-based essays' in the Science how-to section on page 280.
5. Justify the argument you put forward in your answer to Question 4.

Key idea: Form and function

In eukaryotic cells, chromosomes are present within the nucleus. Propose reasons why the form of chromosomes is useful to the function of the DNA it organises.

Success criteria

- I can define the terms 'DNA', 'chromosome', 'gene' and 'allele'.
- I can compare DNA, chromosomes and genes in relation to their roles in genetics.

1·2 ► The structure and function of DNA

Learning intention

At the end of this lesson, I will be able to explain how the structure of DNA is related to the functions it performs.

Key terms

amino acid: an organic molecule that is the building block of a protein

enzyme: a protein that increases the rate of a specific chemical reaction in the body

nitrogenous: containing nitrogen

proteins: long polymer chains made up of amino acid monomers

ribonucleic acid (RNA): a nucleic acid similar to DNA that helps to produce proteins from the genes of DNA

transcription: the process in which a cell makes an RNA copy of a piece of DNA

Key idea: Form and function

Long strands of DNA are arranged very specifically, which is important for how we function, repair ourselves when we get injured, and grow.

DNA is arranged as a double helix

The double helix of DNA is composed of two strands of DNA, each with a sugar–phosphate backbone. **Nitrogenous** bases in the nucleotides on one strand align with their complementary bases on the other strand and join together: adenine with thymine, and cytosine with guanine. A nucleotide consists of a sugar molecule, a phosphate molecule and a single nitrogenous base. In this way, the order of nucleotides on the DNA strands forms specific genes, which code for many different proteins.

Another important molecule is **ribonucleic acid (RNA)**. RNA is very similar to DNA – it also contains nitrogenous bases – but RNA exists as single strands, not double ones. Also, instead of thymine, RNA contains a similar nitrogenous base called uracil. RNA helps DNA to replicate and to use the genetic information in DNA to produce **proteins**.

DNA is held together by chemical bonds

The DNA double helix is held together by different types of chemical bonds. The sugar–phosphate backbone and nucleotides are chemically bonded by covalent bonds, which are strong enough to hold together the nucleotides during replication and protein synthesis. However, the bonds between the nitrogenous bases in the middle of the ladder are weaker hydrogen bonds. This allows the double helix to be opened, so that its genetic codes can be read during protein synthesis.

Figure 1.4: DNA is a double helix, which is composed of two complementary strands that join together to form a 'ladder'.

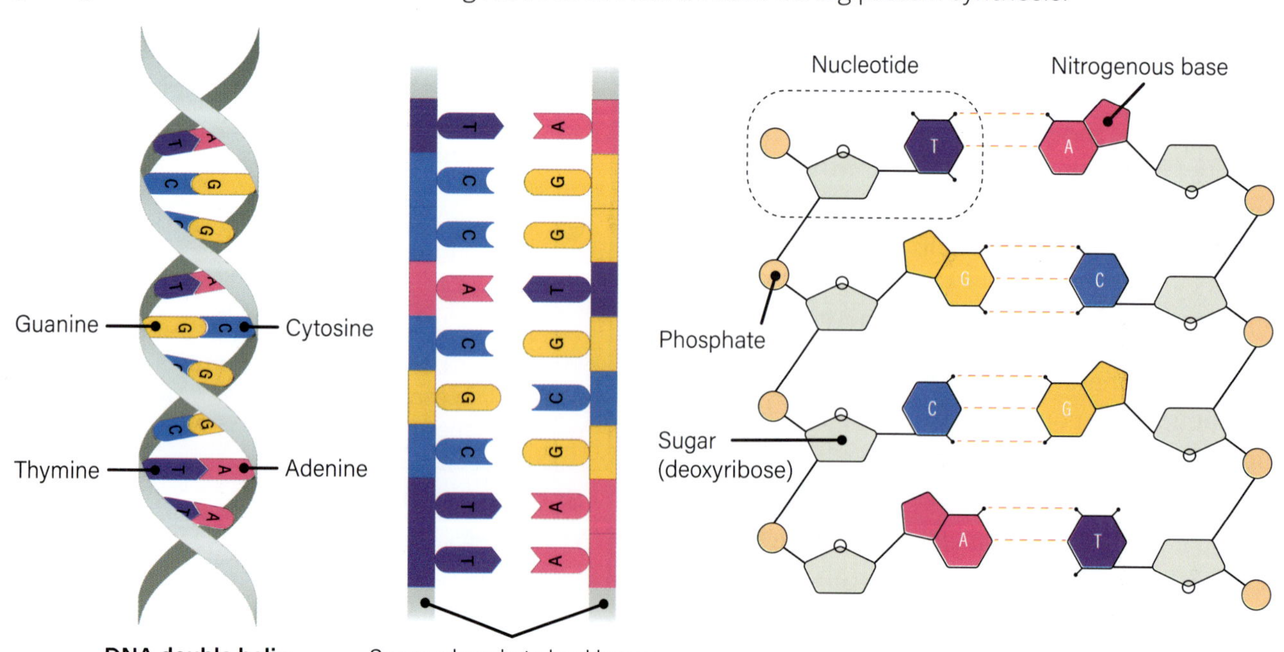

Replicating DNA forms complementary strands

DNA is replicated so that new cells can form. The following steps occur (Figure 1.5).

1 The chromosome containing the DNA 'unwinds', so that the DNA is in long strands.
2 A section of DNA is 'unzipped' by an **enzyme** called DNA helicase, so that the nucleotides on both sides are available.
3 New nucleotides are matched to the existing ones on both sides of the DNA, by another enzyme called DNA polymerase.
4 Each of the new strands is 'glued' back together, making two identical double helixes from the existing one. These are called 'daughter molecules'.

The two new strands of DNA eventually condense into chromosomes. When the cell divides, each of the two new cells receives an identical copy of the replicated DNA. This means that our genetic code is passed on to all new cells.

Figure 1.5: In DNA replication, the existing strands of DNA act as a template to make new complementary ones.

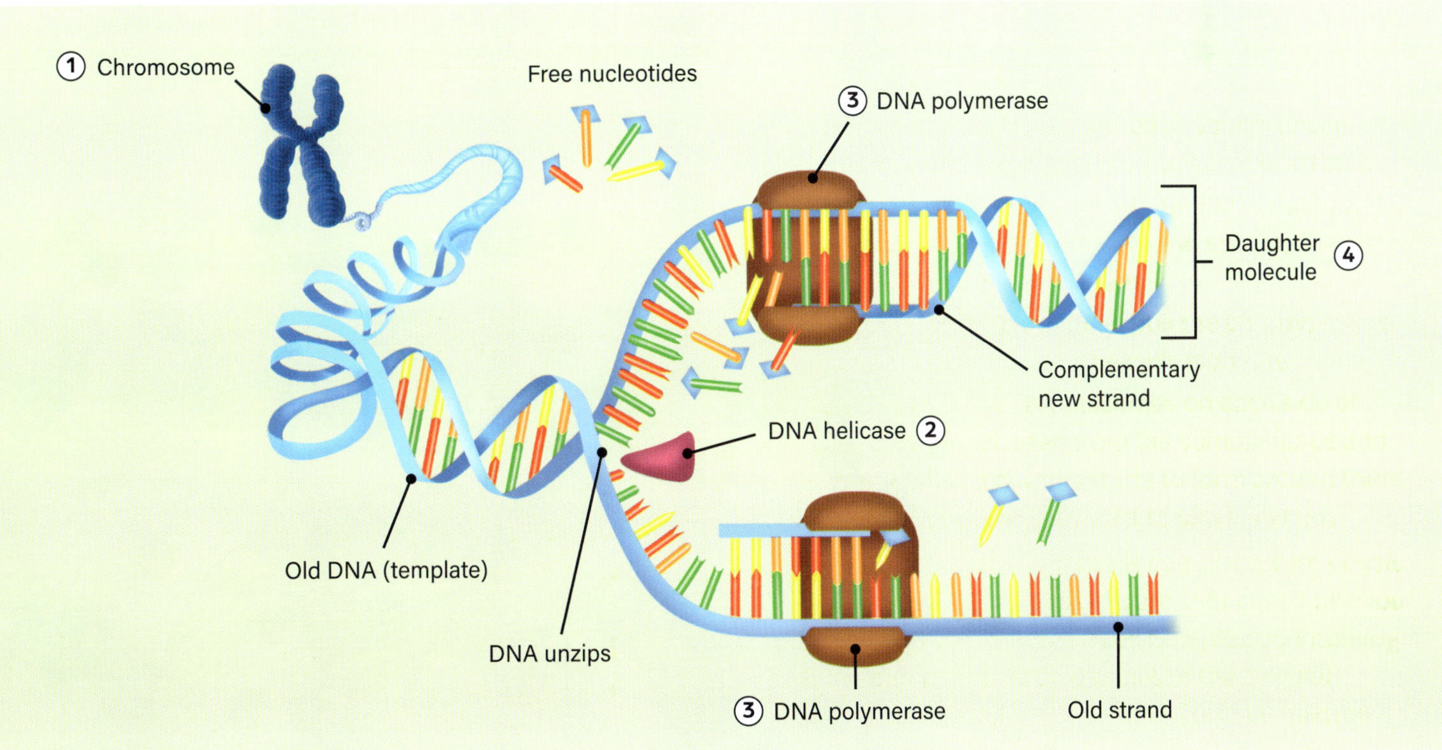

Genes on DNA are used to produce proteins

When cells are not preparing to replicate, DNA has another role – protein production. This is done through two processes: transcription and translation.

During **transcription**, a section of the DNA strand that contains a specific gene unwinds. (This is different from replication, in which the whole length of the DNA strand unwinds.) Genes are specific sequences of nitrogenous bases.

A special type of RNA called messenger RNA (mRNA) reads the unwound section of nucleotides like a barcode and copies the nitrogenous bases that make up the gene. This is called a complementary strand, because, as each base is read, the RNA adds the opposite, matching the base to its chain. The bases are summarised in Table 1.1.

Table 1.1: Complementary DNA and RNA bases

DNA base		RNA base
Adenine (A)		Uracil (U)
Thymine (T)	pairs with	Adenine (A)
Cytosine (C)		Guanine (G)
Guanine (G)		Cytosine (C)

The strand of mRNA then heads to the ribosome, the cell organelle responsible for protein production. Here, the new code is read like a barcode a second time, translating it from nitrogenous bases into a chain of molecules called **amino acids**. These bond together, forming one long chain – a polymer. This step is called translation. The chain is then folded, using complex processes, becoming the final protein we need in our bodies. The longest protein in our bodies is a polymer made up of 34 000 different molecules.

Proteins have many functions

Proteins have many important functions in the body. Many cellular processes rely on specific proteins. These proteins are often in the form of enzymes and assist processes by speeding up important chemical reactions.

Examples of enzymes are amylase, which helps break down starches when we eat food, and lactase, which breaks down lactose (a sugar in milk), in our stomachs. Only some people have the gene for lactase. People who have it can drink milk without any ill effects, but those who lack the gene are lactose intolerant – and milk makes them sick!

Proteins are also important for growth and development. Muscles are mostly made up of protein, which needs to be continually produced to support growth. Proteins can also help us to heal when we are injured. If you cut yourself, the bleeding is stopped by specific blood cells called platelets (Figure 1.7a). Platelets contain special proteins that allow them to stick to the edges of wounds, and to one another, forming clots and allowing us to heal.

Another important role of proteins is in reproduction. Proteins allow humans to produce hormones such as oestrogen and progesterone, which are essential during pregnancy to enable us to produce offspring. Without these proteins, humans would cease to exist.

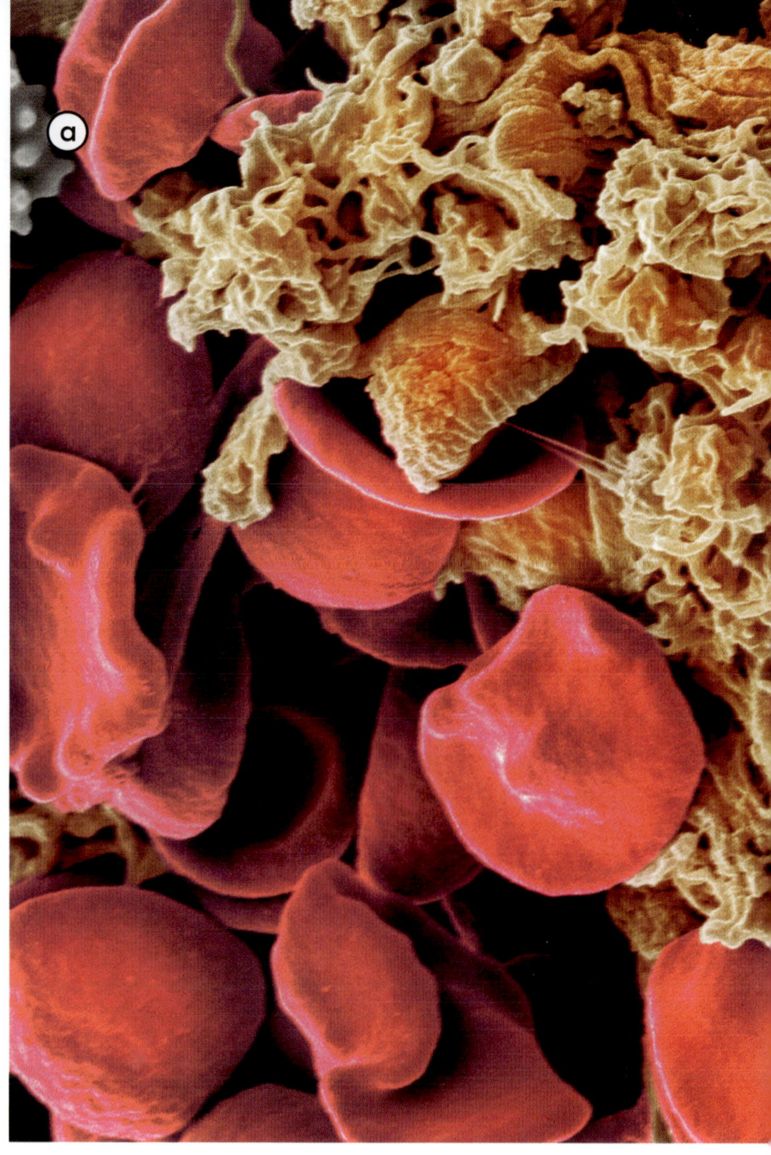

Figure 1.7: Proteins are important for many processes in the body. Platelets (a) allow blood to clot, while keratin makes up fingernails and hair, as seen on the micrograph of a human hair (b).

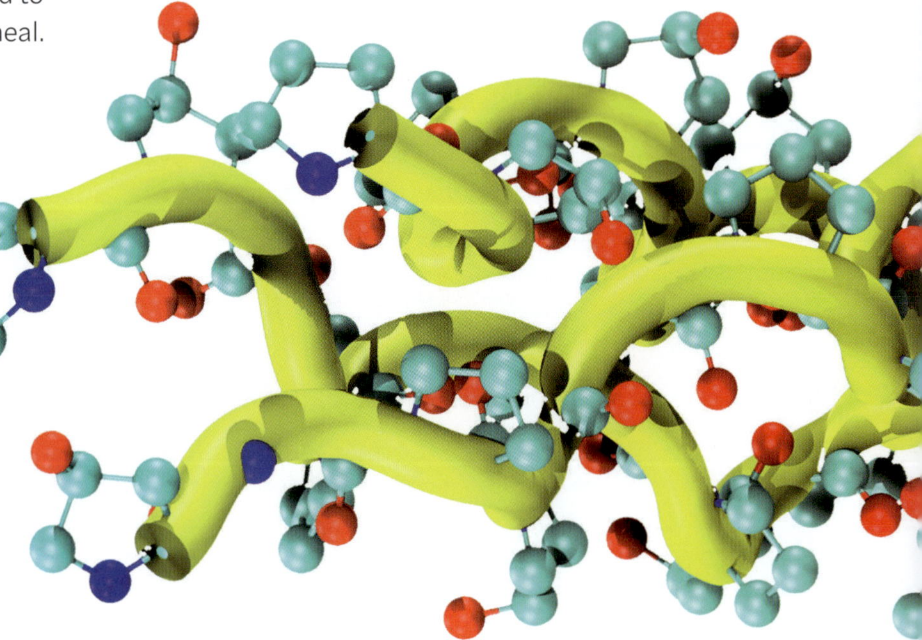

Figure 1.6: Proteins are polymers, meaning they are made of up of many units. The 3D structure shown here is a representation of collagen – a complex protein polymer made up of lots of chains.

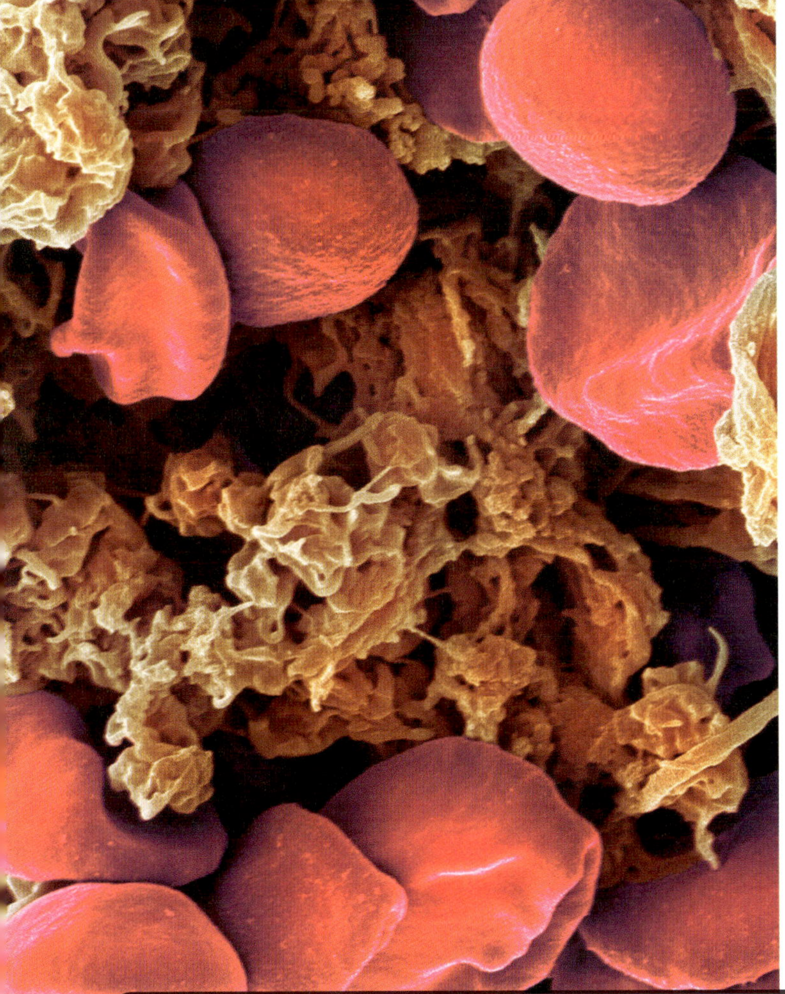

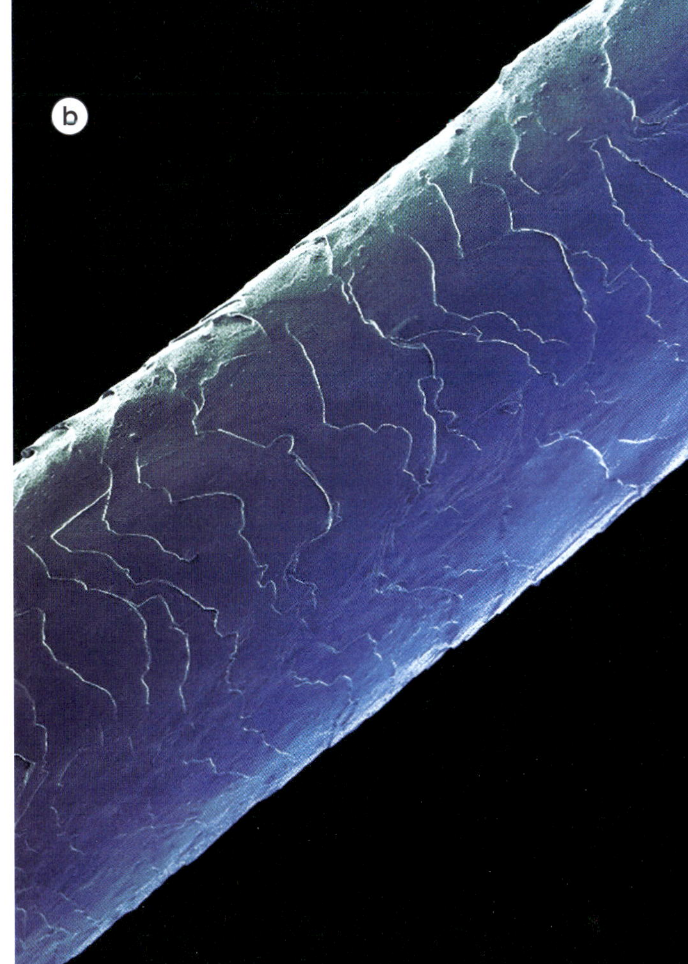

Learning Ladder

Genetics

1. Identify three features of a DNA strand.
2. Describe the arrangement of these components in a DNA strand.
3. Explain the relationship between the structure of DNA and the production of proteins.
4. Discuss the outcome of DNA replication on our genetic material.
5. Evaluate the impact of protein production on the traits and characteristics we possess.

Nature and development of science

1. Identify the technology used to produce the images seen in Figure 1.7 (micrographs).
2. Describe a way that understanding protein production assists scientists in refining our understanding of the different uses of proteins.
3. Explain how understanding DNA replication may help scientists to develop new technologies relating to our genetics.
4. Discuss how the process of DNA replication would need to be reviewed by the scientific community for the information to be considered valid.

Processing, modelling and analysing p. 244

Scientists mapped the DNA of a number of dogs and determined that, on average, 24 per cent of their total nitrogenous bases were adenine, while 26 per cent were guanine.

1. Considering that DNA bases match in set pairs, construct a table comparing the average percentage of all nitrogenous bases in the dogs that were studied.
2. Convert your table from Question 1 into a pie chart. Refer to 'Constructing pie charts' in the Science how-to section on page 297.
3. Explain the pattern in percentages for the nitrogenous bases mapped by scientists.

Key idea: Form and function

Our bodies require hundreds of thousands of proteins to function effectively. Using information from secondary sources, identify three common proteins in our bodies and explain how their form relates to their function.

Success criteria

- I can describe the structure of DNA.
- I can relate the structure of DNA to the processes of replication and protein production.

1·3 ► Cell division basics

Learning intention

At the end of this lesson, I will be able to describe the importance of cell division.

Key terms

cell cycle: the process a cell goes through each time it divides

diploid cell: normal body cells that contain a complete set of chromosomes (2*n*)

haploid cell: sex cells that contain half a complete set of chromosomes (*n*)

Key idea: Form and function

One of the components of cell theory states that 'all cells come from pre-existing cells'. The process that allows this to happen is called cell division, which all cells undergo to ensure that genetic materials can be replicated and passed on.

Cell division is a part of the cell cycle

The **cell cycle** is a continually repeating process that allows cells to replicate their DNA, ready to divide. This replication occurs during the main phase of the cell cycle – interphase – alongside other cellular processes that help the cell to carry out its important functions. Figure 1.8 shows the cell cycle, including the four phases of mitosis.

Cell division produces new cells

After interphase, cells move into one of two types of cell division – mitosis or meiosis. Mitosis, which is carried out by normal body cells, produces genetically identical, or **diploid**, cells. This means that they contain a full set of chromosomes for the organism. Meiosis produces genetically unique sex cells or gametes. These cells are **haploid** cells; they contain half of the total number of chromosomes of a normal body cell, so that when combined in reproduction, the resulting offspring will be diploid.

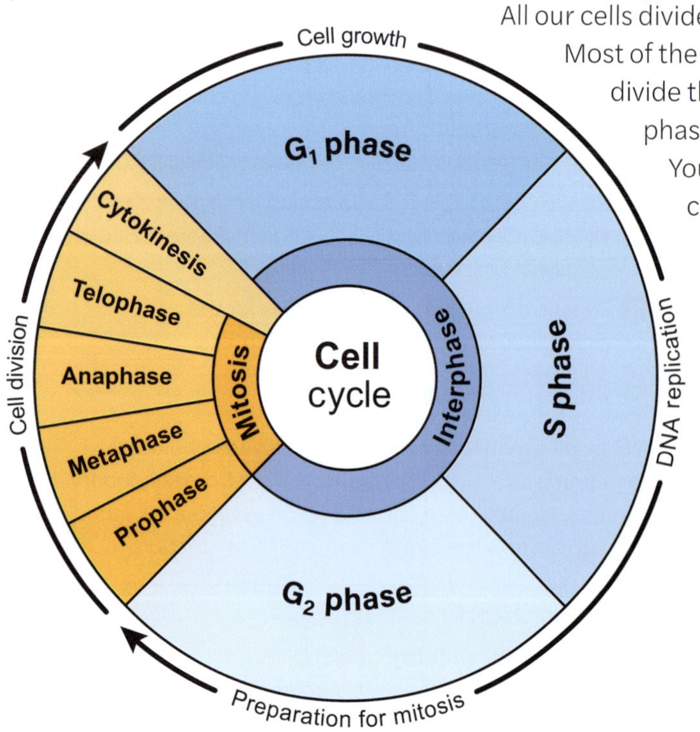

Cell division occurs by mitosis or meiosis

All our cells divide so that we can function as a living organism. Most of the cells of living things, including our body cells, divide through the process of mitosis, which has four phases: prophase, metaphase, anaphase and telophase. You will learn more about these phases across the coming chapter sections. Mitosis is just one stage of the cell cycle.

Unlike other body cells that divide once, gametes divide twice in a process called meiosis. This extra division produces sex cells that have half the usual number of chromosomes found in other cells of living things. The phases of meiosis are similar to mitosis – prophase, metaphase, anaphase and telophase – but they are repeated twice! The first round is called meiosis I, and the second meiosis II. These phases will be discussed in more detail later in the chapter. The first round of division occurs after interphase in the cell cycle.

Figure 1.8: The cell cycle shows all the processes a normal body cell goes through, including mitosis.

Cell growth
G_1 phase
S phase
DNA replication
G_2 phase
Preparation for mitosis
Cell division
Cytokinesis
Telophase
Anaphase
Metaphase
Prophase
Mitosis
Interphase
Cell cycle

Without cell division, organisms would not exist

The processes of mitosis and meiosis are extremely important to all living things. They allow genetic material to be passed from cell to cell as they are produced, ensuring that genes can be inherited – that is, passed from parents to offspring. Without these important processes, living things would cease to exist, as there would be no way for genetic information to be passed on.

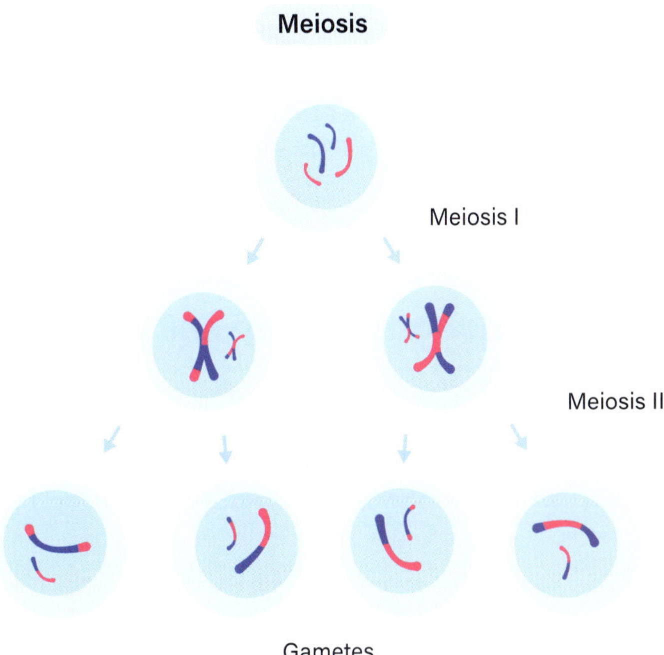

Figure 1.9: Meiosis produces four haploid cells from one parent cell.

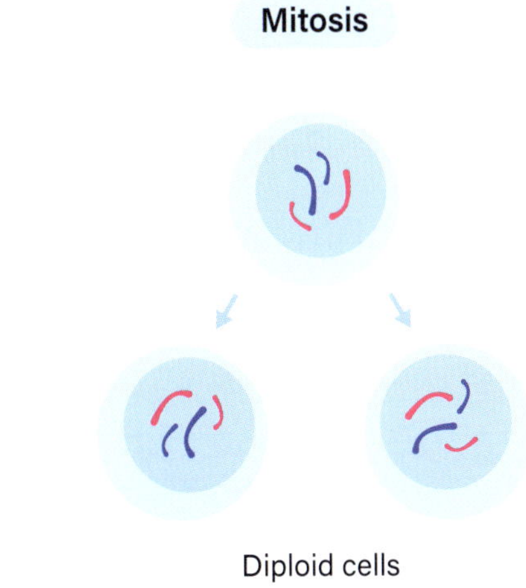

Figure 1.10: Mitosis produces two identical diploid cells from a single parent cell.

Learning Ladder

Genetics

1. Identify the components of the cell cycle.
2. Describe why interphase (the main phase of the cell cycle) is important for our genetics.
3. Explain how meiosis is important to inheritance.

Nature and development of science

When attempting to view cells undergoing division, scientists often use fluorescence microscopy – dyeing cells a bright colour so they can be easily seen and then studying them to isolate visual representations of each of the stages of mitosis and meiosis.

1. Identify the primary technology used to view cells undergoing cell division.
2. Suggest a reason why visualising each stage of division would help scientists to refine their understanding of cell division.
3. Explain how microscopic processes such as cell division have contributed to scientists' development of new technologies, such as the one you identified in Question 1.
4. Discuss why peer review of the stages of cell division would be important when observing and comparing cells under a microscope.
5. Analyse how the advancement you identified in Question 3 has enabled scientists to progress their knowledge of cellular processes.

Communicating p. 262

1. Select an appropriate way to communicate information about the cell cycle and the size of each of its stages.
2. Construct your identified method of communication from Question 1.
3. Construct a digital diagram of mitosis, to show the start and end results of the process.

Key idea: Form and function

Interphase is important not only to cell division, but also to many cellular processes, such as the production of proteins. Explain why interphase needs to occur for much longer periods of time than mitosis.

Success criteria

- I can identify the different types of cell division.
- I can describe the cells produced by cell division and explain their importance.

1·4 ► Cell division: mitosis

Learning intention

At the end of this lesson, I will be able to explain the process of mitosis.

Key terms

cytokinesis: the last stage of mitosis, where a cell divides into two identical daughter cells

mitosis: simple cell division that produces identical cells

Key idea: Form and function

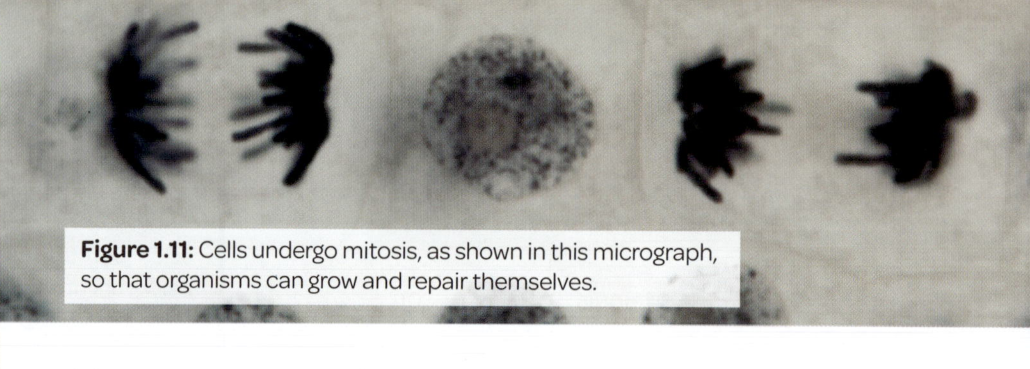

Figure 1.11: Cells undergo mitosis, as shown in this micrograph, so that organisms can grow and repair themselves.

A key component of reproduction is cell division. For organisms that reproduce asexually, this cell division typically occurs as mitosis. **Mitosis** is the division of one cell into two, genetically identical cells.

Mitosis passes chromosomes from cell to cell

All our cells divide so that we can function as a living organism. For our body cells, this division occurs through the process of mitosis. Mitosis is one stage of the cell cycle, which occurs in every cell in our body apart from the sex cells. During the main part of this cycle – interphase – the cell produces the substances needed and does its required job in the body. It also replicates all the DNA in its chromosomes. This is important, as otherwise the new cells would not contain the correct genetic material to carry out their functions! Mitosis passes genetics from cell to cell.

There are four phases of mitosis

Once the cell cycle reaches the stage of mitosis, the cell can divide. This happens as four distinct phases.

1 **Prophase:** The chromosomes condense (coil up), and the nucleus breaks down. Special spindle fibres start to develop.
2 **Metaphase:** The chromosomes line up across the equator (middle) of the cell, and spindle fibres attach to either side of them.
3 **Anaphase:** The spindle fibres pull the chromosomes away from the centre towards either side of the cell, and the cell elongates (becomes longer).
4 **Telophase:** The chromosomes on each side of the cell are released from the spindle fibres and unwind, while a nucleus starts to re-form around them. The middle of the cell starts to pinch inwards.

After telophase, the cell splits in half, down the middle. This splitting of the cell is known as **cytokinesis**. Cytokinesis is the conclusion of cell division, as the one parent cell has now completely divided into two identical daughter cells.

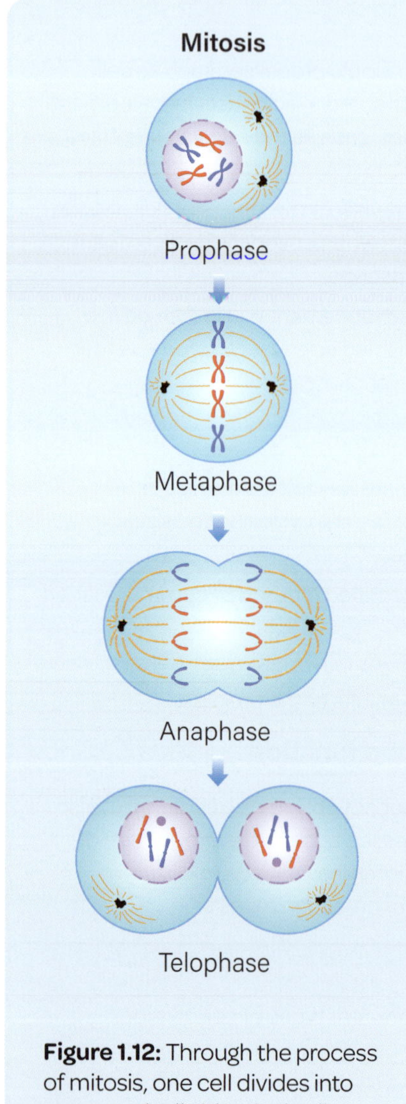

Figure 1.12: Through the process of mitosis, one cell divides into two genetically identical cells.

Growth and repair occur because of mitosis

The process of mitosis is very important because it allows for both growth and repair to occur. Growth happens when body tissues increase in size. This is not due to cells getting larger but to cells dividing to create more cells. As the number of cells in a tissue increases, the size of the tissue also increases, causing the organism to grow.

The growth of an organism is influenced by many factors, but its overall height and structure are determined by its genes. This is the reason humans do not grow as large as elephants.

Cells are constantly dividing to replace cells that are damaged or old. Cell division ensures that cells in the body are healthy and can perform the functions needed for survival. Some cells, such as those in the stomach and intestines, only last a few days because they are exposed to conditions that damage them. Others, such as liver cells, last much longer because they are less likely to be damaged.

Not all body cells can be replaced. For example, nerve and heart muscle cells are unable to divide, so any damage to these cells can be permanent.

Uncontrolled mitosis causes cancer

Our cells contain special genes that control how often they divide by mitosis. If these genes are altered in any way (this is called mutation and will be covered later in the chapter), the cell's instructions for division can be removed entirely. This causes the cell to mutate and divide much more rapidly than normal, forming dangerous tumours which we know as the disease cancer. Changes to these genes are the reason that cancers occur in our bodies

Figure 1.13: This image taken using a microscope shows onion cells during the process of mitosis.

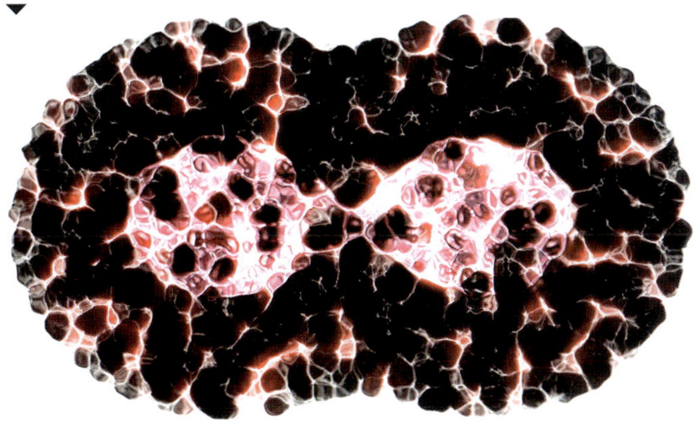

Learning Ladder

Genetics

1. Identify the phases of mitosis.
2. Describe the functions that mitosis allows our body cells to carry out.
3. Explain how mitosis allows us to continue to use our genetics within our bodies.
4. Analyse the outcome of mitosis in relation to inheritance between parents and offspring.
5. DNA profiling is a technology that can map the sequence of genes present within the DNA from our body cells. Evaluate the implications of DNA profiling for our ability to examine our genetic code.

Use and influence of science

1. Identify an issue that our knowledge of mitosis may help to address. (*Hint:* Think about our ability to examine DNA.)
2. Describe a way that the phases of mitosis may be interpreted differently by different scientists when the cell is completing its final separation.
3. Explain how genetic diseases contained within cells may influence the amount of research conducted into mitosis.

Processing, modelling and analysing p. 244

When a single body cell divides, it produces two identical daughter cells. Those cells, after interphase, can then produce four identical body cells, which can then produce eight, and so on.

1. Construct a table to show how many identical cells could be produced after six rounds of mitosis have been completed in a specific area of the body.
2. Convert your table into a scientific graph. (Refer to 'Graphing' in the Science how-to section on page 294.)
3. Explain the trend that appears in this data.
4. Discuss why trends like this would be important within our body when we are injured and need to replace large quantities of cells in order to heal.

Key idea: Form and function

Different cells within the body last for different amounts of time before they need to be replaced. Explain how the location and form of these cells can influence their function, and the reason for these time differences.

Success criteria

- I can outline the phases of mitosis.
- I can explain how mitosis passes genetics between cells.

1·5 ▸ Cell division: meiosis

Learning intention

At the end of this lesson, I will be able to explain the process of meiosis.

Key terms

crossing over: the exchange of genes by homologous chromosomes

gamete: a sex cell – an ovum or a sperm

homologous: chromosomes that have the same genes in the same order

meiosis: complex cell division that produces unique haploid gamete cells

Key idea: Form and function

You started your life when an egg cell was fertilised by a sperm cell. The fertilised cell divided into two, then four, then eight and so on, until you eventually became the approximately 32.7 trillion cells that you are today. None of this would have been possible without meiosis. **Meiosis** is cell division that specifically produces gametes, which are essential for sexual reproduction. They are also the reason you look different from your parents!

Meiosis is different from mitosis

While both mitosis and meiosis are forms of cell division, they occur within different cells and result in different outcomes. Mitosis produces two identical, diploid daughter cells from a single parent cell (as you learnt in Section 1.4). However, unlike other body cells, which divide once, sex cells (**gametes**) divide twice in a process called meiosis. This extra division produces sex cells that have half the usual number of chromosomes found in other cells of the body, making them haploid cells.

Chromosomes rearrange during meiosis

Another important factor in the process of meiosis that occurs at the start of the division is a process called **crossing over**. As the chromosomes of the cell condense, they match up as homologous pairs (see Figure 1.14). **Homologous** chromosomes contain the same combination of genetic material within the same location. The human karyotype is made up of 22 pairs of homologous chromosomes and one pair of sex chromosomes. While these chromosomes contain the same genes, they can have different variations of these genes – different alleles – meaning they are not identical.

Once matched up, each of the homologous pairs undergoes crossing over: random segments, or genes, on the chromosomes break off and switch places. This 'mixes up' the genes present on each of the chromosomes, making them unique. Division then continues, as outlined on the opposite page.

Figure 1.14: Pairs of homologous chromosomes undertake crossing over during the first round of meiosis, swapping random sections of their genetic material. Each gamete ends up with a unique combination of genetics. ▼

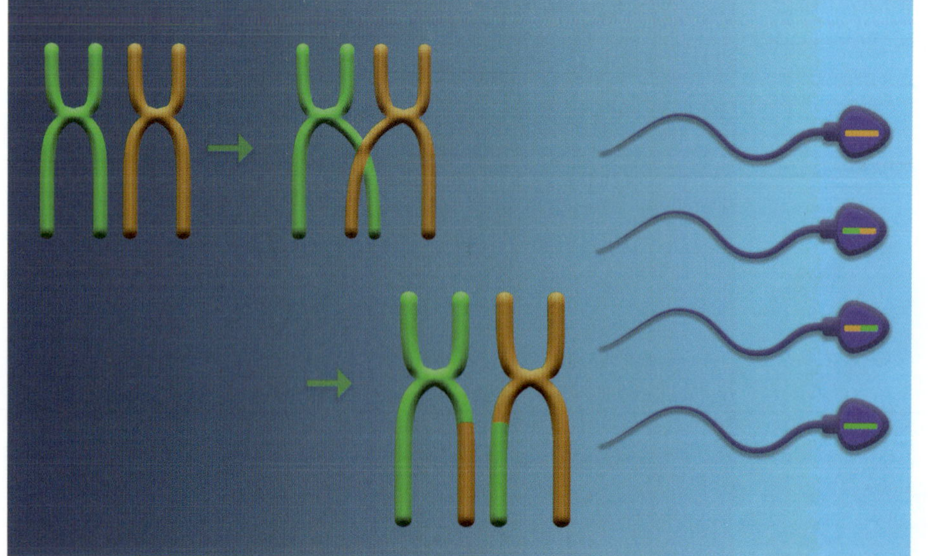

Figure 1.15: The process of meiosis requires two rounds of cell division and produces haploid daughter cells that contain unique combinations of DNA.

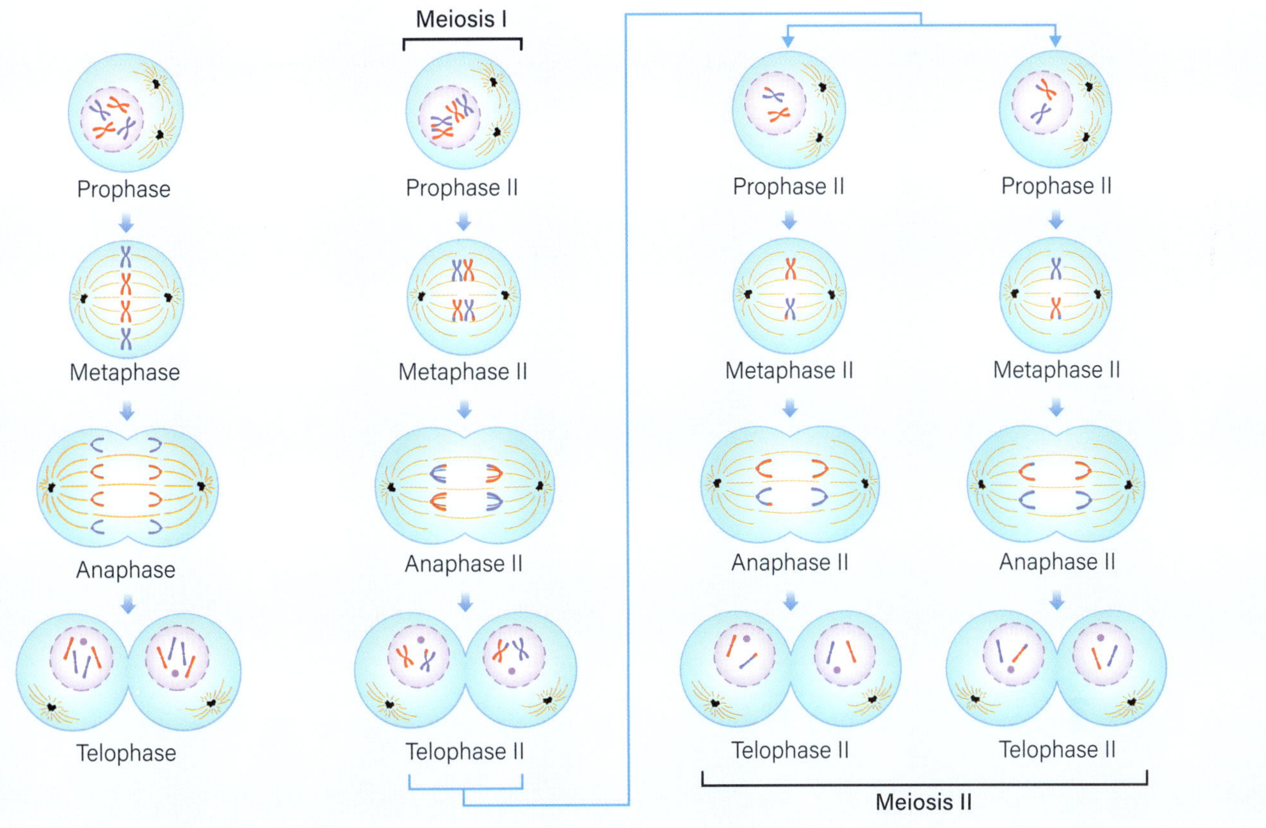

There are eight phases in meiosis

Like mitosis, meiosis occurs in stages. However, meiosis has eight phases, including two rounds of cell division (Figure 1.15).

Meiosis I

1 **Prophase I:** Chromosomes condense (coil up), and pairs of chromosomes that have the same genes in the same order (homologous) match up and randomly switch segments of their genetic material (i.e. they cross over). This mixes up the genes, making the chromosomes different from the original chromosomes. The nucleus breaks down, and spindle fibres appear.

2 **Metaphase I:** Pairs of chromosomes line up across the equator (middle) of the cell, and spindle fibres attach to them.

3 **Anaphase I:** Chromosome pairs separate to the opposite ends of the cell, and the cell elongates (gets longer).

4 **Telophase I:** The chromosomes are released from the spindle fibres, while a nucleus starts to form around each set. The middle of the cell starts to pinch inwards.

After telophase, cytokinesis occurs and the parent cell again splits into two identical daughter cells. However, unlike in mitosis, these cells are unique, and they immediately enter into another round of division: meiosis II.

Meiosis II

1 **Prophase II:** The nucleus of each daughter cell dissolves, and spindle fibres appear in both cells.

2 **Metaphase II:** The chromosomes line up across the equator (middle) of each cell, and spindle fibres attach to either side of them.

3 **Anaphase II:** Spindle fibres pull the chromosomes towards each side, and the cells elongate.

4 **Telophase II:** The chromosomes on each side of both cells are released from the spindle fibres and unwind, while a nucleus starts to form around them. The middles of the cells start to pinch inwards.

When round two of telophase is complete, cytokinesis occurs again, and the two cells are split in half into four unique daughter cells. Each cell is haploid, containing half the number of chromosomes of the original (diploid) parent cell, and contains a unique mixture of the genetic material from its chromosomes.

The result of meiosis is gametes

The four unique daughter cells produced in meiosis are gametes, not normal body cells. In males, they are sperm cells; in females, they are ovum (egg) cells. It is extremely important that gametes are produced in this way – that is, as haploid cells – so that reproduction can occur.

Meiosis and fertilisation both cause variation in genetics

You learnt last year that gametes play an important role in sexual reproduction, which ensures the survival of a species by producing new and unique offspring. But why are they unique?

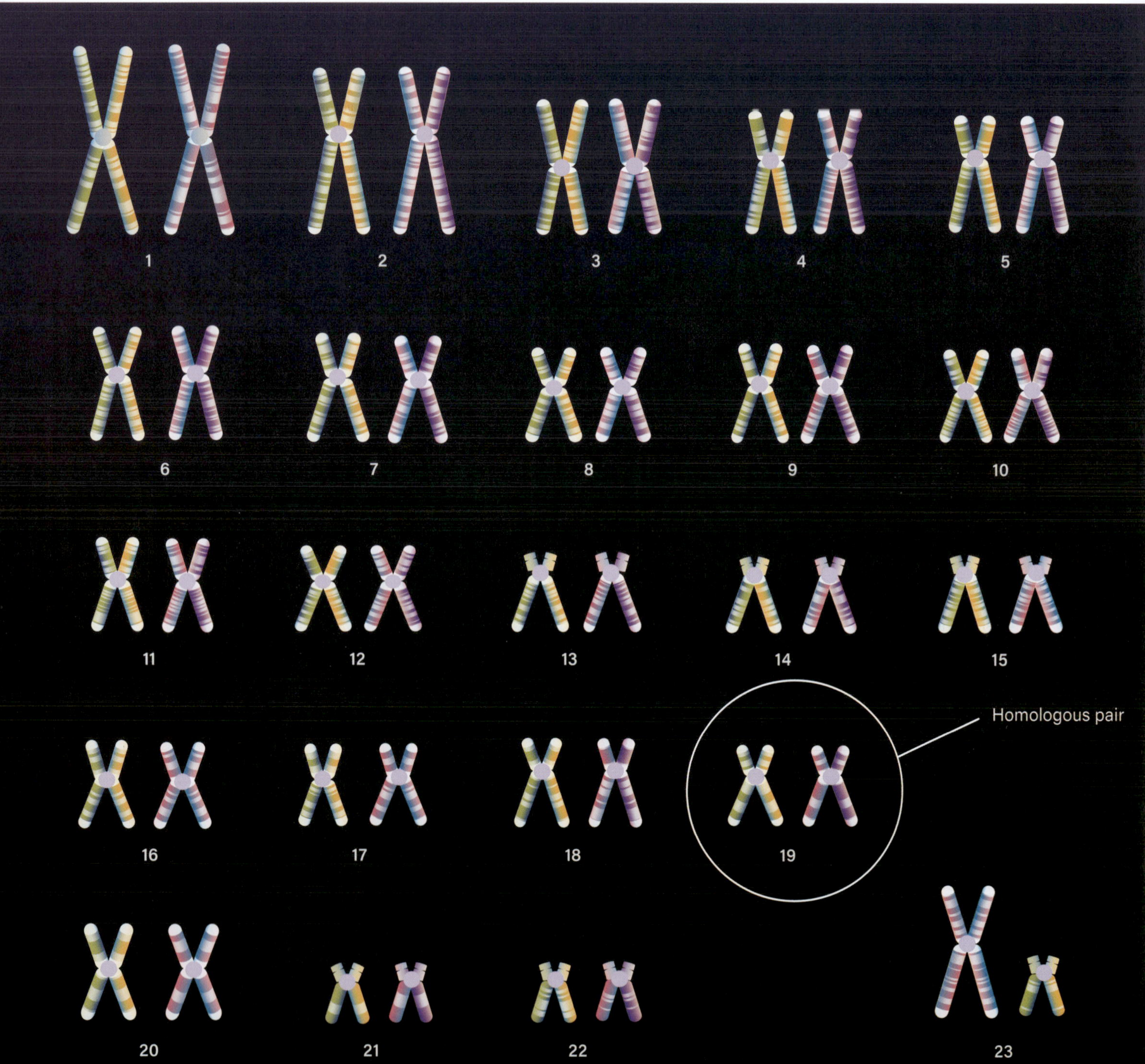

Figure 1.16: During meiosis, homologous chromosomes pair up. Homologous chromosomes contain the same combination of genetic material within the same location. The human karyotype is made up

One reason is meiosis. As discussed earlier in this section, the process of crossing over mixes up the genetics of the cells that are dividing, so that the gametes produced contain unique genetics. In this way, offspring inherit a unique combination of chromosomes from their parents, which they express as their physical characteristics. In humans, because there are lots of different genetic traits, children can look quite different from their parents, depending on which chromosomes they inherit from them. The number of homologous pairs of chromosomes a species possesses is linked directly to the number of possible unique gametes it can produce. This is demonstrated in Table 1.2.

Another reason is the process of fertilisation itself. Fertilisation is the point at which a male and female gamete fuse to form a new offspring. However, the gametes that fuse are randomly determined; there are no set gametes that will combine to produce a new offspring. This means that a new organism inherits a random combination of genetics – half from its biological mother and half from its biological father. The way these genetics show up as the physical characteristics in the new offspring will depend on which versions of different genes were present in the gametes that fused together.

Table 1.2: Meiosis means that different numbers of unique gametes can be produced. The more chromosomes there are, the larger the possible number of unique gametes.

Pairs of homologous chromosomes	Calculation of possible unique combinations	Number of possible unique combinations
2	2^2	4
3	2^3	8
4	2^4	16
5	2^5	32

Learning Ladder

Genetics

1. Identify the stage of meiosis that causes genetic variation in offspring.
2. Describe a way that genetic variation is increased because of the process of meiosis.
3. Explain how the process of meiosis produces gametes that are then used in reproduction.
4. Analyse the outcome of meiosis, in terms of cell production, variation and reproduction.
5. Use secondary research to identify a technology that allows scientists to view and understand the structure of chromosomes. Evaluate the impact of this technology on our understanding of the process of meiosis.

Use and influence of science

Meiosis does not always occur flawlessly. Sometimes, the process of non-disjunction – incomplete separation of chromosomes – can occur during division. This leaves some gametes with extra chromosomes, and others with incomplete sets of chromosomes. Abnormal chromosome numbers can lead to conditions such as Trisomy 21 (Down syndrome).

1. Identify how our knowledge of meiosis could help to address issues that occur within the division process.
2. Describe two interpretations of the way that meiosis causes variation within humans.
3. Explain how scientific research priorities may shift if the number of babies born with Trisomy 21 started to increase.
4. Discuss the ways in which misinformation about the process of meiosis may influence an individual's decision to have children.

Processing, modelling and analysing p. 244

1. Construct a table showing the number of possible unique gametes for species that have between 20 and 25 pairs of homologous chromosomes.
2. Convert your table from Question 1 into an appropriate scientific graph. Refer to 'Graphing' in the Science how-to section on page 294 to help you.
3. Explain the trend that you have observed within your graph.
4. Discuss the relationship between an organism's total number of chromosomes and its ability to produce unique offspring.
5. Analyse the data in the table in relation to its range. Describe the impact of this feature on organisms.

Key idea: Form and function

Meiosis is an integral part of reproduction. Evaluate how the formation of gametes through meiosis is related to the function of reproduction in species.

Success criteria

- I can describe the stages of meiosis.
- I can explain why the process of meiosis is important for variation.

1·6 ▸ Genes and the environment

Learning intention

At the end of this lesson, I will be able to identify how genes and the environment interact when organisms are developing traits.

Key terms

environmental factor: something in the environment that affects gene expression

expression: the process of converting the instructions in DNA into a trait, such as a protein

monogenic: influenced by a single gene

polygenic: influenced by two or more genes

Investigation 1.6

Genetic trait survey, p. 312

Key idea: Form and function

Every physical feature we have is the result of the genes in our DNA. Our hair, eyes, teeth, fingernails and skin colour are all related to our genes. But is that all that controls these traits?

Genes can be varied

Scientists estimate that in the human genome, there are more than 20 000 genes that can be used to produce proteins. Some of those genes are alleles, which are variations of the same gene that can cause slightly different proteins to be produced. These variations occur because the sequence of nucleotides that make up the gene is slightly altered. The alleles you inherit from your parents will influence all your traits.

Almost every gene in your body is one of a pair, which work together to express traits. This combination of two alleles (genes) that influence the same trait is called the genotype.

Some traits are influenced by many genes

Some traits, such as dimples and the ability to roll your tongue, are controlled by a single gene. They are **monogenic**. Other traits, such as height and skin colour, are more commonly **polygenic** – they are influenced by several genes. The way these traits are expressed is called the phenotype, which is what we physically see. Your genotype, which is inherited from your parents, determines your phenotype.

Some traits are controlled by multiple alleles; other traits are controlled by just one pair of alleles. These alleles determine the phenotype of an organism – what is physically expressed; for example, dimples, widow's peaks and freckles in humans, and flower colour in plants.

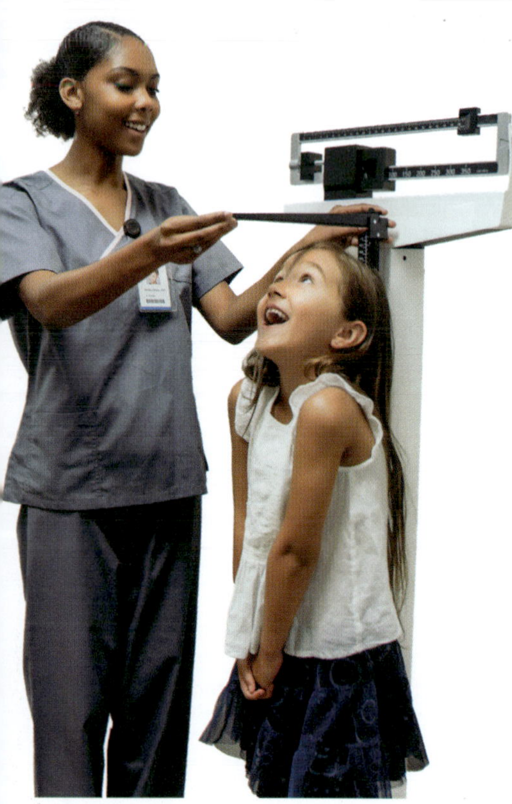

Figure 1.17: Our height is determined by our genes but can also be influenced by environmental factors such as our diet.

Figure 1.18: The darker patches of fur on a Siamese cat are due to the interaction between proteins produced from their genes and cooler temperatures.

Genes are influenced by environmental factors

Gene **expression** is influenced by external **environmental factors**. For example, skin colour is influenced by your genes, but also by sunlight. Your skin will tan, or get darker, the more sunlight (ultraviolet rays) you are exposed to. Height can also be influenced by environmental factors. A diet low in vital nutrients can restrict a child's growth. This can result in shorter individuals or in diseases such as rickets, which causes the bones of the leg to bow.

The genes of all organisms are affected by the environment. Table 1.3 summarises some traits of other organisms that are influenced by their environment.

Table 1.3: Environmental influences on some traits

Organism	Trait	Environmental factor
Hydrangea	Flower colour	Soil pH level
Siamese cat	Fur colour: dark patches around nose, ears, paws and tail	Temperature
Turtle	Sex	Temperature
Banksia	Leaf size	Sunlight exposure

Genes can provide clues about our ancestry

Skin colour is related to how much of the pigment melanin is present; the more melanin you produce, the darker your skin is. In the past, populations that lived close to the equator were exposed to higher levels of sunlight than those who lived closer to the North or South Pole. This meant that they needed higher levels of melanin to protect themselves from harmful ultraviolet rays, which could cause damage and skin cancer. So, if you have more melanin today, chances are your ancestors grew up somewhere with lots of sunlight.

Figure 1.19: The colour of hydrangea flowers is influenced by the pH level in the surrounding soil.

Learning Ladder

Genetics

1. Identify the name given to different variations of the same gene.
2. Describe the difference between monogenic and polygenic traits.
3. Explain how the environment can influence the expression of your genes.

Use and influence of science

1. Identify a piece of knowledge that can be used to help solve problems, such as a lack of growth during early development.
2. Describe two ways that our knowledge of the influence of an environment on animals may be interpreted.
3. Explain how magazines and platforms that display the 'ideal' body type may influence societal research into how height is controlled.
4. Discuss how misinformation about skin colour inheritance may inform an individual's decision-making.
5. Conduct secondary research and analyse some factors that may affect how our knowledge of environmental influences on plants is adopted by society.

Communicating p. 262

A student conducted a survey into the phenotype of everyone in their class. They found that 60 per cent of students could roll their tongue, 70 per cent had dimples, 20 per cent had blue eyes, 50 per cent had brown hair and 10 per cent had a big toe that was shorter than their second toe.

1. Propose an appropriate way to compile and present this information clearly.
2. Construct your presentation using a digital technology.
3. Measure and record the height of everyone in your class and construct a graph showing the data you have collected. Refer to 'Graphing' in the Science how-to section on page 294 to help you.

Key idea: Form and function

Another trait controlled by our genes is hair colour. Considering the number of possibilities, propose why a person's hair colour is more likely to be a function of multiple genes, rather than a monogenic trait.

Success criteria

- I can identify factors that can affect the expression of genes.
- I can identify the relationship between the environment, genes and phenotype.

1·7 ▸ Mutation and genetic variation

Learning intention

At the end of this lesson, I will be able to:

- describe what a mutation is
- explain how mutations affect genetic variation.

Key terms

chromosomal mutation: a mutation caused by a change to a large section of a chromosome

karyotype: a picture of an organism's full set of chromosomes

mutation: an error in DNA that changes the DNA sequence, which can be caused by DNA replication, or cell division

point mutation: a mutation caused by a change to a single nucleotide base

Key idea: Form and function

If you were to put all your DNA end to end, it would reach to the Sun and back over 600 times! With that amount of DNA, it is not surprising that mistakes happen when it is copied or replicated. These errors can become **mutations**.

Errors can occur during DNA replication

During DNA replication, the existing strands of DNA act as templates for new strands to form. Even though the DNA replication process is very tightly controlled, DNA sequences can be disrupted when the bases are being read or synthesised. The resulting changes in the DNA being produced are known as **point mutations**.

There are three main types of point mutation: substitution, insertion and deletion. In substitution, an incorrect base is inserted. In insertion, one or more extra bases are incorrectly inserted. In deletion, one or more bases are removed.

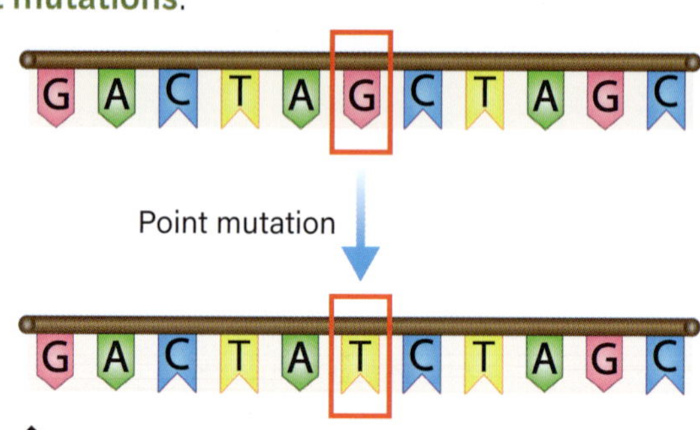

Figure 1.20: A point mutation is a change to a single nucleotide base on a strand of DNA.

Table 1.4: Types of chromosomal mutations

Mutation	Description
Deletion	A section of the chromosome is removed, usually during crossing over in meiosis.
Duplication	A section of a chromosome is replicated twice and inserted into the chromosome.
Insertion	A section from one chromosome is incorrectly inserted into another.
Inversion	A section of a chromosome is inserted upside-down during crossing over.
Translocation	A section of a chromosome breaks off and reattaches to a different section of another chromosome.

Errors can occur during cell division

When cell division occurs, chromosomes move around in the cells, and during meiosis, the chromosomes randomly swap sections of genetic material. If these processes occur incorrectly, it can cause a **chromosomal mutation**. There are several kinds of chromosomal mutation, as shown in Table 1.4.

Sometimes, chromosomes fail to separate correctly during cell division. Some cells may receive additional copies of whole chromosomes, while others may be missing them. This can occur during both mitosis and meiosis. When cells produced by meiosis contain additional chromosomes or some are missing, this mutation can be passed from parents to offspring.

Other factors can cause mutations

Environmental factors that can cause mutations include ultraviolet radiation from the Sun and chemicals such as cigarette smoke.

Mutations can have different effects

The effect of a mutation depends on its type and location. It may be beneficial or harmful, or have no effect (Table 1.5).

Table 1.5: The impacts of mutation

Impact	Description	Examples
Beneficial	The mutation causes a change that improves the organism's survival.	The production of an enzyme that confers the ability to digest lactose (sugars) in milk
Harmful	The mutation causes a change that negatively impacts the organism's survival.	Cystic fibrosis – a disease in humans results in the build-up of sticky mucus in the lungs Sickle cell anaemia – a disease causing red blood cells to be deformed, meaning they cannot transport oxygen effectively
None	The mutation does not cause a change to the organism.	Substitutions of a single base that results in the same protein being formed, producing no change to the organism

An example of a harmful mutation is cystic fibrosis (CF). This disease occurs because of a recessive allele – the child needs to get a copy from both parents to have the disease (more on this later). When both alleles inherited are the CF gene, the child will be born with cystic fibrosis and have sticky mucus in their lungs which affects their ability to breathe.

Karyotypes are pictures of an individual's chromosomes (Figure 1.21). To get this picture, cells containing condensed chromosomes are stained and examined under a microscope. An image of the chromosomes is taken, and each chromosome is cut out, placed next to its pair and numbered. The largest chromosome is chromosome 1. By observing the appearance and number of chromosomes, scientists can check for chromosomal abnormalities.

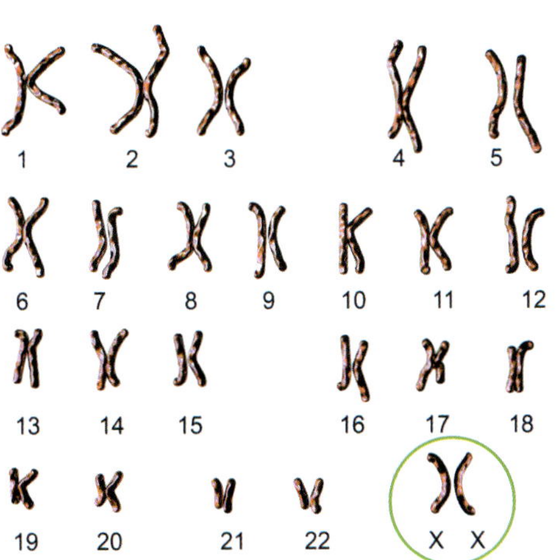

Figure 1.21: The XX chromosome pair in this karyotype shows that the person is female. Males have an XY chromosome pair instead.

Learning Ladder

Genetics

1. Identify the type of cell division that would pass on mutations when they occur.
2. Describe the two main types of mutations.
3. Explain how a chromosomal mutation may have a larger impact on an organism.
4. Analyse the outcome of a beneficial mutation.
5. Evaluate the impact of a technology that could be used to screen for genes such as the ones that cause diseases like cystic fibrosis and sickle cell anaemia.

Nature and development of science

1. Identify the method used to obtain an image of the human karyotype.
2. Describe a way that our knowledge of mutations may have been refined as we continued to study chromosomes and DNA. (*Hint:* Think about the types and impacts of mutations.)
3. Explain how understanding chromosomal mutations has led to the development of technologies, using secondary research as needed.
4. Discuss the role of peer review in relation to our understanding of mutations.
5. A scientist proposed the idea that mutations should be further explored so we can use their benefits. Use your understanding of mutations to justify *or* construct an argument against this proposal.

Communicating p. 262

1. Select an appropriate way to organise information about the types of chromosomal mutations.
2. Construct your suggested method of communication from Question 1.
3. Construct an infographic on the impact of a named mutation.
4. Construct an argument explaining why mutations are not always a bad thing. Refer to 'Writing evidence-based essays' in the Science how-to section on page 280 to help you.

Key idea: Form and function

Sickle cell anaemia impacts the shape of red blood cells. Using secondary research, analyse how the form that red blood cells take influences their function, and link this to the problem with sickle cell disease.

Success criteria

- I can identify a range of mutations.
- I can explain how mutations affect genetic diversity.

1·8 ▸ Genotypes and phenotypes

Learning intention

At the end of this lesson, I will be able to:

- outline the connection between genotypes and phenotypes
- link this connection to Mendelian inheritance.

Key terms

dominant allele: an allele that only requires one copy to be seen in a phenotype

heterozygous: a genotype of two different alleles

homozygous: a genotype of two of the same allele

recessive allele: an allele that requires two copies to be seen in a phenotype

Key idea: Form and function

Genes are responsible for the physical characteristics of all living things. We can use our understanding of genetics to predict patterns of monogenic inheritance.

Alleles can be dominant or recessive

When a gene for a single trait has two alleles, we can often predict the gene's inheritance pattern. One allele of the gene is usually dominant over the other and has more influence on how that gene is expressed.

An organism that is **heterozygous** for a gene has a copy of the **dominant allele** and a copy of the **recessive allele**. They will display characteristics of the dominant allele. An organism that is **homozygous** for a gene has either two copies of the dominant allele or two copies of the recessive allele.

We represent the alleles of a gene as letters – an upper-case letter for a dominant allele and a lower-case letter for a recessive allele (Table 1.6).

Table 1.6: How different genotypes are represented in monogenic inheritance traits

Genotype description	Genotype	Phenotype
Homozygous dominant	AA	Dominant characteristic
Heterozygous	Aa	Dominant characteristic
Homozygous recessive	aa	Recessive characteristic

Figure 1.22: Dimples are a dominant trait, so the person in this image must have a genotype of DD or Dd, not the recessive dd genotype.

Gregor Mendel determined monogenic inheritance patterns

In the 1860s, Austrian biologist Gregor Mendel bred more than 30 000 pea plants to test patterns of inheritance. He manually pollinated the plants by using a paintbrush to control which plants were bred together. He studied the occurrence of specific traits that could be expressed in two ways: height (tall or short), seed shape (round or wrinkled), flower colour (purple or white) and pod colour (yellow or green). By crossing 'pure-breeding' (homozygous) plants with the same traits, he was able to show that when two plants that had only one gene for a trait are crossed, all of their offspring have the same trait.

Mendel's work with cross-breeding plants allowed him to better understand inheritance. When he crossed a homozygous dominant plant with a homozygous recessive plant, all the offspring displayed only dominant traits. When he bred two of the 'hybrid' (heterozygous) plants from this first (F_1) generation together, three-quarters of the offspring displayed the dominant trait and one-quarter of them displayed the recessive trait. This three-to-one (3:1) ratio was consistently displayed when heterozygous offspring were bred together, no matter what trait was being studied. This is shown for pod colour in Figure 1.23.

Table 1.7: Common inheritance patterns in monogenic traits in Mendel's pea plants

Characteristic	Alleles	Dominant phenotype	Genotypes for dominant phenotype	Recessive phenotype	Genotype for recessive phenotype
Pea plant height	Tall (T) Short (t)	Tall	TT, Tt	Short	tt
Seed shape	Round (R) Wrinkled (r)	Round	RR, Rr	Wrinkled	rr
Pod colour	Green (G) Yellow (g)	Green	GG, Gg	Yellow	gg

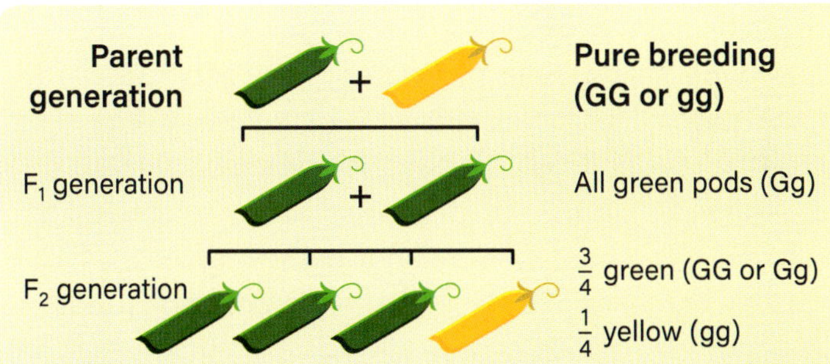

Figure 1.23: Mendel showed that when hybrid plants were cross-bred, the dominant trait was displayed over the recessive trait in a 3:1 ratio. In this experiment, the parent generation are pure breeding, either GG or gg. The F_1 generation is the first generation bred from the pair of parents and are all hybrid (Gg). The F_2 generation is the second generation and contains three GG or Gg (dominant) individuals and one gg (recessive) individual.

Learning Ladder

Genetics

1. Identify:
   a. three characteristics of Mendel's pea plants.
   b. the genotype required for the recessive phenotype to be displayed for each.
2. Describe how a person's genotype can affect whether their phenotype includes dimples.
3. Explain how Mendel's pea plant experiments provide evidence that organisms inherit and pass on genetic characteristics.
4. Two parents who were heterozygous for the characteristic of having dimples decided to have a child. Using the pea plant inheritance as an example, analyse the likely outcome of their child's phenotype. (*Note:* Dimples are a dominant allele.)

Nature and development of science

1. Identify two technologies that would have assisted Mendel with his research.
2. Propose and describe a way that our understanding of genetics may have been refined since Mendel's experiments with pea plants.
3. Explain how Mendel's work has contributed to our understanding of genetics.
4. Discuss the ways that Mendel's work with pea plants could be validated by modern scientists.
5. Analyse how advances in the technologies that scientists use to study genetics can impact our ability to study genes and inheritance.

Processing, modelling and analysing p. 244

A student wanted to replicate Mendel's pea plant experiments into flower colour. She manually pollinated two hybrid pea plants, grew eight offspring plants and observed their flower colours. Six of the plants had purple flowers and two had white flowers.

1. Construct a scientific table that could be used to record the student's results.
2. Convert this data into an appropriate scientific graph, to compare the number of plants produced with different flower colours.
3. State the difference in plant numbers as a ratio. Explain how this pattern relates to Mendel's work.
4. Discuss the relationship between hybrid plants and their offspring when a specific trait is controlled by a dominant and a recessive allele.

Key idea: Form and function

Many characteristics are controlled by a single trait, containing a dominant and a recessive allele. Use secondary research to identify three other characteristics that are controlled by a single dominant/recessive allele combination. Explain how the combination of alleles results in the phenotype seen by others.

Success criteria

- I can outline the relationship between genotypes and phenotypes.
- I can explain Mendelian inheritance patterns.

1·9 ▸ Punnett squares and pedigrees

Learning intention

At the end of this lesson, I will be able to model and predict inheritance patterns using Punnett squares and pedigrees.

Key terms

pedigree chart: a chart that shows the inheritance of a trait through generations of a family

Punnett square: a diagram that shows the possible genotypes of offspring

Investigation 1.9

Investigating proportions of sprout colours, p. 313

Key idea: Form and function

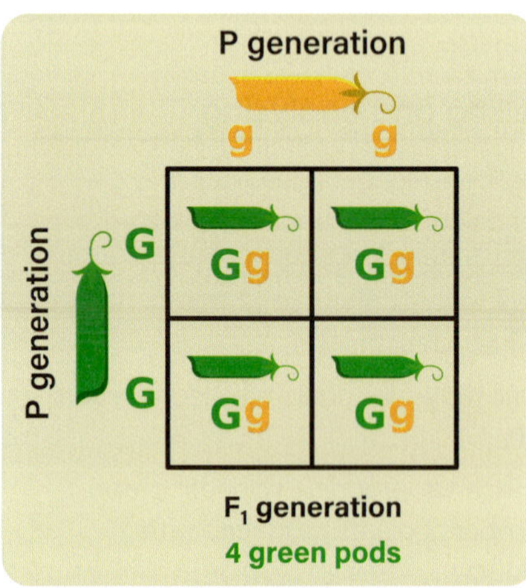

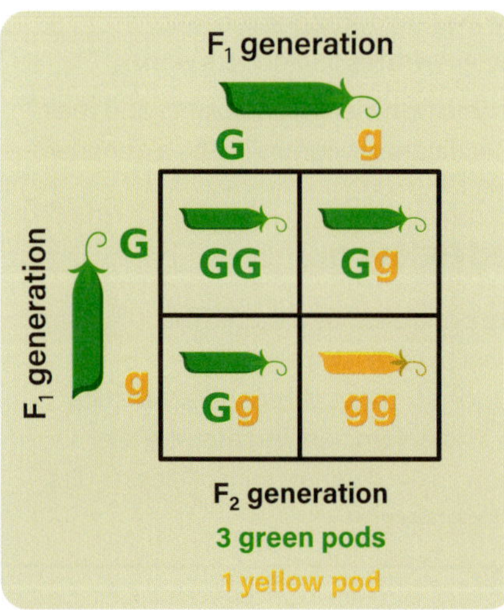

▲ **Figure 1.24:** These Punnett squares show the genotypes of Mendel's pea plant experiment on pod colour.

We can use two methods to map traits in families to predict the likelihood of offspring having specific phenotypes: Punnett squares and pedigrees.

Punnett squares show single crosses

A **Punnett square** is a diagram showing a box with four spaces that can be used to determine the possible genotypes of one generation of offspring from two parents. Up to four possible genotypes will be shown. The information inside the square can then be used to show the expected percentage (%) or ratio of offspring that have a particular genotype. Once the possible genotypes are known, they can be used to determine the possible percentage or ratio of phenotypes of offspring.

The different generations in a cross-breeding experiment are given different names.

- The original pair of parents are the P generation.
- The first generation produced from the P generation is the F_1 generation.
- The generation produced by cross-breeding among the F_1 generation is the F_2 generation.

Punnett squares can predict the genotypes of offspring

To construct a Punnett square, follow these steps.

1 Draw the outline of your Punnett square.
2 Select a trait and assign the allele a letter.
3 Identify the genotype of the two parents (P): are they homozygous dominant, heterozygous or homozygous recessive?
4 Place the alleles of one parent down the left-hand side of the box.
5 Place the alleles of the second parent across the top of the box.
6 Fill in the four boxes in the table by matching up the allele from the left and the allele from the top. Always write the capital letter allele first if there is one.

Once your Punnett square has been drawn, you can determine the percentage or ratio of the possible genotypes, and then phenotypes of offspring from the two parents. For example, consider the Punnett squares in Figure 1.24.

Square 1: P generation

Both parents are homozygous: one for green pods (GG) and one for yellow pods (gg). All potential offspring must inherit a G allele from one parent and a g allele from the other parent, meaning that 100 per cent of the offspring genotypes will be Gg. This means they are all heterozygous and all have the dominant green phenotype.

Square 2: F_1 generation

Both parents are heterozygous Gg – they all have green pods. The offspring will be GG (25%), Gg (50%) and gg (25%). Therefore, the genotypic ratio is 1 GG : 2 Gg : 1 gg.

The phenotypes are green (75%) (remember, heterozygous genotypes always show the dominant trait) and yellow (25%). Therefore, the phenotypic ratio is 3 green : 1 yellow.

Pedigree charts show traits passed down through families

A **pedigree chart** shows which family members over generations have a particular trait or phenotype. Pedigree charts can be used to identify whether a trait is dominant or recessive. Males are represented by a square and females by a circle, and shading indicates the presence of the trait. A horizontal line between them means they are a mating pair, and a vertical line coming from the mating line means they have offspring. Generations are numbered in Roman numerals on the left of the chart. Individuals in each generation are also numbered. For example, in Figure 1.25, the filled-in circle in the first generation is Individual I-1.

Pedigrees can be used to track the inheritance of non-infectious diseases, such as haemophilia (a blood-clotting disease) and colour blindness. These diseases are expressed by a recessive gene that is present specifically on the X chromosome. If a female (with two X chromosomes) has a single copy of the recessive allele, they can be a carrier of the disease, but not have the disease itself because the other X chromosome has the dominant allele. However, males have one X chromosome and one Y chromosome. If their X chromosome has the allele for the disease, then they will have the disease. Males cannot be carriers for recessive conditions that are present on the X and Y chromosomes. That is why these diseases are more commonly seen in males.

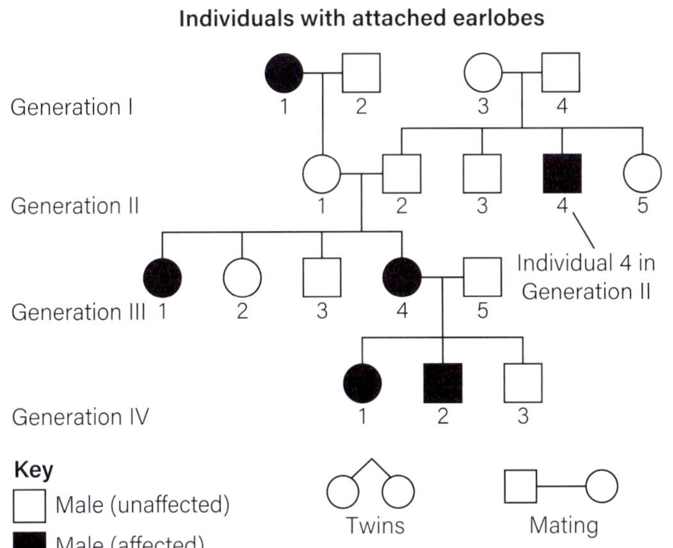

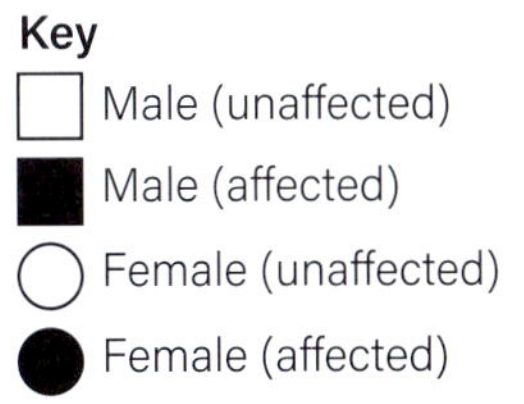

Figure 1.25: A pedigree chart showing the inheritance of attached earlobes through generations of a family

Pedigree charts can give us information about genotypes

In pedigrees, patterns in the occurrence of specific traits can give us hints about whether the trait is dominant or recessive.

For example, if two parents who are both affected by a particular trait (filled-in shapes) produce an offspring who is unaffected (not filled in), then the trait can be predicted to be dominant. Each parent must be heterozygous for the trait and have one copy of the recessive allele. The unaffected offspring received two recessive alleles, one from each parent.

However, if both parents are unaffected by a particular trait and produce an offspring who is affected, then the trait can be predicted to be recessive. Both parents were heterozygous for the trait and carried the recessive allele for it. The offspring received both recessive alleles and so is affected.

If the patterns of pedigree charts look more complicated than this, then a different type of inheritance than just a monogenic trait might be occurring. Sex-linked conditions might appear more often in males, and polygenic trait patterns can be very difficult to distinguish.

Pedigrees and Punnett squares can be constructed from written information

We can use written information based on observations of phenotypes to construct pedigrees showing inheritance through generations of a family. From there, we can predict the likelihood of particular phenotypes occurring, by using a Punnett square to show possible genotypes and linking these to their correct phenotypes.

For example, in Section 1.8 we saw that dimples are a monogenic trait. People with dimples have the dominant (D) allele, while those without dimples have the recessive (d) allele, and *must* have the homozygous recessive genotype (dd).

The following observations were made about a family. Kate was looking at photos to investigate her family history. She noticed that some family members had dimples and others did not. Her mother does not have dimples, and neither does her mother's mother, but her grandfather does. On her father's side, her father and grandmother both have dimples, but her grandfather does not. Kate does not have dimples, but her two brothers do. Kate wondered if her child would have dimples. Her husband and his mother have dimples.

We can construct a pedigree, like the one in Figure 1.28, to show the appearance of dimples in Kate's family by following these steps.

1 Identify the number of generations.
  Three: Kate, her parents and her grandparents.
2 Represent each individual in the information with a symbol – circles for females and squares for males. Align the circles and squares with the generation labels.
3 Draw a line between Kate and her husband, between Kate's parents, and between each set of grandparents, to show that they have had, or may have, offspring.
4 Connect Kate to her two brothers by drawing a line from the horizontal line between her parents and then joining it to a horizontal line that connects to each sibling.
5 Construct a key identifying affected and unaffected males and females.
6 Shade those symbols that represent a person with dimples.
7 Number the circles and squares, so that each individual can be easily identified.

Once the pedigree has been constructed, you can use a Punnett square to determine the likelihood of Kate's offspring having dimples. To do this, determine the genotype of Kate and her husband and place them on the outside of a Punnett square (Figure 1.29).

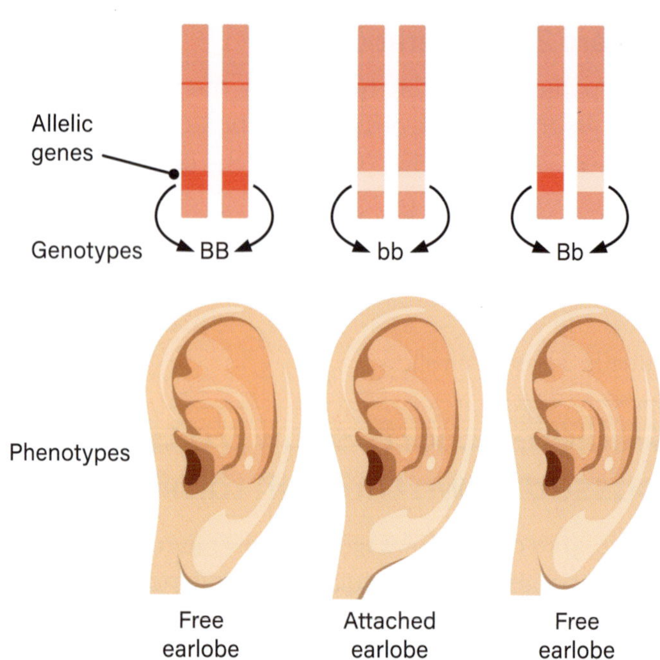

Figure 1.26: Earlobe attachment is controlled by a specific gene. These allele combinations can be tracked and mapped over generations using a pedigree chart, like the one in Figure 1.25.

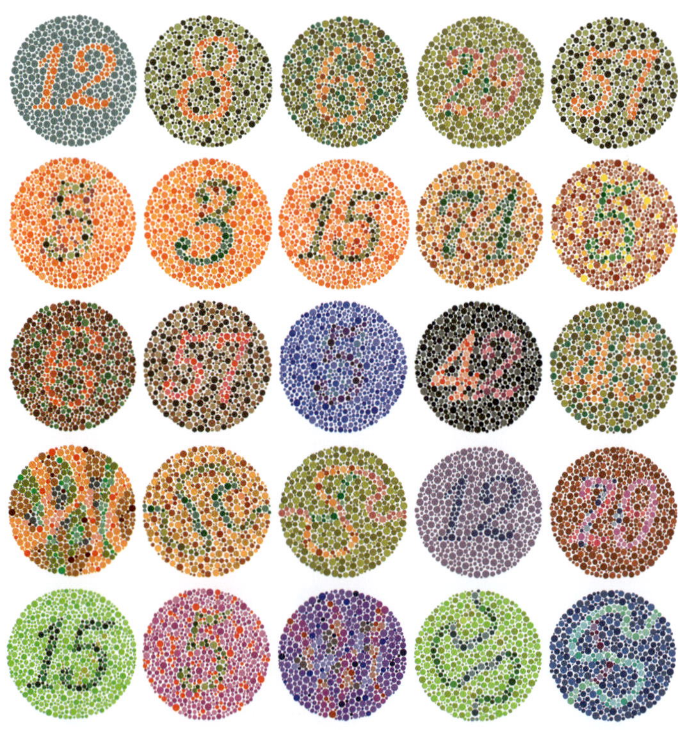

Figure 1.27: Males are more likely to inherit colour blindness than females because it is carried on the X sex chromosome. These charts are used to test for colour blindness; each contains a symbol or an image. People with different types of colour blindness may not be able to see all the different numbers and shapes.

1 As Kate does not have dimples, her genotype *must* be recessive (dd).

2 Kate's husband has dimples, so he must have at least *one* dominant allele (D). However, his father does not have dimples, so his genotype would be homozygous recessive (dd). As Kate's husband would have one copy of the recessive allele from this parent (d), his genotype must be heterozygous (Dd).

3 Construct the Punnett square so that the genotypic and phenotypic ratios can be calculated, and the likelihood determined of Kate's children having dimples.

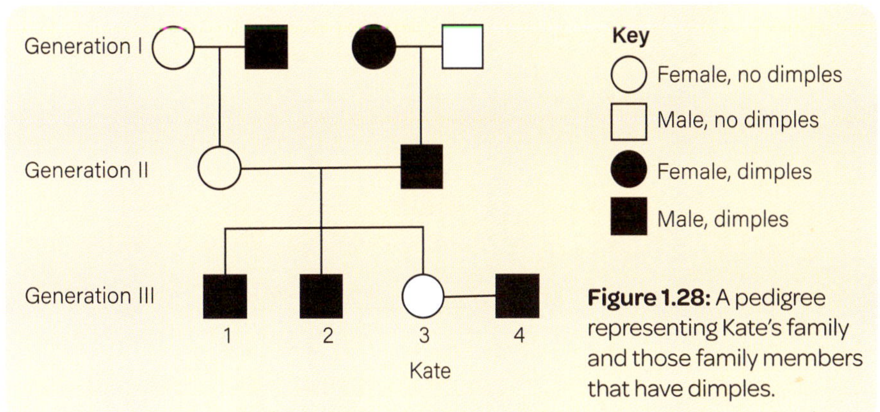

Figure 1.28: A pedigree representing Kate's family and those family members that have dimples.

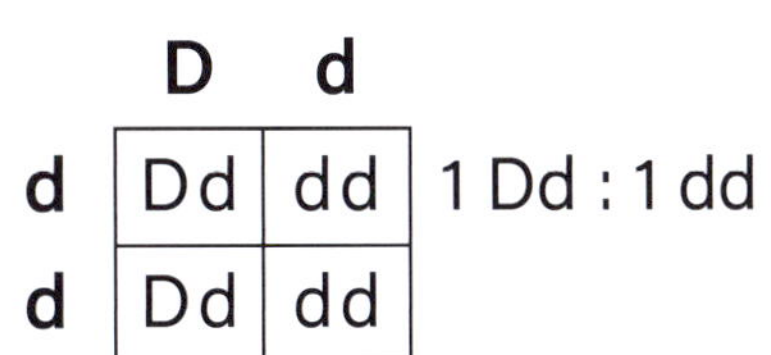

	D	d
d	Dd	dd
d	Dd	dd

1 Dd : 1 dd

Figure 1.29: A Punnett square showing the possible offspring of Kate and her husband. It shows a genotypic ratio of 1 Dd: 1 dd, and a phenotypic ratio of 1 dimples : 1 no dimples. This means there is a 50 per cent chance of Kate's children having dimples.

Learning Ladder

Genetics

1 Identify two mechanisms for tracking genetic patterns of inheritance.

2 Describe a reason why a pedigree could be used if someone wanted to see how they inherited a trait that neither of their parents have.

3 Two parents who are heterozygous for dimples (a dominant trait: D) are planning to have a child. Explain how you could predict the likelihood of the child not having dimples. Use a Punnett square to support your response.

4 Analyse the outcome of using a pedigree to track the inheritance of a specific trait across multiple generations, referring to Figure 1.25 in your response.

5 Evaluate the impact of using Punnett squares to predict the probability of offspring possessing specific genetic traits across a single generation and multiple generations.

Use and influence of science

1 Identify an issue that understanding the use of Punnett squares and pedigrees can solve, when studying how offspring received certain characteristics. (*Hint:* Think about how traits can be mapped.)

2 Describe two ways that the information in a pedigree may be interpreted when looking at a specific trait, in terms of benefit and harm.

3 Explain how research directions may shift if pedigrees were more widely used to track the incidence of inherited diseases.

Processing, modelling and analysing *p. 244*

A student wants to track the inheritance of widow's peaks in his family. (A widow's peak is a V-shaped hairline, with the peak in the centre of the forehead.) He has a widow's peak, but his sister does not. His parents both have widow's peaks. His grandparents on his mother's side both have widow's peaks. However, on his father's side, only his grandfather and an uncle have a widow's peak.

1 Construct a pedigree showing this information. Make widow's peak the affected trait.

2 Calculate the percentage of individuals with and without a widow's peak in the pedigree.

3 Using your pedigree, propose whether widow's peak is a dominant or recessive trait. Give a reason for your choice.

Key idea: Form and function

Haemochromatosis is a recessive genetic condition that impacts iron levels in the body. Conduct secondary research to determine the impact on the materials formed to regulate iron absorption and explain how this condition changed their function. Construct a Punnett square to demonstrate the probability of the condition being inherited by offspring in two parents who are both carriers (heterozygous) for the allele.

Success criteria

- I can construct and use Punnett squares to model and predict inheritance patterns.
- I can construct and use pedigrees to model and predict inheritance patterns.

1·10 ▸ Key idea: Form and function – discovering the structure of DNA

Learning intention

At the end of this lesson, I will be able to discuss the roles of the scientists involved in the discovery of the structure of DNA.

Key term

nuclein: the name given to DNA when it was first discovered

Investigation 1.10

DNA lolly model, p. 315

Key idea: Form and function

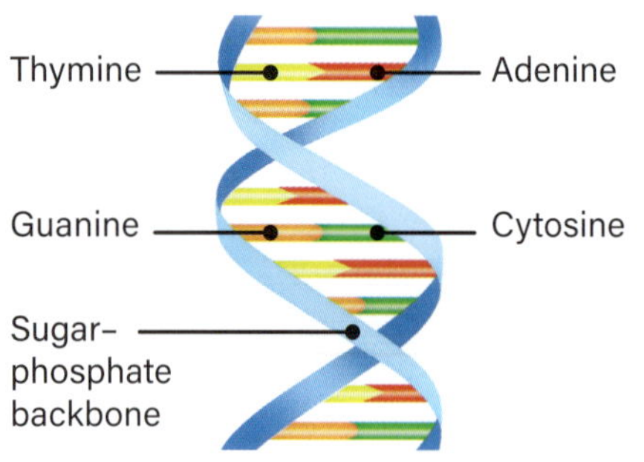

Figure 1.30: The 1953 Watson–Crick model of DNA is still used around the world today.

Although DNA was discovered in the 1860s, its structure was not known until much later. Two scientists – American biologist James Watson and British physicist Francis Crick – presented their model of the DNA double helix in the 1950s. However, their model was based not only on their work but also on the work of other scientists, some of whom remained largely uncredited.

DNA was discovered and investigated across many decades

In 1869, Swiss chemist Friedrich Miescher first found evidence of DNA in cells while investigating white blood cells. He called the molecule **nuclein**, but over time its name changed to deoxyribonucleic acid (DNA) as more discoveries were made.

The next scientists to contribute to our understanding of DNA were Russian biochemist Phoebus Levene and Austrian biochemist Erwin Chargaff. In 1919, Levene proposed that nucleic acids were composed of a series of nucleotides, and that a nucleotide was made up of one of four nitrogenous bases, a phosphate and a sugar molecule, which were always linked in the same order. New evidence quickly modified this idea when Chargaff showed that the order in which nucleotides are linked is highly varied. Chargaff concluded that in every species, the amounts of adenine and thymine were similar, as were the amounts of cytosine and guanine.

Figure 1.31: James Watson and Francis Crick with their famous model of DNA

The Watson–Crick model shows DNA as a double helix

The Watson–Crick model of DNA is the model currently accepted and used around the world. The model is a double-stranded twisted helix (Figure 1.30). The two strands of DNA are antiparallel (parallel, but in the opposite direction) to each other and have a sugar–phosphate backbone. The four nitrogenous bases (adenine, thymine, cytosine and guanine) connect the two backbones and exist in pairs held together by hydrogen bonds between the nitrogenous bases. DNA sequences are written out using letters to represent the bases: A, T, C and G. Watson and Crick presented their model of DNA in 1953.

Franklin and Wilkins played a key role in the Watson–Crick DNA model

Although Watson and Crick were widely credited for their model of DNA, two other scientists also played roles in the discovery. Maurice Wilkins was a New Zealand physicist who worked with English chemist Rosalind Franklin. Wilkins proposed the idea of using X-rays to photograph DNA to observe its structure. Franklin was able to do so, eventually obtaining 'photograph 51' (Figure 1.32), which supported the idea of the double helix structure. Wilkins showed 'photograph 51' to Watson and Crick, who used it as evidence in their published paper without crediting Franklin for it.

Watson, Crick and Wilkins shared a Nobel Prize

In 1962, Watson, Crick and Wilkins received a joint Nobel Prize in Physiology or Medicine for their discovery of the molecular structure of DNA. Although Franklin's work was key, she did not receive the Nobel Prize because she had died four years previously and Nobel Prizes are not awarded posthumously (after death). Also, a Nobel Prize can only be shared by up to three people.

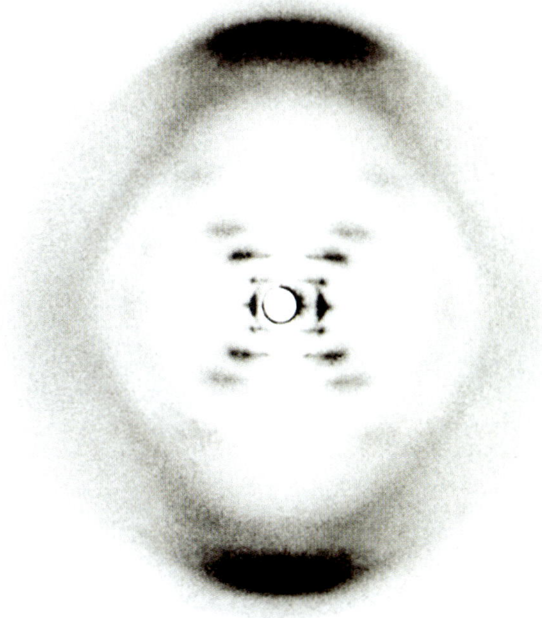

▲ **Figure 1.32:** 'Photograph 51.' This photo, taken by Rosalind Franklin using X-rays, provided supporting evidence for the Watson–Crick model of DNA.

Learning Ladder

Genetics

1. Identify the four scientists involved in the discovery of the structure of DNA.
2. Describe the role Watson and Crick played in presenting the structure of DNA.
3. Explain how Franklin's work contributed to our understanding of the structure of DNA.
4. Analyse the outcome of the combined efforts of the four scientists discussed in terms of our understanding of DNA as a molecule.

Use and influence of science

1. Identify a piece of knowledge linked to DNA's discovery that may support an issue of integrity in scientific research.
2. Describe one way in which the discovery of DNA may be credited to specific scientists.
3. Explain why society may be supportive of new research into the structure of DNA and genes. (*Hint:* Think about the link between DNA and physical health.)
4. Discuss some ways in which misinformation about the discovery of DNA's structure may have influenced which scientists were nominated for a Nobel Prize.
5. Our understanding of structures such as DNA is continually improving. Research and analyse the technologies that are being used today that assist scientists in their understanding of DNA as a molecule. Suggest how this information can be passed on to society.

Communicating *p. 262*

1. Propose two appropriate ways to present information on the structure and function of DNA.
2. Construct one of your suggested methods from Question 1.
3. Construct a digital timeline that shows the roles of the scientists discussed in this section and their contributions to our understanding of DNA.
4. Construct a scientific argument stating which scientists you believe should have been awarded the Nobel Prize for the discovery of DNA, using evidence from the information in the spread.

Success criteria

- I can identify the scientists involved in the discovery of the structure of DNA.
- I can discuss the contributions made by each scientist.

Summary

- Deoxyribonucleic acid (DNA), contained within the nucleus of nearly every cell, contains all of our genetic material.
- Genes are comprised of sections of DNA.
- DNA is packaged and stored as chromosomes, which can become condensed.

Chromosome
Nucleus
Cell
DNA
Gene

- Mitosis is cell division that produces identical body cells. It occurs in four stages: prophase, metaphase, anaphase and telophase.
- The process of mitosis is used for growth and repair in cells.
- Meiosis is cell division that produces unique sex cells (gametes). It completes the stage of mitosis twice in a row, making four daughter cells.
- The process of meiosis is used to produce gametes for reproduction, which increase genetic variation in populations.

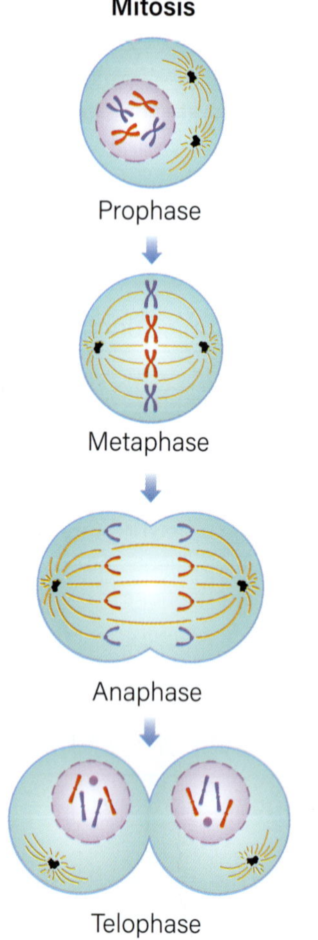

- A molecule of DNA is comprised of four nitrogenous bases and a sugar–phosphate backbone, arranged as a double helix.
- The nitrogenous bases always pair together in the same way: adenine (A) with thymine (T), and cytosine (C) with guanine (G).
- DNA replication occurs to duplicate the DNA contained within a cell, ready for division.
- The bases within DNA allow our cells to produce proteins, using transcription and translation, so that our bodies can function.

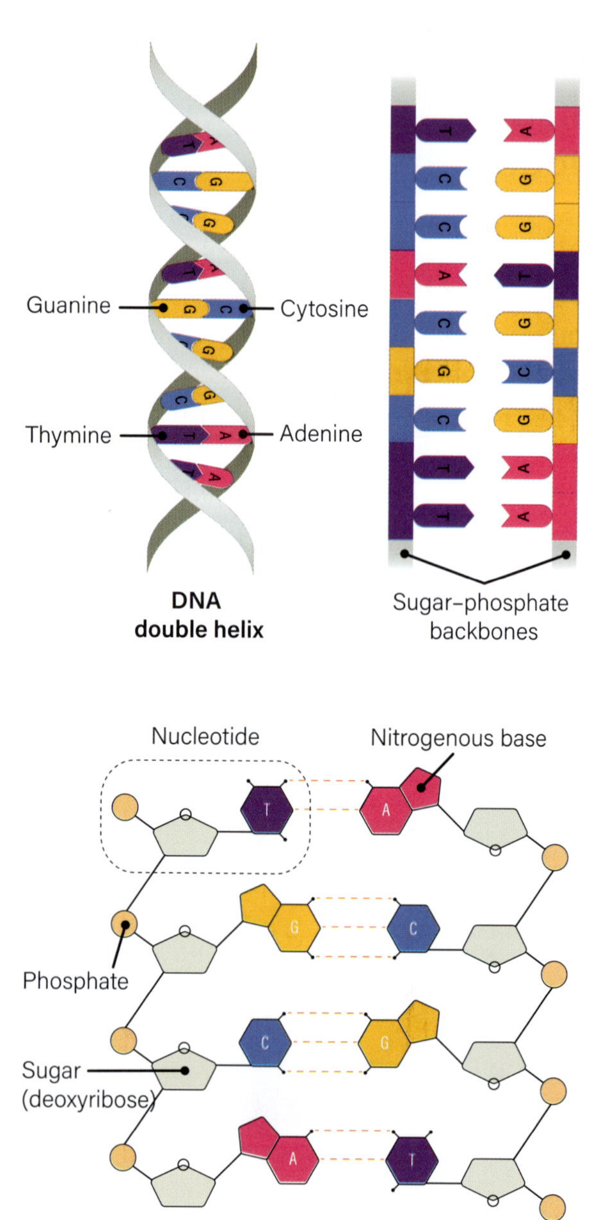

- Variations of a gene are called alleles. You have two copies of each gene in your body.
- The combination of genes is called the genotype, while the physical expression of the trait (what it looks like) is called the phenotype.
- Some traits are controlled by a single gene (monogenic), while others are controlled by multiple genes (polygenic).
- Gregor Mendel studied pea plants and developed a 3:1 ratio of inheritance for monogenic traits.
- Dominant alleles always mask recessive alleles. A dominant trait can be expressed from two genotypes, homozygous or heterozygous (AA, Aa), while a recessive trait will only be seen if the homozygous recessive genotype is present (aa).

- Changes to our DNA sequence or chromosomes are called mutations.
- Point mutations affect only a single nitrogenous base, while chromosomal mutations affect sections or complete chromosomes.
- Mutations can be beneficial (e.g. lactose tolerance), harmful (e.g. sickle cell anaemia) or have no effect.

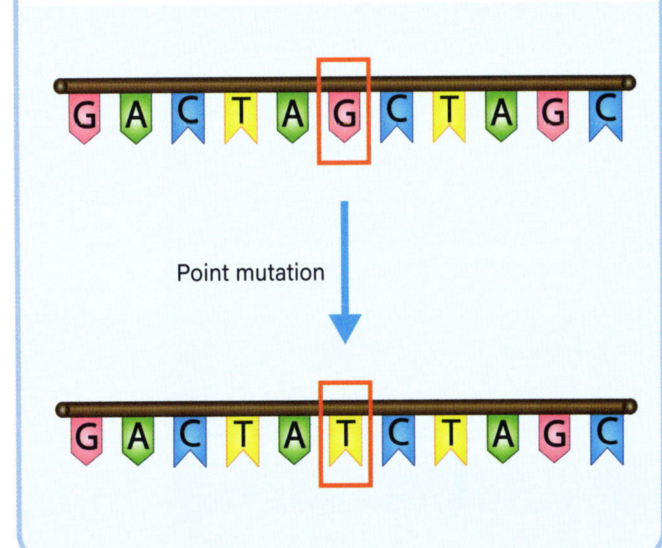

- You inherited your genes from your parents, when gametes fused during fertilisation to create you!
- Inheritance of specific genes can be tracked across single or multiple generations.
- Punnett squares predict the probability of offspring inheriting a specific set of genes across one generation.
- Pedigrees can be used to map the occurrence of a specific trait across multiple generations.

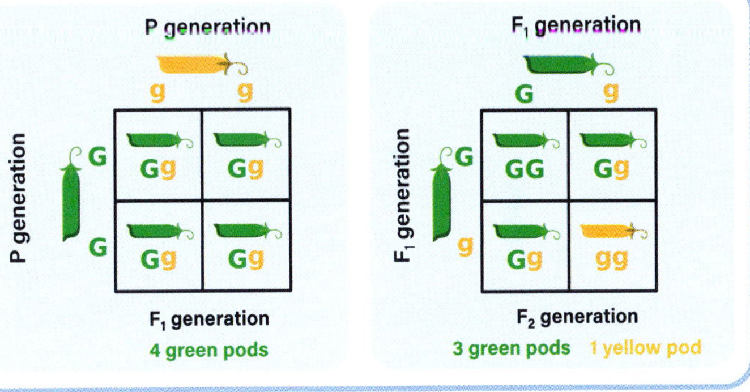

Individuals with attached earlobes

Generation I: 1, 2, 3, 4

Generation II: 1, 2, 3, 4, 5

Individual 4 in Generation II

Generation III: 1, 2, 3, 4, 5

Generation IV: 1, 2, 3

Key

- Male (unaffected)
- Male (affected)
- Female (unaffected)
- Female (affected)
- Twins
- Mating

Key idea: Form and function

- The structure of DNA was discovered by Watson, Crick and Franklin.
- Photos showing the form of DNA as a molecule have helped scientists to understand its function in the body.

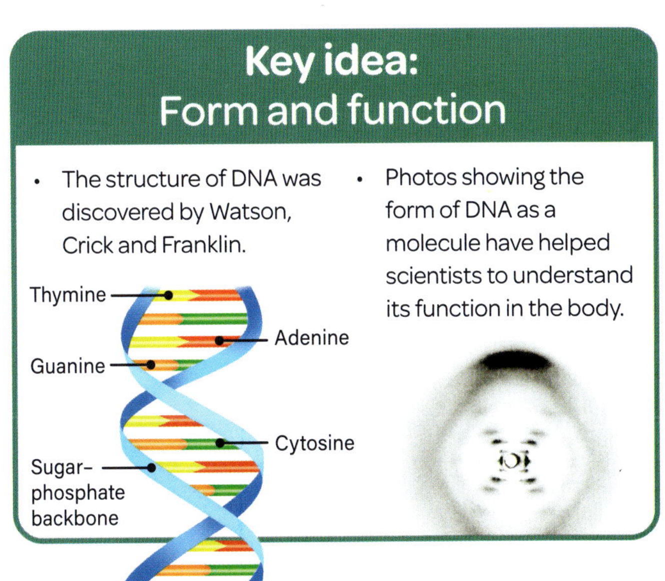

Masterclass

Steps in progression		1	2
Science understanding	Genetics	Identify the feature(s) of cells studied in genetic testing.	Describe a benefit of using genetic testing to conduct population screening for diseases.
Science as a human endeavour	Nature and development of science	Identify a technology in the images showing DNA profiles.	Describe a way that genetic testing has helped to refine our understanding of DNA.
Science as a human endeavour	Use and influence of science	Identify an issue with genetic testing.	Describe two different ways in which required genetic testing may be interpreted by the public.
Science inquiry	Processing, modelling and analysing	Research and construct a table showing the percentage of individuals in the population with an additional five genes that have been linked to cancer risks.	Construct a graph of the data presented in Table 1.8.
Science inquiry	Communicating	Propose a way that you could communicate information about genes linked to increased cancer risks to a rural community.	Construct your suggested communication from Question 1.

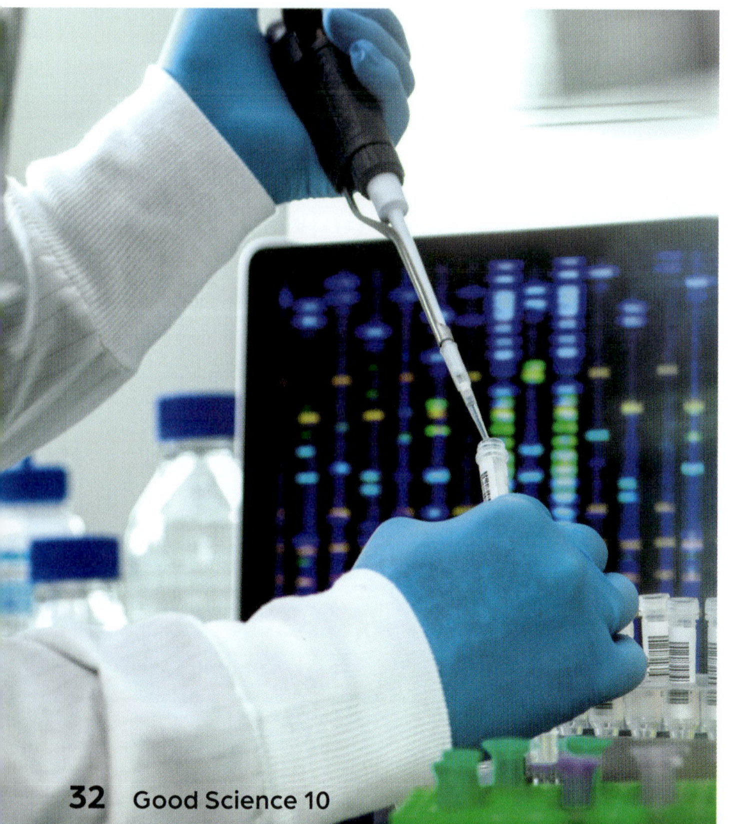

Genetic testing can help to detect diseases

Genetic technologies such as DNA profiling can be used to map the genetic markers of a person from a sample of their DNA. (Blood or saliva are most commonly used for this purpose.) Mapping allows scientists to screen for diseases and/or disease markers such as an increased susceptibility to certain cancers.

These kinds of technologies can be used at a population level to help identify at-risk groups, in order to reduce disease incidence or to monitor people more closely for early detection of disease. While this is advantageous, genetic testing is still new and therefore is limited to those with access to proper health care. It also cannot detect all diseases.

Figure 1.33: Scientists use precise DNA mapping technologies to screen for specific traits and disease markers, which become apparent in DNA profiles.

Demonstrate your understanding

3	4	5	
Explain how genetic testing can be used to provide insight into genetic diseases.	Analyse the outcome of genetic testing for disease markers such as cancer risks.	Evaluate the impact that genetic testing may have on cancer diagnosis and survival rates in a population.	
Explain how the use of genetic testing may help to develop new technologies to prevent or treat diseases.	Discuss how scientists could determine the accuracy of their information on the percentage of the population likely to have a specific gene.	Analyse how advances in genetic testing technologies could enable further advances in scientific studies in additional areas, using secondary research as needed.	
Explain how a gene linked to kidney cancer being found in a large proportion of the population may influence research.	Discuss how misinformation about the purpose of genetic testing may influence an individual's likelihood of participating in studies.	Analyse the factors that have led to information about the link between genes and cancer becoming more well known in society.	
Identify any patterns in the data in Table 1.8. Explain how this may influence an individual's level of worry about cancer-related genes.	Consider any of the genes in Table 1.8 that would be considered anomalies. Discuss your considerations at a population level.	Calculate the range of the data in Table 1.8. Analyse the impact of applying this data to whole countries such as Australia, with a population of approximately 26 million people.	Science how-to p. 244
Use a digital platform to construct an infographic showing what genetic testing is and the diseases it can help to detect.	Construct an argument on the benefits of genetic screening that you could provide to patients at a doctor's surgery.	Conduct secondary research and use the information below to justify your argument, including scientific references.	Science how-to p. 262

Table 1.8: A selection of alleles that increase the risk of certain cancers, which could be detected using genetic testing

Gene name	Found in population (%)	Type(s) of cancers with increased risk of occurrence
ATM	0.4	Breast
BRCA1	0.1	Breast and ovarian
BRIP1	0.2	Ovarian
FH	0.1	Skin, kidney and uterine
MUTYH	2.2	Colorectal
VHL	0.003	Brain and kidney

Figure 1.34: Scientists can read and interpret DNA profiles ▶ to provide them with many different pieces of information.

2.0 Evolution

The world around us is home to an incredibly diverse range of living things: plants, animals, microorganisms and more. But how did this occur? Life on Earth has changed and diversified over time, becoming more complex. The mechanism behind these changes is evolution. Proposed by Charles Darwin and Alfred Russel Wallace, the theory of evolution by natural selection explains how environmental conditions have caused species to change, diversify and increase in number since the first single-celled organisms appeared on Earth. By passing on favourable characteristics over generations, life has been able to survive for billions of years!

Figure 2.1: The unique characteristics of different species of finch (such as the cactus finch – *Geospiza propinqua*) were one factor that led Charles Darwin to propose his theory of evolution by natural selection.

Learning Ladder

The Learning Ladder for each chapter maps the Science Understanding, Science as a Human Endeavour and Science Inquiry strands that will be covered. Each ladder has five levels of progression, called steps. To climb the ladders, you need to develop fluency at each step. This will help you develop the ability to complete tasks that are more complex.

Steps in progression	Science understanding: Biological science: Evolution	Science as a human endeavour: Nature and development of science	Science as a human endeavour: Use and influence of science
5	I can evaluate evidence for how the diversity of life has changed over time	I can analyse how advances in technologies enable advances in science	I can analyse the key factors that contribute to scientific knowledge being adopted more broadly by society
4	I can analyse a range of evidence for evolution	I can discuss how scientific knowledge is validated, including the role of publication and peer review	I can discuss how scientific information and misinformation may inform personal and social decision-making
3	I can explain the theory of evolution using examples	I can explain how science has contributed to developments in technologies and engineering	I can explain how the values and needs of society influence the focus of scientific research
2	I can describe different types of evidence that support the theory of evolution	I can describe how scientific knowledge is refined over time	I can describe scientific knowledge that may be interpreted in different ways
1	I can identify processes involved in natural selection	I can identify technologies that have enabled advances in science	I can identify scientific knowledge that can address socio-scientific issues

Figure 2.2: The Galapagos Islands are home to many unique species of plants and animals that provide supporting evidence of evolution.

Science inquiry

Steps in progression	Questioning and predicting	Evaluating
5	I can develop explanatory models when investigating scientific questions, problems and claims	I can evaluate the validity and reproducibility of investigation methods
4	I can discuss what is needed for a question to be investigable or a prediction to be reasoned	I can construct evidence-based arguments to justify conclusions, address ethical issues or assess claims
3	I can develop a hypothesis that predicts the relationship between investigation variables	I can discuss ways to improve the quality of data and validity of conclusions and claims
2	I can formulate questions to investigate scientific problems	I can describe the impact of assumptions and errors, and propose ways to reduce them
1	I can make simple predictions based on what I know and observe	I can identify assumptions and types of errors in an investigation

2·1 ► Evolution basics

Learning intention

At the end of this lesson, I will be able to outline the process of evolution and describe how it occurs in populations.

Key terms

common ancestor: a previous version of a species that has evolved into two or more recent species

natural selection: the process in which organisms that are better suited to their environment tend to survive and reproduce, passing their traits on to further generations

theory of evolution by natural selection: a theory that says species have evolved adaptations over time that helped them to survive in their environment

Key idea: Stability and change

Figure 2.3: ► Darwin made observations about the platypus when he saw it in Australia. He noted that it behaved a lot like European water rats, despite clearly being a different species of animal.

▲ **Figure 2.4:** Being camouflaged in an environment can enable many species, such as this rabbit, to hide from predators, helping them to survive.

In December 1831, naturalist Charles Darwin set sail from England on the HMS *Beagle*, as part of a crew aiming to map the coastline of South America. This five-year journey took Darwin and the crew to countries all over the world, including Australia and the Galapagos Islands. Importantly, collecting and recording observations of plants and animals while on the trip led Darwin to his most well-known scientific discovery.

Charles Darwin and Alfred Wallace proposed the theory of evolution by natural selection

Scientists Charles Darwin and Alfred Wallace both studied the diversity of life. Throughout the 1800s, they travelled independently to different parts of the world, making observations about living things and collecting evidence. Separately, they came up with the idea that species evolved by passing down genetics for adaptations that helped them to survive in their environment. In 1858, they published a paper together outlining the theory of evolution by natural selection.

Darwin, who also published a book titled *On the Origin of Species*, is much more commonly associated than Wallace with the theory of evolution by natural selection.

Natural selection is how species evolve

An important component of this scientific theory is natural selection. **Natural selection** is the process by which animals evolve, which can eventually lead to the formation of species. For natural selection to occur, there must be a selection pressure present in an environment that impacts on a particular species. Animals with a favourable characteristic are more likely to survive and pass on their genetics (via reproduction) to the next generation. Over time, this means that more of the species will inherit the characteristic. When multiple different characteristics have changed, the organism may even be classified as an entirely new species! Natural selection helps organisms to survive in changing environments. The steps of natural selection will be discussed in more detail later in the chapter.

Evolution is supported by different pieces of evidence

Darwin and Wallace's observations were just one component of the **theory of evolution by natural selection**. Since their original paper, many scientists have studied evolution and how it occurs.

As technologies are developed (such as microscopes, genetic screening, and devices to detect radioactive isotopes), evidence that supports this theory has been put forward. These pieces of evidence are briefly summarised in Table 2.1 and will be discussed in more detail later in the chapter.

Organisms share common ancestors

Populations of animals that exist all over the world are related. Scientists are able to study species that have separated and determine, through their DNA, if they share a common ancestor. **Common ancestors** are previous versions of particular species that have become two or more new species as populations become isolated and experience different selection pressures. By working backwards, scientists were able to determine the last universal common ancestor (LUCA): a single-celled organism that is believed to have lived near thermal vents on the ocean floor approximately 3.9 billion years ago!

Table 2.1: There are many pieces of supporting evidence for evolution.

Evidence	Description	Why it supports evolution
Fossils	Remains of living things that previously lived/existed on Earth	Shows how organisms have changed over time
Similar features	Specific physical traits organisms have that are similar to other organisms	Helps to show relationships between living things; shows features that benefit survival in different environments
Organism distribution	Many living things are found in lots of different places across the world; some are found only in unique locations	Shows that organisms could have lived on previous landmasses and become isolated; demonstrates how an organism's survival is specifically linked to its environment
Genetic similarities	The DNA of all living things is comprised of the same genetic components, and closely related species have more similar genes	Shows that organisms must have shared a common ancestor and that we have genes in common

Learning Ladder

Evolution

1. Identify two processes involved in natural selection.
2. Describe one of the supporting pieces of evidence of evolution by natural selection.
3. Explain the theory of evolution by natural selection. Include an example of an organism Darwin observed.

Nature and development of science

1. Identify two technologies that have helped scientists to process evidence relating to evolution.
2. Describe a way that one of these technologies has refined our understanding of evolution.
3. Explain how scientists trying to explain the process of evolution may have contributed to research into technologies relating to DNA.
4. Discuss how Darwin and Wallace's work has been validated through peer review.

Questioning and predicting p. 234

1. Predict which of the animals below would survive best in a snowy environment. Provide a reason for your response.
   A Lion **B** Camel **C** Polar bear
2. Construct a scientific question that you could use to investigate the impact of an animal's structural features on our understanding of the environment it lives in.
3. Outline a hypothesis that would relate to your constructed question from Question 2.
4. Discuss the features of your constructed question and hypothesis that make them investigable. How might they be improved?
5. Use the information from the text to develop a model that supports the claim: 'Life on Earth has changed significantly over time.'

Key idea: Stability and change

The theory of evolution by natural selection provides a mechanism that explains how species change in response to their environment. There are also many species that do not evolve much over long periods of time. Use secondary research to identify a species whose features have remained stable, and analyse the reasons for this stability.

Success criteria

- I can define the term 'evolution'.
- I can outline how evolution occurs in different populations.

2.2 ▸ Processes of evolution

Learning intention

At the end of this lesson, I will be able to explain how evolution causes changes within and between populations.

Key terms

abiotic: non-living

isolation: when a population is cut off from others

population: a group of organisms of the same species that live and interact within the same area

predation: when an organism preys on another organism within an ecosystem

selection pressure: a factor that can influence which traits are beneficial to the survival of organisms

variation: the difference between individuals in a species, in both genetics and phenotypes

Investigation 2.2

Natural selection in practice, p. 316

Key idea: Stability and change

▲ **Figure 2.5:** Giraffes' long necks are thought to be the result of natural selection.

Have you ever wondered why giraffes have such long necks? Most scientists now think that giraffes got their long necks through the process of natural selection. A giraffe with a longer neck was much more likely to find food (e.g. leaves high up on trees), and therefore to survive, than a giraffe with a shorter neck. So, giraffes with long necks would be more likely to reproduce and pass on the genes for longer necks to the next generation.

Variation can occur through many processes

A key aspect of natural selection is **variation** within **populations**. Variations occur for several reasons, including mutation, meiosis and fertilisation.

Mutation

When a mutation occurs in a point or a section of an organism's DNA, it can have a number of effects. Whether it is a beneficial or a harmful mutation, it can be passed on to offspring, causing variations within the population.

Meiosis

In the previous chapter, we learnt that meiosis is cell division that forms gametes, which allow species to reproduce. During meiosis, chromosomes 'cross over' so that matching chromosomes randomly swap sections of their genetic code, making the resulting cells different from their parent cells. This unique combination of alleles can lead to variations between parents and offspring.

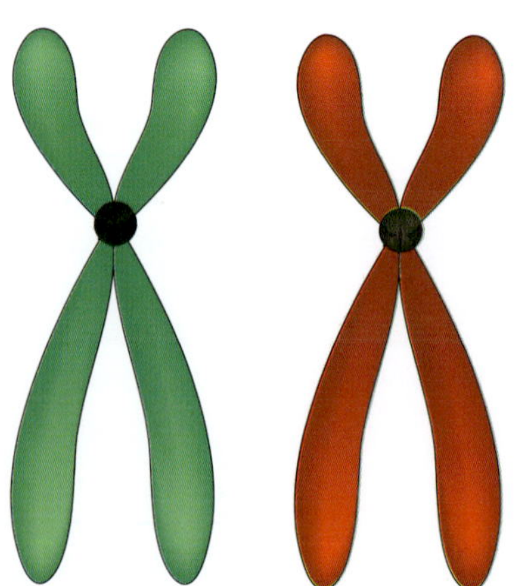

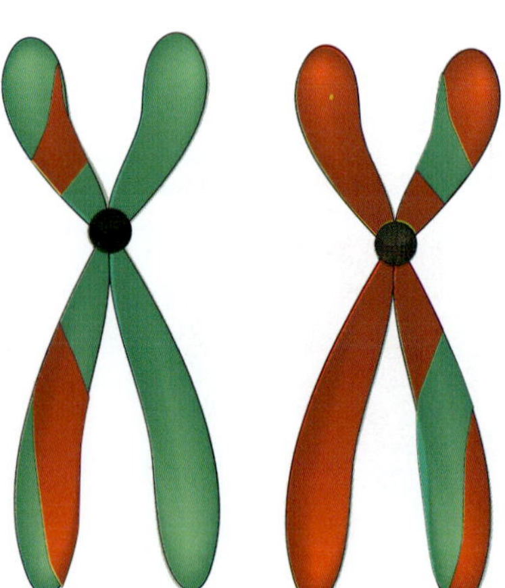

▲ **Figure 2.6:** Crossing over between homologous chromosomes can increase variations within a population.

Fertilisation

Part of the process of sexual reproduction is fertilisation, during which a male and a female gamete randomly fuse. A single new cell forms that then divides and develops into a new organism. The random fusing can lead to variations in offspring, who have a unique combination of alleles from their parents. In a population with a large number of alleles, a variety of phenotypes can be expressed.

Selection pressures affect species' survival

Ecosystems are complex and constantly changing. Various **selection pressures** within ecosystems can influence the survival of populations, as well as of individuals within populations. These factors include competition, predation, temperature changes and isolation.

Competition

All ecosystems contain limited resources. In some ecosystems, multiple species require the same resources, such as a common food source. This results in competition: the species compete to obtain and use the resource they both need. The population of the species that is more successful will rise, while the other population will fall.

Predation

Ecosystems include different plant and animal species. Animals that are omnivorous or carnivorous prey on other animals within the ecosystem. This is known as **predation**. An organism's ability to catch prey – or to avoid being prey – can influence its ability to survive. Prey can use camouflage and other measures to hide within its surroundings.

Temperature changes

Many **abiotic** factors can influence the survivability of an ecosystem, including soil conditions, exposure to sunlight and the availability of water. One abiotic condition is becoming increasingly relevant across many ecosystems – changing temperatures. Continued influences of climate change are causing temperatures to shift in a number of ecosystems – from rainforests and deserts to glaciers and coral reefs. Many organisms, including plant life, struggle to survive against rapidly changing climates.

Isolation

Sometimes populations of a particular species move to different locations and become **isolated**. As a result, the two populations of the species can be exposed to different environmental conditions. Over time, this isolation can cause changes in each population, eventually leading to the formation of entirely distinct species.

Figure 2.7: An organism's ability to camouflage can help it to survive against predators, helping it to pass on its genetics to offspring. In this image, the paler moth is camouflaged against the pale background, so is more likely than the black moth to survive and pass on its genes.

Natural selection passes beneficial traits to offspring

You learnt in Section 2.1 that the process of beneficial traits being passed from parents to offspring is called natural selection. Over time, natural selection can cause changes in the frequencies of certain genes and traits occurring within a population, as those organisms with favourable characteristics survive and pass on the genes to their offspring. Table 2.2 lists the steps involved in natural selection and gives as an example the effects on a population of beetles. The effects of natural selection on the numbers of red and green beetles in the population over time are set out in Table 2.3.

Table 2.2: The five steps of natural selection

Step	Example
1 There is variation within a population.	A population of beetles includes both red and green beetles.
2 A selection pressure in the ecosystem causes one trait to become favourable for the population of organisms.	Bushes that the beetles shelter in lose their red flowers and retain their green leaves.
3 The selection pressure increases the survivability of some of the population, because their traits are favourable.	Green beetles can still camouflage in the bushes.
4 External factors reduce the number of organisms with the non-favourable trait.	Predation from birds reduces the number of red beetles because they are easier to see and prey on.
5 The surviving organisms reproduce, passing on the genes that produce the favourable trait, meaning that over time it becomes the dominant trait in the ecosystem.	The green beetles become more prevalent over generations.

Table 2.3: The change in the numbers of red and green beetles when the environment favours the green beetles

Time (days)	Number of red beetles	Number of green beetles
0	22	22
2	19	26
4	13	31
6	9	34
8	6	39
10	3	42

This process of natural selection happens for all organisms across ecosystems. When populations of the same species become isolated, they can be exposed to different selection pressures, causing independent changes. This will be discussed further in Section 2.5 on evolution.

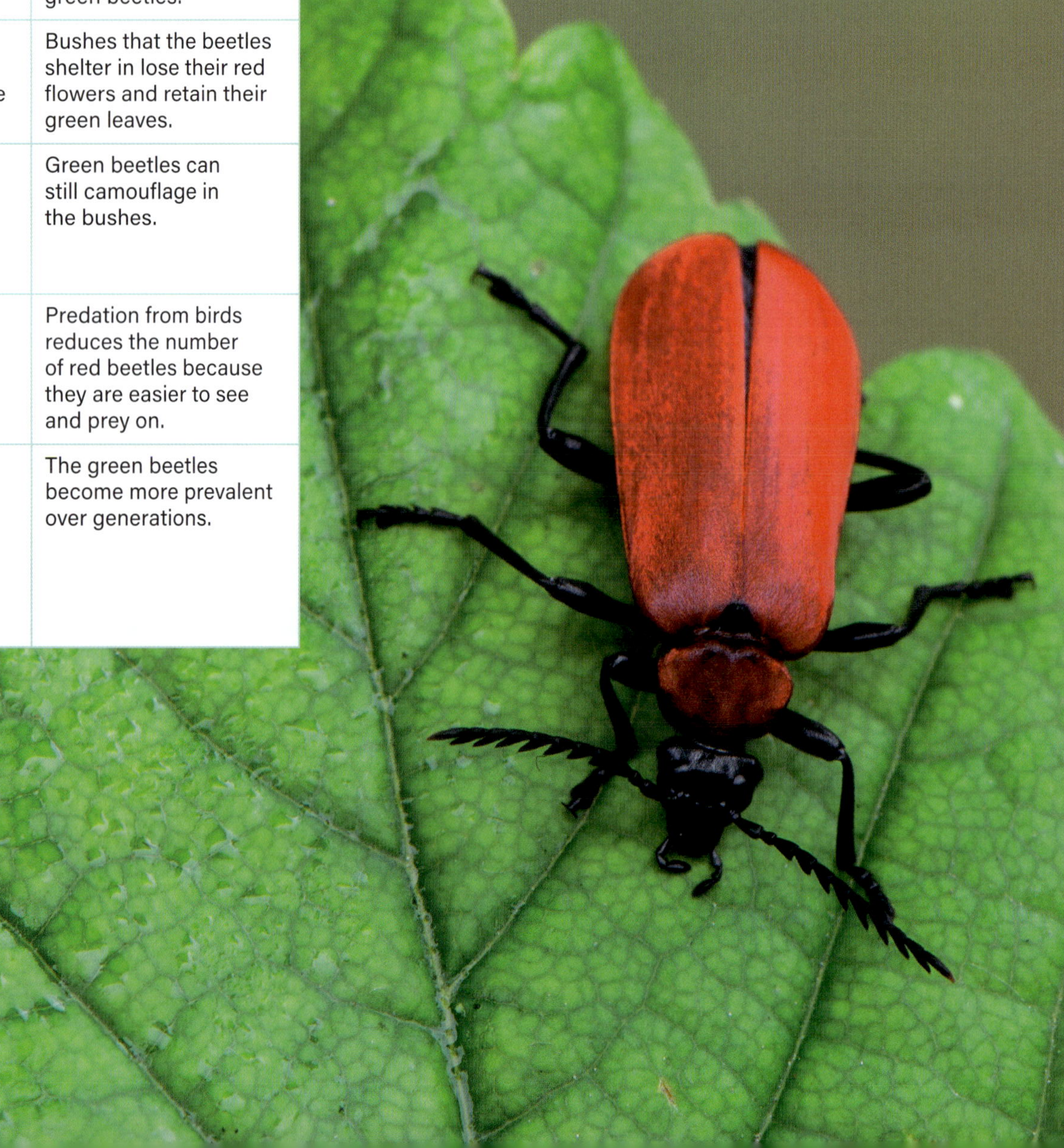

Figure 2.8: A red beetle on a green leaf is more visible than a green beetle on a green leaf, so it is more vulnerable to predation.

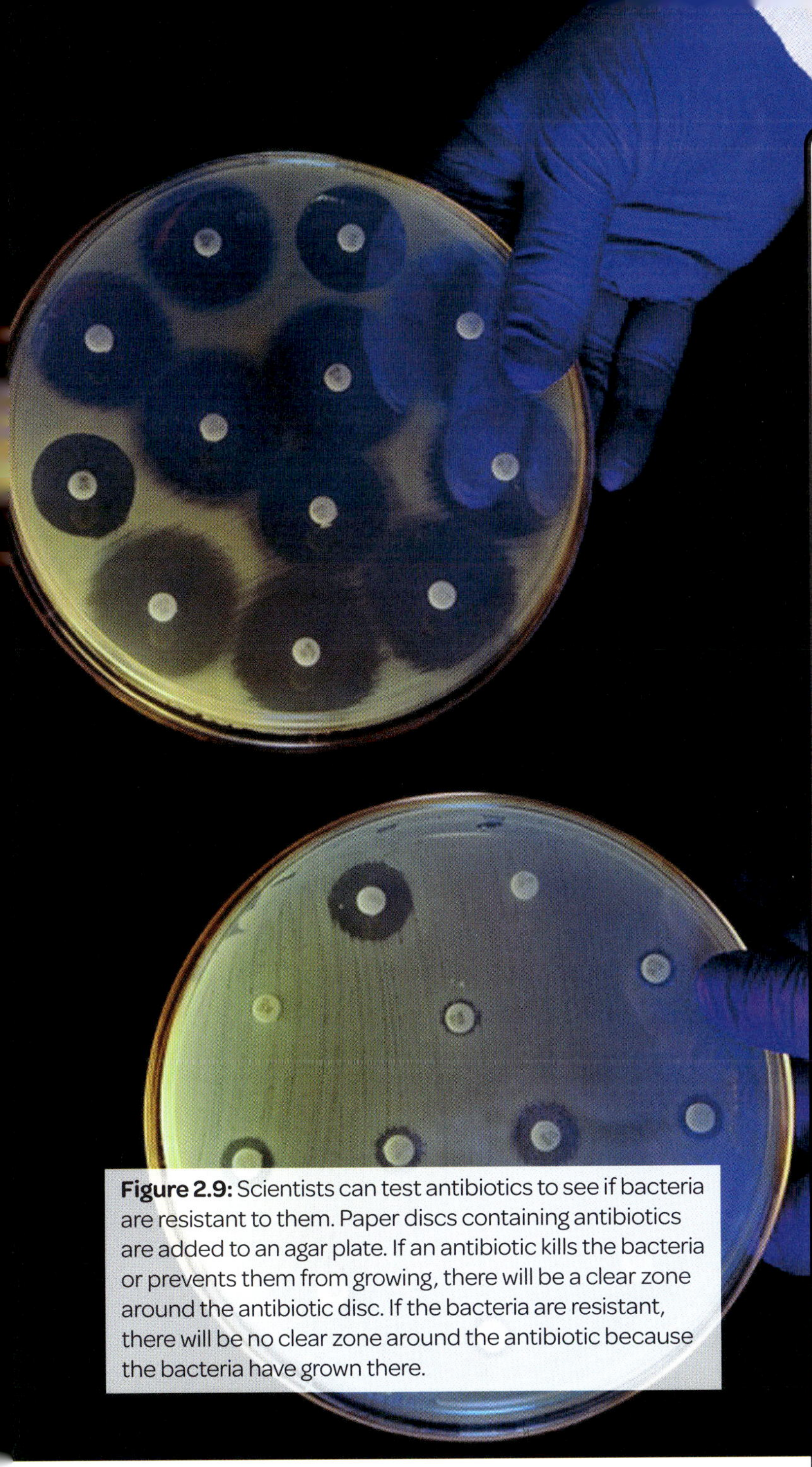

Figure 2.9: Scientists can test antibiotics to see if bacteria are resistant to them. Paper discs containing antibiotics are added to an agar plate. If an antibiotic kills the bacteria or prevents them from growing, there will be a clear zone around the antibiotic disc. If the bacteria are resistant, there will be no clear zone around the antibiotic because the bacteria have grown there.

Antibiotic resistance is an example of natural selection

The overuse of antibiotics has resulted in bacteria having increased resistance to antibiotics – this is an example of natural selection. The resistant bacteria have more chance of surviving and reproducing, and so antibiotic resistance increases throughout bacterial populations. Scientists are in a constant battle to stay one step ahead of the bacteria, synthesising new antibiotics – even though these can become ineffective very quickly.

Learning Ladder

Evolution

1. Identify three processes that can impact the level of variation within species.
2. Describe a way that the information in Table 2.3 provides supporting evidence for evolution.
3. Explain how the colour of an organism can affect its survivability.
4. Analyse how factors such as competition, predation and isolation can provide supporting evidence for the theory of evolution by natural selection.
5. Evaluate the impact of natural selection on the diversity of life on Earth.

Use and influence of science

1. Identify an issue that habitat destruction may cause for populations of species. (*Hint:* Think about their movement.)
2. Describe a way that people may interpret the results of the study shown in Figure 2.9 differently.
3. Explain how people's desire to have allergen-free animals to keep as pets may influence research into the evolution of dog breeds.

Questioning and predicting p. 234

1. Predict how a population of green caterpillars in your garden might change if you suddenly planted bushes that had purple leaves.
2. Construct a scientific question to use to investigate the impact of colour on population numbers. (*Hint:* Think about how an object that comes in various colours could be detected in the open.)
3. Construct a scientific hypothesis stating what the expected outcome of the investigation from Question 2 would be.
4. Discuss the factors in your question and hypothesis that make them investigable.

Key idea: Stability and change

Many species provide evidence of natural selection in the environment. Conduct research into another species and explain how changes in its physical characteristics may have been a result of natural selection. Identify and describe the selection pressures that led to a lack of stability in its environment.

Success criteria

- I can describe the process of evolution.
- I can explain how evolution causes change within and between populations.

2·3 ▸ Evidence of evolution

Learning intention

At the end of this lesson, I will be able to:

- explain how life has diversified over time
- explain how scientists use evidence from different sources to support the idea of evolution.

Key terms

absolute dating: determining the age range of an object, such as a fossil, in numbers of years

biogeography: the study of the past and present distribution of living organisms

comparative anatomy: the study of similar anatomical structures in different species in order to understand their evolution

comparative embryology: the study of the similarity of the embryos of different species as they develop

evolution: the way in which organisms change over generations as a result of adaptations that suit their environment

fossil: the geologically altered remains of a previously living organism

fossil record: a record of all fossils that are on Earth

homologous structures: features in different organisms that are similar in structure, but different in function

relative dating: arranging geological objects and rocks in a sequence from oldest to youngest

vestigial structure: a feature of an organism that has lost some or all of its function through evolution

Investigation 2.3A

Dating fossils, p. 318

Investigation 2.3B

Comparative anatomy, p. 319

Key idea: Stability and change

One of the closest relatives to humans is thought to be the bonobo, a great ape that lives in regions of the Congo in Africa. Bonobos live to about 40 years of age and communicate vocally and through facial expressions. Scientists have used different kinds of evidence to make this conclusion. Similarities between their structures, genetics and distribution provide evidence that many organisms have evolved from a common ancestor and show how life on Earth has diversified over time.

Life on Earth started as single cells

Scientists are still investigating to determine exactly when and where life on Earth started. Current information suggests that the earliest life forms came into existence about 3.5–4 billion years ago, but new discoveries may indicate that it was even earlier than that.

The earliest known life forms were very simple cells in the ocean. Earth's atmosphere had not yet formed, so life could not exist on land. Over time, these organisms expanded across the ocean and gradually became more complex. Structures made of more than one cell emerged, and organisms eventually made their way onto land. This diversification continued and is the reason for all the variety we see today.

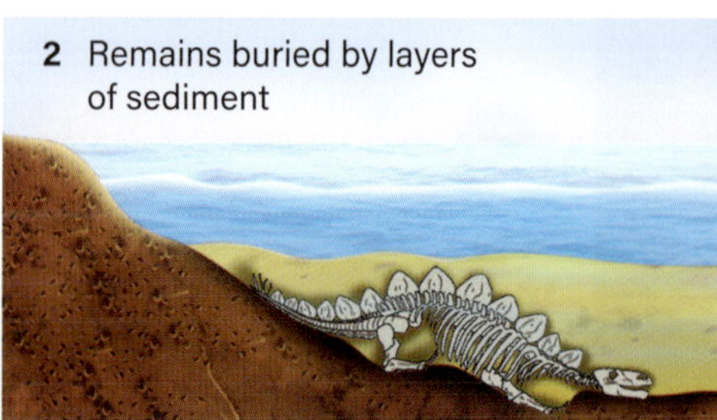

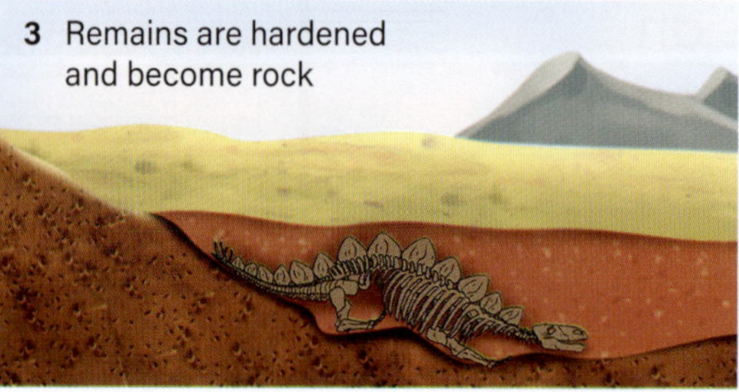

Figure 2.10: Fossils form over millions of years, usually in sedimentary rock.

▲ **Figure 2.11:** Fossils form when layers of sediment cover an organism. After millions of years, the organism can become an impression in solid rock.

Figure 2.12: The fossil record arranges fossils, showing evidence of the diversification of organisms over time. ▶

Fossils form over millions of years

Fossils are the remains, impressions or traces of long-dead organisms formed in rock and kept safe by geological processes. Fossils can be preserved bones or other remnants, shells, footprints, nests or eggs. Fossils are rare because they form only under certain conditions: the temperature, amount of oxygen and even soil acidity must be just right.

For fossils to form in rock, the following steps must occur over time.

1. The organism dies, and its remains settle somewhere they will remain undisturbed (e.g. under water). The soft parts of the organism decay, leaving behind structures such as bones and teeth.
2. The organism is covered with layers of sediment over time, which harden around it into rock.
3. The skeleton dissolves and leaves a cast that is filled with minerals.
4. The minerals crystallise and take the form of the mould left by the organism.

Fossils of complete organisms are also found in amber, which is hardened tree sap that encases small organisms or parts of organisms such as mosquitoes and leaves.

The fossil record provides evidence of evolution

Palaeontologists study the **fossil record**, which provides evidence of evolution. It shows how organisms have changed through geologic time from the simplest single-celled organisms to human beings. Although fossils have been found and dated from all around the world, many species are missing from the record, or their fossils have not yet been discovered. The gaps in the fossil record make it difficult to get an exact picture of all life forms. Also, it is hard to know whether a particular fossil is a good representation of a species.

Scientists date fossils in two main ways: relative and absolute dating. **Relative dating** gives an approximate age of a fossil by comparing it to other fossils or by studying the layers and ages of the surrounding stone. **Absolute dating** is commonly done by studying the radioactive minerals in the rock. Scientists can calculate the age of rocks by investigating how much the radioactive elements in them have decayed. Older fossils are of simpler organisms than younger fossils. This evidence supports the idea that life has diversified over time, because it suggests that much more complex organisms evolved from simpler ones.

Biogeography provides evidence of evolution

Biogeography is the study of the past and present distribution of living organisms. By studying where plants and animals are or were on islands and continents, scientists can start to piece together a complex and ancient puzzle.

Much of the evidence for biogeography comes from fossils. If fossils of one species are found in two separate continents, then this is compelling evidence that they were once connected. Australia has been an isolated island for many millions of years, which is why Australian marsupials and mammals are unique. However, even though Australia has been separated from the South American continent for millions of years, Australian marsupials still have some features in common with those in South America. This means that they probably evolved from a common ancestor millions of years ago. The wildlife in South America and Africa share more features than the wildlife in Australia and South America because that land separation happened later.

Common features provide evidence of common ancestors

Evolution is a gradual change in the physical characteristics of organisms over many generations. The first definite clues of evolution came by comparing the anatomy (especially the bones) of species to find similarities and differences. Scientists using **comparative anatomy** have found evidence to show that different species have evolved from the same ancestor. Parts of the body in different species that show evidence of common features are called **homologous structures**.

Vestigial structures are another type of evidence of common ancestors. These are parts of organisms that have lost their function as a species has evolved. They remain in the organism, sometimes even causing problems. Vestigial structures in humans include wisdom teeth, the appendix, the tailbone, and nipples on men.

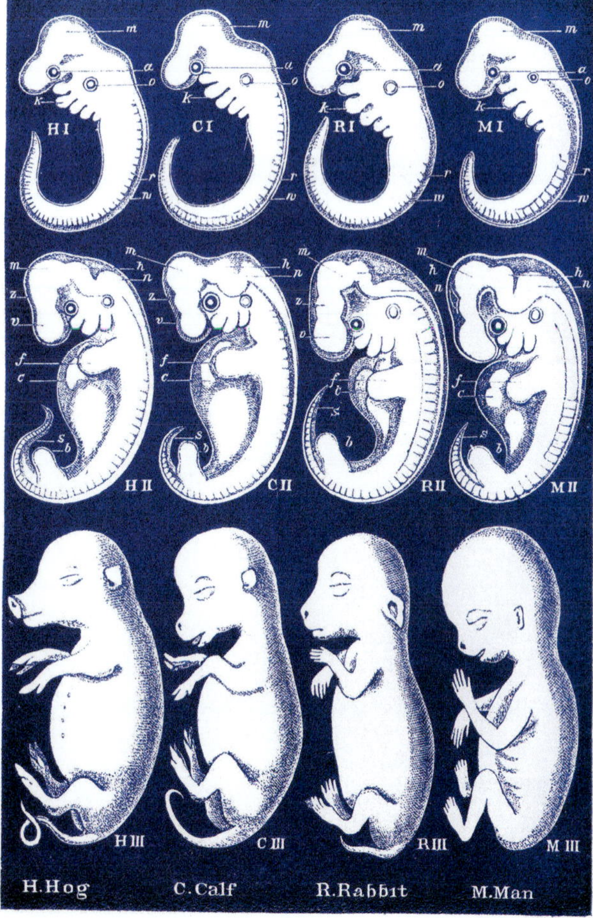

Figure 2.13: Comparative embryology shows the similarity between organisms early in their development.

Comparative embryology is another piece of evidence that supports evolution and the diversification of life over time. Scientists examine the developing embryos of different species to compare what they look like at different stages of development. Early in development, it is nearly impossible to see the difference between a human, bird, turtle and rabbit embryo. This similarity is further evidence that organisms have a common ancestor.

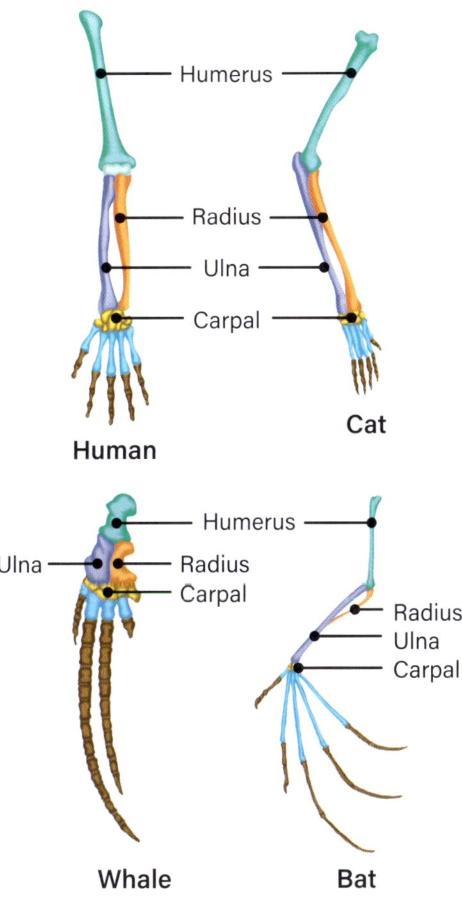

Figure 2.14: The human arm, cat leg, whale fin and bat wing are homologous structures (known as the pentadactyl limb) that show evidence of a common ancestor.

Genetic evidence shows how closely species are related

From 1990 to 2003, teams of scientists all over the world worked together to completely map the DNA of humans in a huge endeavour called the Human Genome Project. Now we can compare the DNA of different species and identify how similar they are. DNA mapping shows that humans and bonobo great apes share many genes, providing evidence that we shared a common ancestor.

Learning Ladder

Evolution

1. Identify two pieces of evidence that are directly linked to the structural features of organisms.
2. Describe three pieces of evidence and explain how each provides support for the idea of evolution by natural selection.
3. Explain how arranging fossils in order can show common ancestors linked to the evolution of new species.
4. Analyse one of the pieces of evidence presented in this section in relation to how strongly it supports the idea of evolution by natural selection.

Use and influence of science

1. Identify an issue with studying the embryos of organisms.
2. Describe some different ways that evidence from comparative anatomy may be interpreted.
3. Explain a potential reason why there may have been a need to study and present evidence on the evolution of species.

Evaluating

p. 253

1. Identify an assumption made when arranging fossils using relative dating.
2. Describe the impact of this assumption if the rock layers were arranged in a different way.
3. Propose a way to improve the quality of the evidence provided by comparative embryology. (*Hint:* Could multiple pieces of evidence be combined?)
4. Construct an evidence-based argument supporting the theory of evolution based on the information in this section.
5. Evaluate the validity of conclusions drawn about evolution from fossils that are arranged using only relative dating.

Key idea: Stability and change

Single-celled organisms such as bacteria have existed for billions of years. Analyse how they could have maintained stability in their populations, even when their surrounding environments changed.

Success criteria

- I can explain how scientific evidence supports the idea that life on Earth has diversified over time.
- I can explain how scientists use evidence from different sources to support the idea of evolution.

2·4 ▸ Evolution and caring for Country

Learning intention

At the end of this lesson, I will be able to understand how First Nations rock art provides important scientific knowledge of changes in plants and animals over many tens of thousands of years.

Key terms

Dreaming stories: complex First Nations stories that hold significant knowledge about history, creation, law, lore, and many other knowledges; have been used by First Nations Peoples for tens of thousands of years to care for Country and to live well

Gondwanaland: the supercontinent that broke from Pangaea approximately 250 million years ago and was made up of the modern-day continents of Africa, South America, Australia and Antarctica

megapode: a class of flightless bird that had large legs and feet and built nesting mounds in which to lay its eggs

Key idea: Stability and change

Continental drift – the splitting of the continents from **Gondwanaland** 250 million years ago – created the conditions and the time required for unique species of plants and animals to evolve on the Australian continent and adjacent islands. This meant that when the first humans arrived, there were many unique species of plants and animals in the environment.

First Nations Peoples hunted megafauna

On the eve of colonisation in 1788, First Nations Peoples had been in Australia for at least 65 000 years. During that time, they had coexisted with many now-extinct animals. There is a growing body of evidence that First Nations Peoples hunted and killed different types of megafauna, including *Diprotodon optatum*, a rhino-sized marsupial wombat, and *Genyornis newtoni*, a giant ground-nesting bird. The **megapode** was a flightless bird that reached more than 2 metres in height. Remains of these animals have been found in the Warratyi Rock Shelter in the Flinders Ranges in South Australia.

◂ **Figure 2.15:** This image of what the diprotodon could have looked like is based on skeletal remains, recorded rock paintings and the physiology of its closest living relative, the wombat. Features such as the colour and patterns on the fur pelt and the shape of the ears cannot be known with certainty.

The Warratyi Rock Shelter is considered to be one of the oldest inhabited places in the Australian arid interior. It had year-round abundant freshwater springs and ample edible plants for gathering and animals to hunt, including the megafauna.

Images of ancient megafauna

Dreaming stories and rock-art images refer to animals that are now extinct on the Australian continent and the adjacent islands. It was once thought that hunting by First Nations Peoples caused the extinction of the megafauna. However, evidence now indicates that a sharp fall in the numbers of megafauna coincided with the last Ice Age, when the drying out of the continent impacted their plant-based food sources.

Some First Nations rock art shows what the extinct megafauna may have looked like. One example is the *Thylacoleo carnifex*, also known as the marsupial lion. This animal lived in forested areas across the Australian continent until its extinction 35 000 years ago. Palaeontologists have excavated several skeletons of the marsupial lion, but it is the rock art that provides insights into the shapes and patterns on the pelt of this animal and the shape of its ears – both parts of the animal that have not been preserved. Interestingly, the climate where these paintings are located has changed significantly, and the rock art suggests what the climate may have been like many thousands of years ago.

Figure 2.16: A rock art painting of a human figure throwing a spear at what is thought to be a *Thylacoleo carnifex* (marsupial lion) estimated to have been created from 15 000 to 22 000 years ago in Western Australia. The fossil remains of the lion tell us very little about the colour and the pattern of the fur pelt (note the stripes in the image); however, this image provides important insights into the appearance of this extinct species. The Drysdale River painting may be the first direct evidence of an interaction between a human and a now-extinct animal in ancient Australia.

Ancient plants are depicted in rock art

There are many rock art sites in the Kimberley region in Western Australia where various plants are depicted. In many cases, the plants are painted in their unprocessed forms and as objects, tools and other items, or as foods. The images are depicted in different styles of art, as listed in Table 2.4.

The various styles show different plants from the local landscape, and their uses, changing patterns of growth and availability, over many tens of thousands of years. These images also provide an important understanding of how the climate and sea-level changes affected the ability of the plants to survive and thrive in the changing landscapes.

Figure 2.17: Water lilies, plum and long yam were common edible plants when the rock paintings were created. First Nations Peoples continue to use these plants today.

Table 2.4: The different styles of art represented at the Kimberley rock art sites

Style	Date	Description
Pecked Cupule	50 000 years ago	Known for its mosaics of rainforest, woodland and grasslands, which reflects the local climate at that time
Irregular Infill Animal	36 000–18 000 years ago	Indicates an increasing aridity in the local landscape
Gwion	18 000–14 000 years ago	Indicates significant changes to sea levels relating to the last glacial maximum of the most recent Ice Age
Static Polychrome	14 000–9000 years ago	Relates to when the monsoon seasonal weather pattern was re-established after the Ice Age and to the loss of coastal plain due to rising sea levels
Painted Hand	9000–5000 years ago	Coincides with the Holocene wet period and the resulting increase in the human population as a result of the environment becoming wetter and better able to sustain a larger population
Wanjina	5000 years ago to present day	Marked by greater volatility in summer rainfall and periods of drought

Learning Ladder

Evolution

1. Identify a characteristic or trait that the marsupial lion, the diprotodon and *Genyornis newtoni* shared that would have contributed to their extinction.
2. Examine Figure 2.15 and describe the traits shown in the image that would have been transmitted through generations of marsupial lions.
3. Explore the images of the native plants in Figure 2.17.
   a Identify a desirable characteristic.
   b Explain how First Nations farmers could have improved this characteristic.
4. Analyse how the art styles in the Kimberley, listed in Table 2.4, support natural selection as a process that influences evolution.
5. Identify a trait of the diprotodon that contemporary wombats do not have. Evaluate how this has ensured that contemporary wombats could survive the increased aridity in the landscape.

Nature and development of science

1. Identify a mechanism that First Nations Peoples used to communicate information about edible plants.
2. Describe a way that traditional rock paintings have helped to refine scientists' knowledge of Australia and its animals.
3. Explain how knowledge from First Nations Peoples' rock paintings may be used to inform farming practices in Australia.
4. Discuss how the painted image of the *Thylacoleo carnifex* in Figure 2.16 can help validate reconstructed images of this animal from its fossilised remains.

Questioning and predicting p. 234

1. Predict which of the following species in Australia would be most likely to become extinct based on its size: the platypus, koala or red kangaroo. (*Hint:* Consider megafauna extinctions.)
2. Construct a scientific question you could use to test the optimal growth conditions of the long yam.
3. Develop a hypothesis for your scientific question from Question 2 and explain why you have predicted the outcome you stated.
4. Discuss the factors needed to make your constructed question and hypothesis investigable.

Key idea: Stability and change

Another extinct Australian native species is the thylacine (Tasmanian tiger). While European colonists only recorded the tiger inhabiting Tasmania, evidence of thylacines has been found in rock paintings by Yolŋu People in far north Australia. Propose reasons for the change in tiger distribution and explain how these paintings provide evidence of change in organisms.

Success criteria

- I can describe how ancient rock art can provide important scientific knowledge from the past.
- I can explain how First Nations Peoples' recorded images can provide supporting evidence of evolution in Australia.

2·5 ▸ The theory of evolution by natural selection

Learning intention

At the end of this lesson, I will be able to:

- discuss the theory of evolution by natural selection
- explain how the theory of evolution by natural selection is supported by evidence.

Key term

biodiversity: the variety of life on Earth

Key idea: Stability and change

Living things have diversified since life began on Earth. How do scientists explain this?

Evolution is caused by natural selection

Biodiversity – the variety of life on Earth – is a result of natural selection. Favourable characteristics are passed from parents to offspring and become more common over generations. As populations separate and are exposed to different conditions, specific traits can become favourable, causing changes between the populations, which eventually become different enough to be seen as two distinct species. The gradual change over many generations is called evolution.

Evolution shows how living things are related

We learnt earlier that Charles Darwin and Alfred Russel Wallace proposed the theory of evolution by natural selection to explain this evolution. The theory, backed by their collected evidence and the work of other scientists, explains that all living things are related and share a common ancestor. By looking at the common ancestors of living things, and at previous organisms using evidence such as the fossil record, the theory of evolution suggests how all species are related. Scientists can directly compare the traits of two similar species and use these to make a first observation about how closely related they are. To strengthen this evidence, technologies such as DNA mapping can now be used to compare genes (and the order in which they appear) in the DNA of species. The greater the overlap, the more closely the species are related.

Galapagos animals helped Darwin develop the theory of evolution

While Darwin was on his sea voyage, he found enormous fossils and many species of animals that forced him to consider new theories for the origins of life on Earth. He also discovered several species of finches on the Galapagos Islands, off the coast of Ecuador. The finches looked very similar, but their beaks were different lengths and shapes. Darwin theorised that the finches had a common ancestor, but that they separated geographically to different islands and had different food sources, so they evolved different beaks through natural selection (Figure 2.18).

Darwin also studied the Galapagos tortoise. He observed that the environment and available food supply affected the tortoise's shell shape (Figure 2.19).

▲ **Figure 2.18:** The finches of the Galapagos Islands evolved different beaks in response to the food sources available to them.

Figure 2.19: The shell shape of the Galapagos tortoise is linked to its evolution. Saddle-shaped shells (a) with a gap at the front help tortoises with a limited food supply to reach food from higher sources, such as cacti, while lower dome-shaped shells (b) are prevalent where the tortoises have a large supply of lush, low-lying vegetation.

Many scientists have contributed ideas on evolution

Before Darwin and Wallace, many scientists contributed to our understanding of evolution. Jean-Baptiste Lamarck suggested that animals adapt to their surroundings. However, unlike Darwin, Lamarck suggested that this adaptation occurred during the life of the animal, rather than adaptations being inherited from parents.

Many scientists have since contributed to our further understanding of evolution. The study of fossils and how to date them, DNA and genetics, and mapping the genome of humans and other animals have all built on the work of Darwin and Wallace. Today, many different processes are used to provide supporting evidence for the theory of evolution by natural selection.

Learning Ladder

Evolution

1. Identify two adaptations of the Galapagos tortoises that remained similar, and one that changed, as they became isolated in different environments.
2. Describe a feature of the Galapagos finches that Darwin observed which supported his theory about the evolution of organisms.
3. Explain how the theory of evolution by natural selection can show the relatedness of species.
4. Analyse the use of the features of the Galapagos tortoise (Figure 2.19) as a piece of evidence that supports the theory of evolution by natural selection.

Nature and development of science

1. Identify a technology that has helped scientists to determine how closely related two species are.
2. Describe a way that the fossil record would allow the theory of evolution to be further refined over time.
3. Explain how scientists studying evolution may have presented ideas about observing evidence that led to developments in technology.
4. Discuss the importance of peer review in relation to scientists contributing information to the theory of evolution by natural selection.
5. Analyse how advances in available technologies led to advances in our understanding of the relationships between living things.

Evaluating

p. 253

1. Identify an assumption made by Darwin when he studied the finches and tortoises of the Galapagos Islands.
2. Describe the impact this assumption may have had on the theory of evolution by natural selection if it had been incorrect.
3. Propose a way that the validity of Darwin and Wallace's theory has been improved over time.

Key idea: Stability and change

Alfred Wallace also visited many places while developing his ideas about evolution. Conduct research to identify three organisms Wallace observed. Explain how changes in these organisms provided supporting evidence for the theory of evolution by natural selection.

Success criteria

- I can explain the theory of evolution by natural selection.
- I can discuss how specific evidence supports the theory of evolution by natural selection.

2·6 ▸ Key idea: Stability and change – evolution of the modern horse

Learning intention

At the end of this lesson, I will be able to apply my understanding of evolution to a specific modern organism.

Key term

ancestral species: an older species from which one or more recent species have evolved

Key idea: Stability and change

Scientists use structural evidence and the fossil record to track the evolutionary change of many animals, but one that can be clearly demonstrated is the evolution of the horse. Horses are mammals of the family Equidae, who live on a diet of vegetation.

The horse evolved across millions of years

After using absolute dating techniques on fossils to arrange them, scientists have been able to determine that the modern horse has evolved from a number of ancestors over the last 50–60 million years. Evidence seen in the horse's complete fossil record demonstrates that its early to recent ancestors travelled across a range of environments, and isolated populations caused a number of branches to form on their evolutionary path, creating multiple **ancestral species**. This led to the eventual emergence of the modern-day horse. Evolutionary changes in the horse that occurred throughout this time include an increasing body size, and changes to the teeth and foot structures.

Figure 2.20: As the horse evolved, its size increased across related species, as did the structure of the bones making up its foot. This image directly compares the size and foot shape of some of the ancestors of the modern horse.

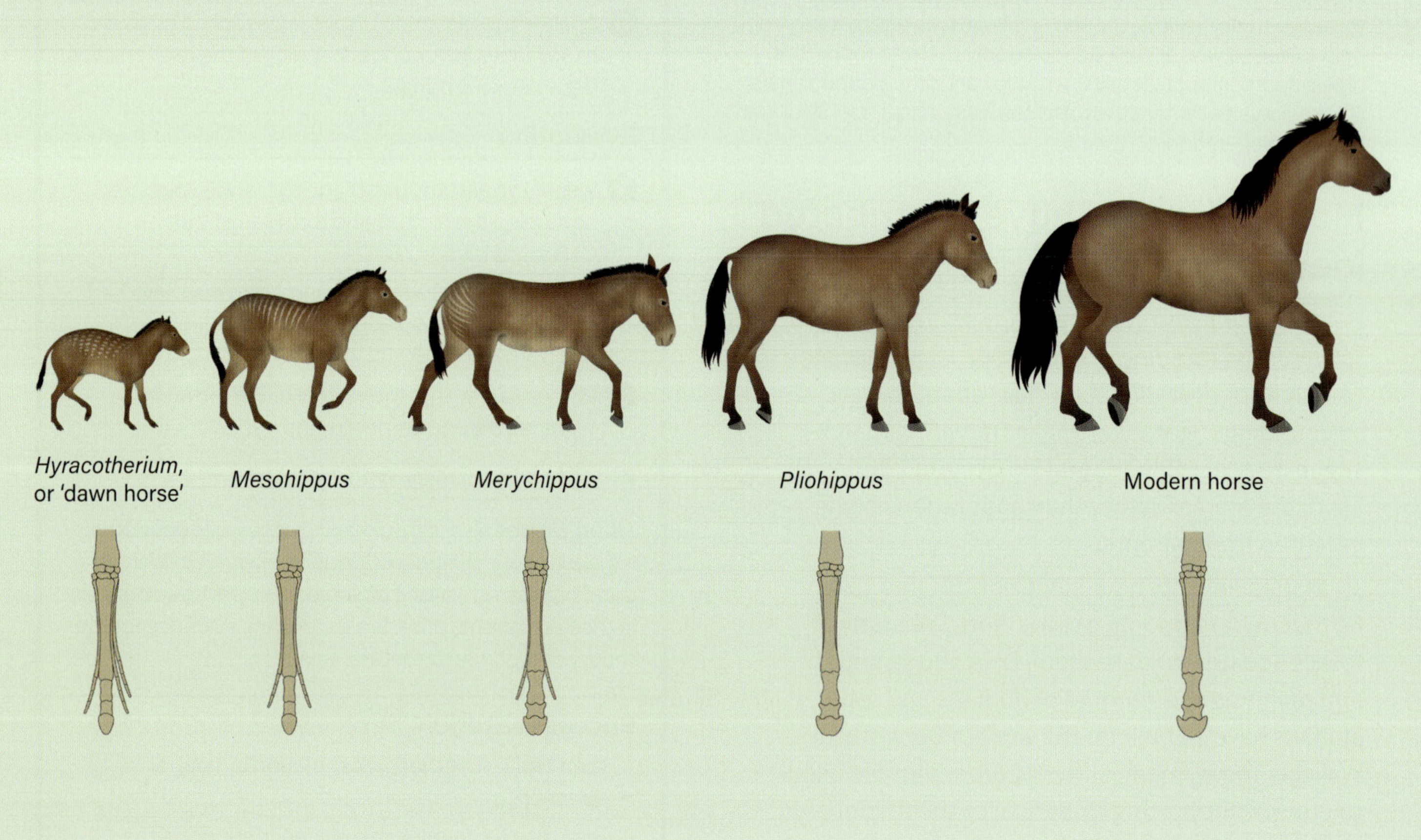

Figure 2.21: The fossil record, such as the fossil pictured here, provides supporting evidence of the evolutionary path of the horse.

Early ancestors of the horse were much smaller

Early ancestors of the horse survived within forests that experienced higher rainfall than environments the modern horse is adapted to. The first species, *Hyracotherium* (known as the 'Dawn horse'), was a small, dog-like creature known to exist around 55 million years ago. It measured only 25–50 centimetres in height at the shoulder (4.9 hands), had four toes on its front feet and three on its back feet, and its teeth were less flat than those of its modern relatives. *Hyracotherium* predominantly subsisted on a diet of soft plants and fruit. The higher rainfall of the time meant that the ground underfoot was often soft as well.

Figure 2.22: A 50-million-year-old fossil of a *Hyracotherium* skeleton. *Hyracotherium* was a similar size to a modern domestic cat.

Ancestors of today's horse responded to selection pressures

Over time, environments became drier, with forest areas giving way to grassland areas. Some of the coexisting early ancestors of the horse started to show adaptations, seen through the fossil record. One of the first noted changes to the structure of early ancestors of the horse was a change in their teeth. Flatter molars were developed, in response to vegetation becoming tougher and needing to be ground down more.

Other changes that are seen in the horse's more recent modern ancestors include increased size and changes in the structure of the foot. While *Hyracotherium* was very small and had four-toed feet in front, these features were not selected for as changes occurred in the climate. It became favourable for horse ancestors to be larger in size and to have a more prominent hoof, made up of fewer toes. Larger bodies made it harder for predators to successfully attack them, while the hoof gave them the ability to run faster and further over drier ground.

A species evidenced in the fossil record approximately 40 million years ago is *Mesohippus*. This ancestor of the horse had only three prominent toes, and was larger, measuring approximately 60 centimetres (5.9 hands) at the shoulder and around 1.2 metres in length. As climate drying continued, ancestral species such as *Merychippus* began to move out onto the grassland and plains environments. *Merychippus* was larger again, measuring about 1 metre at the shoulder (9.8 hands). The animal still had three toes, although the middle one was larger and more prominent than the two external toes.

Closer again to the modern horse was *Pliohippus*, seen in the fossil record approximately 5 million years ago. This species is considered the 'grandfather' of modern horses, standing 1.25 metres tall at the shoulder (12.3 hands), and with a true hoof – a single toe – that had become stabilised for running for longer distances over grasslands and plains. There were many ancestral species of the modern horse that existed within this time period, as populations travelled out and across different areas of the world via land bridges that do not exist today. The stability of this grassland environment meant that there were fewer evolutionary changes noted from this point on.

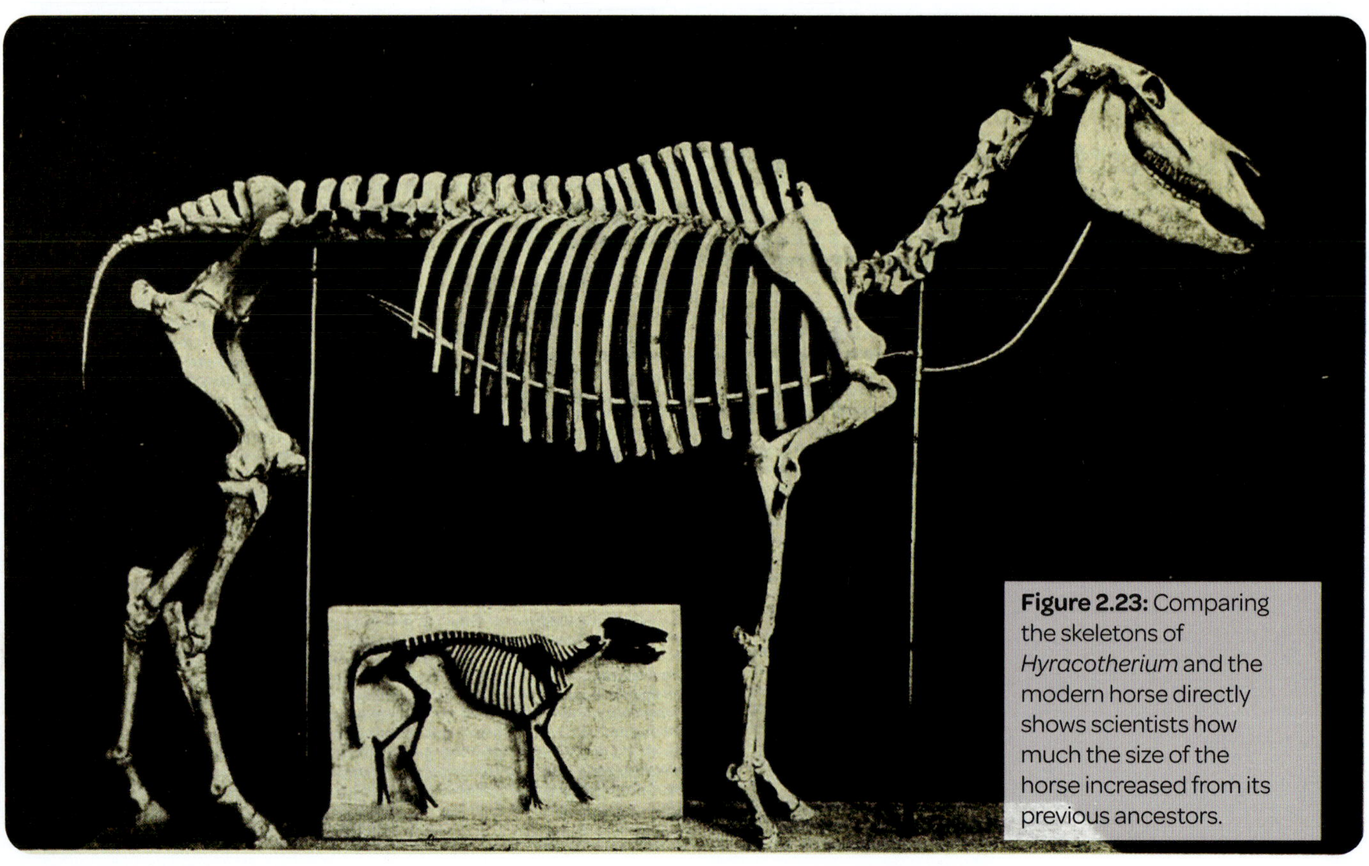

Figure 2.23: Comparing the skeletons of *Hyracotherium* and the modern horse directly shows scientists how much the size of the horse increased from its previous ancestors.

The modern horse of today

One of the new species that branched out from *Pliohippus* and subsequent branching species is *Equus* (the modern horse we know today). *Equus* measures at least 1.4 metres at the shoulder (13.3 hands) and no longer has vestigial remains of previously existing toes; instead, it balances and runs on one large hoof. Species of *Equus* have been found across many continents (excluding Australia and Antarctica). These horses, which humans have since domesticated and bred artificially into different types, continue to exist as wild colonies in grassland environments. Without evidence from the fossil record and their ability to analyse structural features across species, scientists would be unable to trace the ancestry of the modern horse and all of its previous ancestors, including the many branching species that existed before it.

Figure 2.24: Clydesdales are large, modern horses (at least 1.6 metres, or 16 hands, high) whose hooves are well suited to traversing long distances across grasslands.

Learning Ladder

Evolution

1. Identify two factors that caused changes in the anatomy of ancestors of the modern horse.
2. Describe the evidence scientists have used when looking at the evolution of the horse.
3. Explain how this evidence was used to demonstrate the evolutionary path of the horse.
4. Analyse the use of the fossil record to provide evidence of evolution.

Use and influence of science

1. Identify an issue that the use of the fossil record helps to solve when trying to determine the ancestors of a modern species.
2. Describe how the features of ancestors of the modern horse may lead scientists to make different assumptions about the environment it lived in.
3. Explain how the domestication and use of the modern horse may have promoted research into its features and evolution.
4. Discuss ways that misinformation about the collection and use of evidence of evolution may influence society's perspectives on scientists' findings.
5. Analyse the factors that may lead to the equestrian community having a broader understanding of horse evolution than society at large.

Evaluating

p. 253

1. Absolute dating requires scientists to analyse radioactive samples in fossils so they can determine their age. Identify a type of error that could occur when using this process to determine the age of a fossil.
2. Describe an impact that your identified error might have on an investigation into the ancestors of the horse, looking at the time periods they existed in.
3. Explain how the impact described in Question 2 could affect a conclusion drawn by a scientist conducting this investigation. Propose how that situation could be avoided, to improve the quality of the collected data.
4. Construct an evidence-based argument supporting the conclusions scientists have made about the evolution of the horse, using information from this section and secondary research.

Success criteria

- I can outline the evolution of the modern horse.
- I can examine how this evolution provides supporting evidence of change over time.

▶ Summary

- Evolution is the process of organisms changing over time to form new species.
- Charles Darwin and Alfred Russel Wallace developed the theory of evolution by natural selection after studying different species across the world.

Large ground finch
(seeds)

Cactus finch
(cactus fruit and flowers)

Vegetarian finch
(buds)

Woodpecker finch
(insects)

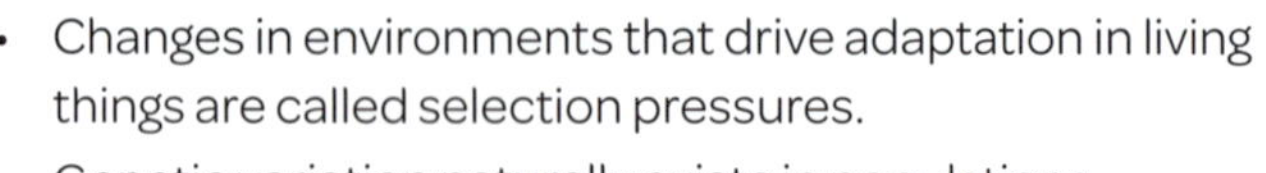

- Changes in environments that drive adaptation in living things are called selection pressures.
- Genetic variation naturally exists in populations, which can influence how they change.
- Isolation of one population from another can cause different changes within each population based on their environments.
- The mechanism of evolution is natural selection.
- In natural selection, organisms with favourable characteristics can survive and pass on their genetics to future generations, which can cause populations to change over time.

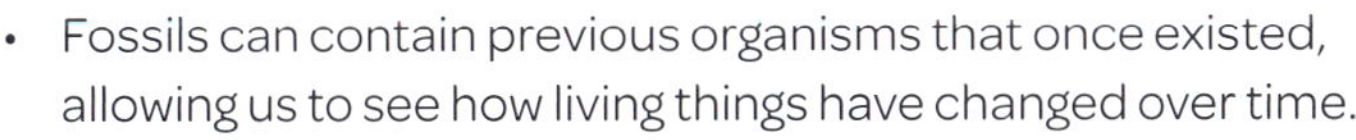

- Fossils can contain previous organisms that once existed, allowing us to see how living things have changed over time.
- Scientists can date and order fossils to show changes over time. This is called the fossil record.
- Biogeography provides information about the distribution of living things over time.
- Comparative anatomy shows the similarities in structural features shared by living things.
- Comparative embryology shows the similarities in embryo development of living things.
- Comparing the DNA of species shows that all living things share a genetic code, meaning they must have shared ancestors in the past.
- Evidence of evolution shows that species are all related and previously shared a common ancestor.

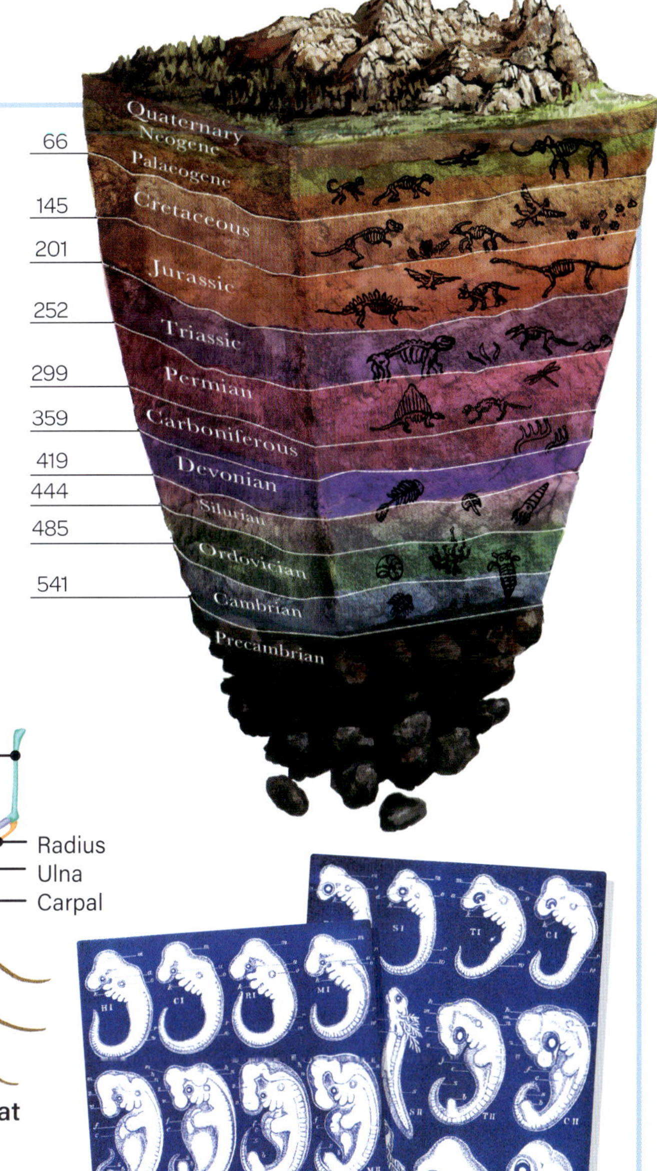

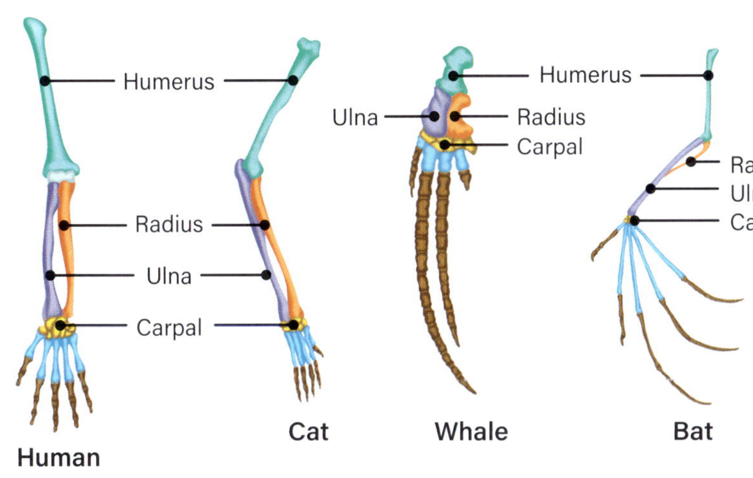

Key idea:
Stability and change

- The modern horse demonstrates how organisms can change over time as their ancestors evolved and formed new species in response to selection pressures.
- Stability within an organism's environment can mean that it undergoes fewer evolutionary changes once it is suited to its environment.

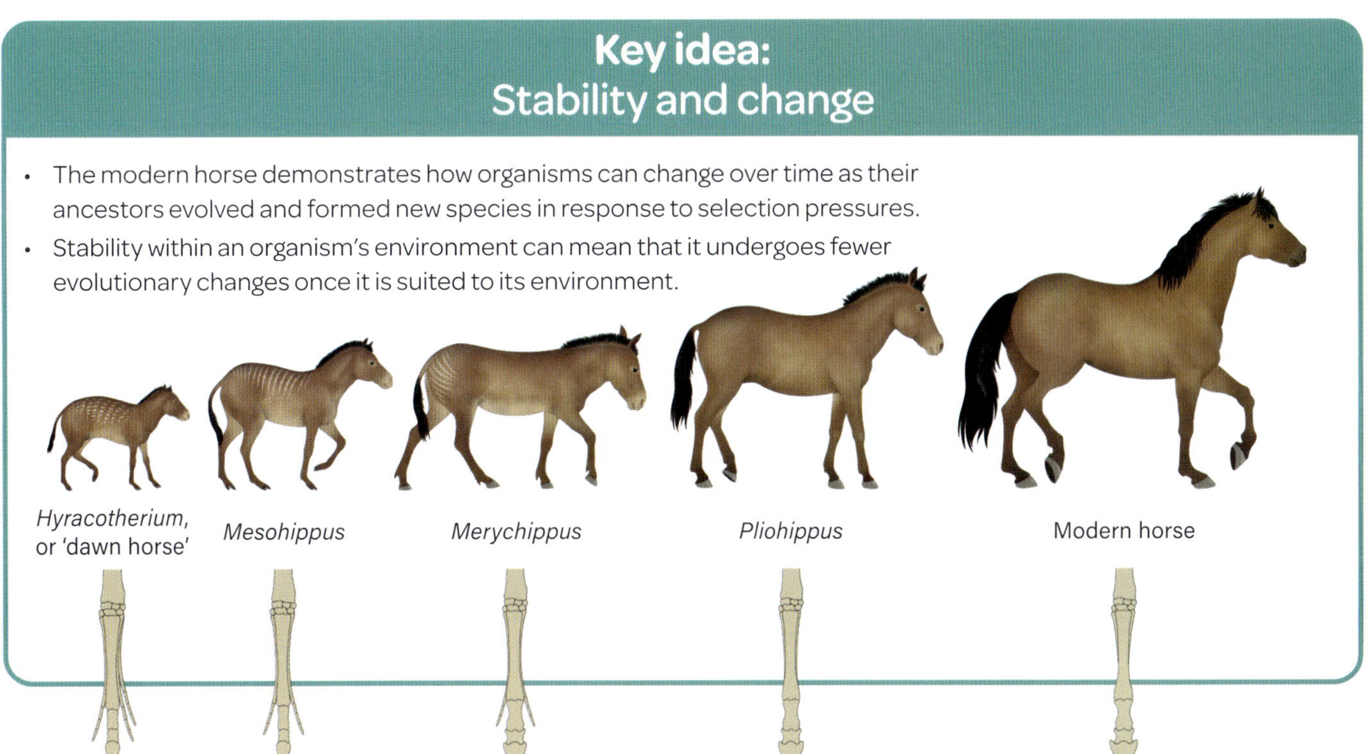

Masterclass

Steps in progression

		1	2
Science understanding	Evolution	Identify two selection pressures that led to the changes in the Galapagos snails.	Describe a feature of the snails that has changed over time as a response to the selection pressures.
Science as a human endeavour	Nature and development of science	Identify a technology that scientists can use to study living things such as snails.	Describe a way that our knowledge of organisms of the Galapagos Islands has changed over time.
Science as a human endeavour	Use and influence of science	Identify a piece of knowledge about the studies of animals of the Galapagos Islands that may influence the issue of ethical treatment of animals.	Describe two ways that information about the Galapagos snails may be interpreted.
Science inquiry	Questioning and predicting	Make a prediction about how the number of snail species may change in the Galapagos Islands over the next 200 years.	Construct a scientific question that could be used to investigate the link between a snail's shell colour and its environment.
Science inquiry	Evaluating	Identify a potential error scientists could make when investigating the number of snail populations in the Galapagos Islands.	Describe an impact this error might have on the outcome of their investigation.

Figure 2.25: A diverse range of habitats and ecosystems means that many species of the Galapagos snail have evolved over generations.

The Galapagos Islands host more than just tortoises and finches

Charles Darwin's research focused on tortoises and finches. However, the Galapagos Islands contain many diverse species, including its snails. Land snails have spread across multiple islands and taken up residence in many different ecosystems. Over 100 distinct snail species have been found in the Galapagos Islands, inhabiting a variety of habitats – from lush forests to damp caves to arid volcanic rocks. The snails have distinct shell lengths, colours and patterns that have evolved over generations as snails in each area passed on these favourable genetics to their offspring. Over time, this has led to the diverse range of snail species found in the Galapagos Islands today.

Demonstrate your understanding

3	4	5	
Explain how the snails provide evidence of evolution by natural selection.	Analyse the strength of this supporting evidence of evolution.	Evaluate how the snails of the Galapagos Islands demonstrate the way in which the diversity of life has changed over time.	
Explain how the diverse range of organisms in the Galapagos Islands may be studied using technologies such as engineering.	Discuss how the snails of the Galapagos Islands help to validate Darwin and Wallace's theory.	Use secondary research to analyse how advances in genetic technologies have helped scientists to study the evolution of organisms in the Galapagos Islands.	
Explain how societal demand for 'designer pets' may influence research into the evolution of organisms.	Discuss how information about evolution may inform the decisions made by society about conservation strategies for organisms.	Analyse the factors that led to the adoption by society of the theory of evolution by natural selection.	
Propose a hypothesis stating the expected outcome of your scientific question from step 2.	Discuss the factors that are needed to form an investigable scientific question and hypothesis, and link these factors to your previous two responses.	Develop a model that addresses the conclusion: 'The animals of the Galapagos Islands continue to provide supporting evidence of evolution.'	Science how-to p. 234
Propose a way that the data collected in these investigations could be kept free of these errors, to improve its validity.	Construct an evidence-based argument that identifies and addresses an ethical issue that may be present when conducting investigations involving living organisms.	Evaluate the validity of the investigations and data that have been collected from the Galapagos Islands over time, based on analysis of secondary sourced resources.	Science how-to p. 253

Figure 2.26: The shells of different Galapagos snail species vary in length and colouring, based on the environments the snails inhabit.

3.0 The periodic table

Our physical world is made up of over 100 different chemical elements. Although these elements each have unique properties, they can be grouped together based on the things they have in common, such as how they look, behave, or react with other substances. To help make sense of this complex world, scientists developed the periodic table; a powerful tool that organises all known elements into a logical system. The table helps us to understand patterns in nature and to predict how elements will interact in the world around us.

Learning Ladder

The Learning Ladder for each chapter maps the Science Understanding, Science as a Human Endeavour and Science Inquiry strands that will be covered. Each ladder has five levels of progression, called steps. To climb the ladders, you need to develop fluency at each step. This will help you develop the ability to complete tasks that are more complex.

Steps in progression	Science understanding: Chemical science: The periodic table	Science as a human endeavour: Nature and development of science	Science as a human endeavour: Use and influence of science
5	I can evaluate how the periodic table has been developed and used over time	I can analyse how advances in technologies enable advances in science	I can analyse the key factors that contribute to scientific knowledge being adopted more broadly by society
4	I can analyse the properties of atoms based on trends in the periodic table	I can discuss how scientific knowledge is validated, including the role of publication and peer review	I can discuss how scientific information and misinformation may inform personal and social decision-making
3	I can explain relationships between the organisation of the periodic table and atomic structure	I can explain how science has contributed to developments in technologies and engineering	I can explain how the values and needs of society influence the focus of scientific research
2	I can describe patterns and trends in the periodic table	I can describe how scientific knowledge is refined over time	I can describe scientific knowledge that may be interpreted in different ways
1	I can identify characteristics of elements based on their position in the periodic table	I can identify technologies that have enabled advances in science	I can identify scientific knowledge that can address socio-scientific issues

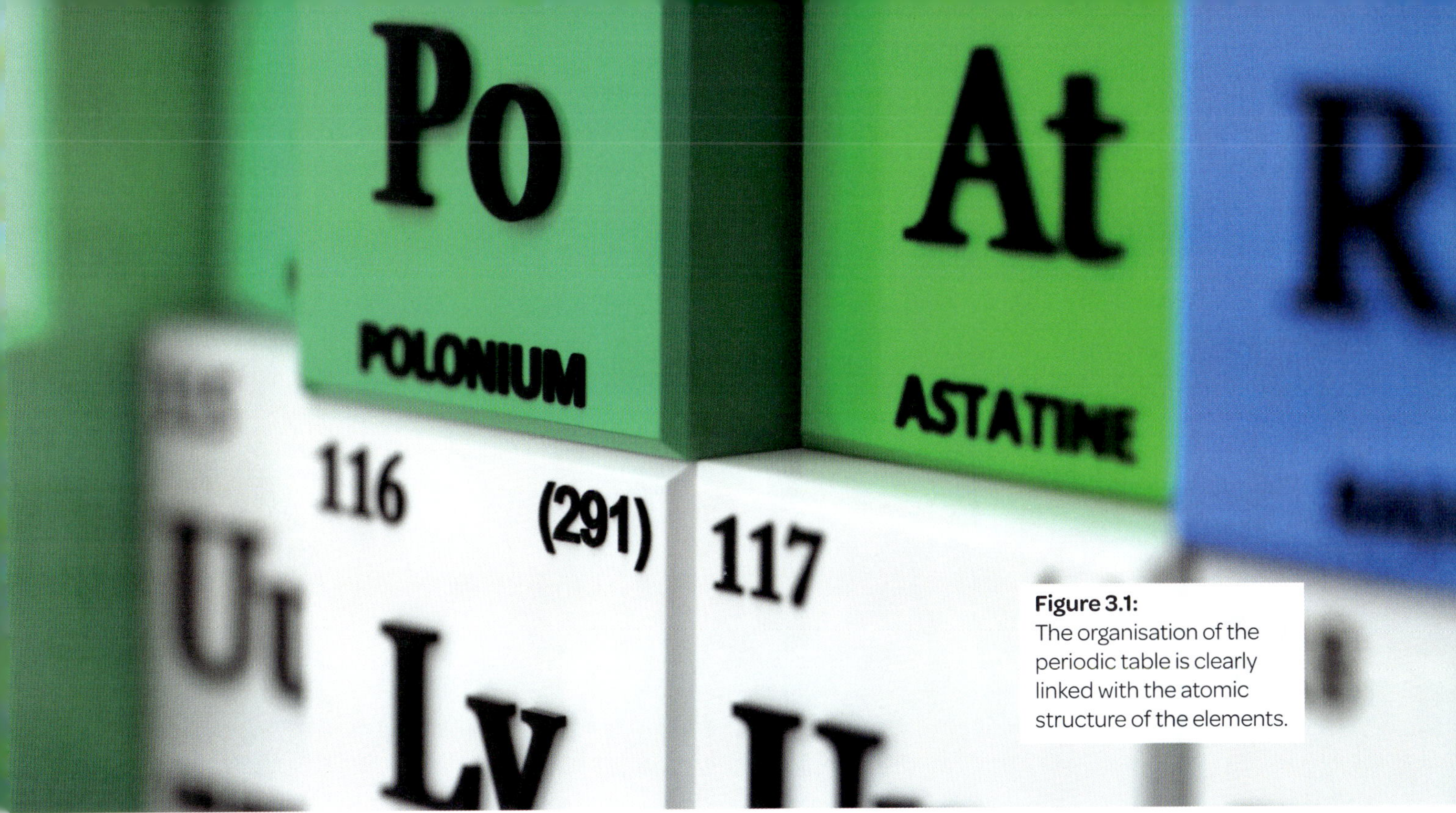

Figure 3.1: The organisation of the periodic table is clearly linked with the atomic structure of the elements.

Processing, modelling and analysing	Evaluating	Steps in progression
I can analyse the quality of data using descriptive statistics	I can evaluate the validity and reproducibility of investigation methods	5
I can discuss relationships and anomalies that emerge in processed data	I can construct evidence-based arguments to justify conclusions, address ethical issues or assess claims	4
I can identify and explain trends and/or patterns in a range of dataset representations	I can discuss ways to improve the quality of data and validity of conclusions and claims	3
I can process data by using mathematical relationships and/or constructing graphs	I can describe the impact of assumptions and errors, and propose ways to reduce them	2
I can organise data and information using tables, keys and/or models	I can identify assumptions and types of errors in an investigation	1

Science inquiry

3·1 ▸ Periodic table basics

Figure 3.2: Light micrograph image of crystallised silver, produced during a metal displacement reaction between copper and silver nitrate

Learning intention

At the end of this lesson, I will be able to describe what the periodic table tells us about each element.

Key terms

atomic mass: the average mass of the atoms in a sample of an element

atomic number: the number of protons in an atom

chemical symbol: a symbol of one or two letters used to represent an element

electron configuration: the number of electrons in each shell of an atom, starting with the smallest – for example, 2, 8, 8, 2

electron shell: the space around an atom's nucleus, in which electrons circulate at different energy levels

element: a pure substance made of only one type of atom

isotopes: atoms of the same element with the same number of protons but a different number of neutrons

periodic table: a table of all known elements and their chemical symbols

Key idea: Patterns, order and organisation

The periodic table includes every known **element**. The table provides a lot of information about each element, including its chemical symbol and information about its atoms. Elements are listed in order by their atomic number from 1 (hydrogen) to 118 (oganesson). The table also groups together elements with similar properties.

Chemical symbols always stay the same

Every known element has a **chemical symbol** (X), sometimes called the element symbol. For example, the chemical symbol for silver is Ag. Scientists throughout the world call this element Ag, regardless of which language they speak. So, while 'silver' is *argent* in French, *plata* in Spanish and *серебряный* in Russian, scientists from these countries can understand each other by referring to silver by its chemical symbol (which originated from its French name). Using 'Ag' as the symbol for silver does not make much sense to English speakers, but it seems a perfect symbol if you are French and use the word *argent* for that element. See Table 3.1 for the origins of some element symbols.

Table 3.1: The origins of some element symbols

Element	Symbol	Origin of symbol
Sodium	Na	The Latin word for sodium is *natrium*.
Iron	Fe	The Latin word for iron is *ferrum*.
Strontium	Sr	Named after a Scottish village called *Sròn an t-Sìthein* in Scottish Gaelic.
Lead	Pb	The Latin word for lead is *plumbum*.
Potassium	K	The Latin word for potassium is *kalium*.

Figure 3.3: Silver has many names in different languages but its chemical symbol (Ag) does not change. This helps scientists to communicate with and understand each other and to collaborate.

The periodic table tells us about the atoms of each element

In the **periodic table**, each element has its own square. This square contains an element's atomic number (Z), chemical symbol, name and atomic mass (see Figure 3.4). There is also often a key that categorises elements for general identification – for example, types of metals and non-metals (see Figure 3.9).

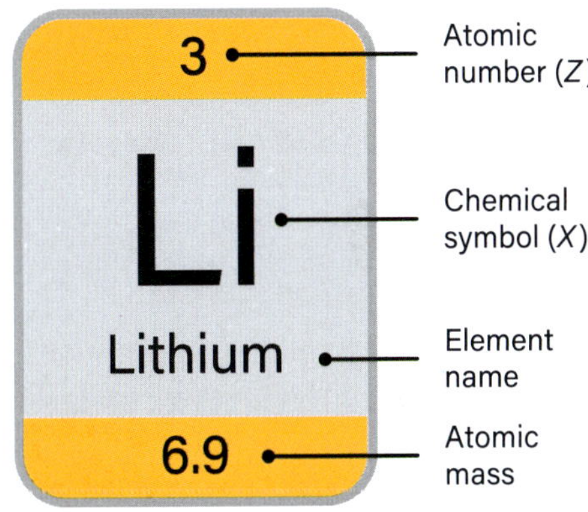

Figure 3.4: The periodic table includes information for every known element.

An atomic number equals the number of protons

The **atomic number** states how many protons are in the nucleus of an atom of the element. For example, every atom of silicon (Si) has 14 protons, so the atomic number of silicon is 14. Similarly, every atom of barium (Ba) has 56 protons, so the atomic number of barium is 56.

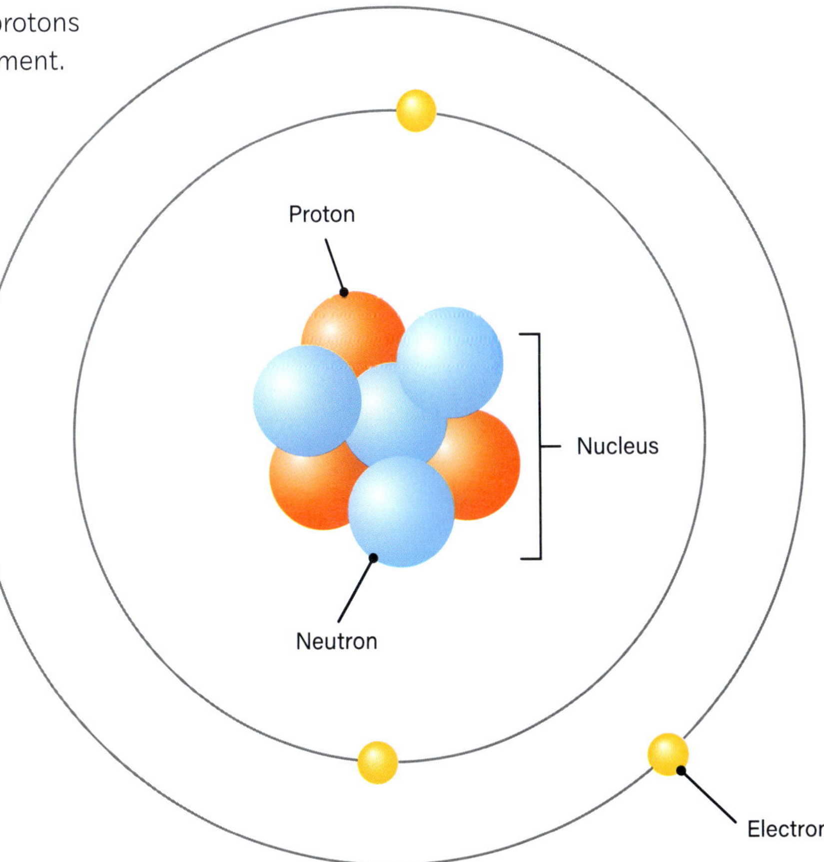

Figure 3.5: Protons and neutrons are located in the nucleus of an atom, while the electrons orbit around the nucleus in different shells. This is a lithium atom because it has three protons, which matches its atomic number ($Z = 3$) in the periodic table.

An atom has the same number of electrons and protons

Atoms are assumed to be neutrally charged unless otherwise stated. Therefore, we can assume that an atom has the same number of protons and electrons. This is because the positive charges of the protons cancel out the negative charges of the electrons, resulting in an overall neutral charge.

Consider calcium (Ca), which has an atomic number of 20 ($Z = 20$). This means that calcium has 20 protons; therefore, a calcium atom has 20 electrons.

Electrons circulate around an atom's nucleus in **electron shells**. An atom's electrons are distributed across the electron shells using the 2, 8, 8, 2 **electron configuration**, starting with the smallest, innermost shell (i.e. the shell that is closest to the nucleus) (see Figure 3.6). The first shell can only hold up to two electrons. Once it's full, the second shell can be filled with eight electrons. The next eight electrons are then distributed into the third shell, followed by two electrons into the fourth shell.

For example, a calcium atom will have 20 protons and 20 electrons. The first two electrons will fill the first shell, the next eight electrons will fill the second shell, followed by eight electrons in the third shell and the remaining two into the fourth shell. This means that a calcium atom has four shells and two electrons on its valence shell (outermost shell), represented by the simple electron configuration: 2, 8, 8, 2. This concept is explored in more depth in Section 3.5.

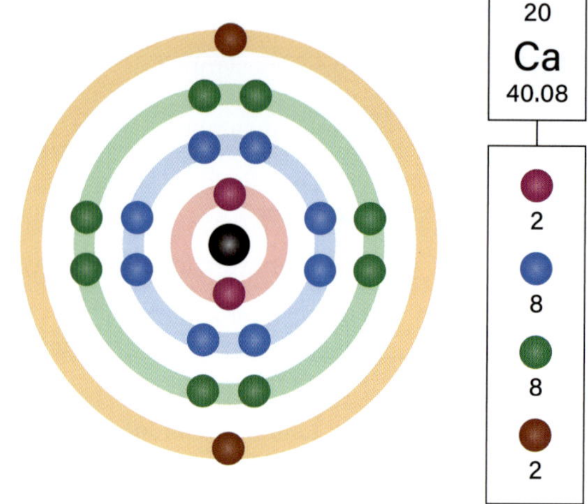

Figure 3.6: A calcium atom has 20 electrons, which are distributed across four electron shells. The first two electrons spread out to opposite sides of the atom. The first four electrons on the second and third shells spread out evenly around the nucleus. The next four form pairs, resulting in four pairs of electrons in the second and third shells when full. There are two electrons in its valence shell, spread out like those in the first shell.

Atomic mass is an average

The **atomic mass** is the average mass of the atoms of a sample of the element. This is measured in atomic mass units (amu). The electrons that orbit an atom's nucleus do not weigh very much, so most of the mass of an atom is in the nucleus, where the neutrons and protons are located. The number of neutrons in an atom is often equal to the number of protons, but it can also vary. This variation leads to many different **isotopes** (types) of atoms of the same element, which is why neutron count is not typically included in the periodic table.

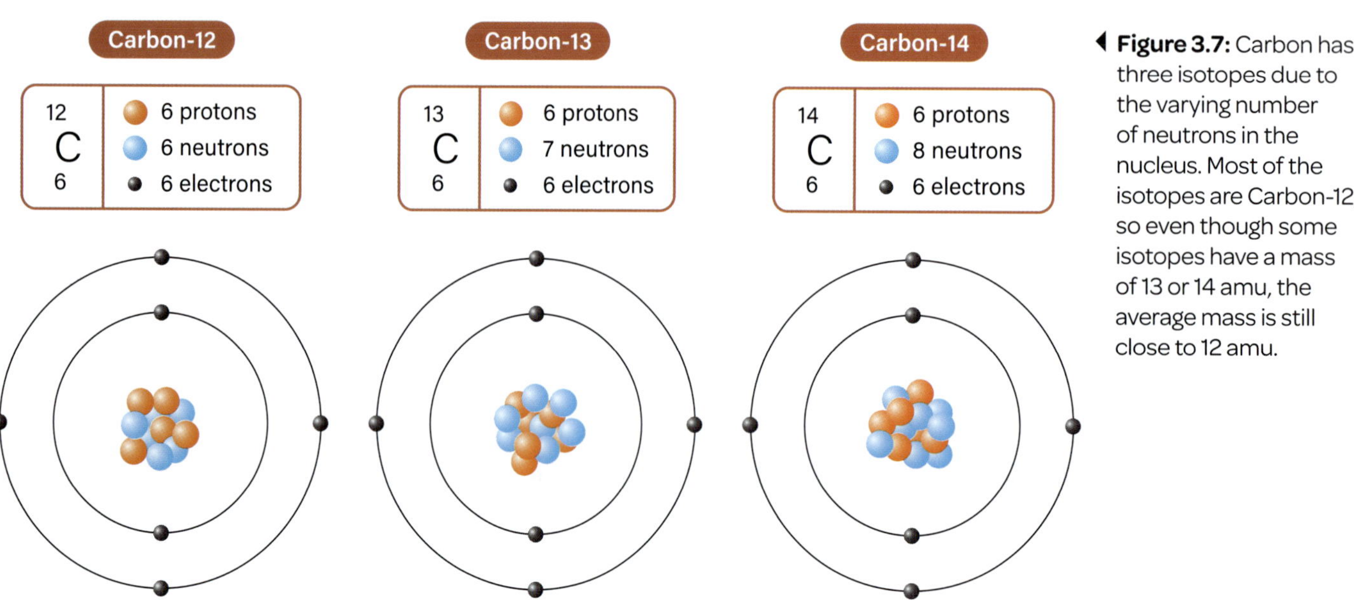

Figure 3.7: Carbon has three isotopes due to the varying number of neutrons in the nucleus. Most of the isotopes are Carbon-12, so even though some isotopes have a mass of 13 or 14 amu, the average mass is still close to 12 amu.

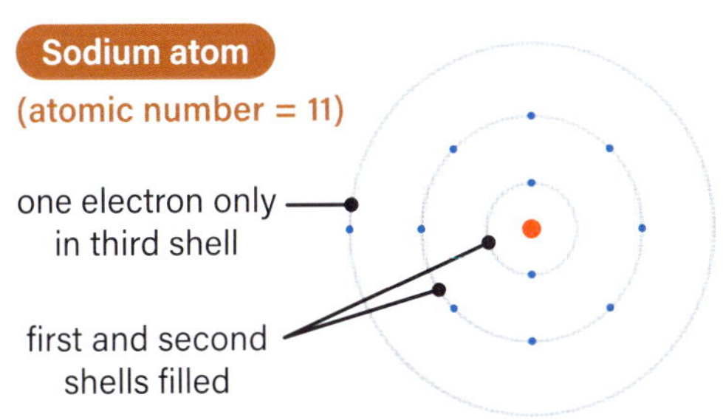

Figure 3.8a: Sodium has 11 electrons distributed across three shells in a 2, 8, 1 configuration.

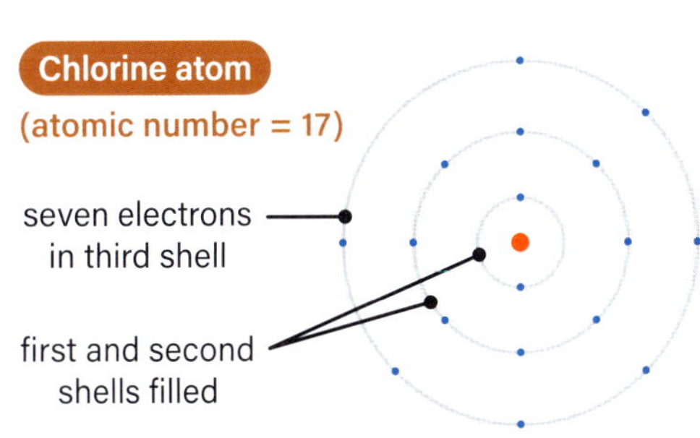

Figure 3.8b: Chlorine has 17 electrons distributed across three shells in a 2, 8, 7 configuration.

Learning Ladder

Refer to the periodic table (Figure 3.9 on pages 66–67) to help you answer the questions below.

The periodic table

1. Copy the paragraph below and fill in the blanks. There are 118 elements in the periodic table, which means there are 118 different types of _______ that make up everything. Boron's atomic number is _______, which means it has _______ protons in its _______. It is next to _______ in the periodic table and _______ aluminium.
2. Locate boron, carbon, nitrogen and oxygen in the periodic table.
   - **a** Describe each of their atomic structures, based on their positions in the periodic table.
   - **b** Identify and describe a trend observed in their structures as atomic number increases.
3. Explain why all isotopes of all atoms are not included in the periodic table.
4. Consider these elements: potassium, magnesium, gold, nitrogen, helium, tin, barium and cobalt.
   - **a** Propose which two elements are likely to have similar properties. Provide a reason.
   - **b** Propose which elements will have the least similar properties. Provide a reason.
   - **c** Conduct an online search to determine if your predictions were accurate. If they were not, reflect on your reasons from parts a and b.
5. Discuss the benefits and drawbacks of using chemical symbols instead of element names to identify elements.

Use and influence of science

1. Identify the feature on each square of the periodic table that minimises communication issues between scientists who speak different languages.
2. 'Atomic mass' and 'mass number' are not the same thing. Propose a reason why it is common for people either to mix up these terms or to use them interchangeably.
3. Explain how organising all known elements into the periodic table addressed needs of the scientific community.

Processing, modelling and analysing

p. 244

1. Copy the table into your notebook and fill in the blanks.

Element	Symbol (X)	Atomic number (Z)
Carbon	C	6
Oxygen		
	Na	
		13
Fluorine		
	Cu	
Calcium		
		80
Tungsten		

2. **a** Construct a table that is similar to Table 3.1. Include 10 elements whose symbols do not match their English names. As well as the columns in Table 3.1, add a column for any interesting facts about the discovery of the elements.
   - **b** Classify the 118 known elements into two groups:
     - **i** Group A: The English name of the element matches its symbol.
     - **ii** Group B: The English name of the element does not match its symbol.
   - **c** Calculate the percentage of elements in Group A and the percentage of elements in Group B. (See 'Percentages' on page 289 of the Science how-to section.)
   - **d** Construct a pie chart to represent the percentages you worked out in part c.

Key idea: Patterns, order and organisation

Use the periodic table to find three elements named after scientists and three elements named after countries. Research the reasons they were named after a scientist or a country. Identify whether there are patterns or rules that determine how an element is named. Discuss your findings with a classmate.

Success criteria

- I can describe what the periodic table tells us about each element.

Periodic table of elements

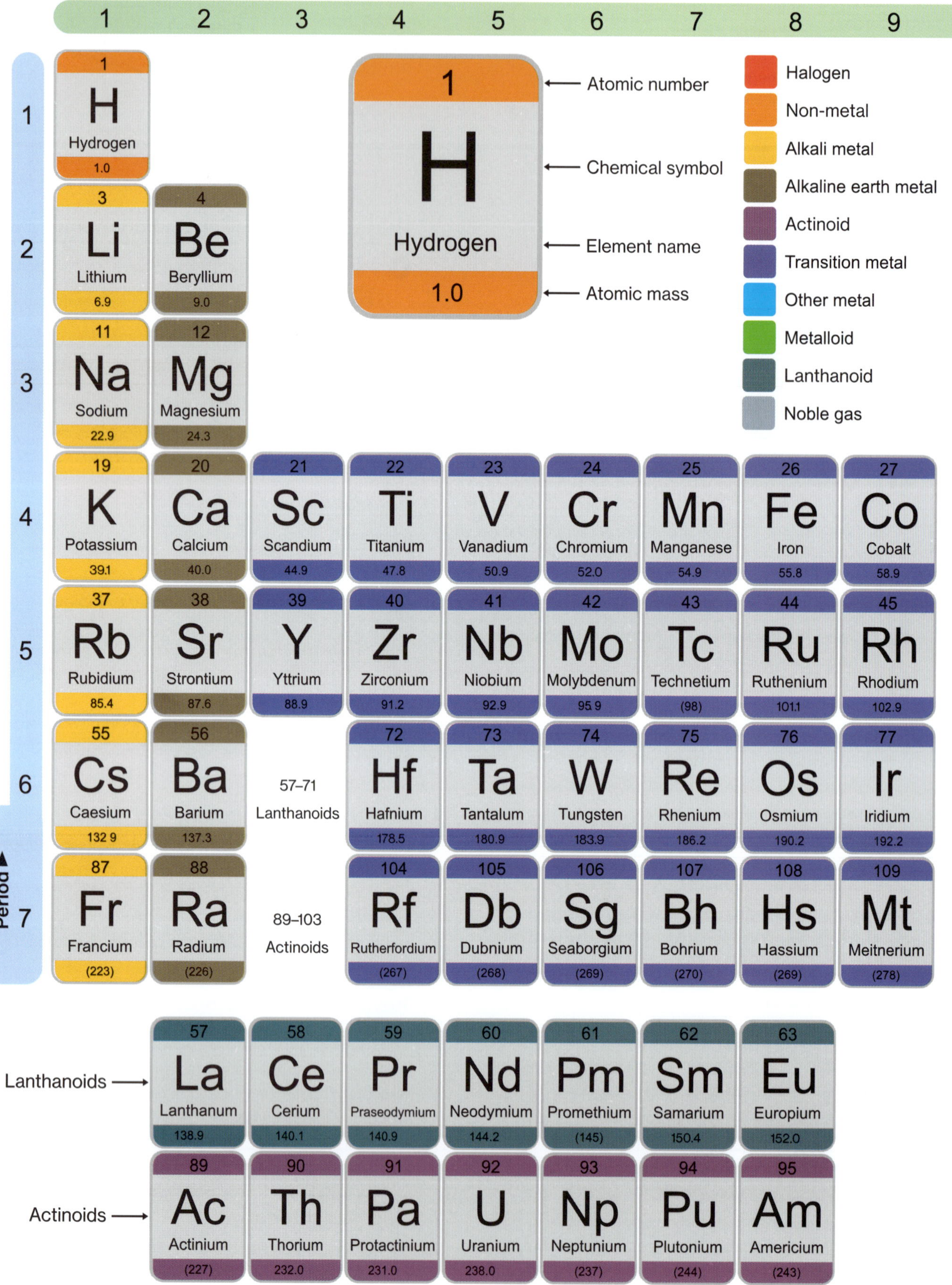

Figure 3.9: The periodic table shows all known elements and their chemical symbols.

10	11	12	13	14	15	16	17	◄ Group 18
								2 He Helium 4.0
			5 B Boron 10.8	6 C Carbon 12.0	7 N Nitrogen 14.0	8 O Oxygen 16.0	9 F Fluorine 19.0	10 Ne Neon 20.1
			13 Al Aluminium 26.9	14 Si Silicon 28.0	15 P Phosphorus 30.9	16 S Sulfur 32.0	17 Cl Chlorine 35.4	18 Ar Argon 39.9
28 Ni Nickel 58.6	29 Cu Copper 63.5	30 Zn Zinc 65.3	31 Ga Gallium 69.7	32 Ge Germanium 72.6	33 As Arsenic 74.9	34 Se Selenium 78.9	35 Br Bromine 79.9	36 Kr Krypton 83.8
46 Pd Palladium 106.4	47 Ag Silver 107.8	48 Cd Cadmium 112.4	49 In Indium 114.8	50 Sn Tin 118.7	51 Sb Antimony 121.7	52 Te Tellurium 127.6	53 I Iodine 126.9	54 Xe Xenon 131.2
78 Pt Platinum 195.1	79 Au Gold 197.0	80 Hg Mercury 200.6	81 Tl Thallium 204.4	82 Pb Lead 207.2	83 Bi Bismuth 209.0	84 Po Polonium (209)	85 At Astatine (210)	86 Rn Radon (222)
110 Ds Darmstadtium (281)	111 Rg Roentgenium (281)	112 Cn Copernicium (285)	113 Nh Nihonium (286)	114 Fl Flerovium (289)	115 Mc Moscovium (288)	116 Lv Livermorium (293)	117 Ts Tennessine (294)	118 Og Oganesson (294)

64 Gd Gadolinium 157.3	65 Tb Terbium 158.9	66 Dy Dysprosium 162.5	67 Ho Holmium 164.9	68 Er Erbium 167.2	69 Tm Thulium 168.9	70 Yb Ytterbium 173.1	71 Lu Lutetium 175.0
96 Cm Curium (247)	97 Bk Berkelium (247)	98 Cf Californium (251)	99 Es Einsteinium (252)	100 Fm Fermium (257)	101 Md Mendelevium (258)	102 No Nobelium (259)	103 Lr Lawrencium 281

3·2 ▸ Organisation of elements in the periodic table

Learning intention

At the end of this lesson, I will be able to:

- describe how the elements in the periodic table are organised
- predict the relative reactivity of elements based on their position in the periodic table.

Key terms

group: a column in the periodic table

period: a row in the periodic table

reactive: undergoes chemical reactions easily

valence electron: an electron in the outermost electron shell

valence shell: outermost electron shell (furthest from the nucleus)

Investigation 3.2

Investigating the reactivity of metals, p. 320

Key idea: Patterns, order and organisation

Figure 3.10: The alkali metals in group 1 of the periodic table are so reactive they must be stored in oil to prevent the metals coming into contact with water or air.

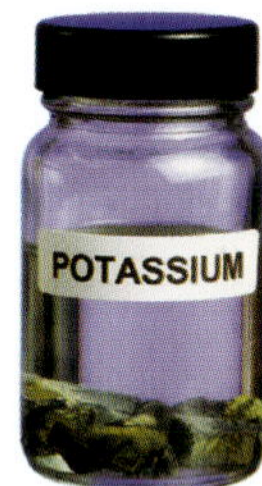

In the periodic table, elements are organised into seven rows and 18 columns. They are listed in order by their atomic number, from 1 (hydrogen) to 118 (oganesson). The table also groups together elements with similar properties.

Elements are organised into groups

The 18 vertical columns in the periodic table are called **groups**. Each group of elements has a number from 1 to 18. These groupings are based on the physical and chemical properties of the elements. For example, certain gases are in one group and certain metals are in another group.

Groups relate to valence electrons

In the periodic table, elements that belong to the same vertical group have the same number of electrons in their outermost shell. The outermost shell is called the **valence shell**, and electrons that occupy it are called **valence electrons**. The group number does not always match the number of electrons in the valence shell; however, a relationship can usually be seen (see Table 3.2). Elements with the same number of valence electrons have similar chemical properties.

The following are some of the groupings in the periodic table.

- **Group 1 (alkali metals):** The group 1 metals – lithium (Li), sodium (Na), potassium (K), rubidium (Rb), caesium (Cs) and francium (Fr) – are the alkali metals. They are very **reactive** (see Figures 3.10–3.12) and are soft, shiny and malleable, and conduct heat and electricity.
- **Group 2 (alkaline earth metals):** The group 2 metals – beryllium (Be), magnesium (Mg), calcium (Ca), strontium (Sr), barium (Ba) and radium (Ra) – are the alkaline earth metals. They have similar properties to the metals in group 1, except they are less reactive.
- **Groups 3–12 (transition metals):** The elements in groups 3–12 are the transition metals. They are all solid in their standard form except for mercury (Hg), which is liquid. Transition metals can combine with other elements to form many different compounds, which are often brightly coloured (see Figure 3.13).
- **Groups 13–16 (metalloids):** To the right of the transition metals in the periodic table are the metalloids (sometimes called semi-metals). The metalloids include boron (B), silicon (Si), germanium (Ge), arsenic (As), antimony (Sb) and tellurium (Te). These elements exhibit properties of both metals and non-metals.

Table 3.2: The relationship between group numbers in the periodic table and valence electrons

Group 1	Group 2	Groups 3–12	Group 13	Group 14	Group 15	Group 16	Group 17	Group 18
1 valence electron	2 valence electrons	Number of valence electrons varies	3 valence electrons	4 valence electrons	5 valence electrons	6 valence electrons	7 valence electrons	8 valence electrons

Figure 3.11: When the alkali metals in group 1 come into contact with water or air, they react violently. In this image, potassium is reacting with water.

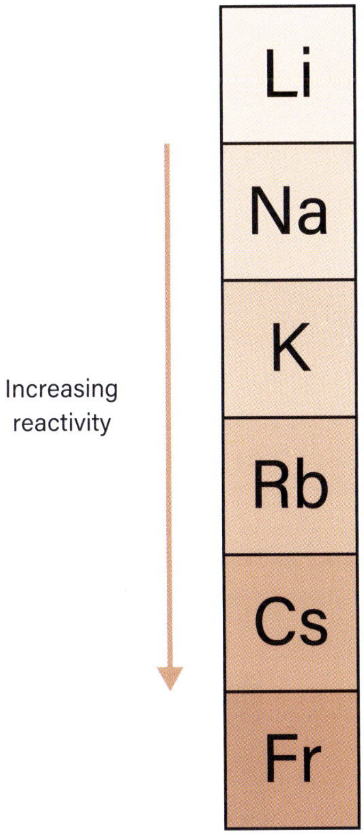

Increasing reactivity

Li
Na
K
Rb
Cs
Fr

Figure 3.12: The reactivity of group 1 metals increases as you move down the group in the periodic table.

The non-metals are mostly found on the right side of the periodic table, with the exception of hydrogen which is in group 1 on the left.

- **Group 17 (halogens):** The six elements in this group – fluorine (F), chlorine (Cl), bromine (Br), iodine (I), astatine (At) and tennessine (Ts) – are the halogens. Halogens are non-metals; they are all gases in their standard form, except for bromine (which is liquid at room temperature) and iodine and astatine (which are solids). The group 17 elements are very reactive; they are the most reactive non-metals.
- **Group 18 (noble gases):** The seven elements in this group – helium (He), neon (Ne), argon (Ar), krypton (Kr), xenon (Xe), radon (Rn) and oganesson (Og) – are the noble gases. These elements are also known as 'inert gases' because they are mostly unreactive. This is because they have a full outer shell of eight electrons, which makes them extremely stable on their own.

Figure 3.13: These brightly coloured salts are all compounds of transition metals.

Elements are also organised into periods

In the periodic table, elements are also arranged into seven horizontal rows (called **periods**), which are numbered from 1 to 7. In the rows, the elements are placed in order of their atomic number.

Periods relate to electron shells and atomic size

The elements in the same period have the same number of electron shells. For example, potassium (K) and bromine (Br) have very different properties, but they both have four electron shells and so are both placed in period 4.

The more electron shells an atom has, the larger its atomic size. In other words, atomic size increases as you go down the group. For instance, sodium in period 3 is larger than lithium in period 2, and potassium in period 4 is larger than sodium, and so on as you proceed down group 1 (see Figure 3.14).

Two rows of the periodic table are separate from the other periods. These rows contain the lanthanoids (57–71) and the actinoids (89–103), which are rare and have very similar properties. These rows are separated so that they do not disrupt the arrangement of the periodic table.

Table 3.3: The relationship between period numbers in the periodic table and electron shells

Period 1	1 electron shell
Period 2	2 electron shells
Period 3	3 electron shells
Period 4	4 electron shells
Period 5	5 electron shells
Period 6	6 electron shells
Period 7	7 electron shells

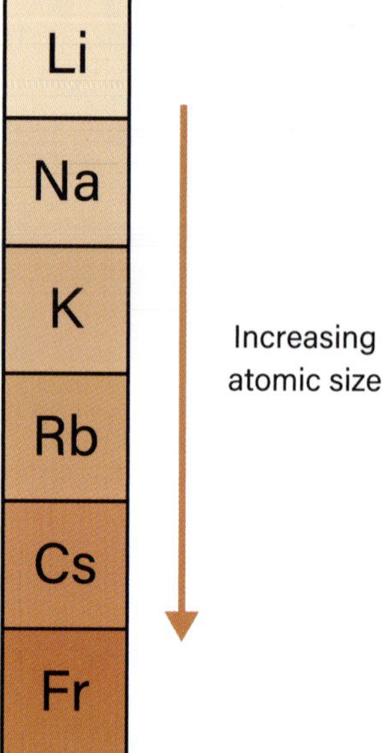

Figure 3.14: The atomic size of group 1 metals increases as you move down the group.

Reactivity trends are observed in the periodic table

The elements in the periodic table are divided into three main sections – metals, non-metals and metalloids – based on their physical and chemical properties.

The reactivity of metals generally increases as you move down a group and left across a period. In other words, the most reactive metals are in the bottom left corner of the periodic table.

The opposite is true for non-metals. They increase in reactivity as you go up a group and to the right; therefore, fluorine (F) is the most reactive non-metal. Noble gases are not included in this trend because they are categorically unreactive (i.e. not reactive at all).

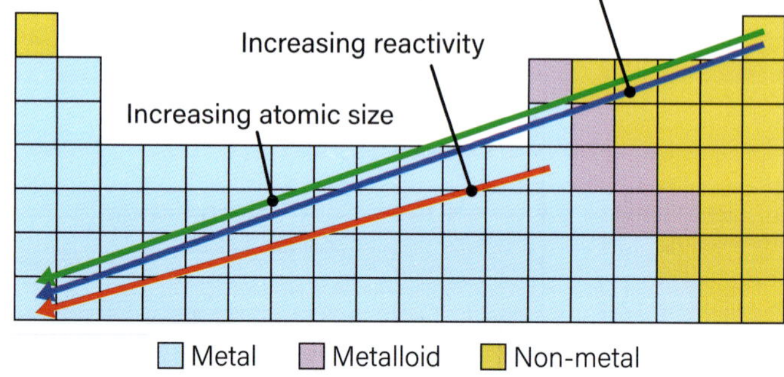

Figure 3.15: The reactivity of metals, metallic properties and atomic size increase towards the bottom left corner of the periodic table.

Figure 3.16: Four metals reacting with hydrochloric acid; from left to right: calcium (Ca), magnesium (Mg), nickel (Ni) and copper (Cu).

Figure 3.17: Neon lights are made from neon, one of the noble gases in group 18 of the periodic table.

Learning Ladder

The periodic table

1 a Identify which subatomic particle is linked to an element's atomic number.
b Identify and describe the features or properties of elements in groups 1 and 2 of the periodic table.

2 Distinguish between *groups* and *periods* of the periodic table.

3 a Identify the type of element found on the far right-hand side of the periodic table.
b Describe what groups and periods tell us about atomic structure.

4 Identify properties that elements on the far right-hand side of the periodic table share.

Use and influence of science

1 Identify a benefit of using the periodic table to determine whether certain elements will react violently in water.

2 Describe your thoughts about the following statement: *No one pays attention to the lanthanoid and actinoid elements because they are not easy to see in the periodic table.*

3 a Explain how the organisation of the periodic table is useful for the scientific community.
b Propose and discuss how the organisation of the periodic table has impacted the wider community.

4 Discuss how the following statement could be problematic:
There are many versions of the periodic table, but the one in this textbook is the one everyone should use.

5 Research and analyse how differences in the organisation of the periodic table across countries have influenced the adoption of scientific knowledge by the global scientific community and society.

Processing, modelling and analysing p. 244

1 a Use the periodic table to identify three elements that are in the same:
i group as potassium. ii period as calcium.
b Construct a table to organise information about the elements you identified in part a.

2 Figure 3.16 shows four metals reacting with hydrochloric acid. Based on the intensity of the bubbles in the photo, create a column graph to show the reactivity (zero, low, medium, high) of each of the metals. Follow the conventions for constructing a scientific graph. (See 'Graphing' in the Science how-to section on page 294.)

3 a Explain how your graph from Question 2 is linked to reactivity trends in the periodic table.
b Propose how you could use your graph to determine the relative reactivity of other metals, such as potassium or zinc.

4 Silver (Ag, $Z = 47$) is more reactive than platinum (Pt, $Z = 78$), but this property does not follow the general trend.
a Conduct a search to explain why this is the case.
b Continue your search to identify and discuss other metals that do not follow the expected reactivity trend.

5 Consider your graph from Question 2.
a Explain why it would not be possible to display descriptive statistics based on the data you have collected.
b Propose how the investigation and/or collection of data would need to be modified in order to allow for a statistical analysis of results.

Key idea: Patterns, order and organisation

Choose one metal, one non-metal and one metalloid. Research the three elements to:

a identify when and by whom each element was discovered.
b describe how each element was found in nature.
c explain the properties and uses of each element.
d identify any patterns that emerge in your findings.
e organise your findings into an orderly chart or poster.

Success criteria

- I can describe how the periodic table is organised, using terms such as 'group' and 'period'.
- I can predict the relative reactivity of elements based on their position in the periodic table.

3·3 ▸ Common substances

Learning intention

At the end of this lesson, I will be able to identify types of substances based on their general properties.

Key terms

chemical bond: an electrostatic force that connects atoms to one another

chemical formula: an expression of the elements that make up a chemical compound, usually presented as a ratio using chemical symbols and numbers – for example, H_2O

compound: a combination of two or more different elements, joined together in a fixed ratio

covalent: made up of non-metals

crystalline solid: a solid with a highly ordered arrangement of particles at the microscopic and macroscopic levels

giant network solid: a continuous network of atoms joined together through covalent bonds (not discrete); also called 'covalent network solid'

ionic: generally made up of metals and non-metals

metallic: made up of metals

Investigation 3.3

Student-designed investigation: properties of common substances, p. 321

Key idea: Patterns, order and organisation

Different elements from the periodic table can join together to form chemical bonds; in doing so, they will form a new substance. Recall that elements are generally classified as either metals or non-metals (see Figure 3.15 in Section 3.2). Depending on what types of elements are bonded, substances can be further classified.

There are different types of chemical bonds

Recall that a **chemical bond** is formed when electrostatic (positive and negative) attractions pull atoms together, so that they are connected. The type of chemical bond that will form between atoms depends on whether metals, non-metals or both are joined to one another. In general, there are three main types of chemical bonding.

- **Metallic bonding:** metals are bonded to other metals.
- **Ionic bonding:** metals are bonded to non-metals.
- **Covalent bonding:** non-metals are bonded to non-metals.

Identifying the elements involved in the bonds and their positions on the periodic table can help you to classify pure substances as metallic, ionic or covalent.

Pure substances can be further classified as elements or compounds

You learnt in previous years that an elemental substance is made of just one type of element, while a **compound** is a substance that contains atoms of two or more different elements. Figure 3.18 shows how the general classification of pure substances – metallic, ionic and covalent – can be further classified as either elements or compounds. By examining the **chemical formulas** of these substances, you can see whether they are made up of metals, non-metals or both. The chemical formula for water is H_2O, so we can see that it is only made up of non-metals, and it is therefore a covalent substance. Mixtures are not included in this classification, as they are not pure substances.

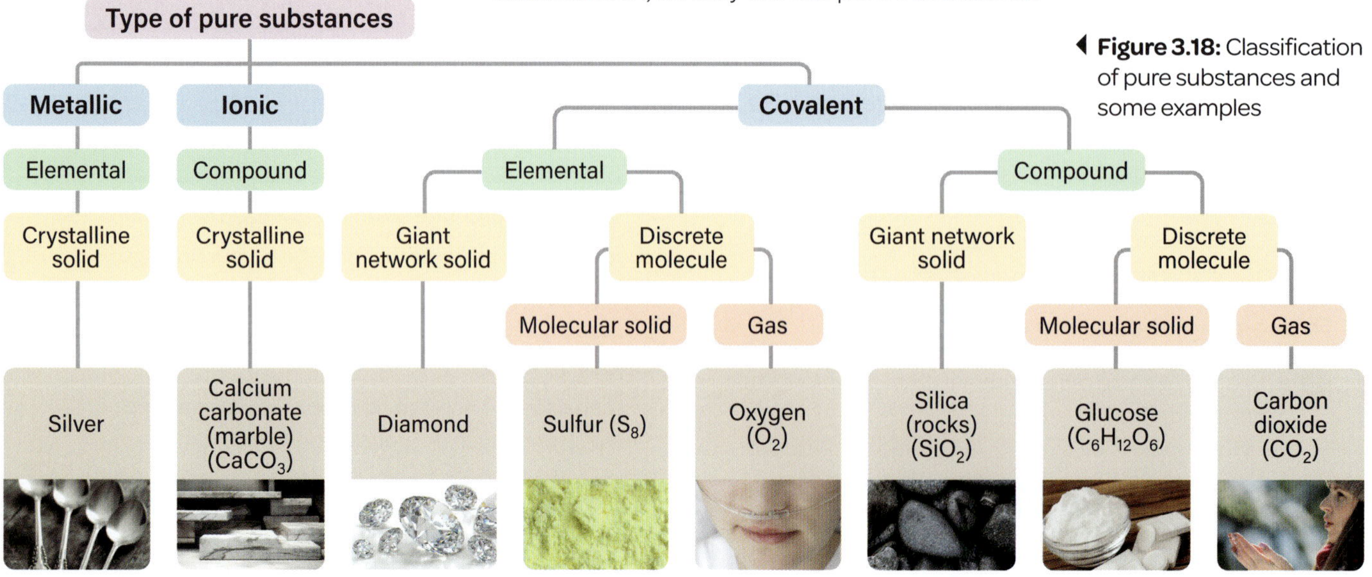

◀ **Figure 3.18:** Classification of pure substances and some examples

Substances have identifying properties

Different substances also have unique properties and characteristics that can be used to classify them. For example, **crystalline solids** have highly ordered arrangements of particles, and **giant network solids** are made up of a continuous network of atoms. Table 3.4 lists the general properties of most substances, which can be related back to their chemical bonding and structure. However, some substances have different properties from those listed in the table. For example, diamond has an extremely high melting point (4500 °C). This is due to the very strong arrangement of covalent bonds between the carbon atoms that repeat to form a giant network solid, similar to a metallic or ionic lattice.

Table 3.4: General properties of metallic, ionic and covalent substances

Property	Metallic substances	Ionic substances	Covalent substances
Melting and boiling points	High	High	Low
Conductivity (electricity)	Good	Conduct when molten or dissolved in water, but not when in a solid state	Not good
Conductivity (heat)	Good conductor	Good insulator	Good insulator
Brittleness	Malleable/ ductile	Brittle	Soft and flexible
Lustre	Lustrous/ shiny	Form crystalline lattice	Dull
Density	High	High	Low

Learning Ladder

The periodic table

1. Classify the following substances as metallic, ionic or covalent, and provide a reason.
   - a Aluminium foil
   - b Sugar ($C_6H_{12}O_6$)
   - c Baking soda ($NaHCO_3$)
   - d Air
   - e Plastic bag
   - f Table salt ($NaCl$)
2. Describe the similarities and differences between metallic, ionic and covalent substances, using a Venn diagram to organise your ideas. In your descriptions, refer to the periodic table, bonding and general properties, and provide examples of common substances.

Use and influence of science

1. Identify a property of metal that makes it suitable for use in frying pans.
2. Describe how unexpected properties of substances, such as diamond's extremely high melting point, can make it difficult for scientific knowledge to be understood by the general public.
3. Single-use plastics have been banned in many places. Based on what you know about their environmental impacts, propose and explain a property of covalent plastics that is not listed in Table 3.4.

Evaluating

p. 253

Read the aim, materials and method for Investigation 3.2.

1. Identify three sources of error in the method. Classify the errors as either random or systematic.
2. For the errors you identified in Question 1, describe specifically how each could impact the results.
3. Propose how the method could be modified to minimise the impacts of each of the errors you described in your response to Question 2.
4. Imagine a classmate has concluded that magnesium is the most reactive metal in the periodic table, based on the results of the investigation showing that it reacted more vigorously than the other metals tested. Construct an evidence-based argument to challenge your classmate's claims, with reference to the investigation method and the periodic table.
5. Evaluate your classmate's claims from Question 4. Propose and explain how the conclusion and/or method could be adjusted to increase the validity of their claims.

Key idea: Patterns, order and organisation

Justify the following statement based on how pure substances can be classified. Use examples of common substances to support your justification.
Metals are never compounds, ionic substances are always compounds, and covalent substances can be compounds.

Success criteria

- I can identify whether substances are metallic, ionic or covalent using their chemical formula and their position in the periodic table.
- I can describe the general properties of metallic, ionic and covalent substances.

3·4 ▸ Development of the periodic table

Learning intention

At the end of this lesson, I will be able to outline how the modern periodic table was developed and refined over time through a process of review by the scientific community.

Key terms

chemical property: a property of a substance that is observed during a chemical reaction

Dalton's atomic theory: the theory that all matter is made up of tiny particles

physical property: a property of a substance that can be observed and measured, such as colour, texture, melting and boiling points, density and hardness

Key idea: Patterns, order and organisation

The periodic table developed over thousands of years and is the work of many scientists. The Ancient Greek philosopher Aristotle believed that matter consisted of four elements: water, fire, earth and air. With continued research and scientific evidence, scientists started to group elements according to their physical and chemical similarities. These groupings were important in shaping the periodic table we have today, which is the work of many scientists from all over the world. Some of the key contributors are Dalton, Döbereiner, Newlands, Mendeleev and Moseley.

Dalton produced an early periodic table of 20 elements

In the early 1800s, English chemist John Dalton proposed that all matter is made of tiny particles that he called atoms (**Dalton's atomic theory**). Dalton produced a table of 20 elements, with symbols to represent each element, although the symbols he used are not the ones used today. As in today's periodic table, Dalton listed an atomic mass for each element, but his calculations for these have been found to be incorrect.

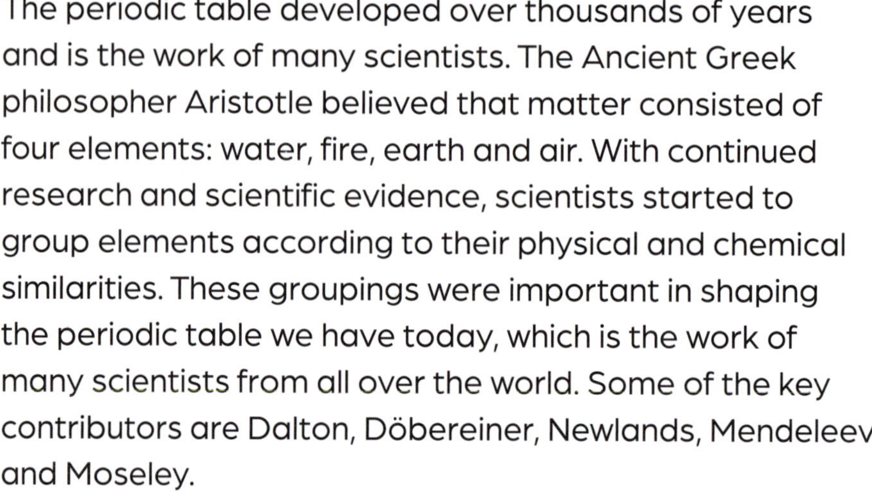

Figure 3.19: ▸ Dalton produced an early table of 20 elements.

Döbereiner grouped elements into triads

As scientists discovered new elements, they expanded Dalton's table. In 1829, German chemist Johann Döbereiner observed patterns or trends in the **physical properties** and **chemical properties** of some elements. He grouped elements with similar properties into triads (i.e. a set or group of three related things) (see Figure 3.20). This grouping of elements that share similar properties laid the foundation for the structure of today's periodic table. To see the relationship between Döbereiner's triads and the modern-day periodic table, find each set of three elements in the periodic table on pages 66–67.

Figure 3.20: Döbereiner's triads shaped the structure of the modern periodic table by grouping elements based on their shared properties. ▾

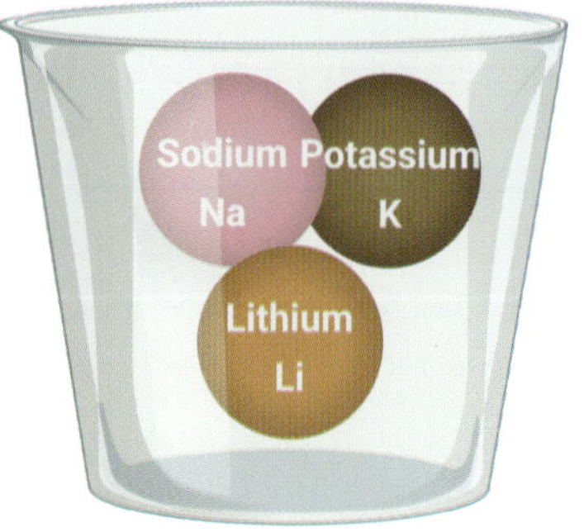

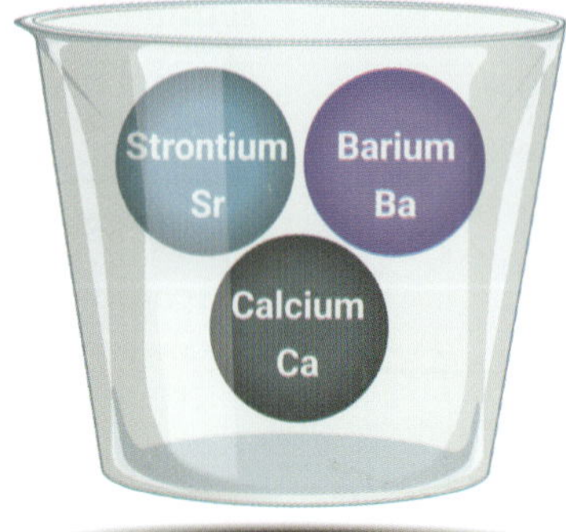

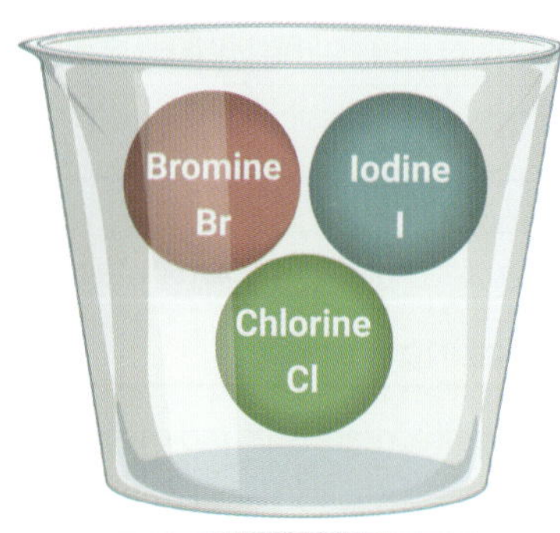

Figure 3.21: Döbereiner identified the relationships between the 55 known elements at the time, and grouped elements based on their common traits.

Newlands arranged elements by atomic mass

In 1864, English chemist John Newlands was the first person to arrange the 60 known elements in order of increasing atomic mass. Newlands is known for his law of octaves. When he arranged elements in seven groups of eight, every eighth element seemed to have similar properties.

Newland's discovery was when the idea of 'periodic' repetition of properties was first considered. However, some elements did not fit the pattern; for example, Newlands put the metal iron into the same group as the non-metals oxygen and sulfur.

Figure 3.22: Newlands arranged elements in order of increasing mass number.

Mendeleev arranged elements into groups and left gaps

In 1869, Russian chemist Dmitri Mendeleev arranged the elements known at the time in order of atomic mass number and put the known element 'families' into vertical columns, which we now refer to as 'groups'.

Mendeleev would group together elements with similar properties, in order of increasing atomic mass. For example, lithium and sodium are both soft, shiny metals that react readily with oxygen and react violently with water. When categorising sodium (Na), Mendeleev placed it in the same group as lithium (Li) on account of their similar chemical characteristics. As sodium has a larger atomic mass, he placed it below lithium.

When sorting elements in this manner, Mendeleev came across elements that had the next-heaviest atomic mass but did not match the characteristics of the next available spot in the table. In Mendeleev's time, gallium (Ga) and germanium (Ge) had not been discovered. After categorising zinc (Zn), arsenic (As) was the next element to be categorised. However, placing arsenic beside zinc would put it in the boron group in his table, and Mendeleev noticed that arsenic matched the properties of phosphorus and nitrogen better than it matched boron (B) or aluminium (Al). Instead of putting it in this group, he placed arsenic below vanadium (V), leaving two gaps in his table for what he claimed were undiscovered elements. He predicted that the unknown elements would have similar chemical properties to those of their 'family'. Later, when the elements were discovered, their properties closely matched Mendeleev's predictions.

Figure 3.23: Mendeleev arranged elements by atomic mass into vertical columns, or 'families', and predicted the properties of unknown elements.

Moseley arranged elements by their atomic number

In 1913, English physicist Henry Moseley suggested that it was an element's atomic number, not its atomic mass number, that was related to the element's physical and chemical properties. Moseley refined the previous periodic table and constructed a more accurate version with fewer elements missing.

Figure 3.24: Moseley refined the periodic table in 1913, suggesting that an element's physical and chemical properties were related to the atomic number, not the atomic mass.

Figure 3.25: Mendeleev's Periodic Law, published in 1869, included all known elements at the time, as well as predictions and gaps for undiscovered elements.

0	I	II	III	IV	V	VI	VII	VIII		
	H 1.01									
He 4.00	Li 6.94	Be 9.01	B 10.8	C 12.0	N 14.0	O 16.0	F 19.0			
Ne 20.2	Na 23.0	Mg 24.3	Al 27.0	Si 28.1	P 31.0	S 32.1	Cl 35.5			
Ar 40.0	K 39.1	Ca 40.1	Sc 45.0	Ti 47.9	V 50.9	Cr 52.0	Mn 54.9	Fe 55.9	Co 58.9	Ni 58.7
	Cu 63.5	Zn 65.4	Ga 69.7	Ge 72.6	As 74.9	Se 79.0	Br 79.9			
Kr 83.8	Rb 85.5	Sr 87.6	Y 88.9	Zr 91.2	Nb 92.9	Mo 95.9	Tc (99)	Ru 101	Rh 103	Pd 106
	Ag 108	Cd 112	In 115	Sn 119	Sb 122	Te 128	I 127			
Xe 131	Ce 133	Ba 137	La 139	Hf 179	Ta 181	W 184	Re 180	Os 194	Ir 192	Pt 195
	Au 197	Hg 201	Ti 204	Pb 207	BI 209	Po (210)	At (210)			
Rn (222)	Fr (223)	Ra (226)	Ac (227)	Th (232)	Pa (231)	U 238				

Döbereiner's triads
Known to Mendeleev
Not known to Mendeleev
Lanthanide series
Noble gases
Actinide series
Gap left for undiscovered element, closely matched Mendeleev's predictions once discovered
Known to Ancients

1	2	3	4	5	6	7	8	9	10	11	12	13	14	15	16	17	18
1 H HYDROGEN																	2 He HELIUM
3 Li LITHIUM	4 Be BERYLLIUM											5 B BORON	6 C CARBON	7 N NITROGEN	8 O OXYGEN	9 F FLUORINE	10 Ne NEON
11 Na SODIUM	12 Mg MAGNESIUM											13 Al ALUMINIUM	14 Si SILICON	15 P PHOSPHORUS	16 S SULFUR	17 Cl CHLORINE	18 Ar ARGON
19 K POTASSIUM	20 Ca CALCIUM	21 Sc SCANDIUM	22 Ti TITANIUM	23 V VANADIUM	24 Cr CHROMIUM	25 Mn MANGANESE	26 Fe IRON	27 Co COBALT	28 Ni NICKEL	29 Cu COPPER	30 Zn ZINC	31 Ga GALLIUM	32 Ge GERMANIUM	33 As ARSENIC	34 Se SELENIUM	35 Br BROMINE	36 Kr KRYPTON
37 Rb RUBIDIUM	38 Sr STRONTIUM	39 Y YTTRIUM	40 Zr ZIRCONIUM	41 Nb NIOBIUM	42 Mo MOLYBDENUM	43 Tc TECHNETIUM	44 Ru RUTHENIUM	45 Rh RHODIUM	46 Pd PALLADIUM	47 Ag SILVER	48 Cd CADMIUM	49 In INDIUM	50 Sn TIN	51 Sb ANTIMONY	52 Te TELLURIUM	53 I IODINE	54 Xe XENON
55 Cs CAESIUM	56 Ba BARIUM	57–71 LANTHANOIDS	72 Hf HAFNIUM	73 Ta TANTALUM	74 W TUNGSTEN	75 Re RHENIUM	76 Os OSMIUM	77 Ir IRIDIUM	78 Pt PLATINUM	79 Au GOLD	80 Hg MERCURY	81 Tl THALLIUM	82 Pb LEAD	83 Bi BISMUTH	84 Po POLONIUM	85 At ASTATINE	86 Rn RADON
87 Fr FRANCIUM	88 Ra RADIUM	89–103 ACTINOIDS	104 Rf RUTHERFORDIUM	105 Db DUBNIUM	106 Sg SEABORGIUM	107 Bh BOHRIUM	108 Hs HASSIUM	109 Mt MEITNERIUM	110 Ds DARMSTADTIUM	111 Rg ROENTGENIUM	112 Cp COPERNICIUM	113 Nh NIHONIUM	114 Fl FLEROVIUM	115 Mc MOSCOVIUM	116 Lv LIVERMORIUM	117 Ts TENNESSINE	118 Og OGANESSON

57 La LANTHANUM	58 Ce CERIUM	59 Pr PRAESODYMIUM	60 Nd NEODYMIUM	61 Pm PROMETHIUM	62 Sm SAMARIUM	63 Eu EUROPIUM	64 Gd GADOLINIUM	65 Tb TERBIUM	66 Dy DYSPROSIUM	67 Ho HOLMIUM	68 Er ERBIUM	69 Tm THULIUM	70 Yb YTTERBIUM	71 Lu LUTETIUM
89 Ac ACTINIUM	90 Th THORIUM	91 Pa PROTACTINIUM	92 U URANIUM	93 Np NEPTUNIUM	94 Pu PLUTONIUM	95 Am AMERICIUM	96 Cm CURIUM	97 Bk BERKELIUM	98 Cf CALIFORNIUM	99 Es EINSTEINIUM	100 Fm FERMIUM	101 Md MENDELEVIUM	102 No NOBELIUM	103 Lr LAWRENCIUM

Figure 3.26: This periodic table colour-codes elements according to when they were discovered.

Before CE
0–1749
1750–1799
1800–1849
1850–1899
1900–1949
1950 onward

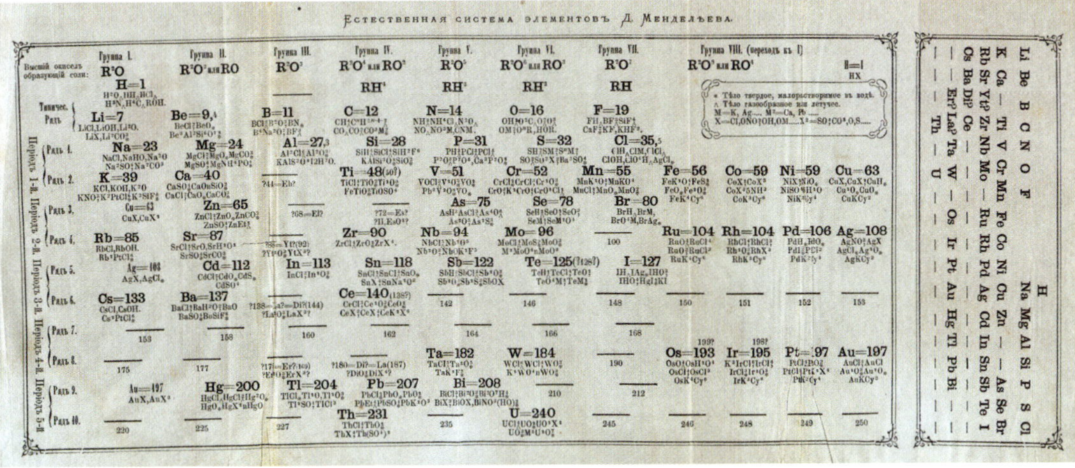

◀ **Figure 3.27:** The first periodic table Mendeleev published, with the elements arranged in vertical columns

Learning Ladder

The periodic table

1. Identify the characteristics or properties that were used to arrange elements by each of the following scientists.
   - **a** John Dalton
   - **b** Johann Döbereiner
   - **c** John Newlands
   - **d** Henry Moseley
2. Describe some similarities and differences between an early periodic table and the modern-day table.
3. Explain how the following scientists contributed to the structure of the modern periodic table.
   - **a** John Newlands
   - **b** Henry Moseley
4. **a** Explain how Johann Döbereiner grouped elements.
   - **b** Select one of Döbereiner's triads and predict which properties they have in common.
   - **c** Research the triad you selected in part b to see if your prediction was accurate.
5. Evaluate how the periodic table has been developed and used over time. Conduct research to discuss if and how these scientists worked together, or if there were other major contributors who have not been included in this section.

Nature and development of science

1. Propose and identify some equipment that may have been used by scientists to explore the properties of unknown substances.
2. Over time, more and more elements have been discovered. Propose some reasons for this.
3. **a** Propose and explain how scientific discoveries related to the periodic table may have contributed to advancements in modern technologies and engineering.
   - **b** Give specific examples of how understanding the periodic table has influenced material selection or development in these fields.
4. Discuss how scientific knowledge about the periodic table was validated over time. In your answer, explain the importance of scientists publishing their findings, and of peer review in confirming and improving the periodic table's accuracy.
5. Research how technological advances have helped to improve the periodic table. Give examples of technologies, and explain their impact on scientific discoveries or on updating the table.

Processing, modelling and analysing p. 244

1. Use Figure 3.26 to construct a timeline that shows the number of elements discovered during each of the seven discovery time periods.
2. Construct a graph to display the information from Question 1.
3. **a** Construct a table with three columns headed 'Period of discovery', 'Elements in order of discovery' and 'Total number of elements discovered'. The periods of discovery could be ancients, 1600s, 1700s, 1800s, 1900s and 2000s.
   - **b** Research all the elements and complete your table.
   - **c** Identify and explain any trends or patterns that you observed.

Key idea: Patterns, order and organisation

Imagine you are a scientist in the 1800s trying to organise the known elements. Using what you know about their properties, describe how you would group and order them. Explain how recognising patterns in element properties can help in predicting the existence and characteristics of elements not yet discovered. If time remains, come up with some pretend elements and properties and swap with a partner to organise.

Success criteria

- I can outline how the modern periodic table was developed over time by different scientists.

3·5 ► Key idea: Patterns, order and organisation – electron configuration

Learning intention

At the end of this lesson, I will be able to:

- describe the structure of atoms in terms of electron shells and energy levels
- use valency to describe the number of electrons an atom needs to gain, lose or share in order to achieve a stable electron configuration.

Key terms

Bohr diagram: a diagram showing the electron configuration of an atom

noble gas configuration: the valence electron configuration that occurs through chemical bonding

octet rule: the tendency of atoms to achieve eight electrons in their valence (outer) shell to be stable

Investigation 3.5

Modelling atomic structure alongside the periodic table, p. 323

Key idea: Patterns, order and organisation

Table 3.5: The maximum number of electrons by shell number

Shell number (n)	Maximum number of electrons ($2n^2$)
1	2
2	8
3	18
4	32

Figure 3.29: These Bohr diagrams represent the electron configurations of lithium and potassium. ▼

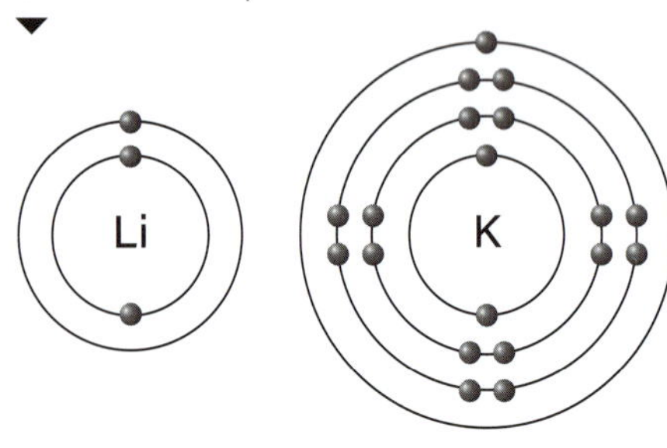

To understand why and how chemical bonding occurs, you must understand that electrons are organised into specific regions around an atom's nucleus.

Electrons exist in energy levels

Recall that protons and neutrons make up the nucleus of an atom, while electrons orbit around the nucleus. The more energy the electrons have, the further they orbit from the nucleus. The orbits are called electron shells.

The number of shells an atom has depends on how many electrons it has. As atomic number (Z) increases, the number of electrons also increases. The period on the periodic table that an element is in corresponds to the number of shells an atom of the element has. For example, sodium is in period 3 of the periodic table, so a sodium atom has three electron shells, or energy levels. The maximum number of electrons in any shell is $2n^2$, where n is the shell number (see Table 3.5). The electron configuration is shown in a **Bohr diagram** (Figure 3.28).

Figure 3.28: A Bohr diagram can be used to show the electron configuration of an atom. ►

X

Shells are drawn as circles around the nucleus.

Electrons are drawn as shaded circles on the shells. If there are more than four in a shell, they are drawn in pairs.

The nucleus, which contains the chemical symbol of the element being shown.

Bohr diagrams show electron configurations

The rules for drawing Bohr diagrams are as follows:

- The electrons fill the lowest energy shells first (i.e. those closest to the nucleus).
- The outer shell (the valence shell) can have only eight electrons. For the first 20 elements, the electron shells follow a 2, 8, 8, 2 pattern (electron configuration). That is, we move to the third shell once the second shell has eight electrons. The pattern for distributing electrons becomes more complex beyond 20 electrons, so you will not be expected to write electron configurations for elements with an atomic number greater than $Z = 20$.

Lithium has three electrons. The first shell can only hold two, so the third electron will be in the second shell. This configuration is written as 2, 1.

Potassium has 19 electrons. The first shell has two, and the second and third shells can each hold eight. We then put the final electron in the fourth shell. This gives potassium a configuration of 2, 8, 8, 1.

Chemical bonding achieves a stable electron configuration

Apart from helium, which has two electrons, noble gases have eight electrons in their valence (outer) shell. We refer to this as 'full', even though from the third shell onwards there is room for more than eight electrons. Having a 'full' outer shell makes them very stable, which is why noble gases tend not to form chemical bonds with other atoms.

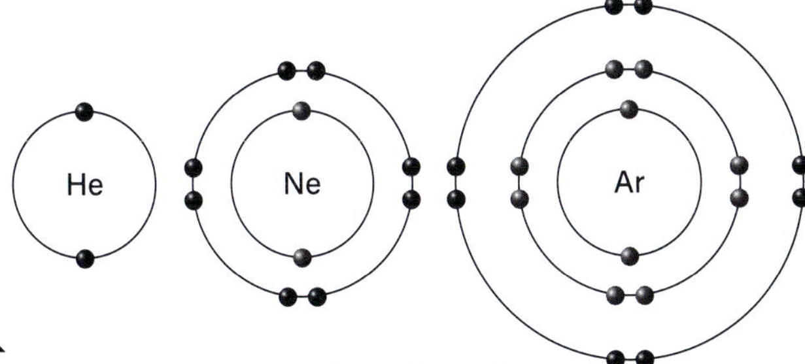

Figure 3.30: Electron configurations of the first three noble gases, showing 'full' and stable valence shells.

All atoms are most stable when they have a full valence shell; therefore, chemical bonding occurs so that atoms can achieve **noble gas configuration** – a full outer shell. This is sometimes called the **octet rule**. Exceptions to this rule include atoms with fewer than five electrons. They can be satisfied with just one shell, which can hold only two electrons. In this case, the atom is stable with just two electrons on the outer shell (e.g. helium).

How an atom achieves a stable octet depends on its number of valence electrons. Atoms with one, two or three valence electrons tend to lose electrons; atoms with four electrons tend to share them, and atoms with five, six or seven electrons gain or share them. Whether electrons are lost, gained or shared will determine whether metallic, ionic or covalent bonds are formed.

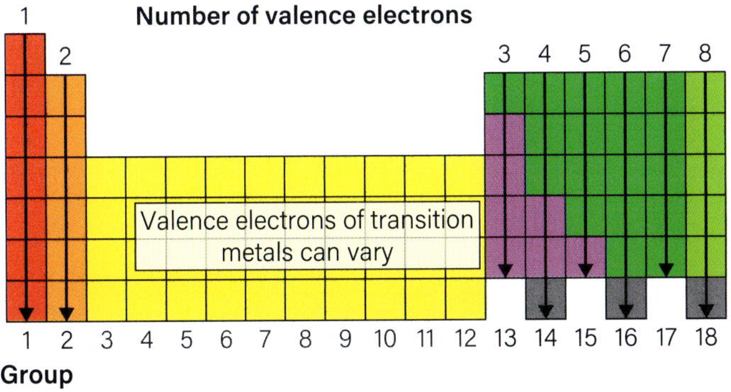

Figure 3.31: The periodic table can be used to determine how many valence electrons the atoms of main group elements have.

Learning Ladder

The periodic table

1. Use the periodic table to determine whether the following elements are likely to lose, gain or share electrons to achieve noble gas configuration.
   - **a** Calcium
   - **b** Carbon
   - **c** Chlorine
   - **d** Aluminium
   - **e** Nitrogen
2. Describe how electron configurations are linked to the following features of the periodic table.
   - **a** Groups
   - **b** Periods
3. Explain why noble gases are unlikely to form chemical bonds with other atoms. Refer to patterns in the periodic table to support your response.
4. Propose and discuss why noble gases are often used to prevent metals reacting with oxygen during welding.
5. Evaluate why noble gases were some of the last elements to be discovered, as seen in Figure 3.26 in Section 3.4.

Nature and development of science

1. Identify materials or equipment you would need to help a friend understand sulfur's electron configuration: 2, 8, 6. Construct a diagram to support your response.
2. **a** Conduct a search to see how the sub-shell electron configuration of sulfur differs from Bohr's model.
   b Describe any similarities or trends between the two representations of scientific knowledge.

Evaluating *p. 253*

Construct an annotated model (digital or physical) of a calcium atom that you could use to teach someone about electron configuration.

1. Identify any assumptions in your model related to the size of subatomic particles, the relative distance between electron shells, and/or the relationship between atomic structure and the organisation of the periodic table.
2. Describe how the assumptions from Question 1 may communicate invalid information about the structure of atoms, electron configuration or the periodic table.

Success criteria

- I can determine the electron configuration of any of the first 20 elements.
- I can explain noble gas configuration and how it relates to chemical bonding.

Summary

- Each known element has a unique chemical symbol (X).
- The periodic table tells us about the atoms of each element.
  - An atomic number (Z) equals the number of protons in the atom's nucleus.
  - Atoms are neutrally charged and therefore have the same number of electrons (–) as protons (+).
  - Atomic mass is the average mass of the different isotopes in a sample of an element.

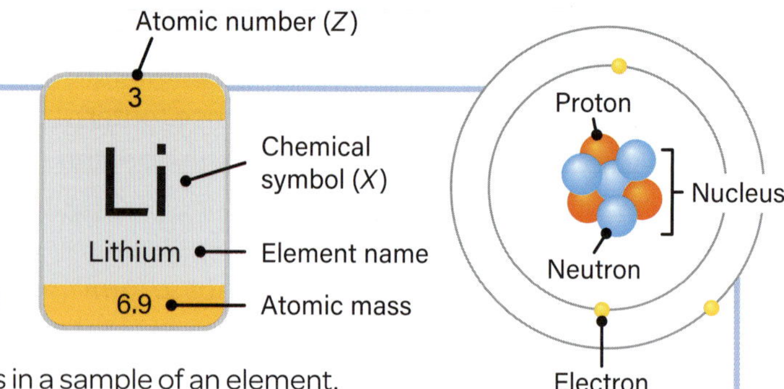

- The organisation of the elements in the periodic table is related to the structure and properties of atoms.
- Elements in the periodic table are organised into groups (columns) and periods (rows).
  - Elements in the same group have similar chemical properties because they have the same number of valence electrons. Exceptions to this rule are the transition elements, whose number of valence electrons can vary.

Group 1	Group 2	Groups 3–12	Group 13	Group 14	Group 15	Group 16	Group 17	Group 18
1 valence electron	2 valence electrons	Number of valence electrons varies	3 valence electrons	4 valence electrons	5 valence electrons	6 valence electrons	7 valence electrons	8 valence electrons

  - The period indicates the number of occupied electron shells.
  - The more shells an atom has that are occupied by electrons, the larger its atomic size.

Period 1	Period 2	Period 3	Period 4	Period 5	Period 6	Period 7
1 electron shell	2 electron shells	3 electron shells	4 electron shells	5 electron shells	6 electron shells	7 electron shells

- In addition to groups and periods, there are many patterns and trends in the periodic table.
  - The reactivity of metals increases towards the bottom left of the periodic table.
  - Atomic size also increases towards the bottom left of the periodic table.
  - Metallic properties gradually change from non-metallic to metallic from the top right towards the bottom left.

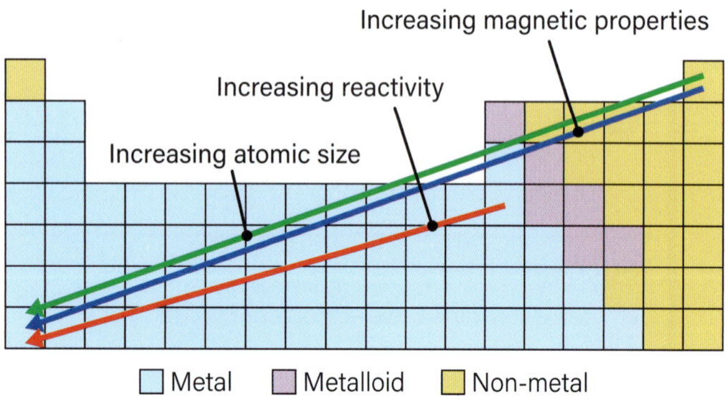

- Different elements from the periodic table can generally join together to form different types of chemical bonds.
  - **Metallic:** metals bonded to metals.
  - **Ionic:** metals bonded to non-metals.
  - **Covalent:** non-metals bonded to non-metals.

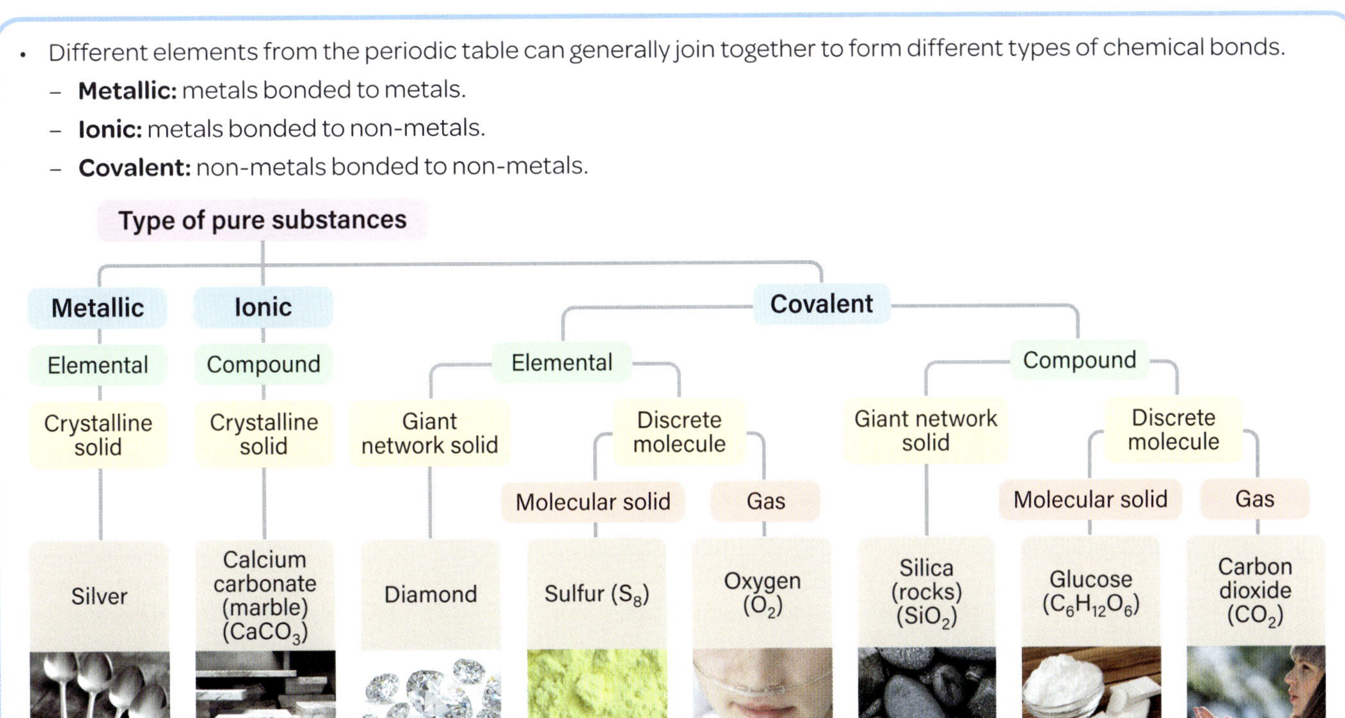

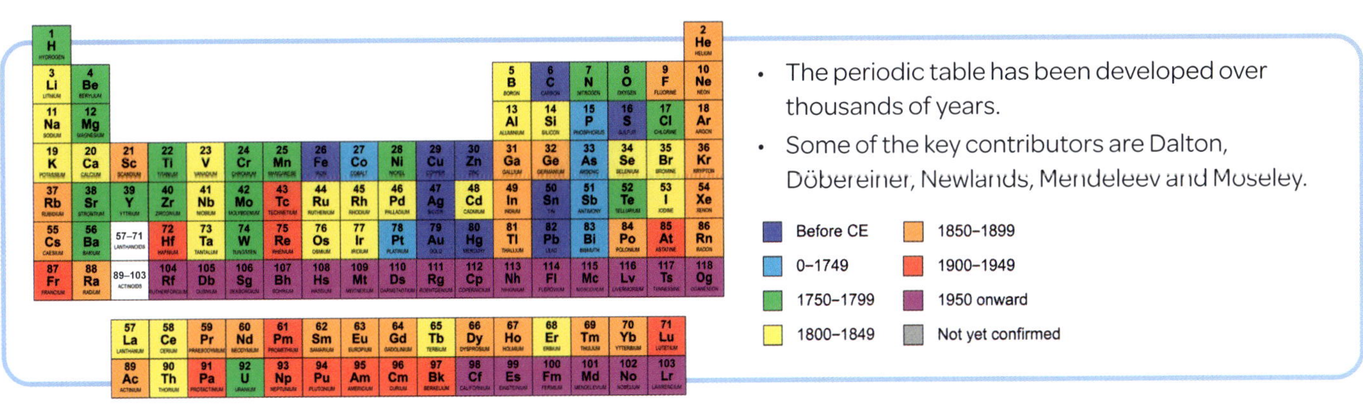

- The periodic table has been developed over thousands of years.
- Some of the key contributors are Dalton, Döbereiner, Newlands, Mendeleev and Moseley.

Key idea: Patterns, order and organisation

- The electrons of an atom exist in energy levels and orbit around the nucleus.
- Bohr diagrams show electron configurations using the 2, 8, 8, 2 distribution pattern.
- The number of valence electrons an atom has will determine the type of bonds it is likely to form in order to achieve a 'full' outer shell.

Number of valence electrons	How it will achieve a 'full' outer shell	Types of bonds it will form
1, 2 or 3	Lose electrons	Ionic or metallic
4	Share electrons	Covalent
5, 6, 7	Gain or share electrons	Ionic or covalent
8	Already has a 'full' outer shell	Does not tend to form bonds

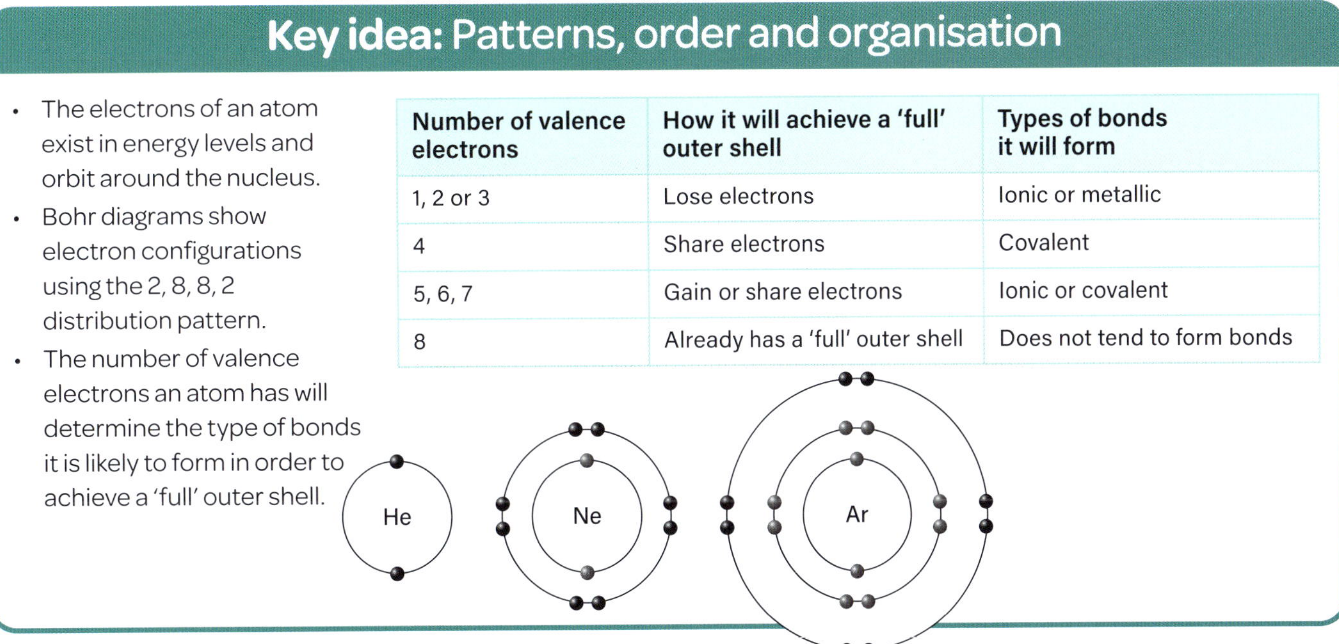

Masterclass

Steps in progression

		1	2
Science understanding	**Periodic table**	Classify the materials used in the *Deepsea Challenger*, Morph folding wheel and Fliteboard.	Describe the type of bonding found between atoms in titanium, a material used in the construction of the *Deepsea Challenger*.
Science as a human endeavour	**Nature and development of science**	Identify a technology that was used in developing the *Deepsea Challenger*.	Propose and describe how knowledge of plastics may have been refined when developing phenolic resin.
Science as a human endeavour	**Use and influence of science**	Propose a problem associated with using a metal for the Fliteboard that is corrosive in salt water.	Carbon fibre is strong, but charcoal – also made of carbon – is weak. Describe how the strength of items made of carbon could be interpreted differently.
Science inquiry	**Processing, modelling and analysing** An investigation is carried out to see how much mass a Morph wheel can hold.	Data indicates that three different Morph wheels can hold 130, 150 and 140 kg, respectively. Organise this information into a table.	Construct an appropriate graph to display the information in your table.
Science inquiry	**Evaluating** An investigation is carried out to see how much mass a Morph wheel can hold by increasing the amount of weight from 100 kg in 10 kg increments until it collapses.	Identify possible assumptions or sources of error in the method.	Give specific examples of how the assumptions or errors could impact the results.

The periodic table and material innovation

Understanding the periodic table is key in selecting the right materials for engineering, as it helps in predicting properties of elements such as strength, reactivity and conductivity. These properties directly affect performance, safety and efficiency.

The *Deepsea Challenger*, a submersible designed to explore the deepest parts of the ocean, is constructed of titanium (a strong, corrosion-resistant metal) and composite polymers, or plastics. Its core safety material is phenolic resin, a hard-setting polymer. These choices were guided by the need for materials that resist extreme pressure – properties that are predictable based on the positions of these elements in the periodic table.

Morph folding wheels, used in portable mobility devices, are made from glass-filled nylon polymer, chosen for its balance of strength, flexibility and light weight. These traits stem from the structure of elements such as carbon, nitrogen and oxygen, which dominate polymer chemistry.

The Fliteboard, an electric hydrofoil surfboard, is constructed of carbon fibre, for carbon's strong bonding, and marine-grade aluminium, for its light weight and resistance to corrosion – qualities that are predictable based on aluminium's position in Group 13 of the periodic table.

Innovations such as these show how knowledge of the periodic table helps engineers to choose elements and compounds with the desired properties, thereby enabling safe, efficient and high-performance designs.

Demonstrate your understanding

3	4	5	
Explain why aluminium is a good choice of material for the Fliteboard.	Analyse the properties of the atoms that make up the substances used to form phenolic resin.	Discuss how understanding the organisation of the periodic table can help engineers select appropriate substances for material innovation. Refer to examples.	
Explain how the materials used to construct the *Deepsea Challenger* are related to effects such as safety.	Discuss how sharing information about early prototypes of innovations helps in developing ideas.	Analyse the significance of there being dozens of prototypes of a product before the design is finalised.	
Explain the links between the criteria for developing a folding wheel to be used in a wheelchair and the materials selected.	Discuss how understanding the materials that a product is made from can influence a person's desire to purchase it.	Propose and analyse factors that contribute to society's acceptance or rejection of material innovation.	
Explain any trends or patterns observed in your graph.	**a** Explain why it would not be possible to explain anomalies in your graph. **b** Give an example of a data result that would be an anomaly.	**a** Add error bars to your graph from step 2, based on information from step 4 of the 'Evaluating' ladder. **b** Discuss what this indicates about experimental uncertainty versus true value.	Science how-to p. 244
Suggest specific amendments to the method to ensure the quality of the data collected.	The values from step 1 of 'Processing, modelling and analysing' are averages of five trials with an experimental uncertainty of 10, 5 and 10 kg, respectively. Write a conclusion, justifying your comments.	Discuss the validity of the conclusion based on the quality of data observed in the graph you constructed in step 2 of 'Processing, modelling and analysing'.	Science how-to p. 253

Figure 3.33: Engineers explored numerous prototypes, using a variety of materials, in developing the Fliteboard. The board is designed to tolerate submersion in salt water, as well as heat generated from the motor.

Figure 3.32: Canadian film director James Cameron carefully considered the purpose and impact of the materials selected to construct the *Deepsea Challenger*, an ocean submersible that was designed to reach the deepest-known part of the ocean.

Figure 3.34: The Morph folding wheel incorporates glass-filled polymer nylon, which is strong enough to hold the weight of the user while still allowing some flexibility.

4.0 Chemical reactions

Reactions are powerful processes that transform substances into something new. In this chapter, you will explore different types of chemical reactions, including synthesis, decomposition and displacement. You will learn how these reactions can be classified based on energy changes as either exothermic or endothermic. You will also investigate the factors that influence how fast a reaction occurs. Through practical and theoretical work, you will deepen your understanding of how chemical reactions work and why they are essential to both natural processes and human technology.

Learning Ladder

The Learning Ladder for each chapter maps the Science Understanding, Science as a Human Endeavour and Science Inquiry strands that will be covered. Each ladder has five levels of progression, called steps. To climb the ladders, you need to develop fluency at each step. This will help you develop the ability to complete tasks that are more complex.

Steps in progression	Science understanding: Chemical science: Chemical reactions	Science as a human endeavour: Nature and development of science	Science as a human endeavour: Use and influence of science
5	I can evaluate the impact our understanding of reactions has in various industries	I can analyse how advances in technologies enable advances in science	I can analyse the key factors that contribute to scientific knowledge being adopted more broadly by society
4	I can analyse the effect of changing reactant and reaction conditions on chemical reactions	I can discuss how scientific knowledge is validated, including the role of publication and peer review	I can discuss how scientific information and misinformation may inform personal and social decision-making
3	I can explain how different factors affect reactions	I can explain how science has contributed to developments in technologies and engineering	I can explain how the values and needs of society influence the focus of scientific research
2	I can describe and represent changes that occur in a range of chemical reactions	I can describe how scientific knowledge is refined over time	I can describe scientific knowledge that may be interpreted in different ways
1	I can identify different types of chemical reactions	I can identify technologies that have enabled advances in science	I can identify scientific knowledge that can address socio-scientific issues

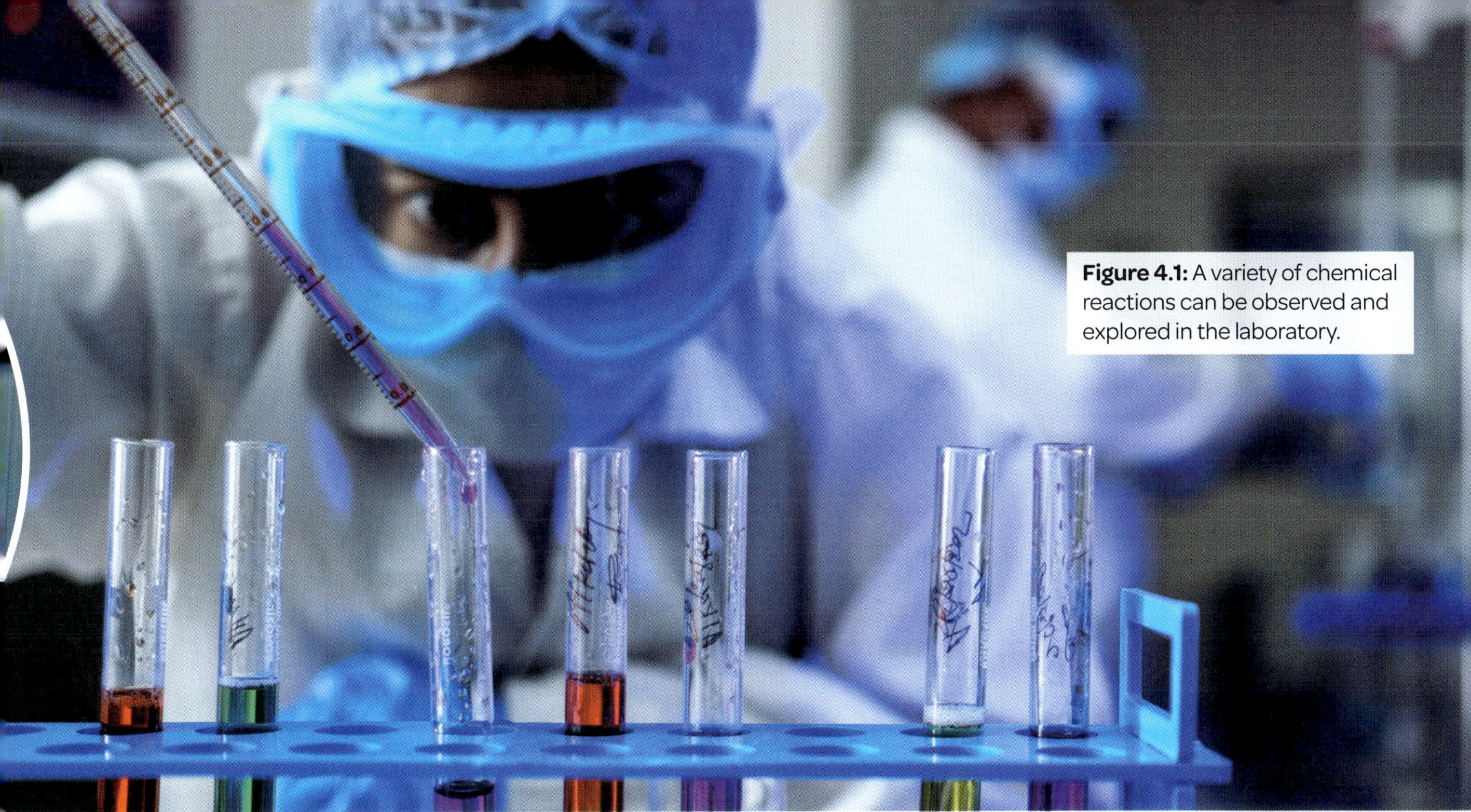

Figure 4.1: A variety of chemical reactions can be observed and explored in the laboratory.

Science inquiry

Steps in progression	Questioning and predicting	Planning and conducting
5	I can develop explanatory models when investigating scientific questions, problems and claims	I can evaluate scientific methods regarding safety, ethical and procedural considerations
4	I can discuss what is needed for a question to be investigable or a prediction to be reasoned	I can design and conduct reproducible investigations that consider safety, ethical and procedural factors
3	I can develop a hypothesis that predicts the relationship between investigation variables	I can use equipment to generate and record data with precision, to obtain replicable data, using digital tools as appropriate
2	I can formulate questions to investigate scientific problems	I can develop and follow risk assessments that consider safety and ethical issues
1	I can make simple predictions based on what I know and observe	I can select appropriate equipment to collect precise data for scientific investigations

4·1 ▸ Chemical reaction basics

Learning intention

At the end of this lesson, I will be able to:

- identify types of chemical reactions
- describe features of chemical reactions in terms of energy changes and reaction rates.

Key terms

chemical reaction: a process in which one or more substances are changed to form new substances; a rearrangement of the way atoms are joined

endothermic: a reaction that absorbs energy in the form of heat

exothermic: a reaction that releases energy in the form of heat or light

Key idea: Matter and energy

Figure 4.2: Explosions are exothermic chemical reactions that release large amounts of energy very quickly.

Chemical reactions are the foundation of chemistry, transforming substances into new forms through various processes. In order to grasp how matter behaves in the world around us, it is essential to understand the different types of reactions, how they involve energy changes and what affects their speed.

There are many types of chemical reactions

All **chemical reactions** involve the rearrangement of atoms to form new substances. The reactants undergo chemical changes, which results in new products being formed. Although this is true for all chemical reactions, there are many types of chemical reactions, each with unique characteristics.

In *synthesis* reactions, two or more substances combine to form a new compound, while *decomposition* reactions break down a compound into simpler parts. *Displacement* reactions involve one element replacing another in a compound, and *neutralisation* reactions occur when acids and bases react to form water and a salt. By recognising these reaction types, based on given information about the reactants and products, we can predict and understand chemical behaviour. Each of these reaction types (see Table 4.1) will be explored in more depth in this chapter.

Table 4.1: Classifications of chemical reactions

Type of reaction	Description	Example
Synthesis	More than one reactant combines to form a single product.	Iron rusting with oxygen to form iron oxide
Decomposition	One reactant breaks down into multiple products.	Water breaking down into hydrogen gas and oxygen gas
Displacement	A more reactive element replaces a less reactive one in a compound.	Copper metal replacing silver in solution to form silver metal
Neutralisation	An acid and a base react to form a neutral salt and water.	Hydrochloric acid and marble chips reacting to form calcium chloride salt, water and carbon dioxide

Chemical reactions can release or absorb energy

Chemical reactions are accompanied by energy changes (represented by a triangle symbol). **Exothermic** reactions release energy, usually as heat, making the surroundings warmer. **Endothermic** reactions absorb energy from their surroundings, causing cooling effects. Thermometers can be used to measure the area that surrounds a chemical reaction, to determine whether it is exothermic or endothermic. The concept of energy changes in chemical reactions is covered in detail in Section 4.7.

Figure 4.3: Endothermic reactions absorb energy, while exothermic reactions release energy.

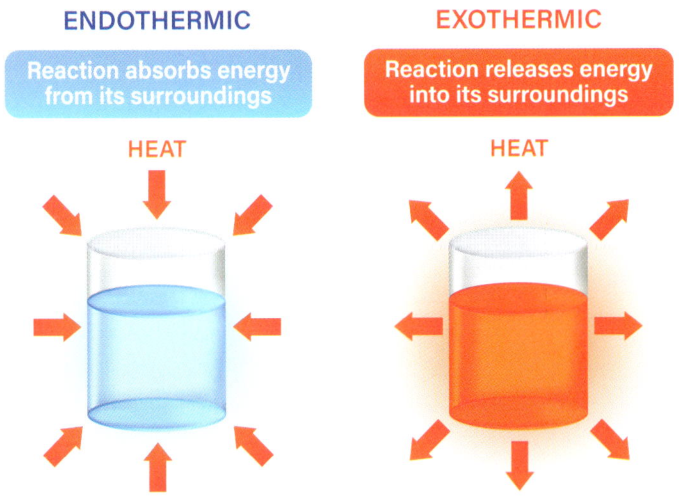

ENDOTHERMIC

Reaction absorbs energy from its surroundings

HEAT

EXOTHERMIC

Reaction releases energy into its surroundings

HEAT

Chemical reactions can occur at different rates

The speed of chemical reactions, also known as the reaction rate, depends on how often and how effectively the particles of reactants collide. Several key factors influence the collisions of particles and therefore affect how quickly a reaction happens. These factors include the surface area of solid reactants, the concentration of reactants, the temperature, and the presence of catalysts. Each of these factors will be explored in Section 4.7. Understanding these factors is important for controlling reaction rates in both experiments and everyday situations.

Figure 4.4: Many chemical reactions are common in everyday life – for example, when a bike rusts slowly over time.

Learning Ladder

Chemical reactions

1. List at least four types of chemical reactions.
2. Describe the difference between a synthesis reaction and a decomposition reaction.
3. Explain how the speed of a reaction is related to the collisions that occur between reactant particles.

Nature and development of science

1. Identify a technology that can be used to determine whether a reaction is exothermic or endothermic.
2. Describe the scientific knowledge that has developed beyond the 'types' of reaction and is used to further classify reactions.

Planning and conducting p. 239

Imagine you are conducting an investigation to determine whether the reaction between baking soda and vinegar is exothermic or endothermic.

1. Identify three pieces of equipment you will need to conduct this investigation and to collect data.
2. **a** Identify the hazards that would be present when conducting the investigation, based on the equipment identified in Question 1 and the reacting substances.
   b Develop a risk assessment to minimise the safety hazards identified in part a.
3. Explain how the equipment identified in Question 1 will collect precise data. Consider whether the equipment should be adapted to ensure the data is replicable.
4. Design a step-by-step method for this investigation that a Year 7 student would be able to follow.
5. Share your method with a classmate and read their method.
   a Evaluate your classmate's method for safety, ethical and procedural factors.
   b Prepare an evaluation summary that includes recommendations for how their method could be improved.

Key idea: Matter and energy

Discuss how all chemical reactions involve changes in energy and matter, no matter the type.

Success criteria

- I can differentiate between types of chemical reactions.
- I can describe how chemical reactions involve energy change and can happen at different rates.

4·2 ▸ Synthesis and decomposition reactions

Learning intention

At the end of this lesson, I will be able to describe the features of synthesis and decomposition reactions.

Key terms

carbonate: a substance containing the elements carbon and oxygen

chemical equation: a chemical reaction represented using chemical formulas of reactants and products

corrosion: the degradation of a metal due to its reaction with its environment

decomposition reaction: a reaction in which one reactant breaks down to form multiple products

degradation: deterioration of physical properties of a material

rusting: the corrosion of iron to form iron oxides

synthesis reaction: a reaction in which two or more reactants combine to form a more complex product

word equation: a representation of a chemical reaction using the names of the reactants and products

Investigation 4.2A

Synthesis of iron oxide, p. 325

Investigation 4.2B

Decomposition of copper(II) carbonate, p. 327

Key idea: Matter and energy

Figure 4.5: The iron in this old car has corroded to produce orange–brown rust.

In some chemical reactions, reactants combine to form a single product; in others, a complex reactant might break down into simple parts.

Synthesis reactions combine substances

A chemical reaction that involves two or more reactants combining to form a single, more complex product is called a **synthesis reaction**. In general, a chemical equation for a synthesis reaction looks like this:

Reactant A	+	reactant B	→	product AB
A	+	B	→	AB

An example is the formation of water from the gases of hydrogen and oxygen. Below are the **word equation** and the **chemical equation** for this synthesis reaction.

Hydrogen gas	+	oxygen gas	→	water
$2H_2(g)$	+	$O_2(g)$	→	$2H_2O(l)$

Corrosion is a synthesis reaction

Corrosion is a natural process that involves the gradual **degradation** of metals by chemical reactions with substances in their environment. When metals react with oxygen, compounds form on the surface of the metals. The oxygen can be in the air, water or salt water. Corrosion affects the properties of a metal structure, such as its strength and appearance.

Iron corrodes to create rust

Rusting is the corrosion of iron. The scientific name for rust is iron(II) oxide (Fe_2O_3), and it forms so easily that pure iron is rarely found in nature. Rust forms as a flaky red–brown solid on iron structures. For rusting of iron to occur, oxygen is required.

Iron	+	oxygen	→	iron(II) oxide
$4Fe$	+	$3O_2$	→	$2Fe_2O_3$

Rust causes a lot of damage to buildings, cars and ships because it does not form a protective layer on the metal's surface. Water can get through to the metal underneath, leading to further corrosion. Little bits of rust flake off, leaving the rest of the metal exposed to oxygen. Eventually, all the metal corrodes.

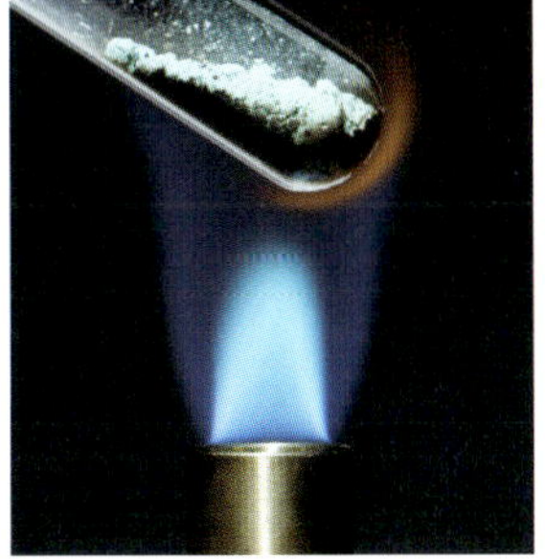

Figure 4.6: Heating a metal carbonate causes it to decompose and produce the metal oxide and carbon dioxide gas.

Decomposition reactions break down substances

In **decomposition reactions**, one substance breaks down into two or more simpler substances. The opposite of a synthesis reaction, a chemical equation for a decomposition reaction looks like this:

Reactant AB → product A + product B

AB → A + B

Most decomposition reactions require energy to get them started. Thermal decomposition is started by heat energy. This is an easy reaction to see in the laboratory if you heat a reactive substance over a Bunsen burner. For example, if you heat copper(II) carbonate, it decomposes into copper(II) oxide and carbon dioxide gas. (A **carbonate** contains the elements carbon and oxygen.)

Copper(II) carbonate $\xrightarrow{\text{heat}}$ copper(II) oxide + carbon dioxide

$CuCO_3(s) \xrightarrow{\text{heat}} CuO(s) + CO_2(g)$

Electrical decomposition – electrolysis – is usually performed by passing an electrical current through a liquid reactant. For example, if you apply electricity to water, it decomposes into hydrogen and oxygen.

Water $\xrightarrow{\text{electricity}}$ hydrogen + oxygen

$2H_2O(l) \xrightarrow{\text{electricity}} 2H_2(g) + O_2(g)$

Photochemical decomposition is a very slow reaction triggered by light energy. For example, when silver chloride is exposed to light, it slowly decomposes into silver and chlorine ions.

Silver chloride $\xrightarrow{\text{light}}$ silver ions + chlorine ions

$AgCl(aq) \xrightarrow{\text{light}} Ag^+(aq) + Cl^-(aq)$

Learning Ladder

Chemical reactions

1. Identify the following reactions as synthesis or decomposition.
   - **a** $2H_2O_2(l) \rightarrow O(l) + O_2(g)$
   - **b** $CaCO_3(s) \rightarrow CaO(s) + CO_2(g)$
   - **c** $2NaHCO_3(s) \rightarrow Na_2CO_3(s) + CO_2(g) + H_2O(g)$
   - **d** $2Mg(s) + O_2(g) \rightarrow 2MgO(s)$
2. **a** Describe corrosion as a synthesis reaction.
   b Write word equations for each reaction above.
3. Corrosion, thermal decomposition, electrolysis and photochemical reactions require more than just the reactants to start a reaction.
   - **a** Identify the factors required for each reaction.
   - **b** Propose why these factors are needed.

Use and influence of science

1. Identify the general equations used to represent synthesis and decomposition reactions.
2. **a** Describe the difference between a word equation and a chemical equation.
   b Propose why a word equation might be more suitable when communicating scientific knowledge to the general public.

Planning and conducting p. 239

Imagine you wanted to compare how long it takes for different metal carbonates to decompose on heating.

1. Propose which piece of equipment would be the most appropriate vessel for decomposing metal carbonates: test tubes, beakers or conical flasks. Provide a reason for your selection.
2. Describe how to minimise the risks involved in decomposing metal carbonates over heat.
3. Identify and explain materials that would be required in this investigation to generate and record precise data, including digital tools if applicable.
4. Propose an aim and a method for the investigation.
5. Identify the features of the method from Question 4 that make it safe, ethical and reproducible.

Key idea: Matter and energy

You are given a gold coin, an iron nail, a piece of copper plate and some aluminium foil. Which of these materials do you think will corrode the fastest, and why? Write a method to test your hypothesis.

Success criteria

- I can distinguish between and provide examples of synthesis and decomposition reactions.

4·3 ▸ Displacement reactions

Learning intention

At the end of this lesson, I will be able to describe the features of displacement reactions.

Key terms

anion: negative ion

cation: positive ion

displacement reaction: a reaction in which an element replaces (displaces) another element from a compound

dissociate: to split apart into ions in water; to dissolve ionic compounds

double-displacement reaction: a reaction in which parts of two compounds replace each other to form two new compounds

ion: an atom that has lost or gained an electron to become charged (positive or negative)

ionic salt: a compound made up of a cation and an anion

net ionic equation: a chemical equation that only includes the ions and precipitate involved in the reaction; not spectator ions

precipitate: an insoluble product that forms a solid in solution

precipitation reaction: a type of double-displacement reaction that forms a precipitate when two solutions are combined

single-displacement reaction: a reaction in which a more reactive element replaces a less reactive element from a compound

spectator ion: an ion that does not take part in the reaction

Investigation 4.3A

Growing metallic crystals: metal displacement reactions, p. 329

Investigation 4.3B

Precipitation reactions, p. 331

Key idea: Matter and energy

In some reactions, an element displaces another element in a compound. A single element might replace another element, or two elements of two different compounds may replace each other.

In single-displacement reactions, one element replaces another element

Displacement reactions occur when an element replaces another element in a compound. When one element displaces another element in a compound, it is called a **single-displacement reaction**. This typically happens when a more reactive element replaces a less reactive one in a compound. In the general equation in Figure 4.7, element A is more reactive than element B, so A forms a chemical bond with C, causing B to be displaced from the compound.

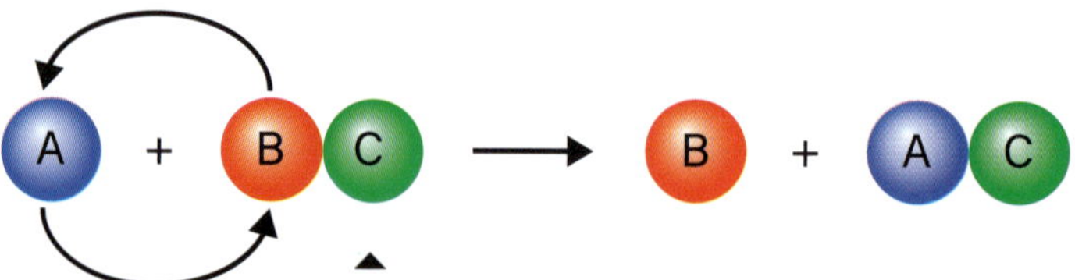

Figure 4.7: A single-displacement reaction

Single-displacement reactions often occur between metals and ionic salts. (An **ionic salt** is a compound made up of a **cation** – a positive **ion** – and an **anion** – a negative ion.) For example, when a piece of zinc metal is placed in copper(II) sulfate solution, zinc displaces copper to form zinc sulfate and solid copper.

$$Zn(s) + CuSO_4(aq) \rightarrow ZnSO_4(aq) + Cu(s)$$

In this case, zinc is more reactive than copper and, therefore, displaces it from the ionic solution.

Figure 4.8: Zinc is more reactive than copper, so when zinc metal is placed in a solution of copper(II) sulfate, the zinc replaces the copper in solution and solid copper forms.

The activity series can be used to predict displacement reactions

To determine whether a displacement reaction will occur, we can use the activity series of metals, as shown in Figure 4.9. If pure, solid metal is more reactive than the metal cation in solution, then a reaction will occur. For example, zinc is above copper on the activity series, indicating that it will displace copper from its compound, forming solid copper.

Figure 4.9: This simplified activity series of metals can be used to determine if a displacement reaction will occur when a metal is combined with an ionic compound in solution.

Lithium (Li)
Potassium (K)
Barium (Ba)
Calcium (Ca)
Sodium (Na)
Magnesium (Mg)
Aluminium (Al)
Zinc (Zn)
Chromium (Cr)
Iron (Fe)
Tin (Sn)
Lead (Pb)
Copper (Cu)
Silver (Ag)
Platinum (Pt)
Gold (Au)

Increasing activity

Double-displacement reactions involve two elements of two different compounds swapping

When parts of two compounds replace each other to form two new compounds, this is called a **double-displacement reaction**. In general, a double-displacement reaction looks like Figure 4.10.

Figure 4.10: A double-displacement reaction

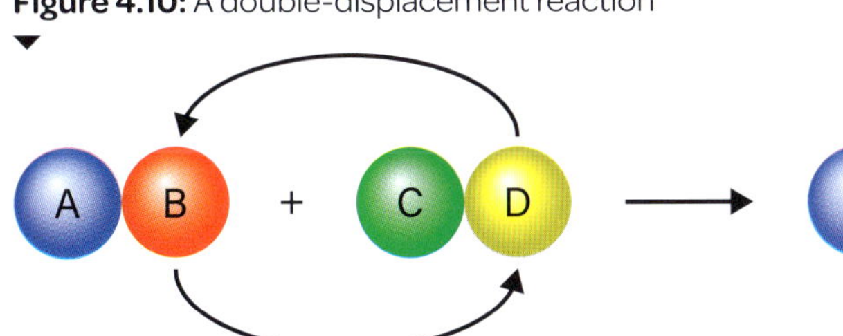

These reactions typically occur in aqueous solutions and ions are exchanged between the reacting compounds. For example, when sodium chloride and silver nitrate solutions are mixed, a double-displacement reaction occurs, forming sodium nitrate and silver chloride.

$$NaCl(aq) + AgNO_3(aq) \rightarrow AgCl(s) + NaNO_3(aq)$$

Precipitates can form in double-displacement reactions

A **precipitation reaction** occurs when two salt solutions are mixed, and two new products are formed (Figure 4.10): an insoluble solid and a soluble salt. The insoluble solid, called a **precipitate**, may be suspended for a while but eventually settles to the bottom of the container (Figure 4.11). Precipitation reactions can be used to identify the presence of certain compounds or ions.

Figure 4.11: Solutions are clear liquids. When a precipitate forms, a solid drops out of the solution.

Ionic compounds can form precipitates

Most ionic compounds are soluble in water, so when they are mixed with water, the water molecules pull them apart and they **dissociate** into freely moving ions. The ions are then available to form new compounds with other ions. If the new compound is insoluble, a solid precipitate will form.

Consider what happens when solutions of potassium iodide and lead(II) nitrate are mixed together. The potassium iodide solution is made up of positive potassium ions (K^+) and negative iodide ions (I^-). The lead(II) nitrate solution is made up of positive lead ions (Pb^{2+}) and negative nitrate ions (NO_3^-). Once these ions are mixed, a precipitation reaction occurs, and the insoluble lead and iodide ions combine to form a solid. This produces a solution of potassium nitrate, with a precipitate of insoluble yellow lead(II) iodide forming within it (Figure 4.12).

The potassium cation and the polyatomic nitrate anion do not take part in the reaction. They are called **spectator ions** because they are not involved in the reaction. They are shown in the complete ionic equation for the reaction but not in the **net ionic equation**.

Potassium iodide	+	lead(II) nitrate	→	lead(II) iodide	+	potassium nitrate
$2KI(aq)$	+	$Pb(NO_3)_2(aq)$	→	$PbI_2(s)$	+	$2KNO_3(aq)$

Complete ionic equation:

$$2K^+ + 2I^- + Pb^{2+} + 2NO_3^- \rightarrow PbI_2 + 2K^+ + 2NO_3^-$$

Cancel out the spectator ions.

$$\cancel{2K^+} + 2I^- + Pb^{2+} + \cancel{2NO_3^-} \rightarrow PbI_2 + \cancel{2K^+} + \cancel{2NO_3^-}$$

Net ionic equation:

$$Pb^{2+}(aq) + 2I^-(aq) \rightarrow PbI_2(s)$$

Solubility rules help to identify precipitates

If a reaction produces a precipitate, you may want to identify it. The solubility rules (see Table 4.2) are rules for ionic compounds that tell you whether certain ions will form precipitates with other ions or stay dissolved in solution.

For example, the rules tell you that in the previous reaction of potassium iodide with lead(II) nitrate, all nitrates are soluble but lead(II) iodide is insoluble. This helps to identify the yellow precipitate as lead(II) iodide. Similarly, you can see that sodium chloride (table salt) is soluble but silver chloride is not soluble.

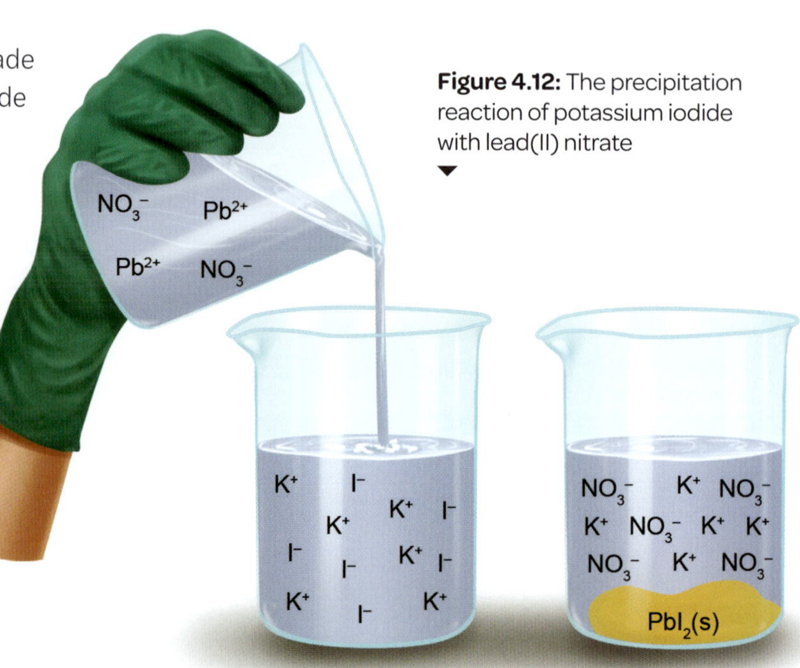

Figure 4.12: The precipitation reaction of potassium iodide with lead(II) nitrate

Table 4.2: The solubility rules

Ions	Solubility
Group 1 salts (Na^+, K^+)	All soluble
Ammonium salts (NH_4^+)	All soluble
Nitrates (NO_3^-)	All soluble
Chlorides (Cl^-)	All soluble except for $AgCl$, $PbCl_2$ and Hg_Cl_2
Iodides (I^-)	All soluble except for AgI, PbI_2 and HgI_2
Sulfates (SO_4^{2-})	All soluble except for $PbSO_4$, $CaSO_4$ and $BaSO_4$
Carbonates (CO_3^{2-})	All insoluble except for group 1 carbonates (e.g. Na_2CO_3, K_2CO_3) and $(NH_4)_2CO_3$
Hydroxides (OH^-)	All insoluble except for group 1 hydroxides (e.g. NaOH, KOH), $Ca(OH)_2$, $Ba(OH)_2$ and NH_4OH

Figure 4.13: When a solution of potassium iodide is added to a solution of lead(II) nitrate, a yellow precipitate of lead(II) iodide is formed.

Learning Ladder

Chemical reactions

1 Classify the following reactions as single or double displacement.

a Copper metal + silver nitrate → silver metal + copper(II) nitrate

b Sodium chloride + copper(II) nitrate → sodium nitrate + copper(II) chloride

c Magnesium chloride + potassium fluoride → potassium chloride + magnesium fluoride

d Zinc metal + iron(II) sulfate → iron metal + zinc sulfate

2 Describe general observations that indicate a displacement reaction has occurred.

3 Explain how the observations of a reaction would help you to distinguish between single- and double-displacement reactions. Provide examples.

4 Write the net ionic equations, including state symbols, for the following reactions. You may use the solubility rules to help you.

a $NaCl + AgNO_3 \rightarrow NaNO_3 + AgCl$

b $CaCl_2 + Na_2CO_3 \rightarrow CaCO_3 + 2NaCl$

c $2NaOH + CuSO_4 \rightarrow Na_2SO_4 + Cu(OH)_2$

d $BaCl_2 + K2SO_4 \rightarrow BaSO_4 + 2KCl$

e $FeI_2 + 2KOH \rightarrow 2KI + Fe(OH)_2$

5 Propose how displacement reactions could be used to remove the following contaminants from waterways. (Include reference to the type of displacement reaction, activity series and solubility rules.)

a Lead

b Sulfate

c Chromium

d Phosphate

Use and influence of science

1 Identify the tool that provides knowledge for investigators to predict whether displacement reactions will occur between metals and compounds in solution.

2 Propose and describe how the tool identified in Question 1 could be used in different ways.

3 Consider and explain how precipitation reactions could be used to remove heavy metals such as lead from contaminated waterways.

4 Propose and discuss how a misunderstanding of solubility rules could impact the ability of industries to make suitable decisions.

5 Analyse how advances related to using the activity series and solubility rules have been embedded in industry and accepted by society.

Questioning and predicting p. 234

1 Predict which metal will displace the other in the following pairs.

a Lithium and aluminium

b Iron and sodium

c Silver and gold

2 Identify the question that would be most suitable to investigate scientifically.

A How can I tell if copper is more reactive than zinc?

B Does temperature affect the rate of the formation of ionic precipitates?

C How does the concentration of ionic solutions affect the rate of metal displacement?

D Construct a scientific question that could be used to investigate the relative reactivities of copper, zinc, silver and iron.

3 Several different metals were mixed with several different ionic compounds dissolved in solution. Develop a hypothesis to predict whether a reaction will take place, based on the activity series.

4 Discuss your response to Question 2 in relation to what is needed for the question to be scientific.

5 Evaluate Figure 4.9 on the previous spread. Discuss the pros and cons of using a simplified version of the activity series of metals.

Key idea: Matter and energy

Conduct a search to learn about how a metal (e.g. copper or iron) is mined and refined. Summarise the process, including details of any displacement reactions used or energy required. Also make reference to the reactivity series and/or solubility.

Success criteria

- I can use the activity series to predict if a displacement reaction will occur.
- I can use solubility rules to predict the formation of a precipitate, based on the combined solutions.

4·4 ▸ Neutralisation and other reactions of acids

Learning intention

At the end of this lesson, I will be able to:

- identify whether a substance is acidic or basic, based on its pH and the use of indicators
- describe the features of neutralisation reactions and other reactions of acids.

Key terms

acid: a substance with a pH of less than 7

alkali: a base that is dissolved in water

base: a substance with a pH of more than 7

caustic: able to burn or corrode organic tissue through chemical action

concentration: the amount of a substance in a volume of solution

corrosive: highly reactive and damaging or destructive to another substance

indicator: a substance used to determine the acidity of a solution

neutralisation reaction: a reaction involving an acid and a base to produce water and a salt

neutralise: to make something chemically neutral; neither acidic nor basic

pH: a figure expressing the acidity or alkalinity of a solution

Investigation 4.4A

Neutralising hydrochloric acid, p. 333

Investigation 4.4B

The effects of indicators on acids and bases, p. 335

Investigation 4.4C

Reactions of acids with metals, p. 337

Investigation 4.4D

Reactions of acids with carbonates, p. 339

Key idea: Matter and energy

Two of the most commonly occurring groups of compounds are acids and bases. They are found in many everyday items and places, such as cleaning products, swimming pools and kitchens. Many of the foods we eat are also either acidic or basic, and these properties help with digestion. Acids can be dangerous because they are corrosive, and they can react with some metals to produce hydrogen gas and metallic salts. However, when an acid and a base react together, they neutralise each other, forming products that are usually much safer.

▼ **Figure 4.14:** Weak acids and bases occur naturally in some foods.

Acids produce hydrogen ions

An **acid** is a **corrosive** chemical substance that produces hydrogen ions (H^+) when mixed with water. The hydrogen ions can react with the other substances to produce water and hydrogen gas, as well as ionic salts, which are made up of cations and anions.

The higher the **concentration** of hydrogen ions produced by an acid, the higher its acidity. Strong acids are very dangerous, especially when they contact skin and eyes. That is why you must always wear protective clothing and eyewear when working with acids in the laboratory.

Weak acids are important in our diet. Citrus fruits contain a weak acid, called citric acid, which contributes to their sour flavour. Soft drinks, coffee and vinegar also contain weak acids.

Bases produce hydroxide ions

A **base** is a substance that reacts with an acid to **neutralise** it. Bases that dissolve in water are called **alkalis**. When mixed with water, bases produce hydroxide ions (OH^-) – an atom of oxygen bonded to an atom of hydrogen, with an overall negative charge.

Strong bases are described as **caustic**, meaning they cause burns by chemical action; therefore, they can be just as dangerous as acids. Bases such as sodium hydroxide and ammonia react with oils and fats and so are used in many household cleaners. Weak bases are found in toothpaste, conditioners, antacid tablets and baking powder.

Three common types of substances act as bases.

- Metal hydroxides (e.g. potassium hydroxide) contain metals bonded with hydroxide (OH^-).
- Metal oxides (e.g. zinc oxide) contain metals bonded with oxide ions (O^{2-}).
- Metal carbonates (e.g. copper carbonate) contain metals bonded with carbonate ions (CO_3^{2-}).

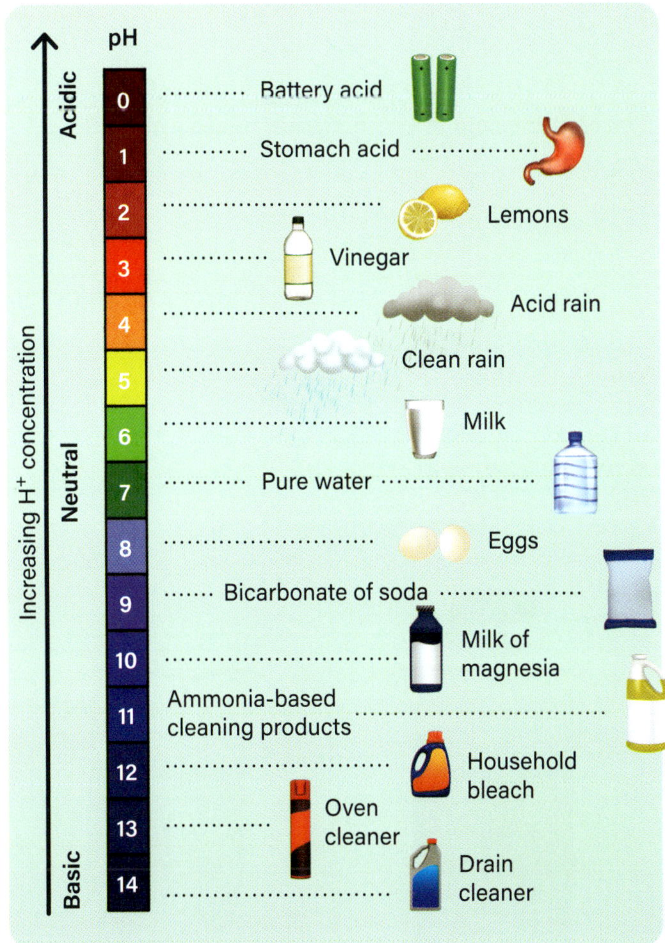

▲ **Figure 4.15:** The pH scale

The pH scale measures acidity

The acidity of a solution is measured on a scale called the **pH** scale (Figure 4.15). The higher the proportion of H^+ ions, the lower the pH. Therefore, the lower the proportion of H^+ ions in a solution compared to hydroxide ions (OH^-), the higher the pH. Acids have a low pH, while bases have a high pH. A solution with a pH of 7 is neutral, a solution with a pH of less than 7 is acidic, and a solution with a pH greater than 7 is basic. Pure water is neutral because it has a pH of 7 at 25 °C.

You can measure pH by using a digital pH meter, or by adding an indicator to a solution and matching its colour to a chart. **Indicators** change colour depending on whether they are mixed with an acid or a base. A common indicator is litmus, which turns red when mixed with acids and blue when mixed with bases.

Indicators can be used to measure pH

Different indicators show a wide range of colour changes depending on the pH of a solution. For example, a universal indicator can show multiple colours to indicate the pH level from 1 to 14, from red in strong acids to violet in strong bases (Figure 4.16).

Some indicators display only two or three colours throughout the pH range. This can be useful when we are only concerned about whether a substance is above or below a certain pH. Figure 4.17 shows a variety of different indicators and their colour changes at specific pH values. For example, bromocresol green changes from yellow to blue at pH 4–5, so it would not be able to identify with certainty if a solution is basic (alkaline), though it could identify if its pH is above or below 4–5.

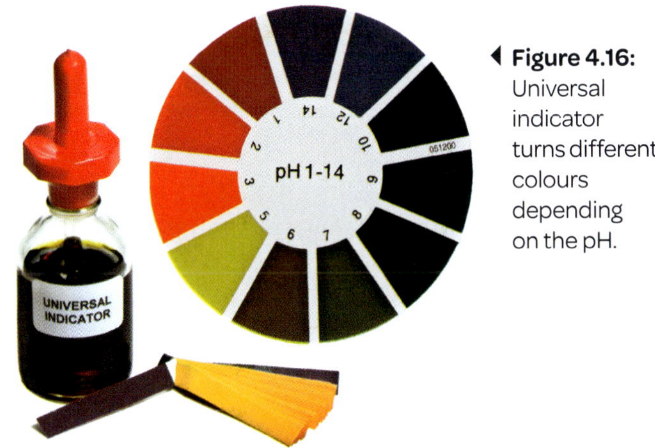

◀ **Figure 4.16:** Universal indicator turns different colours depending on the pH.

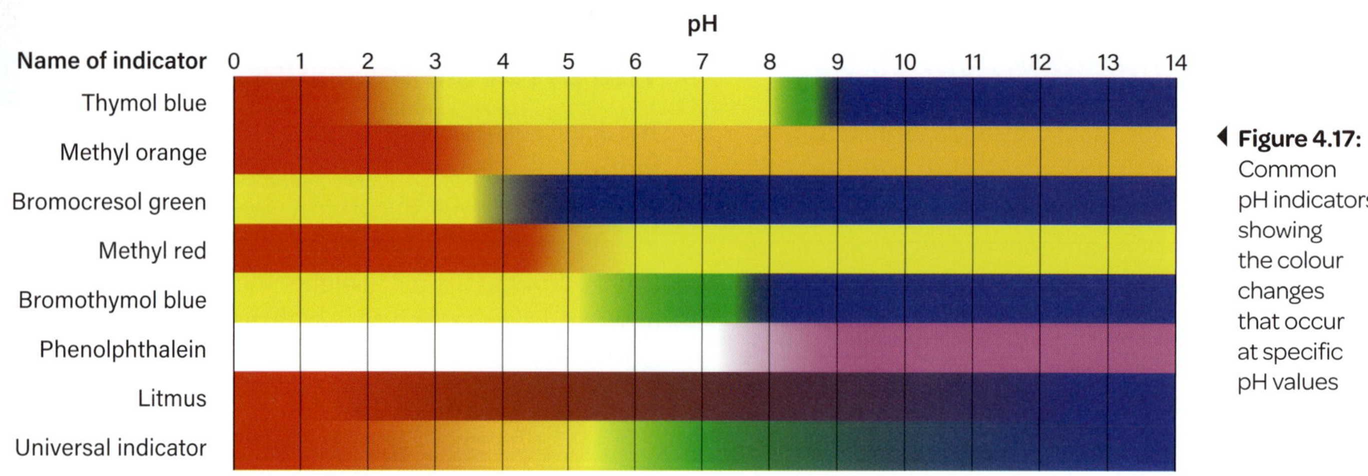

Figure 4.17: Common pH indicators showing the colour changes that occur at specific pH values

Acids and bases neutralise each other

In a **neutralisation reaction**, an acid and a base react to form a salt and water.

Acid + base → salt + water

This happens because acids are a source of hydrogen ions (H^+), while bases are a source of hydroxide ions (OH^-). In an acid–base neutralisation reaction, the H^+ from the acid and the OH^- from the base combine to form water, which is neutral, having a pH of 7 at 25 °C.

Hydrogen ion	+	hydroxide	→	water
$H^+(aq)$	+	$OH^-(aq)$	→	$H_2O(l)$

The other parts of the acid and base combine to produce a salt. The reaction of a strong acid with a strong base results in a neutral solution with a pH of 7 and a neutral ionic salt. For example, nitric acid reacts with lithium hydroxide to form the salt lithium nitrate ($LiNO_3$) and water.

Nitric acid	+	lithium hydroxide	→	lithium nitrate salt	+	water
$HNO_3(aq)$	+	$LiOH(aq)$	→	$LiNO_3(aq)$	+	$H_2O(l)$

Acids react with metals to produce salts and hydrogen gas

Acids react with metals. Some metals are more reactive than others. You can see this in the activity series of metals in Table 4.3. The metals at the top of the reactivity series, such as potassium and sodium, react violently, while those at the bottom react very little. This is one reason that gold and silver are used to make jewellery, rather than iron and zinc. Gold and silver are fairly unreactive, so they keep their shiny surface even when exposed to acids and oils in the air or on your skin.

Acids react with metals to form salts and hydrogen gas. More reactive metals react faster, which you can see by how quickly the hydrogen gas bubbles are released.

Acid + metal → salt + hydrogen

The salt formed depends on the acid the metal reacts with. For example, magnesium reacts with hydrochloric acid to produce magnesium chloride and hydrogen gas.

Hydrochloric acid	+ magnesium	→	magnesium chloride	+ hydrogen
$2HCl(aq)$ +	$Mg(s)$	→	$MgCl_2(aq)$ +	$H_2(g)$

Figure 4.18: Universal indicator is red in a solution of 0.1 M hydrochloric acid. If calcium hydroxide ($Ca(OH)_2$) is added to the solution, the acid and base neutralise each other and the universal indicator turns green.

Figure 4.19: Samples of magnesium, zinc, iron and lead each react differently if placed in a hydrochloric acid solution. Magnesium ribbon (left) reacts vigorously to produce magnesium chloride and hydrogen gas, visible as bubbles. Zinc and iron react less vigorously with the acid, and release hydrogen gas bubbles more slowly than the magnesium. Lead (right) is relatively unreactive, so no gas bubbles are seen.

Table 4.3: The reactivity of different metals with acids

Metal	Reactivity
Potassium (K)	Most reactive
Sodium (Na)	
Calcium (Ca)	
Magnesium (Mg)	
Aluminium (Al)	
Zinc (Zn)	
Iron (Fe)	
Tin (Sn)	
Lead (Pb)	
Copper (Cu)	
Silver(Ag)	
Gold (Au)	Least reactive

Figure 4.20: Potassium is further up the activity series, indicating it is more reactive than calcium.

(a) Potassium reacting with water

(b) Calcium reacting with water

Acids react strongly with metal carbonates

A metal carbonate is a compound containing a metal cation and the carbonate anion (CO_3^{2-}). Acids react with metal carbonates to form salts, carbon dioxide and water.

Acid + metal carbonate → salt + carbon dioxide + water

For example, calcium carbonate (a compound used to settle an upset stomach) reacts with hydrochloric acid to produce calcium chloride, carbon dioxide gas and water.

Hydrochloric acid + calcium carbonate → calcium chloride + carbon dioxide + water

$$2HCl(aq) + CaCO_3(s) \rightarrow CaCl_2(aq) + CO_2(g) + H_2O(l)$$

People burp when they take calcium carbonate to settle an upset stomach. The carbonate reacts with the stomach acids to produce carbon dioxide, which fills up the stomach and has to be released by burping. To test if the gas released in a reaction is carbon dioxide, bubble the gas through limewater. Carbon dioxide turns limewater milky or cloudy.

Figure 4.21: Calcium carbonate reacts with hydrochloric acid to produce calcium chloride and carbon dioxide, which you can see as bubbles.

Acids react readily with metal oxides

A metal oxide is an ionic compound that contains a metal cation and oxide ions (O^{2-}). When acids react with metal oxides, they typically undergo a neutralisation reaction, producing a salt and water. Hydrogen ions (H^+) from the acid combine with the oxide ions (O^{2-}) from the metal oxide to form water (H_2O), while the remaining ions form a salt.

$$\text{Acid + metal oxide} \rightarrow \text{salt + water}$$

For example, when hydrochloric acid (HCl) reacts with copper(II) oxide (CuO), the products are copper(II) chloride ($CuCl_2$) and water.

Hydrochloric acid	+	copper(II) oxide	→	copper(II) chloride	+	water
2HCl(aq)	+	CuO(s)	→	$CuCl_2$(aq)	+	H_2O(l)

This type of reaction is often used to illustrate how acids can neutralise basic compounds, as metal oxides are generally basic in nature. This process not only highlights the principles of acid–base chemistry but also demonstrates practical applications, such as the production of salts in a laboratory setting.

Figure 4.22: Copper(II) oxide is a black powder but forms a greenish solution when it reacts with HCl.

Summary of reactions of acids

As we have seen in this section, acids can react in a number of ways to form neutral products. It is important to be familiar with these reactions, and to be able to predict the products, based on what the acid is reacting with. Table 4.4 is a summary of the reactions of acids.

Table 4.4: Reactions of acids

Reacts with acid to form →	Products
Acid + base	Salt + water
Acid + metal	Salt + hydrogen gas (H_2)
Acid + metal carbonate	Salt + water + carbon dioxide
Acid + metal oxide	Salt + water

When experimenting with reactions of acids in the laboratory, it is important to maintain safety at all times. Safety glasses and protective clothing should always be worn, and if the acid is strong and/or concentrated, gloves should be worn to prevent contact with skin. Look for the corrosives pictogram on the container.

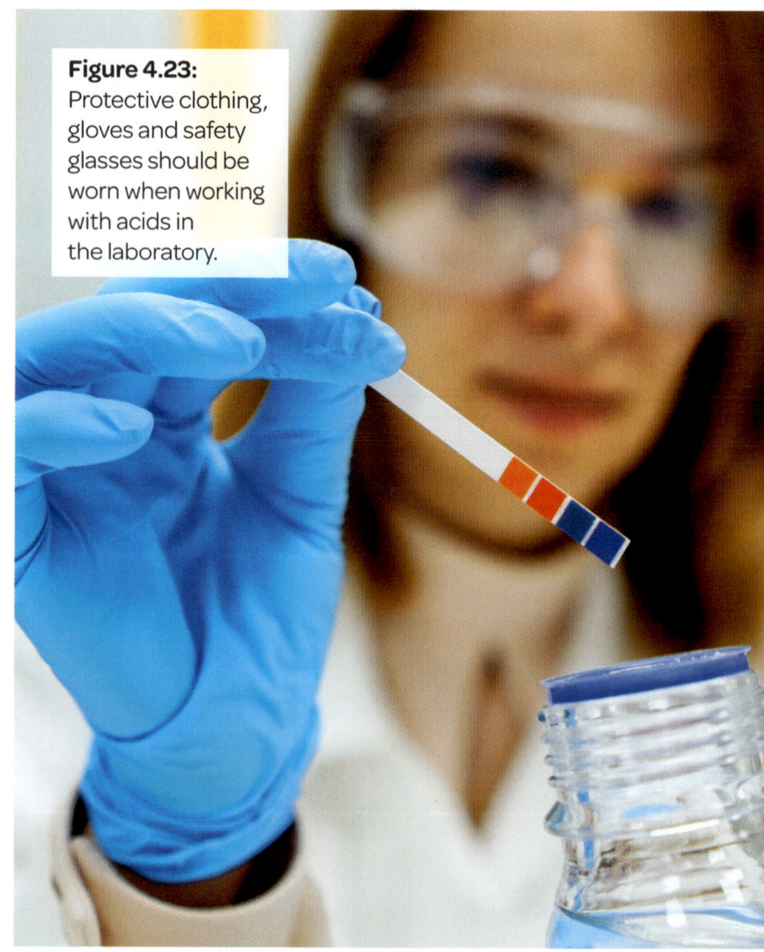

Figure 4.23: Protective clothing, gloves and safety glasses should be worn when working with acids in the laboratory.

Figure 4.24: The corrosives and hazard pictograms could be indicated on acids, depending on the level of risk they present. If an acid is labelled with a corrosives pictogram, it is important to wear gloves.

Learning Ladder

Chemical reactions

1. a Copy and complete the following sentences.

   Acids have a pH range of ____________.

   Bases have a pH range of ____________.

   b Write the general word equation for an:

   i acid reacting with a base.

   ii acid reacting with a metal.

   iii acid reacting with a carbonate.

2. a Describe two observations that could be used to identify:

   i an acid.

   ii a base.

   b Describe what happens when an acid reacts with a metal.

3. a Predict the products of a reaction between sulfuric acid and zinc metal.

   b Explain how you could determine that the gas formed is not carbon dioxide.

4. a Propose what you might do to increase the pH of an acidic solution.

   b Write a complete and balanced chemical equation to represent the reaction in Question 3a.

5. a Discuss how calcium carbonate can help an upset stomach.

   b Conduct a search to explore one way that indicators are used in industry. Share your findings with a classmate.

Nature and development of science

1. Copy and complete the following sentence.

   Universal ________ turns a range of colours which represent values on the ________ scale between 0 and ________.

2. Indicators have been refined over time. Describe how indicators with only one or two colour changes can be more useful than universal indicators.

3. Propose and explain the pros and cons of technologies such as digital pH probes, compared to simple colour indicators, for measuring acidity.

4. Common pH indicators are published in science books for public use. Discuss how peer review may have contributed to the variety of indicators used for a variety of purposes.

5. Conduct a search to identify technologies related to reactions of acids not already included in this section. Analyse how these technologies have contributed to advances in science.

Planning and conducting p. 239

1. a Propose two types of equipment or materials that could be used in an investigation to measure the acidity of a solution.

   b Describe two pieces of equipment that increase safety when handling strong acids and bases.

   c Propose two indicators that would be suitable to detect a pH change of between 7 and 9.

2. a Identify a hazard present when acids react with metals.

   b Describe two ways to minimise the hazard you identified.

Key idea: Matter and energy

Many types of matter can serve as natural indicators. Investigate a range of edible indicators such as cabbage, cherries, blueberries, curry powder, radishes, plums, tomatoes and turnips. Prepare an indicator chart, similar to Figure 4.17, that shows the colour changes that occur at specific pH values for the natural indicators you have explored.

Success criteria

- I can describe the pH scale, including the ranges for acids and bases.
- I can explain how pH indicators are used to measure the pH change of neutralisation reactions.
- I can describe what happens when an acid reacts with a base, a metal and a carbonate.

4·5 ► Rate of reaction: collision theory

Learning intention

At the end of this lesson, I will be able to describe the features of successful collisions required for chemical reactions to occur.

Key terms

activation energy (E_a): the energy required for successful collisions and a chemical reaction to occur

collision theory: a theory which states that for a chemical reaction to occur, particles must collide with the correct orientation and with enough energy; can be used to predict the rate of reactions

Investigation 4.5A

The effect of concentration on the rate of reaction, p. 341

Investigation 4.5B

The effect of temperature on the rate of reaction, p. 343

Key idea: Matter and energy

Chemical reactions can happen quickly, like when a campfire burns or a missile is launched, or they can happen slowly, such as when milk spoils or a ship rusts. The rate of a chemical reaction refers to how quickly reactants are converted into products over a period of time. This rate depends on how particles interact and react with each other to form new substances.

Collision theory helps to predict reaction rates

For a reaction to occur, atoms or molecules must collide successfully. Collisions happen between particles all the time, but not all of them are successful. **Collision theory** states that there are two requirements for a successful collision:

- sufficient energy
- correct orientation of particles.

A collision must occur with enough force to provide the energy required to break the bonds of the reactants so that new bonds are able to form. This is called **activation energy (E_a)**. Particles must also be positioned in such a way that when a collision occurs, the atoms are aligned properly to allow new products to form, as shown in Figure 4.26.

If the proportion of successful collisions in a given amount of time is high, the rate of the reaction is high. If the proportion is low, the rate of the reaction is low. Both the frequency of collisions and their success rate contribute to the rate of chemical reactions.

The rate of a chemical reaction is usually measured by how much of a reactant is used up, or how much of a product is formed, over a certain period of time. This is often expressed in units such as grams per second (g^{-1}) or millilitres per second ($mL\ s^{-1}$), depending on what is being measured. If time is the only thing being measured, we use 1/s as the unit for reaction rate, which represents the change in reaction progress per second.

Figure 4.25: (a) Wood burning is a fast chemical reaction that may be completed in a few hours. (b) Iron rusting is a slow chemical reaction that can take many years.

Many factors can affect reaction rate

Every chemical reaction makes products from the reactants at a certain rate, but this rate is not fixed. It is affected by:

- concentration
- surface area
- pressure
- temperature
- catalysts
- agitation (stirring).

It is possible to adjust many of these factors, either to reduce the reaction rate to make the reaction safer to handle, or to increase the reaction rate so that products are made more quickly. Increasing the reaction rate is very common in the chemical industry, to reduce costs and maximise profits.

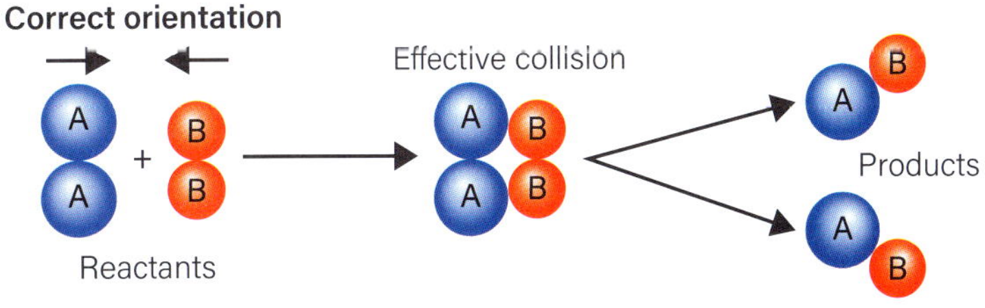

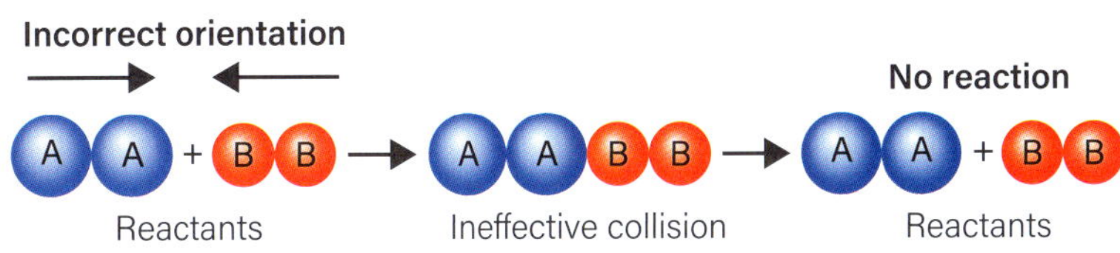

Figure 4.26: Reacting particles must collide with the correct orientation for the collision to be successful.

Concentration, temperature, surface area and catalysts will be explored in more detail in the following section.

Learning Ladder

Chemical reactions

1. Identify two things that must occur for a chemical collision to be successful.
2. Propose what would happen to the rate of a reaction if you were to increase the pressure within a given volume.
3. Justify your response to Question 2 with reference to collision theory.
4. When H^+ and OH^- ions collide with enough force, it might result in a successful collision to form water (H_2O). Using labelled diagrams, show how the ions would be aligned during a:
   - **a** successful collision.
   - **b** unsuccessful collision.

Use and influence of science

1. Identify the theory that, if understood, can help to address socio-scientific issues related to chemical reactions that happen too quickly.
2. The orientation of particles is an aspect of collision theory that determines whether a chemical reaction will be successful. Describe how this aspect could be interpreted in different ways.
3. Some medicinal tablets can take a long time to work because they react slowly in the body. Explain how this could influence whether scientific research will work to improve this.

Questioning and predicting p. 234

1. Predict, giving a reason, which form of medicine will last longer in the body:
   A Liquid **B** Tablet
2. **a** Select the question below that is most suitable to be investigated scientifically:
   - **A** How fast is a reaction between sulfuric acid and:
     - **i** marble chips?
     - **ii** sodium carbonate?
     - **iii** baking powder?
   - **B** How does increasing the concentration of sulfuric acid affect how quickly it reacts with marble chips?

   b Rewrite the scientific question not selected in part a so it can be investigated scientifically.
3. Propose a hypothesis for each of the suitable scientific questions from Questions 2a and 2b, with reference to collision theory.
4. For the hypotheses you constructed in Question 3, ask yourself the following questions.
   - **a** Did I include the independent and dependent variables, as well as the predicted outcome?
   - **b** Can my hypotheses be tested scientifically in the laboratory?
5. Evaluate your hypotheses in Question 3 based on your responses to Question 4. If you answered 'no' to Question 4a or 4b, adjust your hypotheses to follow the set conventions.

Key idea: Matter and energy

Explain how collision theory incorporates the role of particle collisions as well as energy in chemical reactions.

Success criteria

- I can state the two requirements for a successful chemical collision to occur.
- I can describe the rate of a reaction in terms of collision theory.

4·6 ▸ Factors that affect rate of reaction

Learning intention

At the end of this lesson, I will be able to describe how different factors affect the rate of chemical reactions.

Key terms

catalyst: a substance that increases the rate of a chemical reaction without being used up

enzyme: a biological catalyst that increases the rate of reactions in cells

kinetic energy: energy of motion

surface area: the area of the outermost layer of an object

Investigation 4.6A

The effect of surface area on the rate of reaction, p. 344

Investigation 4.6B

The effect of a catalyst on the rate of reaction, p. 346

Investigation 4.6C

Student-designed investigation: investigating the rate of chemical reactions, p. 348

Key idea: Matter and energy

Many factors influence the rate of reaction, including concentration, temperature, surface area and the presence of a catalyst. Temperature and concentration are relatively easy to control. We adjust these factors every day in our kitchens. Catalysts and surface area may require special equipment or materials to control.

Concentrated substances react more quickly

The concentration of a substance is a measure of how many particles are in the volume of a liquid solution or a container of gas. Increasing the concentration of the reactants increases the number of particles within a given volume. The more crowded the particles are, the more collisions occur. When particles collide more frequently, the proportion of successful collisions increases, leading to an increase in the rate of the reaction.

Ⓐ **LOWER concentration**

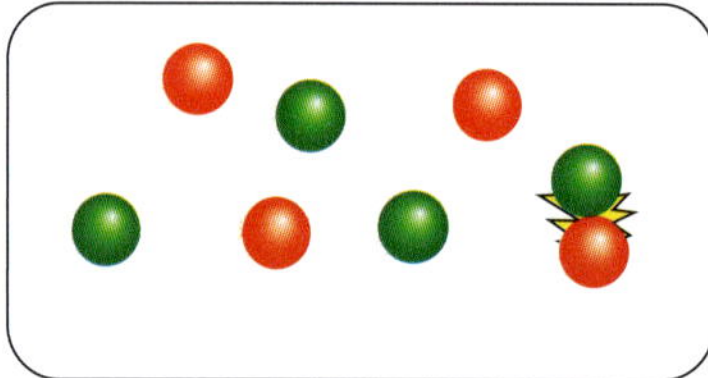

Ⓑ **HIGHER concentration**

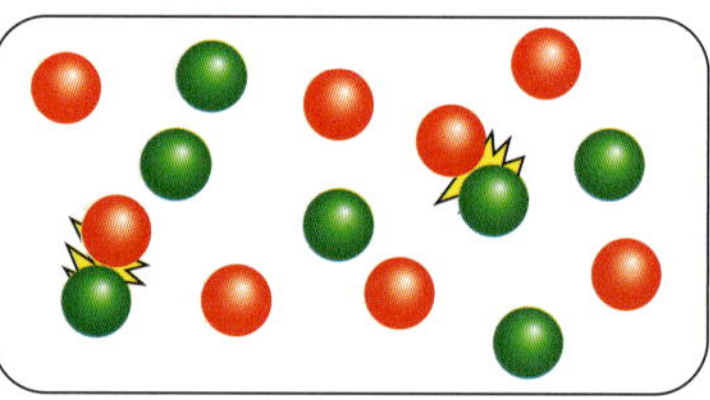

Figure 4.27: (a) Fewer reactants in a specific amount of space will have fewer opportunities to collide in a given amount of time, as the particles are further away from each other. (b) More reactants in the same amount of space will have more successful collisions in the same amount of time, as there are more opportunities for collisions to occur, which will increase the rate of the reaction.

The concentration of a gas can be adjusted by changing the pressure. The pressure of a system increases when the volume of the reactant vessel decreases. This means that there are now the same number of particles in a smaller volume, leading to more collisions between particles.

Figure 4.28: Concentrated sulfuric acid reacts dramatically with sucrose (table sugar).

Raising the temperature increases the reaction rate

Increasing the temperature of a system increases the average **kinetic energy** of the particles in the system. As the average kinetic energy increases, the particles move faster, so they collide more frequently and have more energy when they collide.

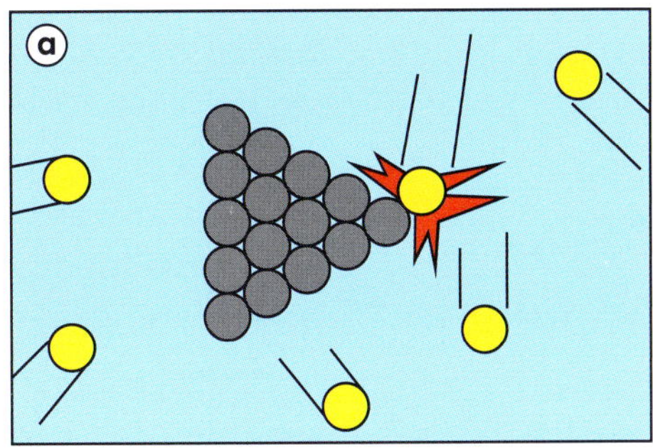

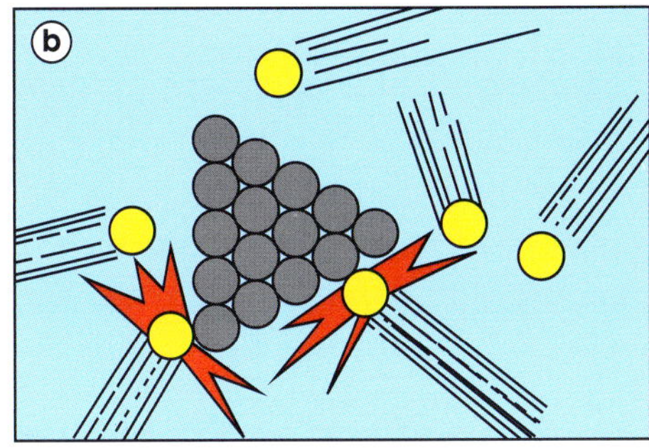

Figure 4.29: Adding heat to a reaction provides the particles with higher levels of kinetic energy. This means they are able to move around more quickly and to collide with more energy. Increased kinetic energy means that the activation energy requirements of the reaction can be reached.

By increasing the temperature of reactants, more particles will have enough energy and collide more often. This means that the hotter the reactants, the faster the reaction. Likewise, the cooler the reactants, the slower the reaction.

We change temperature to adjust reaction rates every day. You boil water before making a cup of tea because the tea dissolves faster in hot water. If you make a cup of tea with water at room temperature, it will take much longer for the tea to flavour the water. Similarly, we keep food in the refrigerator to slow down chemical reactions that cause it to spoil over time.

Increasing the surface area increases the reaction rate

Surface area is the exposed area of a solid substance. If you increase the surface area of a substance, there are more particles at the surface that have the opportunity to react with another substance, so the reaction rate increases.

Consider the reaction of zinc metal with an acid (Figure 4.30). If the zinc is in the form of a large cube, the total surface area of the cube is the sum of the areas of the faces of the cube. If the zinc cube is broken up into many smaller pieces, the total surface area greatly increases, and more area is exposed to react with the acid. If the zinc is crushed into a powder, its surface area increases even further. This form of zinc reacts the fastest.

Small pieces of food cook faster than larger pieces. These smaller pieces of food have a large surface area (compared to their volume) exposed to heat. When you take tablets or pills, different tablets are digested at different rates, depending on their surface areas.

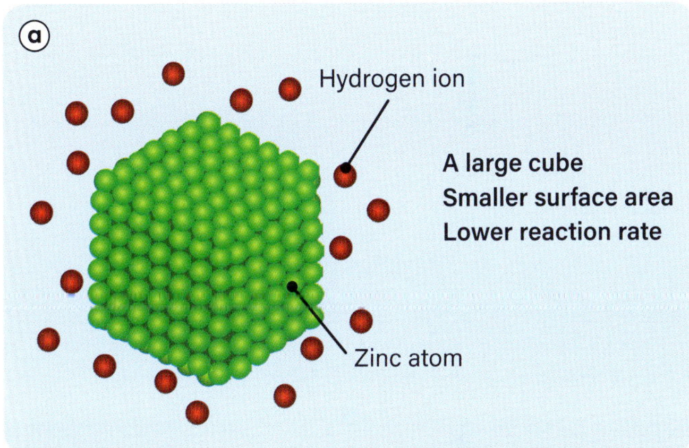

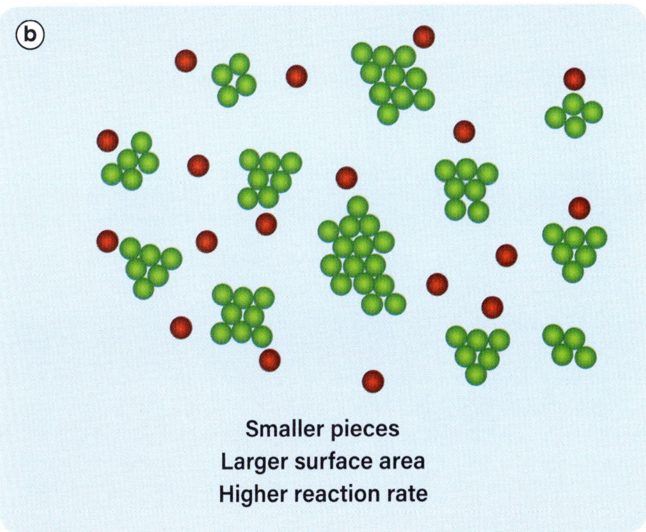

Figure 4.30: (a) Hydrogen ions from the acid can collide only with the outer layer of zinc atoms of the large cube of zinc. The reaction rate is slow. (b) When the zinc block is broken into smaller pieces, more surface is exposed for the hydrogen ions to collide with and the rate of the reaction increases.

Catalysts increase the rate of reaction

A **catalyst** is a substance that increases the rate of a chemical reaction without being used up itself. When writing a chemical equation, we often write the name of the catalyst above the reaction arrow; this indicates its presence but also its lack of involvement in the reaction itself.

Catalysts lower the activation energy of a reaction, which is the energy required to start a reaction. This enables the reaction to occur at a lower temperature.

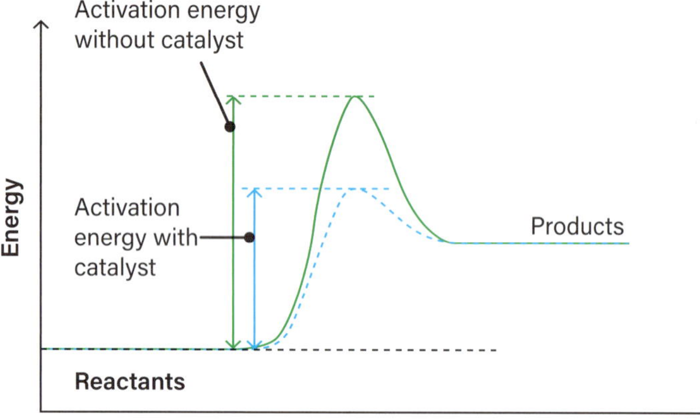

Figure 4.31: Adding a catalyst to a chemical reaction lowers the activation energy of the reaction. More collisions will have enough energy to overcome the smaller energy barrier.

Catalysts are specific, which means that a catalyst increases the rate of some reactions but not others. Catalysts are important in industry because they reduce reaction times and enable chemical processes to take place at lower temperatures and pressures. Therefore, they reduce the cost of production of valuable products.

Enzymes are biological catalysts. They are proteins that help speed up chemical reactions in the body. For example, amylase in saliva catalyses the breakdown of carbohydrates in your mouth.

Here are some common uses of catalysts.

- Catalytic converters in cars contain platinum, which catalyses the conversion of toxic carbon monoxide into carbon dioxide.
- Biological catalysts in detergents remove protein stains, such as egg yolk, which are otherwise hard to dislodge.
- Catalysts in the paper industry break down paper pulp to produce smooth paper for magazines.
- Catalysts turn milk into yoghurt and petroleum into plastic milk jugs, CDs and bicycle helmets.

Figure 4.32: Saliva contains enzymes that start the chemical reactions of digestion.

Figure 4.33: UV light energy from the Sun catalyses many chemical reactions, including biochemical reactions such as photosynthesis.

Learning Ladder

Chemical reactions

1. Identify the following statements as true or false.
   - **a** A crushed substance has a large surface area.
   - **b** Diluting an acid with water will make it react faster with baking soda.
   - **c** Decreasing the volume of a reaction container of gas will decrease the rate of reaction.
   - **d** Stirring oats and water over heat will cook porridge faster than without stirring.
2. Describe two things you could do to decrease how long it takes to cook vegetables.
3. Explain:
   - **a** how temperature increases the rate of a chemical reaction.
   - **b** how surface area influences the rate of chemical reactions.
   - **c** what a catalyst is, and give an example of one.
   - **d** how enzymes are a type of catalyst.
4. Sunlight is a catalyst for the breakdown of hydrogen peroxide into hydrogen gas and oxygen gas. Write a word equation to represent this reaction, remembering that a catalyst is not used up in a reaction.
5. Research why reaction rate needs to be controlled in industry. Analyse the compromises that are made regarding temperature and concentration, and why these might be necessary.

Nature and development of science

1. Identify technologies that commonly utilise catalysts.
2. Propose and describe how the concepts of concentration and pressure have been linked together over time.
3. Research and explain how understanding factors that affect reaction rates has contributed to developments in technologies and engineering.

Planning and conducting

p. 239

You wish to investigate whether a whole or a crushed tablet of antacid eases symptoms of heartburn the fastest.

1. Prepare a list of materials you may require to conduct this investigation.
2. Performing tests on animals (including humans) poses a high risk. Propose how you can eliminate this risk when planning your investigation. (*Hint:* Stomach acid that causes heartburn is primarily hydrochloric acid.)
3. Explain how the materials you listed in Question 1 will enable the collection of accurate, precise and replicable data.
4. **a** Construct an aim for this investigation that considers a wider context.
   - **b** Summarise the method you could follow to address the aim of the investigation. Be sure to consider safety, ethics and reproducibility.

Key idea: Matter and energy

With respect to matter and energy, discuss how catalysts are important in our lives.

- **a** Research catalysts used in industry. Identify a catalyst and state where it is used.
- **b** Explain how this catalyst helps to improve the rate of this reaction.
- **c** Write the word and chemical equations for the reactions that take place.
- **d** What would happen if this catalyst was not used in the reaction?

Success criteria

- I can describe the effect of concentration, temperature, surface area and catalysts on the rate of some common chemical reactions.
- I can explain how factors affect the rate of reactions, with reference to collision theory.

4·7 ▸ Key idea: Matter and energy – exothermic and endothermic reactions

Learning intention

At the end of this lesson, I will be able to:

- describe the features of exothermic and endothermic reactions
- classify chemical reactions as exothermic or endothermic, based on observed or given energy changes.

Key term

energy profile diagram: a diagram that shows energy changes as a reaction progresses from the reactants to products, including activation energy and whether energy is released or absorbed

Investigation 4.7

Predicting energy changes for exothermic and endothermic reactions, p. 350

Key idea: Matter and energy

Chemical reactions are just as much about energy being transformed as they are about atoms rearranging into new substances. The transfer of energy, especially in the form of heat, is central to understanding how and why reactions occur. These energy changes are directly related to how chemical bonds break and form, linking the behaviour of matter to the flow of energy.

Whether a chemical reaction releases energy or absorbs it depends on the total energy involved in the bond-breaking and bond-making processes. This energy transformation is what classifies a reaction as either exothermic or endothermic; a key concept in understanding how matter and energy interact.

Chemical reactions involve energy changes

Every chemical reaction involves energy. This is because bonds between atoms must be broken and new bonds formed to create new substances. These changes always involve a transfer of energy between the reacting particles and their surroundings.

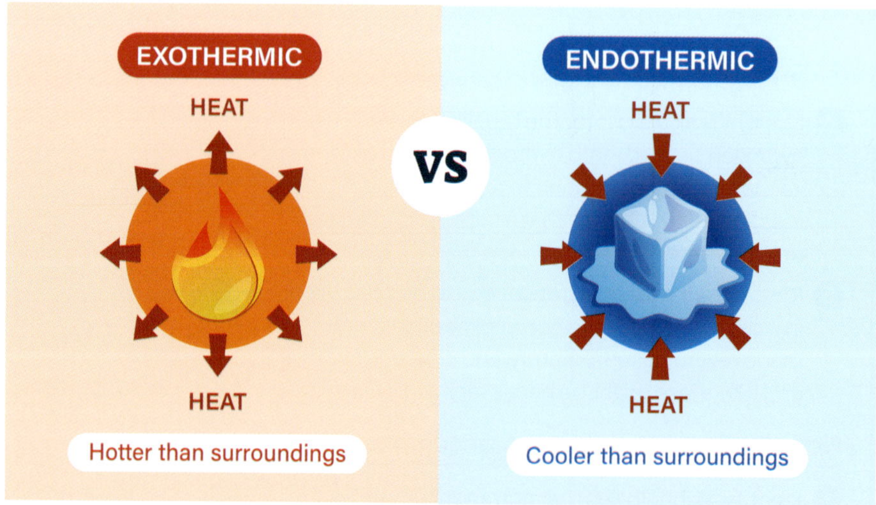

Figure 4.34: Endothermic reactions absorb heat energy, whereas exothermic reactions release heat energy. The area surrounding an endothermic reaction gets colder, whereas the area around an exothermic reaction gets hotter.

In *exothermic* reactions, energy is released, often as heat or light, and the temperature of the surroundings increases. In contrast, *endothermic* reactions absorb energy from their surroundings, causing a temperature drop (see Figure 4.34). Common exothermic examples include neutralisation and many displacement reactions. Combustion is also an example of an exothermic process.

Endothermic reactions can be seen in processes such as photosynthesis and some decomposition reactions.

The total energy before and after the reaction stays the same (energy is conserved), but the way it is stored or transferred can change. This is why understanding the energy flow in a reaction helps us to predict and explain real-world chemical processes.

Bond-breaking and bond-forming determine energy changes

The energy change in a chemical reaction comes from two main steps: breaking bonds in the reactants and forming new bonds in the products. Breaking bonds always requires energy. This energy is needed to separate atoms that are chemically joined. Forming new bonds, on the other hand, releases energy as atoms come together to create more stable arrangements.

Whether a reaction ends up releasing or absorbing energy overall depends on the balance between these two steps.

- If more energy is released when bonds form than is needed to break the original bonds, the reaction is exothermic.
- If more energy is required to break bonds than is released when new ones form, the reaction is endothermic.

So, the overall energy change depends on how much energy goes in and how much comes out during the reaction. This can be tricky to understand, which is why we often use energy profile diagrams.

Energy profile diagrams can be used to represent exothermic and endothermic reactions

Energy profile diagrams are useful tools for showing how energy changes during a chemical reaction. They show energy (E) on the vertical axis and the progress of the reaction on the horizontal axis.

At the start of a reaction, the reactants must gain energy to break bonds and begin reacting. This is the activation energy (E_a), which you learnt about in Section 4.5. Activation energy is the minimum energy needed to initiate the reaction, shown as the 'hump' in Figure 4.35. The bigger the hump, the more energy is required to start the reaction.

- The diagram shows that in an exothermic reaction, the products are at a lower energy level than the reactants. The difference in height represents the energy released to the surroundings.
- In an endothermic reaction, the products are at a higher energy level than the reactants. The gap shows the energy absorbed from the surroundings.

In other words, if the overall energy change is negative, the reaction is exothermic (energy is released). If the overall energy change is positive, the reaction is endothermic (energy is absorbed).

These diagrams help us to visualise how energy flows into or out of the reaction system and give insight into how easily a reaction can start and whether it releases or absorbs heat. Knowing whether a reaction releases or takes in energy helps us to understand how it works and how it can be used, such as in factories or in the human body.

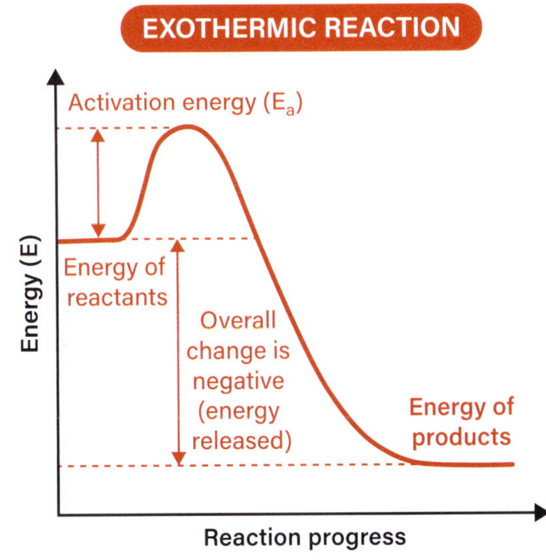

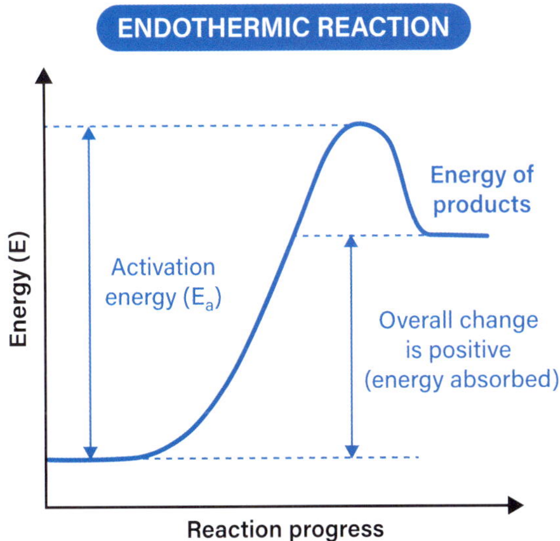

Figure 4.35: Energy profile diagrams show energy changes during chemical reactions. If the difference in energy between reactants and products is negative, the reaction is exothermic. If the difference is positive, it is endothermic.

Instant cold and hot packs are endothermic and exothermic reactions

The instant cold and hot packs found in first aid kits use endothermic and exothermic reactions to function.

Instant cold packs work by absorbing heat from their surroundings in an endothermic reaction. The packs usually contain a bag of water and a solid chemical, such as ammonium nitrate. When the pack is squeezed, the inner water bag bursts and the ammonium nitrate dissolves into the water. As it dissolves, the ammonium nitrate particles split into positive ammonium cations and negative nitrate anions. This process requires energy in order to break the ionic bonds between the ions, and the reaction absorbs this energy as heat from its environment – including your skin!

Instant heat packs work in a similar way: these often contain a mixture of ingredients, such as iron powder, salt, water and activated carbon. When airtight packaging is opened, the oxygen in the air reacts with the iron powder, salt and water, causing the iron to oxidise.

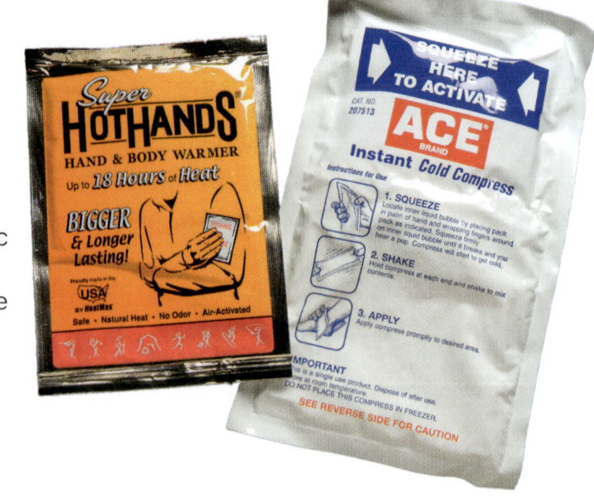

Figure 4.36: Instant cold and hot packs use endothermic and exothermic reactions to rapidly produce a cooling or heating effect.

To do this, the iron atoms lose electrons, which bond with the oxygen atoms to form iron(III) oxide (rust). These new iron(III) oxide bonds are more stable than the original bonds in the oxygen and iron atoms, and so they have less energy. The energy released when these bonds form is also less than the amount of energy required to break the original bonds. This means the reaction is exothermic, and this excess energy is released into the surroundings as heat.

Figure 4.37a: An instant cold pack involves an endothermic reaction.

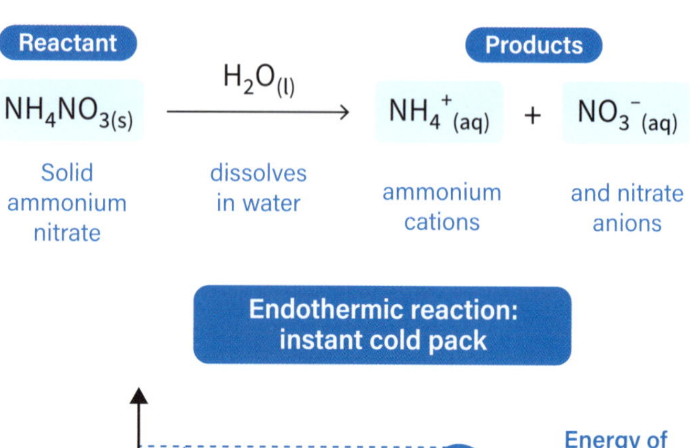

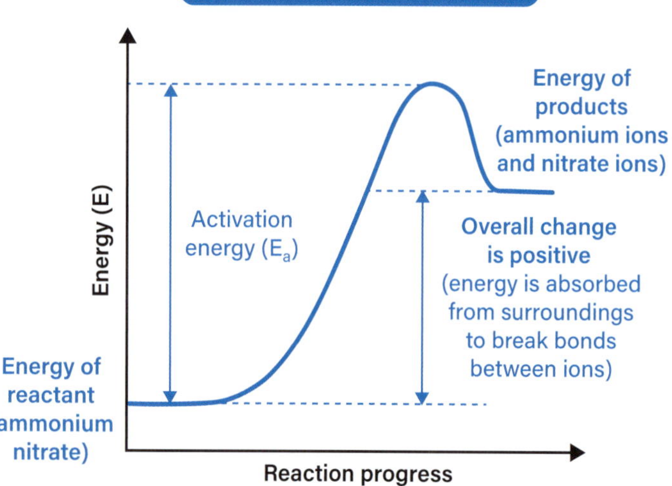

Figure 4.37b: An instant hot pack involves an exothermic reaction.

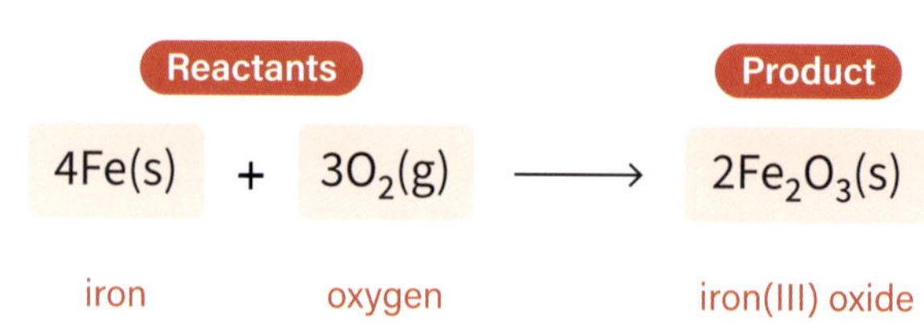

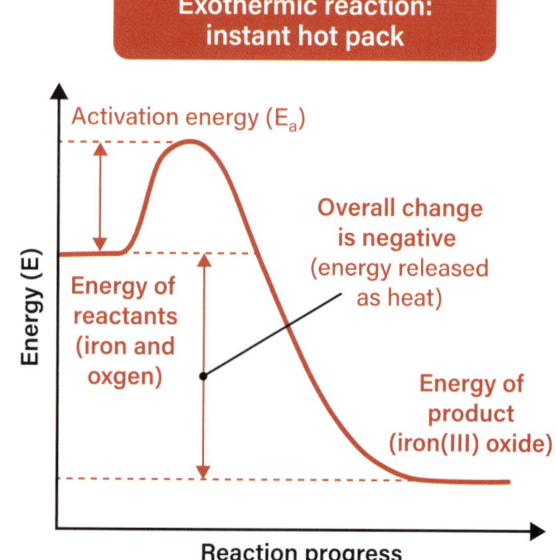

Learning Ladder

Chemical reactions

1. Classify the following reactions as exothermic or endothermic, based on the given information.
   - **a** The reactants have more energy than the products.
   - **b** The products have more energy than the reactants.
   - **c** The temperature surrounding the reactions decreases.
   - **d** The test tube feels warm as the reaction occurs.
2. Consider Figures 4.37a and 4.37b.
   - **a** Describe why energy is required for both exothermic and endothermic reactions.
   - **b** Classify and describe each figure as representing either a synthesis reaction or a decomposition reaction.
3. **a** Construct two labelled energy profile diagrams to represent the reactions and factors described in Questions 1a and 1b.
   - **b** Explain how ice melting is an endothermic process, even though the temperature of the ice would increase.
4. Analyse how a catalyst would impact the energy changes that occur in a chemical reaction. Consider the following in your discussion.
   - **a** Energy required to break bonds
   - **b** Energy required to form bonds
   - **c** Total energy released or absorbed
5. Consider the following energy profile diagrams for two possible processes in industry. If both processes form the desired product, propose and evaluate which one will be more cost effective. Justify your response with reference to features of the energy profile diagrams.

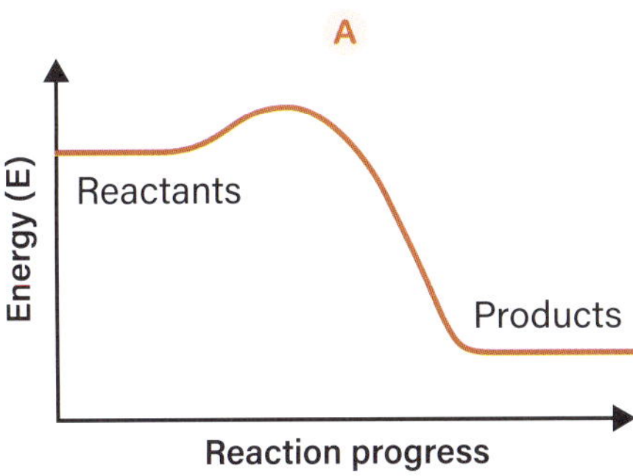

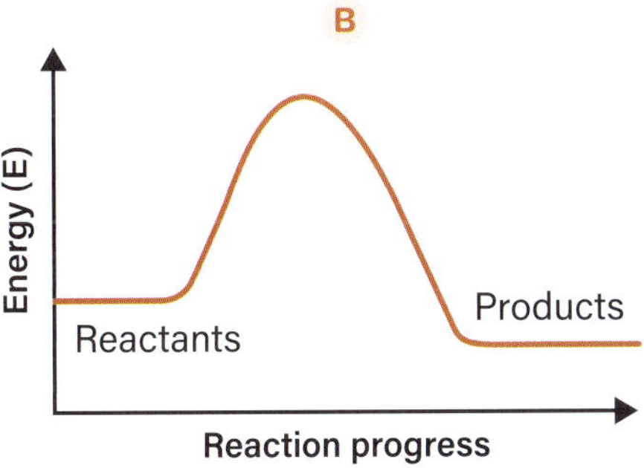

Nature and development of science

1. Identify a technology used in the laboratory that allows scientists to measure energy changes during chemical reactions.
2. Copy and complete the following sentences:

   Energy __________ diagrams allow scientists to visualise chemical __________ as either exothermic or __________, which enables scientists to refine concepts over time. The total energy __________ is determined based on the difference between the energy of the __________ and the __________ of the products, irrespective of the __________ energy.
3. Developments in chemical technologies can involve selecting the safest, most efficient and cost-effective reaction pathways for a particular industrial purpose.
   - **a** Explain how scientific understanding of the features of endothermic and exothermic reactions may have contributed to advancements in chemical technologies.
   - **b** Explain how energy profile diagrams help contribute to our understanding of these features. *Hint:* Consider diagrams A and B from 'Chemical reactions' Question 5 in your response.

Questioning and predicting p. 234

1. A chemical reaction you observe lets off a bright light. Predict whether this is an exothermic or endothermic process.
2. You wish to conduct an investigation to determine whether dissolving different types of salts (sodium hydroxide, potassium nitrate and ammonium nitrate) in water is an exothermic or endothermic process. Formulate a scientific question that would guide this investigation in the laboratory.
3. You conduct research which indicates that potassium nitrate and ammonium nitrate can be used in instant cold packs, but that sodium hydroxide cannot. Use this information to develop a hypothesis that predicts whether these salts dissolving in water will be exothermic or endothermic reactions.
4. Discuss the features of your hypothesis from Question 3 that indicate it is a reasoned hypothesis. *Hint:* This may relate to the 'because' portion of your hypothesis.
5. Analyse how energy profile diagrams can be used as an explanatory model that addresses scientific questions, problems and claims.

Success criteria

- I can describe the features of exothermic and endothermic reactions.
- I can classify chemical reactions as exothermic or endothermic, based on observed or given energy changes.

► Summary

- There are many types of chemical reactions.

Type of reaction	Description	General equation	Example
Synthesis	More than one reactant combines to form a single product.	A + B → AB	Iron rusting with oxygen to form iron oxide Iron + oxygen → iron(II) oxide $4Fe + 3O_2 \rightarrow 2Fe_2O_3$
Decomposition	One reactant breaks down into multiple products.	AB → A + B	Water breaking down into hydrogen and oxygen Water $\xrightarrow{\text{electricity}}$ hydrogen + oxygen $2H_2O(l) \xrightarrow{\text{electricity}} 2H_2(g) + O_2(g)$
Displacement	A more reactive element replaces a less reactive one in a compound.	A + BC → B + AC	Zinc metal replacing copper in copper(II) sulfate solution to form solid copper metal $Zn(s) + CuSO_4(aq) \rightarrow ZnSO_4(aq) + Cu(s)$
Neutralisation	An acid and a base react to form a neutral salt and water.	Acid + base → salt + water	Hydrochloric acid and copper(II) oxide react to form copper(II) chloride salt and water Hydrochloric acid + copper(II) oxide → copper(II) chloride + water $2HCl(aq) + CuO(s) \rightarrow CuCl_2(aq) + H_2O(l)$

- A precipitation reaction is an example of a double-displacement reaction where a solid forms in solution.

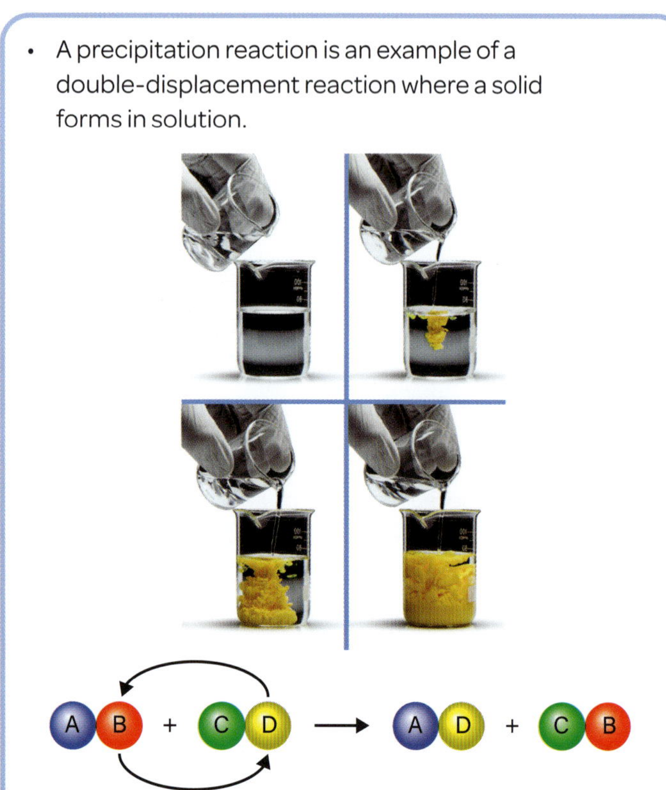

- Chemical reactions can be classified as exothermic or endothermic, depending on whether they release or absorb energy.

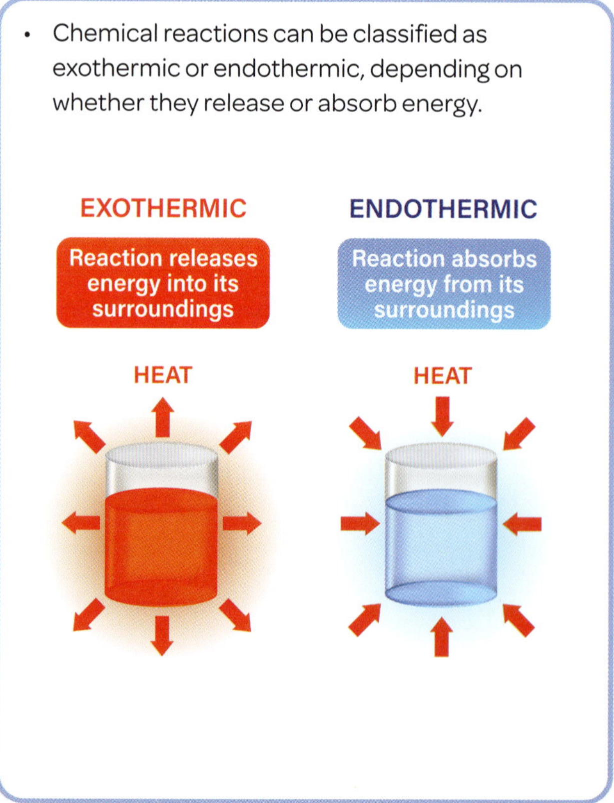

- Collision theory helps us to predict the relative rates of reactions.
  - Sufficient energy and correct orientation of particles is required for successful collisions.
  - The higher the proportion of successful collisions in a given amount of time, the higher the rate of the reaction.
- Factors that affect reaction rate include concentration, temperature and surface area.

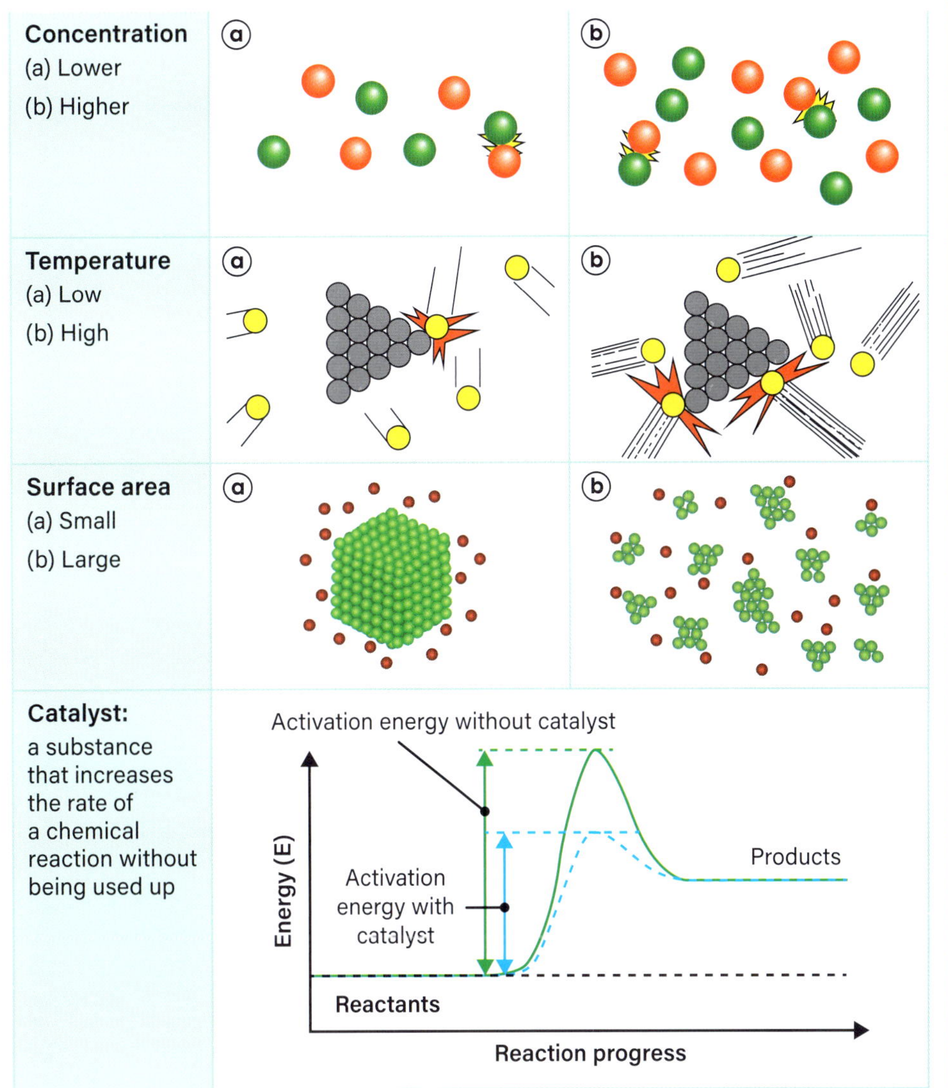

Key idea: Matter and energy

- Bond breaking and bond forming determine energy changes.
  - Forming new bonds releases energy (an exothermic process).
  - Breaking bonds requires energy (an endothermic process).
- Energy profile diagrams can be used to represent the energy changes that take place during a chemical reaction.

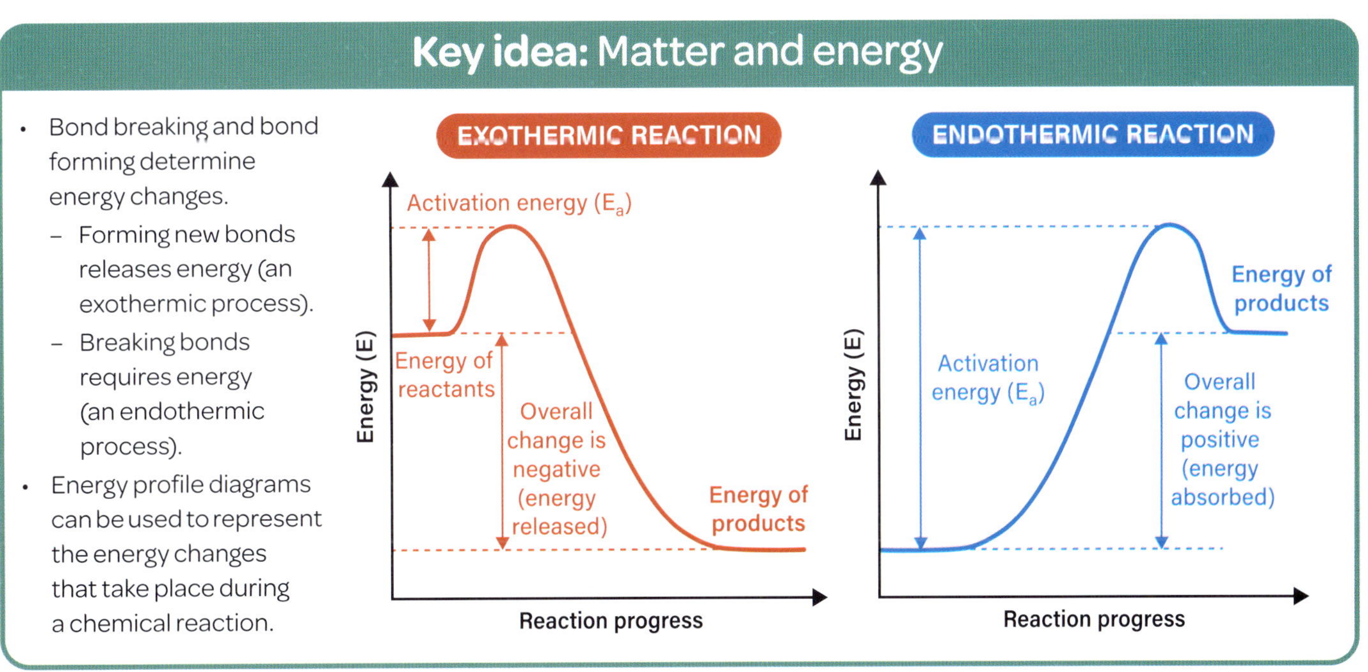

Masterclass

Steps in progression

		1	2
Science understanding	**Chemical reactions**	Identify the type of reaction used to convert sulfur dioxide into sulfur trioxide.	Summarise the contact process in a few sentences.
Science as a human endeavour	**Nature and development of science**	Identify industrial machinery or equipment used in the contact process (see Figure 4.38).	Complete the following sentence: When it was found that water undergoes a violent ________ reaction with ________, a step to form ________ was added to reduce ________.
Science as a human endeavour	**Use and influence of science**	Identify two dangers that would be present if water reacts directly with sulfur trioxide.	Describe why a general understanding of 'acid' could impact how the contact process is viewed.
Science inquiry	**Questioning and predicting**	Predict what will happen to the rate of sulfur trioxide production if the temperature is increased during the contact process.	Formulate a scientific question to investigate how the surface area of sulfur crystals will affect the rate of step 1 of the contact process (sulfur dioxide formation).
Science inquiry	**Planning and conducting**	Make a list of equipment you would select to measure the temperature and volume of gas during a small-scale laboratory version of the contact process.	Research and develop a risk assessment to outline the safety precautions that should be considered when heating sulfur dioxide and using a vanadium (V) oxide catalyst.

The contact process: using chemical reactions to make sulfuric acid

Sulfuric acid is one of the most important chemicals used today. It is essential for making fertilisers, detergents and batteries, for refining crude oil and for metal production. Its manufacture involves several key steps, known as the contact process. Each step includes chemical reactions that connect to what you have learnt in this chapter.

The first step is a synthesis reaction, where pure sulfur burns in oxygen to produce sulfur dioxide.

$$\text{Sulfur} + \text{oxygen} \rightarrow \text{sulfur dioxide}$$
$$S(s) + O_2(g) \rightarrow SO_2(g)$$

Sulfur dioxide is converted to sulfur trioxide (SO_3) using a vanadium (V) oxide (V_2O_5) catalyst, which speeds up the reaction without being used up. When heated, sulfur dioxide passes over the catalyst and reacts with oxygen to form sulfur trioxide. Without the vanadium pentoxide catalyst, the reaction is too slow for industry.

$$\text{Sulfur dioxide} + \text{oxygen} \xrightarrow{\text{vanadium (V) oxide}} \text{sulfur trioxide}$$
$$2SO_2(g) + O_2(g) \xrightarrow{V_2O_5} 2SO_3(g)$$

In the final stage, sulfur trioxide reacts with water to form sulfuric acid (H_2SO_4). This reaction is highly exothermic, releasing a lot of heat. To make it safer, sulfur trioxide is first absorbed into concentrated sulfuric acid to create oleum ($H_2S_2O_7$), which reduces

Demonstrate your understanding

3	4	5
State and explain the factor that increases the rate of the reaction that forms sulfur trioxide.	Discuss the importance of creating oleum as an additional step before producing sulfuric acid.	Evaluate the suitability of the contact process with regard to industry practice.
Explain how sulfuric acid is important in industry, with respect to scientific developments.	Research the historical progression of sulfuric acid being used in industry, including identifying the contributing scientists.	Analyse the link between the need for industrial equipment such as roasting towers and the demand for sulfuric acid in industry.
Explain how the need for sulfuric acid in industry may have influenced the development of the contact process.	Sulfuric acid can be very dangerous. Discuss how this could impact a person's choice to donate funds to manufacture more sulfuric acid.	Propose and analyse key factors that influence whether society will support continued research related to the contact process.
Develop a hypothesis that predicts the outcome of the question you formulated in step 2.	Discuss why it would be difficult to test the effect of temperature on the contact process in a school laboratory.	Develop an annotated diagram or explanatory model that represents the relationship between factors that affect the rate of reactions and the contact process. Science how-to p. 234
Describe how you would ensure that the measurements of sulfur dioxide and oxygen volumes during the reaction are accurate and repeatable.	Research and design a safe, ethical and reproducible method to test how temperature affects the rate of sulfur trioxide production in a school laboratory.	Evaluate a classmate's method from step 4. How could it be improved to better address safety, ethical and procedural factors? Science how-to p. 239

the risk of heat and acid mist. Water is then added to the oleum to produce sulfuric acid.

Formation of oleum:

Sulfur trioxide + sulfuric acid → oleum

$$SO_3(g) + H_2SO_4(l) \rightarrow H_2S_2O_7(l)$$

Dilution of oleum to form sulfuric acid:

Oleum + water → sulfuric acid

$$H_2S_2O_7(l) + H_2O(l) \rightarrow 2H_2SO_4(aq)$$

The contact process produces sulfuric acid used in metalworking for pickling, which removes rust and impurities by soaking metal in acid. This cleans the surface and prepares it for painting or welding, improving quality.

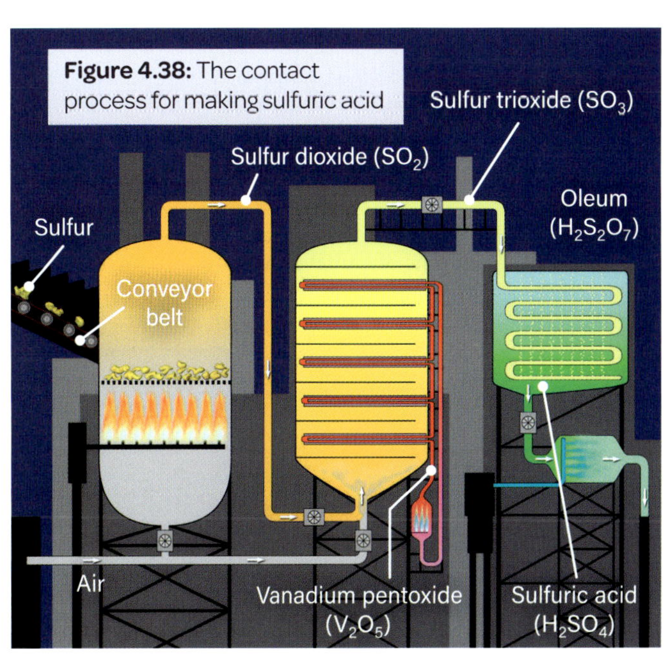

Figure 4.38: The contact process for making sulfuric acid

5.0 Climate change

Earth's climate is changing. This change is being driven by human activity, which has increased the amount of greenhouse gas in the atmosphere. By understanding that everything on Earth is interconnected, and how one thing can affect another, we can better understand the impacts of a changing climate on Earth's systems. The scientific community is working together to develop solutions to address climate change that will allow us to slow – even to reverse – the changes that humans have caused.

Learning Ladder

The Learning Ladder for each chapter maps the Science Understanding, Science as a Human Endeavour and Science Inquiry strands that will be covered. Each ladder has five levels of progression, called steps. To climb the ladders, you need to develop fluency at each step. This will help you develop the ability to complete tasks that are more complex.

Steps in progression	Science understanding	Science as a human endeavour	
	Earth and space science: Climate change	Nature and development of science	Use and influence of science
5	I can discuss how humans impact Earth's systems and evaluate strategies to mitigate climate change	I can analyse how advances in technologies enable advances in science	I can analyse the key factors that contribute to scientific knowledge being adopted more broadly by society
4	I can analyse interactions between greenhouse gas emissions and energy exchanges within and between Earth's systems	I can discuss how scientific knowledge is validated, including the role of publication and peer review	I can discuss how scientific information and misinformation may inform personal and social decision-making
3	I can explain patterns of global climate with reference to causal factors	I can explain how science has contributed to developments in technologies and engineering	I can explain how the values and needs of society influence the focus of scientific research
2	I can describe trends and patterns of global climate	I can describe how scientific knowledge is refined over time	I can describe scientific knowledge that may be interpreted in different ways
1	I can identify and differentiate between the features of Earth's systems	I can identify technologies that have enabled advances in science	I can identify scientific knowledge that can address socio-scientific issues

Figure 5.1: World Heritage-listed Ningaloo Reef sits just off the coast of Western Australia and is more than 260 kilometres long. Home to thousands of species of fish, corals and other marine life, the reef is also a vital breeding ground for endangered turtles. Humpback whales, dugongs, and dolphins pass through the reef during their annual migration. Since 2024, an intense marine heatwave has driven unprecedented coral bleaching across the reef, with some areas affected for the first time. Warming oceans due to climate change increase the frequency and intensity of these heatwaves, so the reefs do not have enough time between bleaching events to recover.

Steps in progression	Evaluating	Communicating
5	I can evaluate the validity and reproducibility of investigation methods	I can justify scientific ideas, findings, arguments and proposals for diverse audiences
4	I can construct evidence-based arguments to justify conclusions, address ethical issues or assess claims	I can communicate scientific findings and arguments effectively for specific purposes to specific audiences
3	I can discuss ways to improve the quality of data and validity of conclusions and claims	I can use digital technologies and/or scientific representations to communicate data and information
2	I can describe the impact of assumptions and errors, and propose ways to reduce them	I can prepare a variety of presentation formats to communicate ideas and findings
1	I can identify assumptions and types of errors in an investigation	I can select appropriate formats, content and vocabulary to communicate scientific ideas and findings

Science inquiry

5·1 ▸ Earth's systems

Figure 5.2: Particles and gases from volcanic eruptions ejected high into the atmosphere can lead to vibrant sunrises and sunsets but also contribute to cooling, as they can reflect sunlight.

Learning intention

At the end of this lesson, I will be able to identify and describe the atmosphere, biosphere, hydrosphere and lithosphere.

Key terms

atmosphere: the layer of gas that surrounds Earth

biodiversity: the variety of organisms in an ecosystem

biosphere: all the living things on Earth

hydrosphere: all the water on Earth

lithosphere: Earth's rigid outer zone (crust and upper mantle), made up of tectonic plates

Key idea: Stability and change

We depend on the natural environment for our survival. Our food is sourced from plants and animals, aquatic ecosystems provide clean drinking water, and photosynthesis creates oxygen. Bees pollinate our crops, coral reefs and mangroves prevent erosion of coastal areas, and humans obtain a sense of wellbeing from spending time in nature. However, human activity is disrupting the natural environment and its ecosystems. Earth scientists think of the Earth as being divided into four systems: the atmosphere, biosphere, hydrosphere and lithosphere. When we think of Earth as being made up of interconnected systems, we can better understand how the planet functions and how to protect and restore it. By understanding how these systems have changed over time, we can predict how they may change in the future.

The atmosphere is Earth's protective layer

The **atmosphere** is a system that incorporates the layer of gases that surrounds the planet. Earth's atmosphere has evolved over its 4.6-billion-year life span. It did not exist at the very beginning; however, as volcanic eruptions and chemical reactions on Earth's surface produced gases, they became trapped by Earth's gravitational pull. Over time the composition of the atmosphere stabilised, becoming something close to what we know today. It did this with the help of the evolution of photosynthetic life, which utilised the carbon dioxide emitted by volcanoes and returned oxygen as a by-product. The 600-kilometre-thick layer of gases that now comprises Earth's atmosphere protects the planet from harmful cosmic and ultraviolet radiation. It keeps the average surface temperature warm enough to support life, and allows us to experience sunrises, sunsets and weather.

The biosphere is home to life's diversity

The biosphere is the system that incorporates all the living things on Earth, from blue whales cruising through the oceans to bacteria living in volcanic vents on the ocean floor. It is difficult for biologists to estimate, but it is thought that there are around 10 million species alive on Earth today, with most of them being unknown to science.

Since the first single-celled organisms arose more than 3 billion years ago, hundreds of millions of species have evolved to be specifically adapted to their ecosystem and the role they play in it. As ecosystems have changed, species that could not survive became extinct and new ones evolved.

Biodiversity is key for ecosystem stability. A biodiverse ecosystem contains many species. In an ecosystem, all living things are connected to each other, and the loss of one thing can result in the loss of them all. The more biodiverse an ecosystem is, the more stable it is and the more resistant it is to change. It is important to conserve biodiversity because different species provide us with different benefits: bacteria and fungi act as decomposers to return nutrients to the soil; insects pollinate food crops; and many species of fish are important food sources. We also source many medicines from nature: antibiotics can be synthesised from chemicals produced by fungi, and painkillers can be derived from plants. If we lose species, we also lose these products.

The hydrosphere is vital for life on Earth

The hydrosphere is the system that contains all the water on Earth – whether in the form of ice, liquid water or water vapour. About 70 per cent of Earth's surface is covered by oceans, with a further 10 per cent covered by ice, which includes sea ice, glaciers, and the Greenland and Antarctic ice sheets.

Earth is the only known location in the solar system where water exists in solid, liquid and gas forms. Its properties mean that in its liquid form, it can absorb a large amount of heat energy before it evaporates into a gas, while in its solid form – ice – the molecular structure means it is less dense than the liquid and so ice floats on top. These properties of water allow it to cycle through the spheres in processes that make up the water cycle.

Water can be found in all the other spheres. It makes up a significant portion of the bodies of living things; it can be found as water vapour, which can condense into clouds in the atmosphere; and it has the ability to change the lithosphere through erosion or to percolate through soil and rocks to be stored as groundwater.

The location of water has changed on Earth over time. Ocean basins open and close due to the movement of tectonic plates. Periods of global cooling have seen ice caps extend from the poles and even cover the globe, whereas periods of global warming have seen ice caps disappear.

◀ **Figure 5.3:** Biodiversity is important for healthy ecosystems. Natural ecosystems are more biodiverse than those changed by humans.

Figure 5.4: Oceans cover 70 per cent of Earth's surface, with ice covering a further 10 per cent.

Figure 5.5: The Himalayas, the world's tallest mountain range, were formed when the Indian and European continents collided.

The lithosphere is always changing

The **lithosphere** is the system that incorporates Earth's crust and the top layer of the mantle. From an aeroplane, you can see the Earth's surface stretched, broken and folded into landforms such as plains, valleys and mountains, covered by plant life or often by water. The lithosphere is broken up into 15 major tectonic plates made up of a combination of continental crust and oceanic crust. The slow movement and interaction of the tectonic plates over time has caused the continents to move around, changing their own shape and that of the oceans. Today the continents are spread apart, but in the past they formed supercontinents such as Pangaea.

Processes such as weathering and erosion help to form soils. The type of soil, and landscape features such as mountains, are key factors that determine the type and amount of plant life that will be able to grow.

Learning Ladder

Climate change

1. a Identify a major feature of the atmosphere.
   b Identify a major feature of the biosphere.
   c Identify a major feature of the hydrosphere.
   d Identify a major feature of the lithosphere.
2. Copy and complete the sentences below.
   a Winds blow sand across a desert to form dunes. This is an interaction between the ______ and the ______.
   b Tree roots crack rocks. This is an interaction between the ___________ and the ___________.
   c Ocean currents bring nutrients up from the depths to the surface so that phytoplankton can grow. This is an interaction between the ___________ and the ___________.
   d Plants photosynthesise, using carbon dioxide and water to form glucose and oxygen. This is an interaction between the ___________, the ___________ and the ___________.
3. Explain how the hydrosphere and biosphere are interconnected.
4. Analyse how changes in the biosphere can lead to changes in the atmosphere.
5. Discuss why knowing how the atmosphere, biosphere, hydrosphere and lithosphere are connected helps us to understand how humans are impacting Earth.

Nature and development of science

1. Propose one technology that scientists can use to gather information on the:
   a atmosphere. b biosphere.
   c hydrosphere. d lithosphere.
2. Describe, using an example of one of Earth's systems, how scientific knowledge has been refined over time.

Communicating

p. 262

You have been asked to create a product that will teach Year 5 students about the atmosphere, biosphere, hydrosphere and lithosphere.

1. a Brainstorm some different products that you could create.
   b Write one to two sentences that describe the atmosphere, biosphere, hydrosphere and lithosphere.
   c Select or create some images, diagrams and/or video that you could use in your product.
2. Select one of the products you identified in Question 1a. Use your responses to Questions 1b and 1c to create it.
3. Brainstorm how you could use digital technologies to improve how you communicate to your audience.

Key idea: Stability and change

Conduct further research into either the atmosphere, biosphere, hydrosphere or the lithosphere to find out how it has changed since Earth was formed.

Success criteria

- I can identify and describe the atmosphere, biosphere, hydrosphere and lithosphere.

5·2 ▸ Earth's climate system

Figure 5.6: Earth's climate system is driven by incoming solar radiation.

Learning intention

At the end of this lesson, I will be able to describe factors that influence Earth's climate system.

Key terms

albedo: the amount of light reflected from a surface

convection: the transfer of energy by movement of a liquid or a gas

Earth's energy budget: the amount of energy coming into and leaving Earth's climate system

greenhouse effect: the trapping of the Sun's heat by Earth's atmosphere

greenhouse gas: a gas that traps the Sun's heat energy in Earth's atmosphere

thermohaline circulation: the movement of ocean currents due to differences in temperature and salinity in different regions of water

Investigation 5.2

The albedo effect, p. 352

Key idea: Stability and change

Earth's climate system is driven by incoming solar radiation and depends on interactions between the atmosphere, hydrosphere and lithosphere. Small changes in these interactions have the potential to change the climate.

Earth's energy comes from the Sun

All of the energy for processes that happen on the surface of Earth comes from the Sun. If Earth had been able to absorb all the incoming solar energy (solar radiation) or reflect it all back into space, our planet would have been inhospitable. Instead, incoming solar radiation is balanced with energy that is being reflected, absorbed and radiated through various processes. If the amount of solar radiation coming in is higher than that going out, then the planet will be warmer. If the amount of radiation coming in is lower than that going out, then the planet will be cooler. This is referred to as **Earth's energy budget** and is a key driver of its climate. Small changes in Earth's energy budget can have significant impacts.

There are two key factors in maintaining Earth's energy budget.

- **Composition of the atmosphere.** Incoming radiation can be reflected by particles, gas molecules and clouds in the atmosphere. The radiation can also be absorbed by atmospheric gas molecules and then be reflected either back to Earth or out into space. This means that a cloudy day will be colder than a clear day, as the clouds will prevent some solar radiation reaching the surface. It also means that some of the outgoing radiation is trapped by gases in the atmosphere, keeping Earth warm (the **greenhouse effect**).

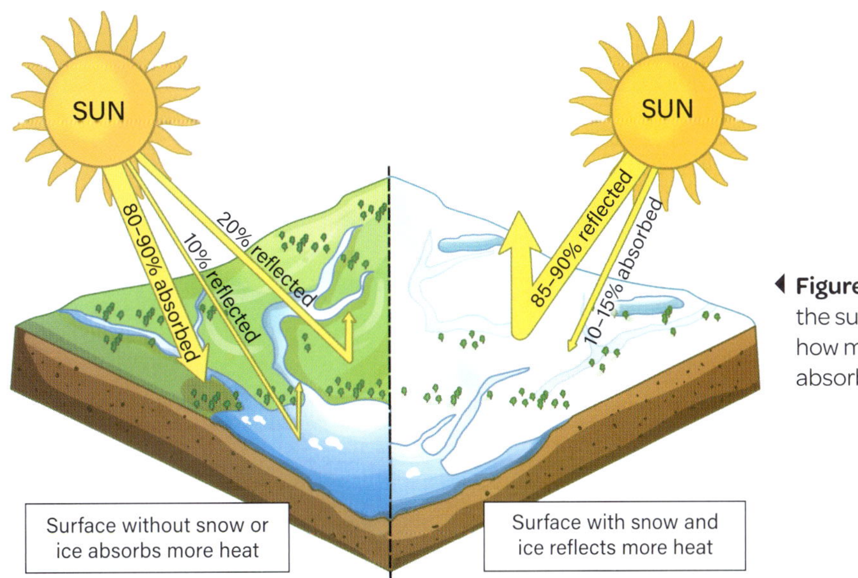

Figure 5.7: The albedo of the surface determines how much solar energy is absorbed and reflected.

- **Colour of the surface.** Have you ever experienced snow blindness or been sunburnt at the snow? This is because light colours reflect more light – we say they have a high albedo. Lighter-coloured regions (such as ice caps) will reflect solar radiation, whereas darker-coloured regions (such as oceans and forests) will absorb solar radiation. If the whole of Earth's surface was a dark colour, the planet would absorb more radiation, increasing the temperature; by contrast, if it was all a light colour, more radiation would be reflected, cooling it down.

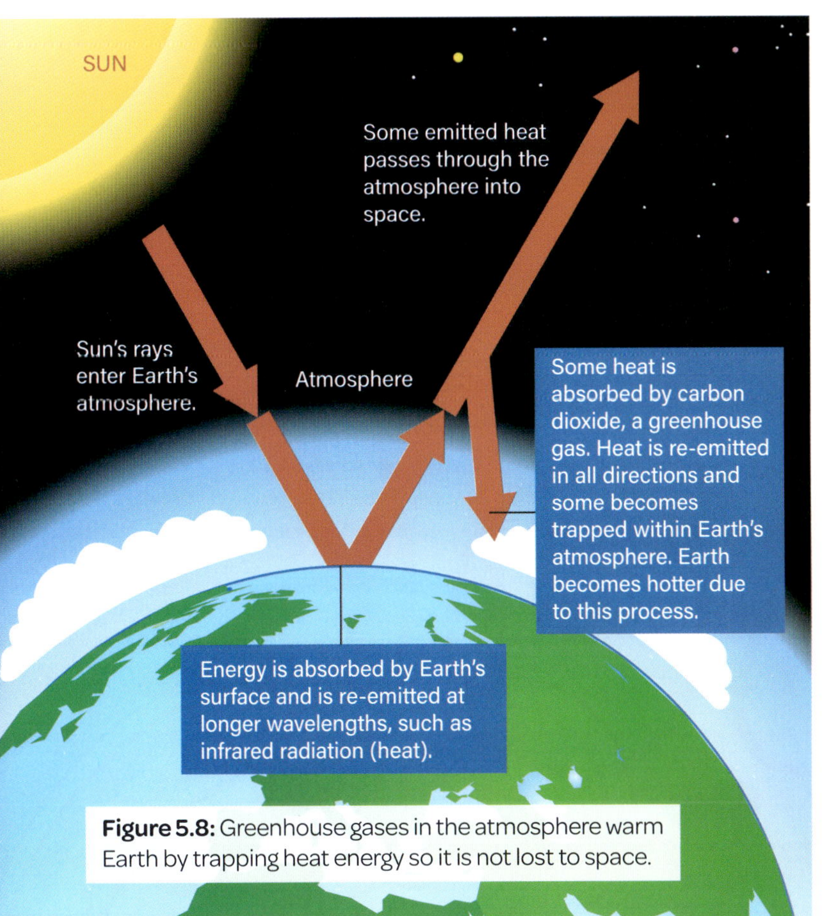

Figure 5.8: Greenhouse gases in the atmosphere warm Earth by trapping heat energy so it is not lost to space.

The greenhouse effect keeps Earth's surface warm

The greenhouse effect is the process where greenhouse gases in Earth's atmosphere absorb and re-emit infrared radiation, heating the surface of Earth so it is warm enough to sustain life. Without this process, Earth's surface would be −18 °C! Greenhouse gases warm Earth's surface and the lower atmosphere by trapping heat emitted by the surface and preventing it from travelling back out into space.

The main greenhouse gases in the atmosphere are water vapour (H_2O), carbon dioxide (CO_2), methane (CH_4) and nitrous oxide (N_2O). These gases are only present in the atmosphere in very small quantities (Table 5.1), but it is enough to trap energy to keep Earth's surface warm – an average of about 13.6 °C in pre-industrial times. The molecules of some greenhouse gases, such as nitrous oxide, can remain for a long time – more than 100 years. Others, such as water vapour, stay in the atmosphere for only a short time – up to a few days.

Table 5.1: Atmospheric concentration of common greenhouse gases

Greenhouse gas	Pre-industrial concentration (parts per million, ppm)
Water vapour, H_2O	1–3
Carbon dioxide, CO_2	280
Methane, CH_4	0.7
Nitrous oxide, NO_2	0.27

Incoming solar radiation is not spread evenly

Earth is a sphere that is tilted at 23.5° on its axis. This means that different parts of the planet will receive different amounts of incoming solar radiation, heating its atmosphere and surface unevenly. The equator receives a higher amount of energy compared to the poles, and this heat imbalance drives circulation in the atmosphere and the oceans, distributing the heat more evenly around the globe.

Figure 5.9: Solar radiation is more concentrated at the equator compared to the poles, resulting in a heat imbalance.

Low angle of incoming sunlight
Sunlight strikes most directly
Low angle of incoming sunlight
60°N
30°N
Tropic of Cancer
0° (Equator)
Tropic of Capricorn
30°S
60°S

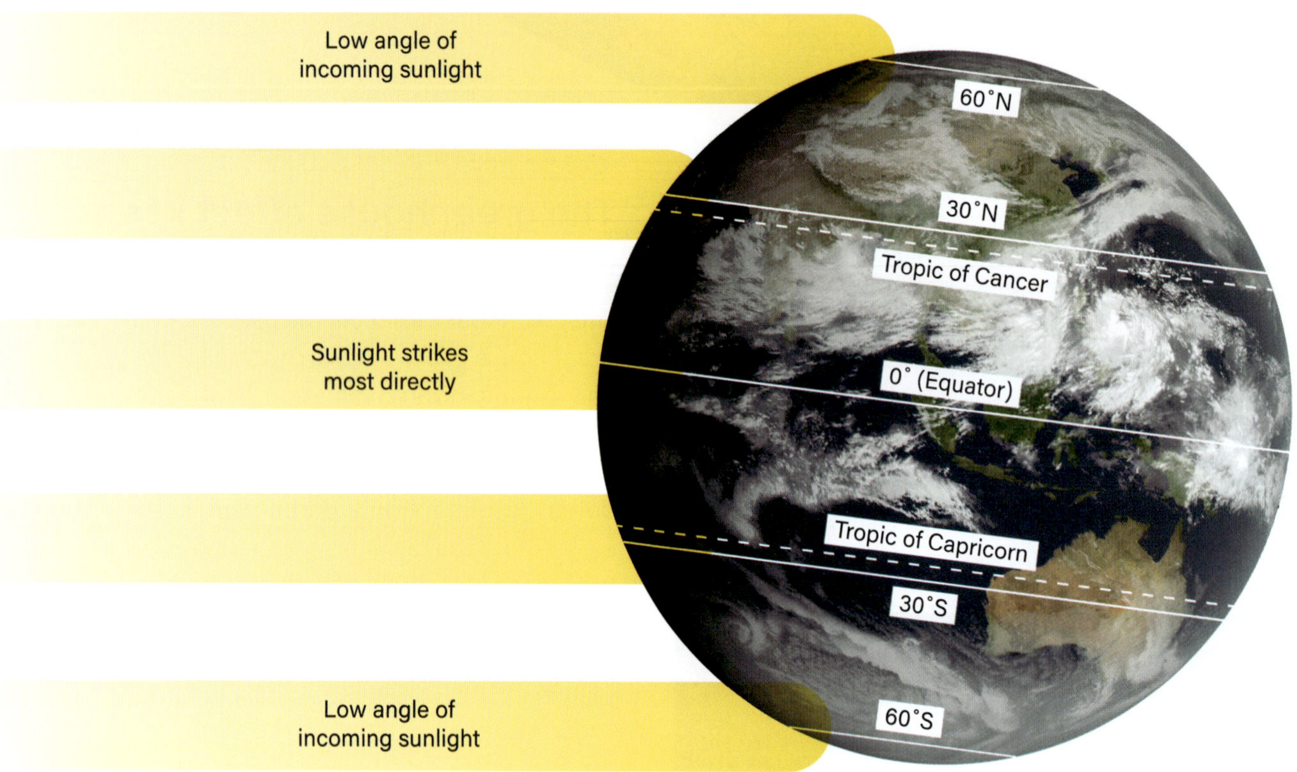

Figure 5.10: The movement of air across the planet is determined by the uneven spread of solar radiation across Earth's surface. Atmospheric circulation redistributes heat by moving rising hot air from the surface and replacing it with cooler, denser air.

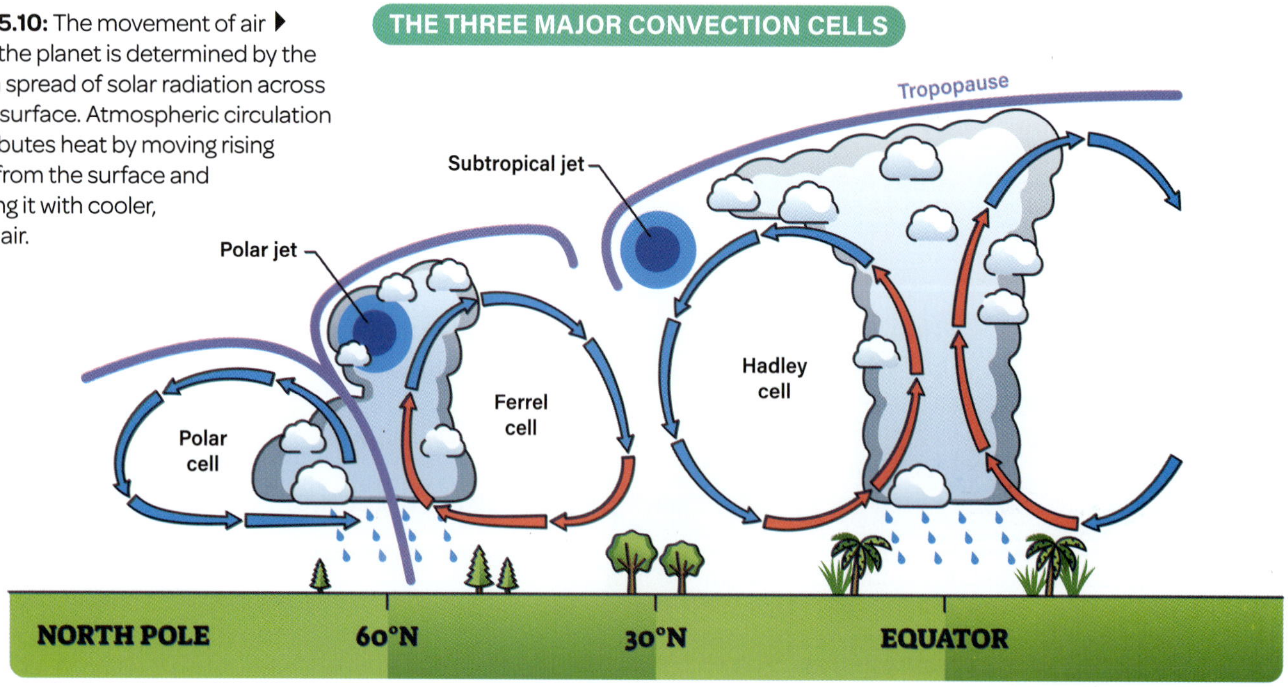

THE THREE MAJOR CONVECTION CELLS
Tropopause
Subtropical jet
Polar jet
Polar cell
Ferrel cell
Hadley cell
NORTH POLE
60°N
30°N
EQUATOR

Atmospheric circulation balances global temperatures

There is a significant temperature difference between the poles and the equator, and atmospheric circulation helps to redistribute this heat around the planet. When air is heated, the molecules move further apart, making it less dense. As a result, warm air rises and cooler, denser air sinks – creating convection currents. Rising warm air leads to areas of low pressure, while sinking cold air creates high-pressure regions. On Earth, this process forms three major convection cells in each hemisphere. These cells play a key role in balancing global temperatures and driving weather patterns.

- **Hadley cells** are located at the tropics. Warm air rises at the equator and starts to travel towards the poles. As the air rises, it begins to cool, and any water vapour will condense and precipitate. This is why there is high rainfall in the tropics. Once the warm air reaches about latitude 30°, it cools and begins to sink, becoming warmer and drier as it flows back towards the equator. This is why most of the world's deserts are located at around latitude 30°.
- **Ferrel cells** are located between latitudes 30° and 60°. Warm air flows along the surface towards the poles, picking up moisture as it moves. When it hits cooler air from the poles at latitude 60°, it will rise because it is less dense.
- **Polar cells** are located at the poles, where cold air sinks and moves along the surface towards the equator. It hits warmer air at latitude 60° and rises. This collision of warm and cool air results in a lot of unstable weather, including high winds and rainfall.

If the temperature difference between the poles and equator changes, this will have an impact on the strength of atmospheric circulation. A larger difference will increase the strength; a smaller difference will decrease the strength.

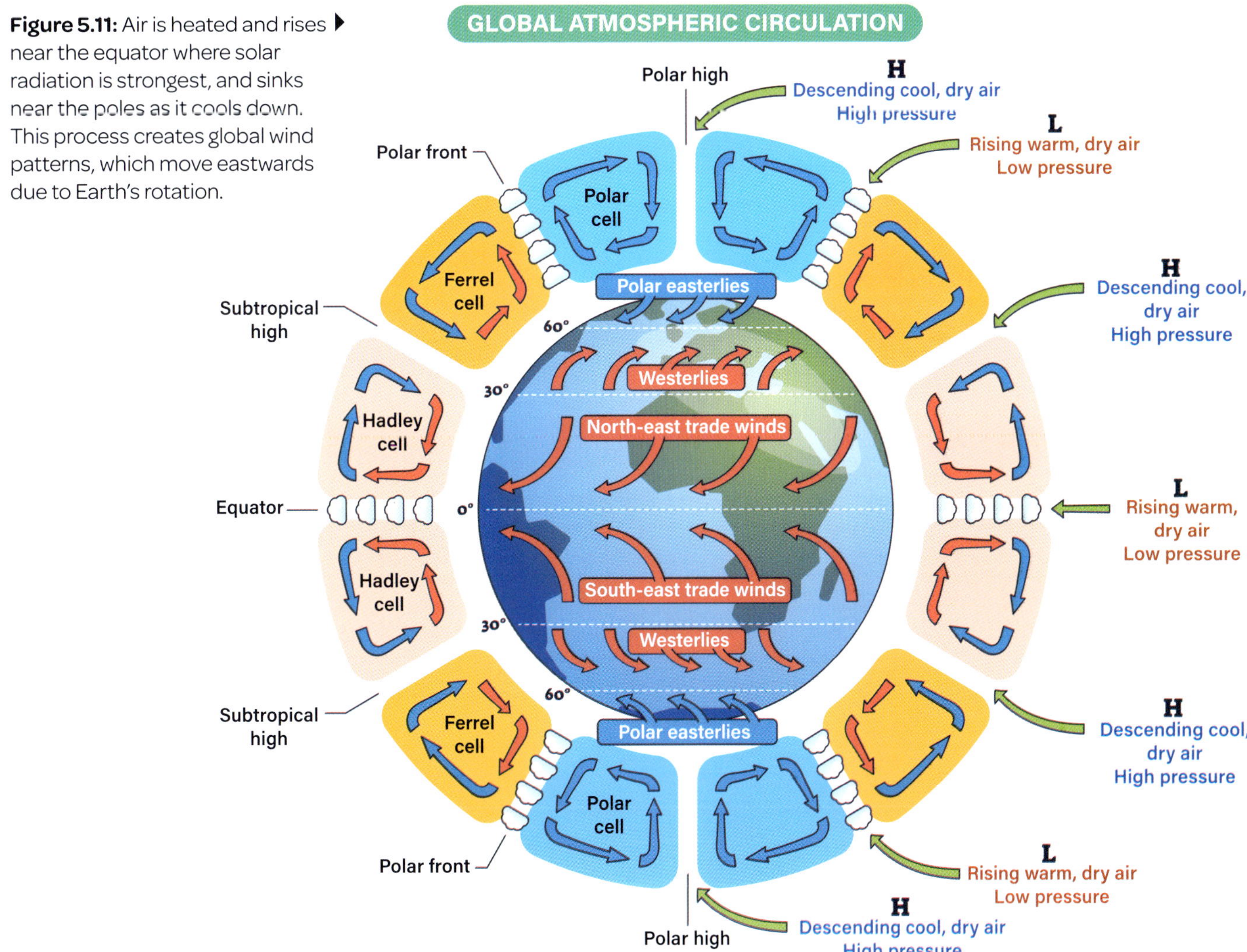

Figure 5.11: Air is heated and rises near the equator where solar radiation is strongest, and sinks near the poles as it cools down. This process creates global wind patterns, which move eastwards due to Earth's rotation.

Figure 5.12: Tropical cyclones form when warm ocean water evaporates and rises rapidly upwards and cooler air rushes in to replace it. The warm, moist air cools in the atmosphere and condenses into large convective cloud systems, which are given their spiralling shape by Earth's rotation.

Ocean circulation distributes heat energy and nutrients around the planet

The ocean absorbs a large amount of incoming solar radiation due to its low albedo. Ocean currents work to redistribute this heat energy. Winds caused by atmospheric circulation patterns drive surface currents, but below 100 metres the larger, deeper and slower-moving currents are driven by differences in density that are caused by differences in temperature and salinity. This is known as **thermohaline circulation**, or the great ocean conveyor belt.

As water at the poles cools and forms ice, it leaves the salt behind. This cold, salty, dense water sinks to the bottom of the ocean. Warmer water from the equator will move to fill this space, in turn becoming denser and sinking. This water will travel slowly towards the equator until it rises, often bringing nutrients with it. The warm surface waters at the equator travel towards the poles, completing the loop – releasing heat into the atmosphere as it starts to cool at around latitude 60°. The movement of warm surface water towards the poles brings heat to regions that would otherwise be cold; for example, the Gulf Stream brings warm water from the equator to northern Europe.

The Antarctic Circumpolar Current is the only place on Earth where the deep-water currents and warm-water currents can flow without being interrupted by land. This current is important, as it connects the Atlantic, Indian and Pacific oceans.

Thermohaline circulation relies on the temperature at the poles being cold enough to form sea ice and the resulting cold, dense, salty water. Without the formation of this water, circulation will begin to slow.

Figure 5.13: The great ocean conveyor belt transports nutrients and heat energy around the globe. The movement of the ocean's water through this cycle happens very slowly: one complete trip around Earth on the conveyor belt takes about 500 years!

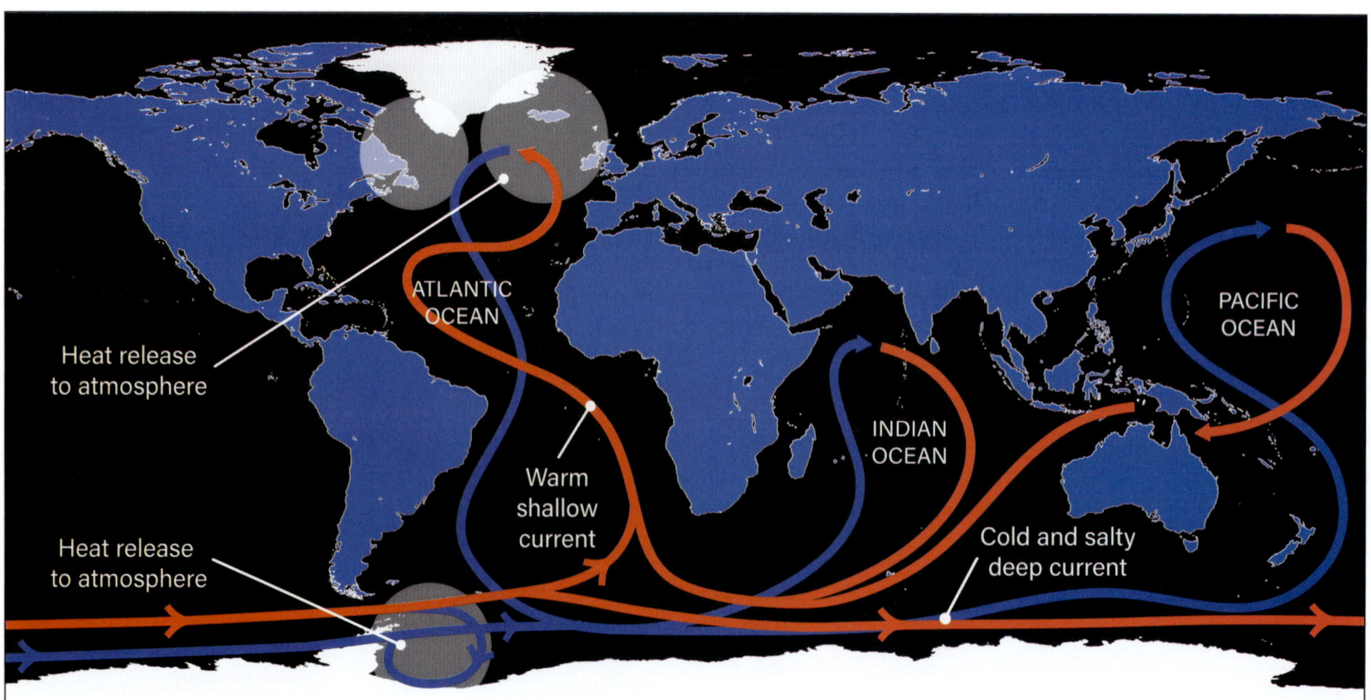

Figure 5.14: Equatorial areas are warmer than polar areas, and this heat is distributed around the planet via atmospheric and oceanic circulation. Differences in temperature and moisture drive the variety of weather and climate conditions across different regions of the Earth (see Section 5.3).

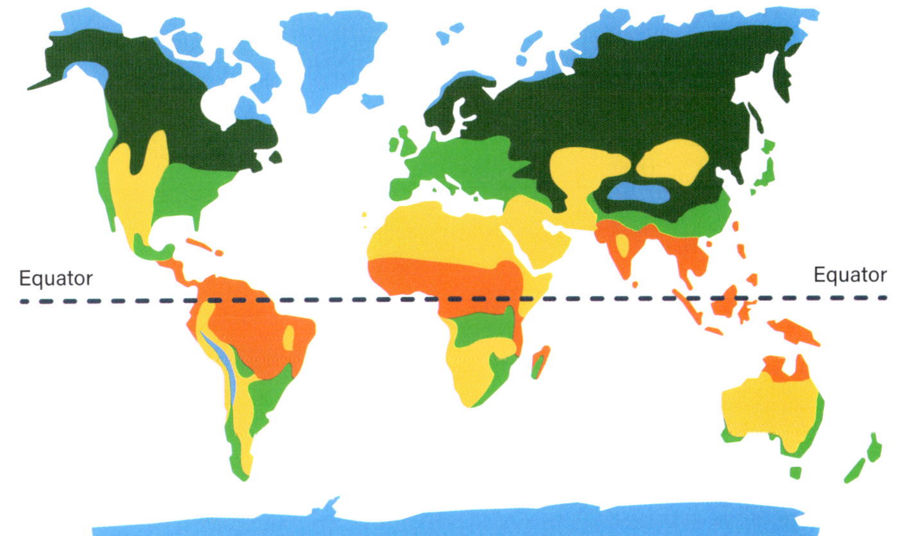

Learning Ladder

Climate change

1. Use these words to complete the sentences below: *absorbs, atmosphere, equator, low, poles.*
   a Circulation in the __________ redistributes hot air from the __________ and cold air from the __________.
   b The ocean __________ a large amount of solar radiation due to its __________ albedo.
2. Describe how Earth's energy budget influences how warm or cold the planet is.
3. Propose and explain what would happen to the temperature of Earth if:
   a a large amount of dust entered the atmosphere due to a meteorite impact.
   b sea ice melted, leaving the dark surface of the ocean exposed.
   c sea ice, glaciers and ice caps extended over vast areas beyond the poles.
4. It is thought that the entire Earth was covered with ice caps about 720–630 million years ago – a period commonly referred to as 'snowball Earth'.
   a Discuss what this would have meant for Earth's energy budget.
   b Discuss how this might have impacted oceanic and atmospheric circulation.
5. Evidence indicates that the Gulf Stream is slowing down. Analyse the impact this could have on Europe's climate.

Use and influence of science

1. a Suggest one way that scientific knowledge about the albedo effect could be used to understand climate change.
   b Suggest one way that scientific knowledge about thermohaline circulation could be used to understand the impacts of warmer oceans at the poles.
2. Describe how scientific knowledge of the influence of ice caps on Earth's climate could be interpreted in different ways.
3. Explain how the needs of society have influenced research into changing oceanic circulation.

Evaluating

p. 253

Read through Investigation 5.2 on pages 352–53.

1. Identify a type of error in the method.
2. Describe how the error you identified in Question 1 could impact the data gathered in the investigation.
3. Propose a way you could improve the investigation to mitigate the impact of the error.
4. Use scientific evidence to justify the conclusion you reached from this investigation.

Key idea: Systems and change

Describe what would happen to Earth's surface temperature over a 24-hour period if there were no atmosphere.

Success criteria

- I can describe the impact on Earth's climate system of:
  - Earth's energy budget
  - atmospheric circulation
  - the greenhouse effect
  - ocean circulation.

5·3 ▸ Weather and climate

Learning intention

At the end of this lesson, I will be able to distinguish between weather and climate and identify global climate trends.

Key terms

climate: trends in weather over a period of at least 30 years

climate zone: a region with a specific climate

meteorologist: a scientist who studies the atmosphere and its effects on Earth, including weather patterns

ocean circulation: the movement of water in the oceans due to major currents

weather: what is happening in the atmosphere at a specific place and time

Investigation 5.3

Observing the weather, p. 354

Key idea: Stability and change

Figure 5.15: The term 'weather' describes what is happening in the atmosphere at a specific place and time.

The difference between weather and climate is a matter of time. Studying weather and climate can prepare us for extreme weather events and allow us to observe how climate trends are changing.

Weather happens over short periods

The term '**weather**' describes what is happening in the atmosphere at a specific place and time – it can be minutes, hours, days or weeks. The weather is always changing because it depends on factors such as temperature, air pressure, wind, humidity and cloud formation. It can also be influenced by landforms and **ocean circulation**.

Meteorologists study the weather and use complex computer models to make predictions. Short-term forecasts are reasonably accurate, but long-term forecasting is more challenging.

Extreme weather events include cyclones, storms, floods and heatwaves. Being able to predict where and when they will happen allows communities to be prepared to respond.

Climate happens over longer periods

The term '**climate**' describes the average weather patterns at a specific location or region over a period of at least 30 years. The climate for a region can be described using mean temperature and rainfall data at different times of the year, but it also includes data such as the frequency of extreme weather events. For example, Melbourne has higher temperatures during summer and lower temperatures during winter, and consistent rainfall throughout the year, and is known for experiencing 'four seasons in one day' due to winds that can come from the south or the west. Northern Australia has higher rainfall over summer (the wet season) than in winter (the dry season). Tropical cyclones are more common in these regions between November and April.

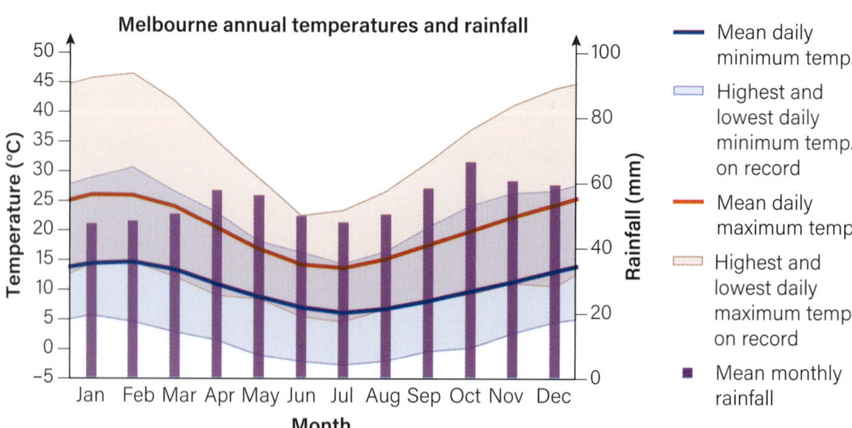

Figure 5.16: This graph shows climate data for Melbourne using mean daily temperatures and mean monthly rainfall data taken from the Bureau of Meteorology's Melbourne Regional Office.

There are trends in climate across the world

Climate zones are regions with a specific climate. The Köppen climate classification system (see Figure 5.14 on page 125) is based on temperature and rainfall. In this system, there are five major climatic zones, as described in Table 5.2.

Table 5.2: Köppen climate classification

Zone	Description
Tropical	Located at the equator High temperatures and high rainfall
Dry	Located north and south of the tropics Much less rainfall than in the tropics Includes deserts
Temperate and continental	Lower temperatures and high rainfall Coastal regions influenced by ocean currents Continental climates highly variable
Polar	Low temperatures, permanent ice caps

The global climate is changing

Long-term data shows that Earth's climate is changing, driven by an increase in atmospheric temperature. This influences rainfall, ocean temperatures and ocean circulation.

Australia's average temperature has risen by 1 °C since 1910 (Figure 5.17). In this time, we have observed more extreme heatwave events, prolonged droughts and longer bushfire seasons. Cyclones are increasing in strength and are forming further away from the equator and outside of the expected season.

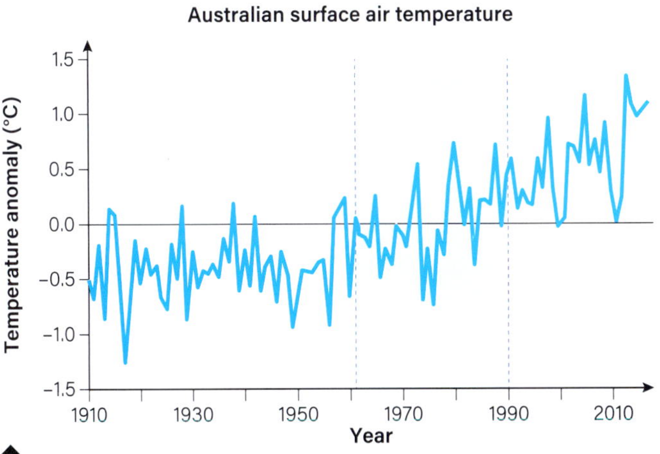

Figure 5.17: Australia's average temperature has increased by 1 °C on average since 1910. This data shows how far the average temperature has varied from the 1961–90 average.

Learning Ladder

Climate change

1. Make an observation to describe the weather today.
2. Describe the characteristics of the climate zone that Victoria is part of.
3. Explain the difference between weather and climate.

Use and influence of science

Communities in Australia need to be able to adapt to more extreme weather events.

1. Identify at least one way that scientific knowledge could be used to assist communities to plan to adapt to more extreme weather events.
2. Conduct some research to find an example of where scientific knowledge of extreme weather events has been interpreted in different ways.
3. Explain how the focus of scientific research to predict extreme weather events has been influenced by the values and needs of society.
4. Discuss how scientific information and misinformation about extreme weather events may inform personal and social decision-making in preparing for an event.
5. Analyse the key factors that contribute to scientific knowledge of extreme weather events being adopted more broadly by society.

Communicating p. 262

Use the data in Figures 5.16 and 5.17 to answer the following questions.

1. Identify an alternative way that you could represent or describe the trends shown in each figure.
2. Create one of your responses to Question 1.
3. Use digital technologies to help you represent data that supports the statement, 'Australia's average temperature is rising.'

Key idea: Stability and change

The Australian continent experiences three major climatic zones.

a Predict how this influences the weather from the north to the south of the continent.

b Propose and analyse how this influence has changed over time as the Australian continent has drifted north from Antarctica.

Success criteria

- I can distinguish between weather and climate.
- I can identify global climate trends.

5·4 ▸ The enhanced greenhouse effect

Learning intention

At the end of this lesson, I will be able to evaluate scientific evidence for the effect that human activity has on the greenhouse effect.

Key terms

enhanced greenhouse effect: an increase in the greenhouse effect due to human greenhouse gas emissions

Industrial Revolution: a period in the late 1700s when manufacturing transformed to large-scale factories that were powered by the burning of fossil fuels

Investigation 5.4

The enhanced greenhouse effect, p. 355

Key idea: Stability and change

Humans have added more greenhouse gases into the atmosphere, which has artificially enhanced the greenhouse effect, raising the average temperature of Earth by about 1 °C.

Humans have increased the greenhouse effect

Since the **Industrial Revolution**, human activity, such as the burning of fossil fuels and factory emissions, has increased the levels of greenhouse gases in the atmosphere. Because there are more gas particles, more infrared radiation stays in Earth's atmosphere and less is radiated back into space. This is called the **enhanced greenhouse effect**.

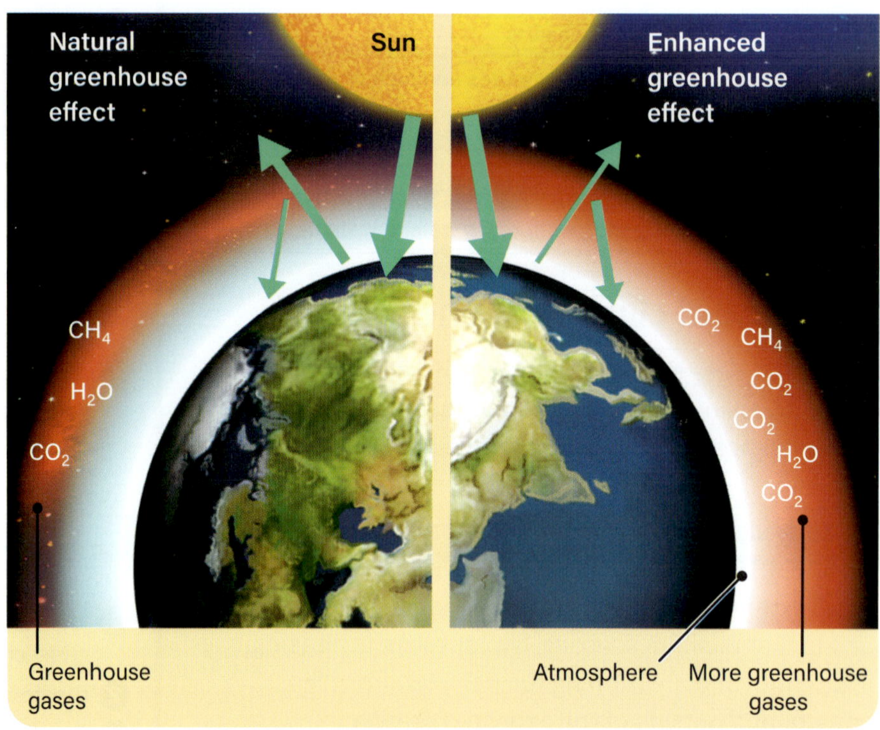

Figure 5.18: A comparison of the natural greenhouse effect and the enhanced greenhouse effect

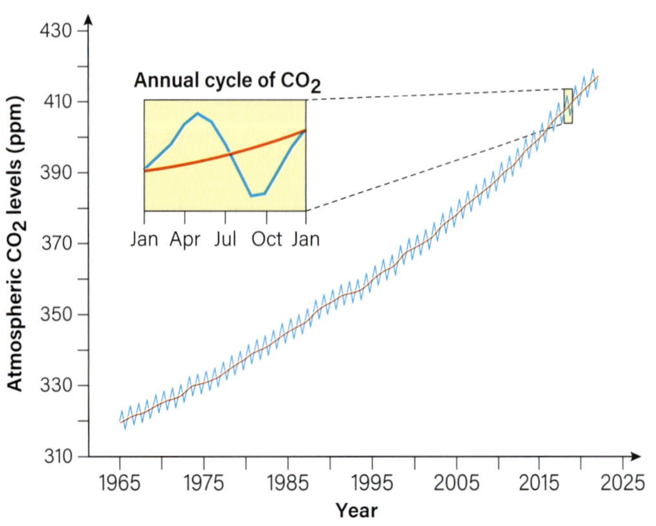

Figure 5.19: The Keeling curve shows atmospheric carbon dioxide concentration (in ppm) from 1965 to today.

Atmospheric carbon dioxide levels are increasing

Scientists can measure the amounts of greenhouse gases in the atmosphere and correlate this with average temperatures to demonstrate the relationship between the two. A global network of stations samples the atmosphere regularly, including Mauna Loa Observatory on the Big Island of Hawaii and the Kennaook/Cape Grim Baseline Air Pollution Station in north-west Tasmania. Measurements taken at Kennaook/Cape Grim show that carbon dioxide levels in the atmosphere are increasing. In 2024, the atmospheric concentration of carbon dioxide was about 425 ppm (parts per million).

The Keeling curve plots carbon dioxide levels in the atmosphere

In 1958, US scientist Charles Keeling started monitoring carbon dioxide concentrations in the atmosphere at Mauna Loa, in Hawaii. His research showed that the concentration of carbon dioxide in the atmosphere was steadily increasing and changed throughout the year.

The annual cycle is because most of Earth's land mass and plant life are in the northern hemisphere. Over the northern spring and summer, plant growth and photosynthesis increase, which reduces carbon dioxide in the atmosphere. During autumn and winter, plant growth and photosynthesis decline and levels of carbon dioxide in the atmosphere increase. When plotted as a graph, this data is known as the Keeling curve (Figure 5.19).

Ice core data also shows carbon dioxide levels are increasing

Ice cores are cylinders of ice drilled from ice sheets. Scientists use ice cores that are up to 800 000 years old to find out more about Earth's past temperature and climate. Importantly, they can analyse air bubbles trapped in the ice to determine the concentration of gases in Earth's atmosphere over time.

Data from ice cores shows that the recent increase in carbon dioxide levels due to human activity is unprecedented (Figure 5.20).

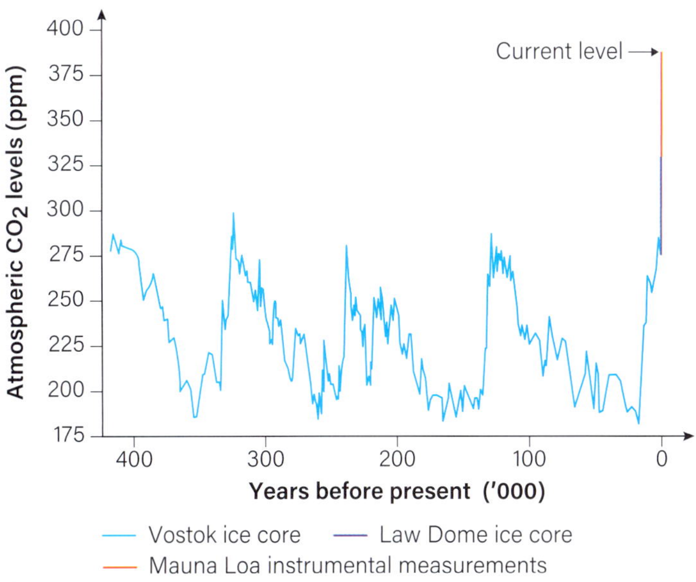

Figure 5.20: Combined data from the Vostok and Law Dome ice cores in Antarctica and from Mauna Loa show how the carbon dioxide concentration in the atmosphere has varied over the last 400 000 years.

Learning Ladder

Climate change

1. Identify the feature(s) of the atmosphere that cause the enhanced greenhouse effect.
2. Describe the difference between the natural and enhanced greenhouse effect.
3. Explain a major cause of the enhanced greenhouse effect.
4. Discuss a potential impact of increased concentrations of greenhouse gases in the atmosphere.
5. Analyse the data on atmospheric carbon dioxide levels to explain the impact human activity has had on atmospheric temperature. *Note:* You may need to refer to the Australian temperature trends in Figure 5.17 in Section 5.3.

Nature and development of science

1. Identify the location of the key observatories where observations are made about the carbon dioxide content of the atmosphere.
2. Describe how Keeling's atmospheric data has helped to refine scientific knowledge of the atmosphere over time.

Evaluating *p. 253*

1. Identify an assumption that scientists may have made about the atmospheric data gathered at Mauna Loa and Kennaook/Cape Grim.
2. Describe the impact of the assumption you identified for Question 1 on the atmospheric data.
3. Some students use data loggers to obtain carbon dioxide concentration from the atmosphere at their school in inner Melbourne. Propose ways that they could improve the quality of their data.
4. Use evidence from Figures 5.19 and 5.20 to support the following claim: 'Carbon dioxide levels in the atmosphere have increased at unprecedented levels since 1965.'

Key idea: Stability and change

Create an infographic to explain how human activity has changed the atmosphere and the greenhouse effect in the last 200 years.

Success criteria

- I can describe how human activity has enhanced the greenhouse effect.
- I can discuss some of the evidence for the enhanced greenhouse effect.

5·5 ▶ Climate change

Learning intention

At the end of this lesson, I will be able to describe some of the impacts of increasing atmospheric temperatures on Earth.

Key terms

glacier: a slowly moving mass of ice formed by the accumulation of snow

precipitation: liquid or solid water that forms in the atmosphere and falls to Earth's surface

thermal expansion: the increase in volume of a substance due to an increase in temperature

water cycle: the cycle of processes by which water circulates between Earth's oceans, atmosphere, land and biosphere

Investigation 5.5A

Modelling thermal expansion, p. 356

Investigation 5.5B

Melting ice and sea-level rise, p. 358

Key idea: Stability and change

Earth's climate is changing. Long-term data shows that Earth's average temperature has increased since 1880. This is influencing weather patterns, sea levels, ocean temperatures and the presence of ice at the poles.

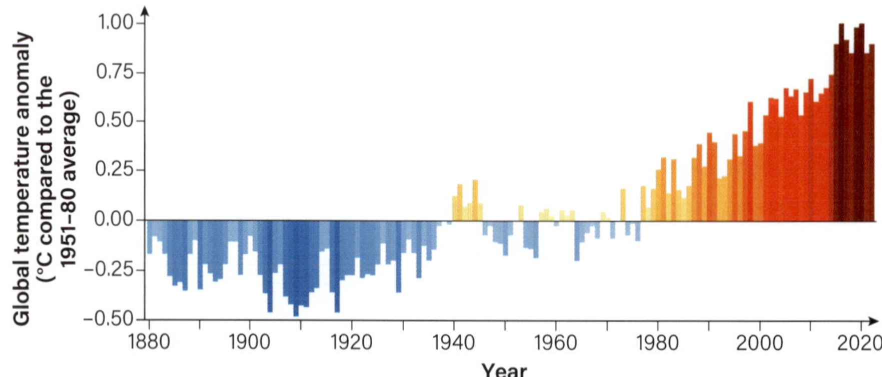

Figure 5.21: The world's average temperature has increased by more than 1 °C since 1880. This graph shows how far the average temperature has varied from the 1951–80 average.

Evaporation and precipitation will increase

Water moves around Earth in the **water cycle**. Increasing temperatures are causing some parts of the water cycle to speed up. As the oceans get warmer, more water evaporates, which means that more water moves into the atmosphere. This results in more cloud formation, more **precipitation** and larger storms. Climate modelling shows that rainfall patterns will also change, with a lot more precipitation over the oceans or along the coastlines of continents, while the interior of the continents will become dryer. If there is more water in the atmosphere, rainfall events can become larger, resulting in flooding. Higher temperatures, and therefore evaporation in regions that receive low rainfall, will result in more extreme droughts.

Figure 5.22: The number of days when mean temperatures were in the top 1 per cent of records has increased in Australia since 1910.

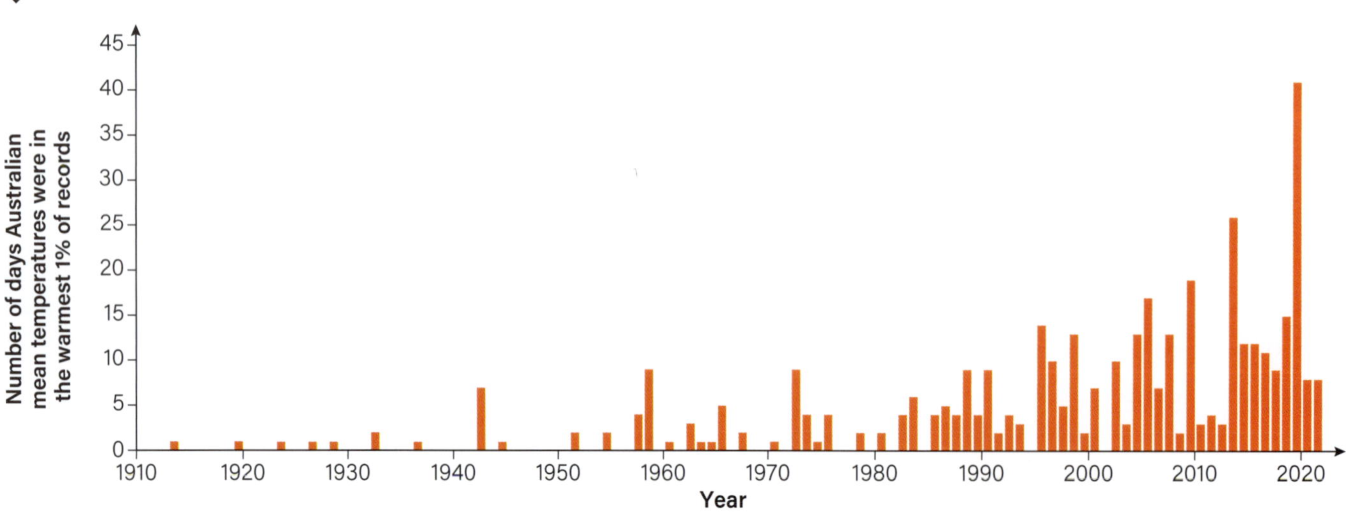

Melting ice contributes to rising sea levels

The increase in global temperatures has led to a reduction of ice at the poles, as well as in glaciers on continents. When glaciers and other continental ice melts, the water runs off into the oceans, contributing to sea-level rise. Figure 5.23 shows how the Franz Josef Glacier in New Zealand has changed since 2009.

When sea ice melts, it leaves dark-coloured sea water behind, which absorbs incoming solar energy instead of reflecting it, like ice does. This increases the sea temperature and melts the remaining sea ice faster. Satellite imagery has shown that there has been a 13 per cent reduction in the permanent summer sea ice in the Arctic each decade since the late 1970s (Figure 5.24). It is predicted that the Arctic Ocean will have ice-free summers by the 2050s.

Figure 5.23: ▸ The Franz Josef Glacier, on the west side of the South Island of New Zealand, retreated about 800 m between 2009 and 2020. It has retreated about 3 kilometres since the late 1800s.

2021

▲ **Figure 5.24:** The extent of the permanent summer sea ice in the Arctic Ocean has decreased by 13 per cent every decade since 1979.

Heatwaves and bushfires will become more frequent

A heatwave is a period of three or more days of higher-than-normal temperatures for a particular region. Not only are temperatures expected to rise, but heatwave events are expected to become more frequent and of longer duration. This will cause challenges for communities, as they need to develop infrastructure to support people who are vulnerable to the extreme temperatures.

The high temperatures and dry conditions produced by climate change also increase the risk of bushfires. Bushfires move rapidly through dry vegetation. In Australia, high temperatures are causing the bushfire season to extend into spring and autumn and the number of extreme fire danger days to increase.

Figure 5.25: Firefighters in Moolort in Central Victoria. The 2019–20 bushfire season was the worst ever recorded in Victoria and across Australia. It was brought about by severe drought combined with intense fire behaviour. Millions of hectares were burned, homes were destroyed and lives were lost. ▼

Sea levels are rising

Sea levels have risen 0.10–0.20 metres around the world in the last century. Some of this rise is attributed to melting glaciers, but at least half is caused by thermal expansion. The amount by which sea levels rise varies around the world because the strength of winds and ocean currents varies, and this affects how much heat can be stored in the deeper parts of the ocean. As the water in the ocean absorbs more heat energy, this energy is transferred to the water molecules, which vibrate more rapidly and move further away from each other, increasing the volume of the water.

As sea levels rise, this will put pressure on coastal communities and threaten infrastructure such as roads and buildings.

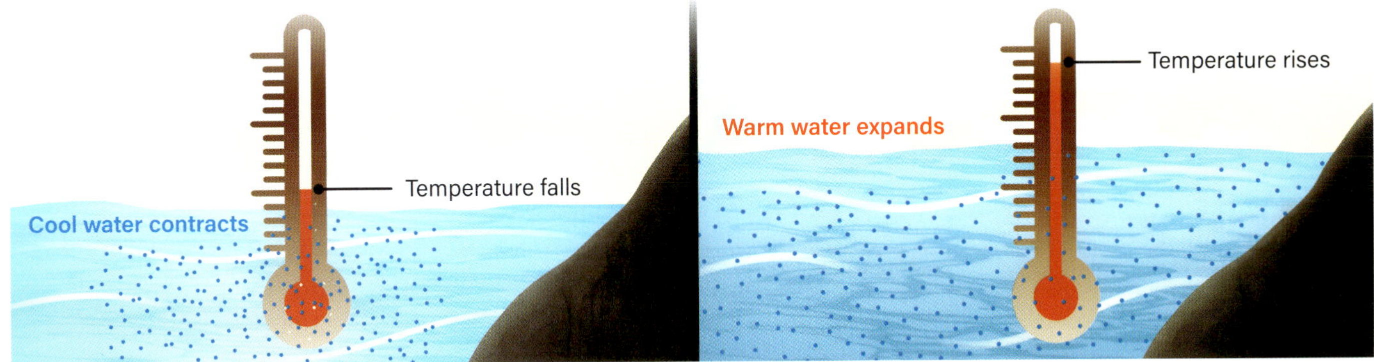

▲ **Figure 5.26:** As the temperature of the water increases, the molecules push each other further apart and so the volume increases.

Learning Ladder

Climate change

1. Identify one way that climate change will impact the:
   - **a** atmosphere.
   - **b** biosphere
   - **c** hydrosphere.
   - **d** lithosphere.
2. Describe the effect that increasing atmospheric temperatures has had on sea-level rise.
3. Explain how higher atmospheric temperatures are impacting ice in the Arctic Ocean.
4. Analyse the effect of greenhouse gas emissions on the frequency and severity of bushfires.
5. Discuss the impact that the burning of fossil fuels has had on the natural world. Evaluate how moving to renewable energy sources can mitigate this impact.

Use and influence of science

1. Identify an example of how scientific knowledge about bushfires can be used to help Australian communities adapt to a changing climate.
2. Conduct research to find and describe an example of how scientific knowledge about bushfires could be interpreted in different ways.
3. Explain how the values and needs of Australian society have resulted in increased scientific research on bushfire behaviour.

Evaluating p. 253

Read through Investigation 5.5B on page 358 before answering the questions below.

1. Identify an error or assumption that is present in this investigation.
2. Describe the impact of the error or assumption from Question 1 on the investigation.
3. Propose how you could modify the method to control the error or assumption you identified in Question 1.
4. Use the data gathered in this investigation to assess the claim that 'all melting ice contributes to sea-level rise'.
5. Evaluate the validity and reproducibility of the investigation.

Key idea: Stability and change

Ask your parents and grandparents about how they have experienced climate change during their lifetime.

Success criteria

- I can describe how increased atmospheric temperatures are affecting evaporation and precipitation.
- I can describe how increased atmospheric temperatures are increasing the risk of heatwaves and bushfires.
- I can describe how increased atmospheric temperatures are reducing the size of glaciers and permanent sea ice at the poles.

5·6 ▸ Effects of climate change

Learning intention

At the end of this lesson, I will be able to:

- describe some of the consequences of climate change
- discuss how data can be used to evaluate the impact of climate change.

Key terms

alpine: areas of high elevation

carbon cycle: the natural cycling of carbon through ecosystems

coral atoll: a ring-shaped island formed by a coral reef

coral bleaching: a phenomenon where corals lose their colour due to the absence of symbiotic algae

groundwater: water that flows underground in spaces between rocks and within soils

ocean acidification: a decrease in the pH of the oceans due to the absorption of more carbon dioxide

Investigation 5.6

Modelling ocean acidification, p. 359

Key idea: Stability and change

Climate change is affecting ecosystems worldwide as natural processes are disrupted. Increased levels of atmospheric carbon dioxide are causing ocean acidification, and high temperatures are leading to habitat loss. Scientific data can be used to create models that predict how a changing climate will affect communities in the future.

Oceans are becoming more acidic

Oceans absorb carbon dioxide from the atmosphere as a natural part of the **carbon cycle**, but the amount they are absorbing is increasing as the amount of carbon dioxide in the atmosphere increases.

When carbon dioxide dissolves in water, it undergoes a chemical reaction to form carbonic acid (H_2CO_3). The carbonic acid then breaks down, releasing hydrogen ions (H^+) that react with free carbonate ions (CO_3^-) in the water to form bicarbonate. This decreases the pH of oceans, leading to **ocean acidification**.

The chemical reactions also mean that less calcium carbonate ($CaCO_3$) is available for zooplankton, coral and shellfish, which need it to build their shells and skeletons. As well, the carbonate in their shells reacts with hydrogen ions in the water, so these organisms cannot grow or they produce thinner shells that are more easily damaged. This will have flow-on effects in the ocean ecosystem, as important organisms in food webs start to disappear.

Climate change is causing habitat loss

Across the world, we are observing habitat loss due to climate change. The Arctic is warming almost twice as quickly as anywhere else in the world, affecting the level of sea-ice coverage. This is limiting the habitat available for polar bears.

Figure 5.27: Ocean acidification is caused by higher amounts of carbon dioxide being absorbed by oceans.

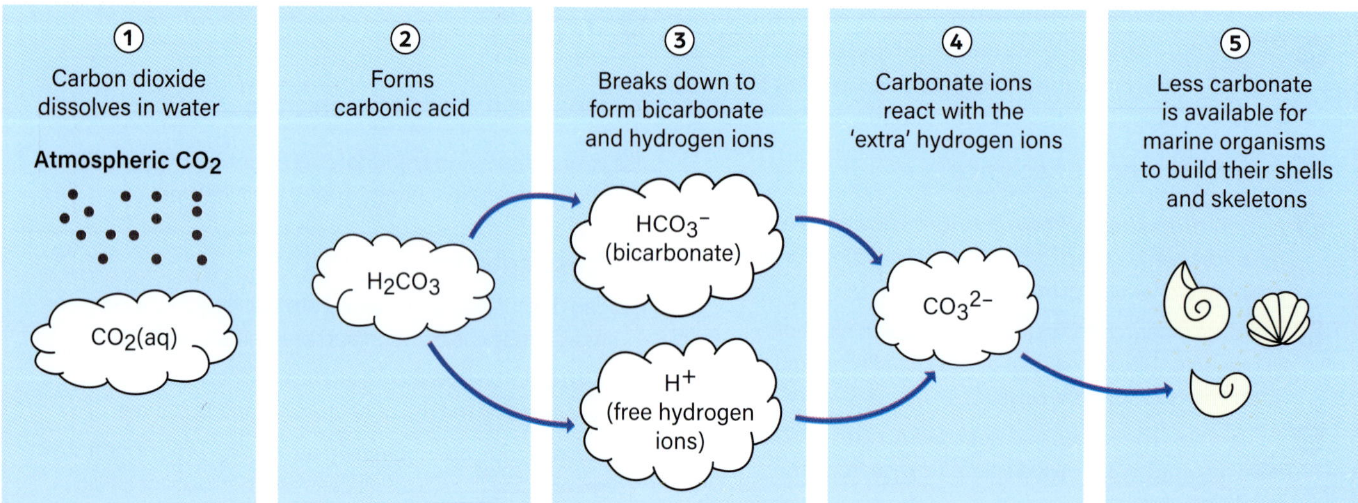

Figure 5.28: A female polar bear hunts a ringed seal to feed her cubs.

Polar bears have evolved to rely on the sea ice as a platform for hunting seals. They do most of their hunting over winter and spring, storing energy for summer and autumn. In the southern areas of the polar bears' habitat, such as Hudson Bay in Canada, the sea ice now melts earlier in spring and forms later in autumn, reducing the time available to hunt. Previously, polar bears would have rested for the short period of time when there is no sea ice, but now their time on land is extended. For every week the ice breaks up earlier, the bears can be about 10 kilograms lighter and in poorer condition. This is leading them to scavenge for food on land and can even lead to starvation. Habitat loss due to climate change is the biggest threat to the survival of the polar bear.

Many Australian species are also experiencing habitat loss due to climate change. There are thought to be fewer than 2000 mountain pygmy possums left in the wild, living in small populations in the **alpine** regions of northern Victoria and southern New South Wales. Their primary habitat is boulder fields – the rocks provide shelter from the extreme winter cold and the summer heat.

Mountain pygmy possums are one of the few Australian mammals that hibernate during winter when food is scarce. Snow in winter provides an insulating layer on top of the boulder fields, preventing the temperatures from dropping below freezing and keeping the mountain pygmy possums warm. If there is no snow cover, the unprotected animals can freeze to death. Changing snowfall patterns and early snowmelt are contributing to population decline, as the mountain pygmy possums' habitat is reduced to smaller and smaller areas that receive the required snow cover over winter.

Figure 5.29: Mountain pygmy possums live in alpine boulder fields. They rely on winter snow cover to provide insulation for their habitat.

Reduced precipitation will result in shorter ski seasons

Warmer winters and reduced precipitation will result in less snowfall in the alpine regions of southern New South Wales, Victoria and Tasmania. This will affect the fragile alpine ecosystems that rely on the snow cover and the waterways that are fed by the spring snowmelt. It will also affect the ski season, an important contributor to the economies of these regions.

Scientific modelling has shown that the length of time when temperatures are low enough for snow to fall and stay on the ground is decreasing. A ski resort needs to be able to maintain about 30 centimetres of snow on the ground of the ski runs, but the depth of natural snowfall has been declining since the 1950s. To continue operating, ski resorts will need to rely more on expensive snowmaking activities to supplement natural snowfall. However, this process still requires low temperatures. If greenhouse gas emissions continue to increase at their current rate, the length of the ski season will halve by 2050 – from an average of 110 days to 55 days. Precipitation will reduce by 4–25 per cent over that time.

Snowfall will be restricted to higher altitudes, where it is colder. This will significantly change the alpine ecosystems at lower altitudes and reduce these important habitats. The winter snowfall also contributes a significant 29 per cent of the water flowing into the Murray–Darling Basin. Less precipitation in the alpine regions will have a significant impact on ecosystems downstream.

Figure 5.30: Sparse snow cover at Mount Hotham, Victoria

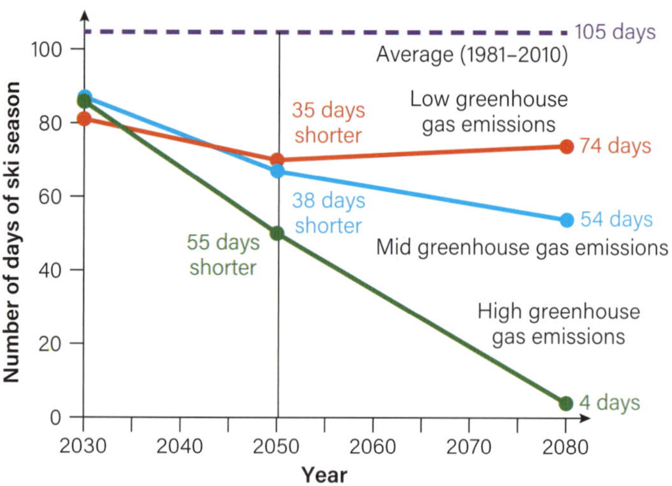

Figure 5.31: Projected trends in the length of ski seasons, based on the amount of greenhouse gas emissions

Low-lying islands will disappear

Many Pacific Island nations consist of low-lying islands or **coral atolls** that are already experiencing adverse effects of climate change.

- By 2030, sea levels in the Pacific will rise 0.09–0.18 metres compared to 1986–2005 levels. Sea levels may rise by 0.20–0.36 metres by 2050 and by 0.33–0.63 metres by 2070.
- Pacific Islands will experience more hot days, and an increase in sea temperatures, which will affect **coral bleaching**.
- Climate change will result in fewer cyclones, but they will be more intense.

Tuvalu is halfway between Australia and Hawaii. It is the second-lowest-lying nation in the world, with most of its nine islands sitting only about 3 metres above sea level. By 2050, much of Tuvalu will be below the high-tide mark, and the islands will become uninhabitable in the next 50–100 years.

Tuvalu has experienced a higher sea-level rise than anywhere else: sea levels have risen more than 5 millimetres a year since 1950, compared to the global average of 1.8 millimetres. During storms, waves crash over the islands, contributing to coastal erosion and damaging crops and infrastructure. The rising sea water has infiltrated **groundwater**, leaving the population reliant on rainfall for fresh water.

Algae from bleaching coral reefs can be toxic for people who eat fish that have ingested the algae (see Section 5.9).

Modelling shows the impacts of different emissions scenarios

Scientific modelling shows different scenarios. It considers what will happen if emissions continue to rise (high), if immediate action is taken to significantly reduce emissions (low) and scenarios in between the two.

Modelling shows that reducing greenhouse gas emissions will have an impact on climate change. If we reduce emissions, we will see some recovery in alpine ecosystems and the ski seasons and less impact on Pacific Island nations.

Figure 5.32: Tuvalu's capital Funafuti is located on Fongafale Island, a coral atoll that sits only about 3 metres above sea level.

Learning Ladder

Climate change

1. Identify a change that can be expected due to climate change in:
   - **a** marine (ocean) ecosystems.
   - **b** Australian alpine ecosystems.
   - **c** Pacific Island nations such as Tuvalu.
   - **d** the Arctic ecosystem.
2. **a** Describe the impact that excess carbon dioxide in the atmosphere is having on the oceans.
   - **b** Describe the effects that rising sea levels are having on Pacific Island nations such as Tuvalu.
   - **c** Describe the impact that climate change will have on Australian alpine ecosystems.
3. Explain the relationship between increasing greenhouse gas emissions and:
   - **a** a decrease in the health of polar bears.
   - **b** the rising sea levels being observed in Tuvalu.
   - **c** shorter ski seasons in Australia.
   - **d** the population decline in mountain pygmy possums.
4. Analyse how the reduction in Arctic sea ice, changes in snowfall patterns in the Australian alpine regions and sea-level rise in the Pacific provide reliable evidence for global climate change.
5. Discuss the impact of reducing greenhouse gas emissions on one of the following.
   - **a** Alpine ecosystems
   - **b** Arctic ecosystems
   - **c** Marine ecosystems
   - **d** Pacific Island ecosystems

Nature and development of science

Satellite data measuring the extent of sea ice in the Arctic is important to quantify the impact of warming oceans on the Arctic ecosystem.

1. Identify one technology that has enabled this data to be gathered.
2. Describe how gathering data over many decades can help to refine scientific knowledge of the impact of warming oceans on sea ice.
3. Explain how wanting to gather more data on Arctic sea ice could lead to the development of new technologies.
4. Conduct research to find out how the scientific community has validated and reviewed studies on Arctic sea ice.
5. Analyse how satellites have enabled advances in science that allow us to better understand Earth's systems.

Evaluating p. 253

Read through Investigation 5.6 on pages 359–60 before answering the questions below.

1. Identify one way that an error could be introduced into the method of this investigation.
2. Describe the impact that the error you identified above could have on the results of the investigation.
3. Propose ways the method could be modified to improve the quality of data that is collected.

Key idea: Stability and change

Construct an argument you could present as part of a class debate on this topic: 'We must change how we use fossil fuels by 2035.'

Success criteria

- I can describe how increased atmospheric carbon dioxide levels lead to ocean acidification.
- I can describe how climate change is causing habitat loss.
- I can discuss how data can be used to evaluate the impact of climate change.

5·7 ▸ Mitigating climate change by reducing greenhouse gas emissions

Learning intention

At the end of this lesson, I will be able to:

- identify strategies to reduce greenhouse gas emissions
- suggest some strategies I can use to reduce my carbon footprint.

Key terms

carbon footprint: a measure of the amount of carbon dioxide emitted as a result of the activities and choices of an individual, organisation or community

carbon sequestration: the process of capturing and storing atmospheric carbon dioxide

energy efficiency: how much usable energy is produced compared to how much energy has been supplied

greenhouse gas emission: the production of a greenhouse gas

landfill: disposal of waste by burying it

net zero emissions: greenhouse gas emissions that are produced are balanced out by those that are absorbed by other processes

Key idea: Stability and change

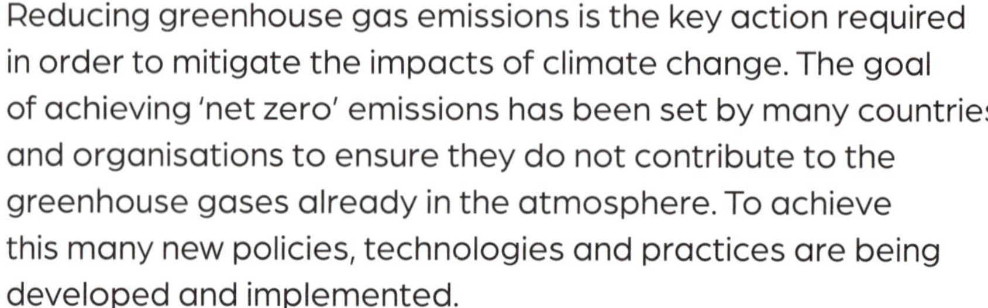

Reducing greenhouse gas emissions is the key action required in order to mitigate the impacts of climate change. The goal of achieving 'net zero' emissions has been set by many countries and organisations to ensure they do not contribute to the greenhouse gases already in the atmosphere. To achieve this many new policies, technologies and practices are being developed and implemented.

The international community has agreed to reduce greenhouse gas emissions

The Intergovernmental Panel on Climate Change (IPCC) is a United Nations group that has stated that reducing emissions of carbon dioxide and other greenhouse gases will reduce the concentration of these gases in the atmosphere and thus reduce global temperature increases.

Governments have agreed to address these issues and to reduce the **greenhouse gas emissions** of their countries. Major agreements include the Kyoto Protocol (1997) and the Paris Agreement (2015).

Australia plans to reduce its emissions to 43 per cent below its 2005 levels by 2030 and to reach **net zero emissions** by 2050.

We are changing how we power ourselves

Electricity to power homes, workplaces, factories and cities contributes the highest amount to Australia's greenhouse gas emissions. Phasing out coal and gas in favour of renewables such as solar and wind, accompanied by storage facilities such as batteries and pumped hydro, will reduce emissions. Switching the manufacturing industry to be fully electric, instead of relying on fossil fuels, will significantly reduce emissions.

Figure 5.33: Installing solar panels and switching to more energy-efficient appliances will reduce greenhouse gas emissions.

Figure 5.34: Generating electricity from renewable sources such as wind instead of fossil fuels will reduce greenhouse gas emissions.

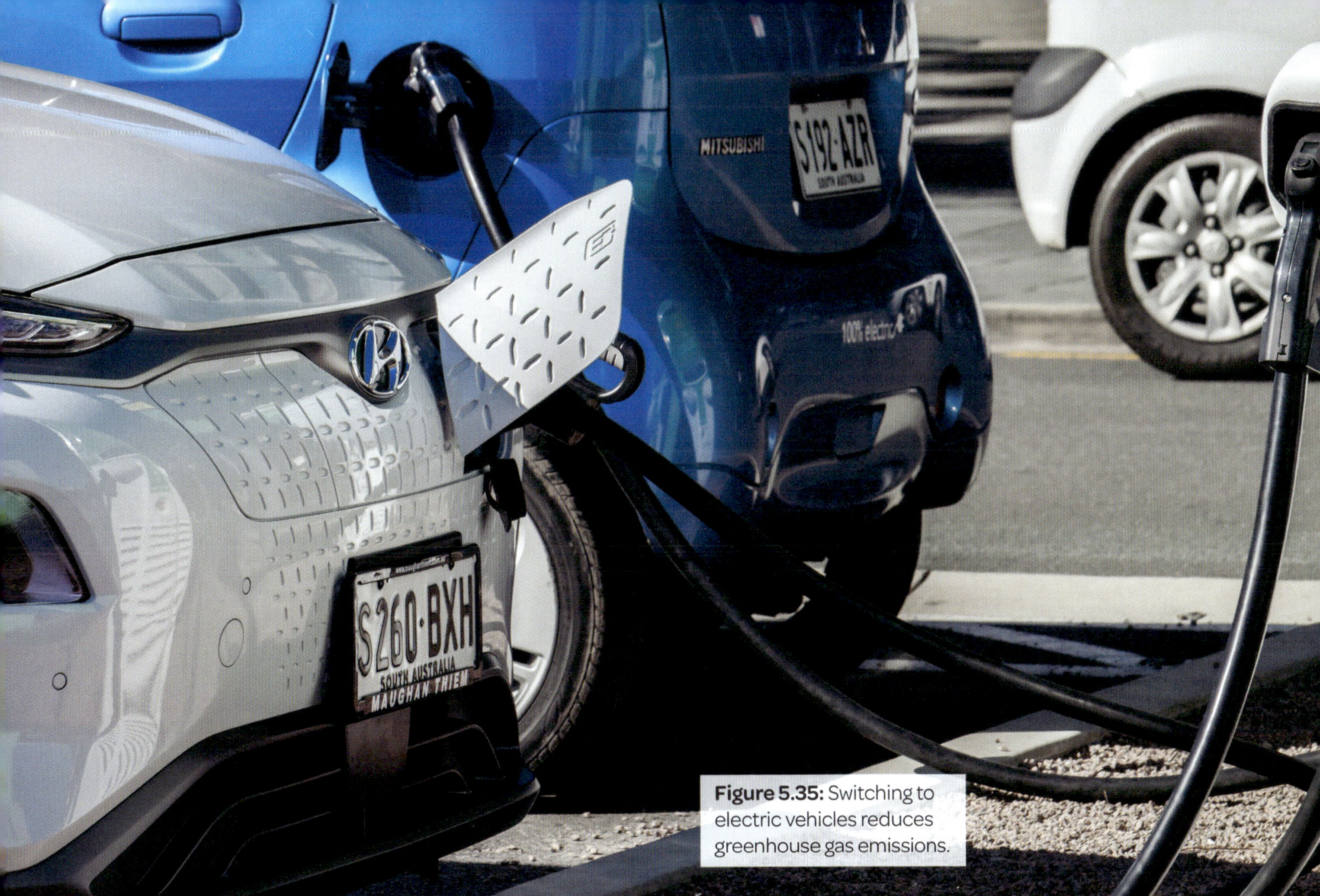

Figure 5.35: Switching to electric vehicles reduces greenhouse gas emissions.

We can reduce our personal greenhouse gas emissions by:

- switching from gas to electrical appliances
- switching to more energy-efficient appliances
- being mindful of energy use, such as switching appliances off at the wall when they are not in use
- installing solar panels and battery storage so that households can run off the energy they produce.

Changing transport will reduce emissions

After power generation and the manufacturing industry, transport is the third-largest source of greenhouse gas emissions in Australia. Switching the transport industry to use rail for freight and using electric vehicles will significantly reduce emissions.

We can reduce our personal transport greenhouse gases by:

- using public or active transport
- switching to hybrid or electric vehicles
- reducing our air travel.

Changing how we use land will reduce emissions

Our land management and agricultural sectors also contribute to greenhouse gas emissions. Clearing native forests removes an important carbon sink, and livestock produce high levels of methane. Changing agricultural practices to encourage the capture and storage of atmospheric carbon dioxide in plants and soils (carbon sequestration), reducing methane production by changing the food provided to livestock, and stopping the clearing of native forests can help to bring the carbon emissions produced by the land-management and agricultural sectors to net zero.

Changing how we manufacture and consume resources will reduce emissions

The manufacturing industry is the second-largest source of greenhouse gas emissions in Australia. Factories often have very high energy needs and rely on fossil fuels for their electricity. The transportation of raw and manufactured materials also contributes to greenhouse gas emissions in the manufacturing process. Continued consumption of manufactured items is unsustainable, not only from an emissions point of view but also because many resources, such as metals, are non-renewable.

Figure 5.36: Australians buy more new clothing annually per person than any other country. A lot of that clothing is fast fashion that is designed not to last. Buying only what we need and shopping in second-hand outlets helps to reduce waste and greenhouse gas emissions.

Australian consumption habits are very wasteful. As of 2024, Australians buy more clothing per person than any other country in the world, with the average person buying 56 new items a year. A significant amount of this is fast fashion – inexpensive clothing that is produced quickly and cheaply as part of a business model that uses novelty and impulse-purchasing, so that consumers buy new clothing more often. This industry has an enormous environmental impact through its consumption of fossil fuels, production of significant greenhouse gas emissions, and use of large amounts of water, chemicals and energy. The clothing produced is often made from materials such as polyester, which is derived from fossil fuels and takes up to 200 years to break down. Despite this, Australians dispose of more than 200 000 tonnes of clothing into landfills each year. By changing our purchasing habits to reduce the amount of new clothing we buy, and reusing and repairing the items we already own, we can reduce both textile waste in landfills and emissions produced from the processing of raw materials. We can reduce our personal greenhouse gas emissions and waste contributions by:

- reducing the amount of clothing we buy
- buying second-hand clothing where possible
- reusing and repairing items to extend their useful life
- avoiding buying clothing made from fossil-fuel–based materials
- recycling items so the material can be used for another purpose.

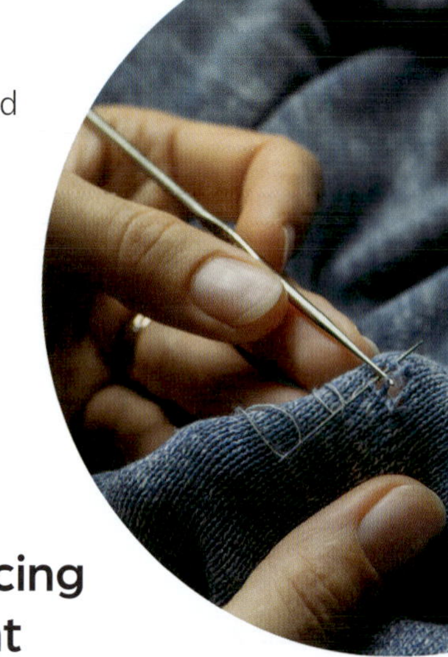

Figure 5.38: Repairing items, instead of replacing them, extends their usable life and reduces our overall consumption of resources.

Strategies for reducing our carbon footprint

Your **carbon footprint** can be thought of as the amount of carbon and other greenhouse gas emissions that are emitted to sustain your lifestyle. Things like housing, transportation, what you eat, what you wear and other items that you buy and use contribute to your carbon footprint. Strategies people use to reduce their carbon footprint include:

- buying locally produced and low-processed food
- opting for smaller and more energy-efficient houses
- choosing low-emission transport options
- selecting appliances and clothing that have long life spans and are repairable and recyclable.

Figure 5.37: The fast fashion industry produces a lot of greenhouse gas emissions through their manufacturing operations, and transport emissions through shipping their products to shops or customers. Business models that encourage consumers to buy lots of cheap clothing result in large amounts of textiles going to landfills, often in developing countries.

Challenges exist in reducing greenhouse gas emissions

Barriers to reducing greenhouse gas emissions are mostly economical. For some countries, fossil fuels are still the cheapest form of energy. Many countries, including Australia, rely heavily on the income from the mining and sale of fossil fuels. New technologies can be expensive and require rare earth metals, the processing of which can cause other environmental damage. While charging infrastructure for electric vehicles is expanding, it is not yet widespread in Australia, creating a barrier to the use of these vehicles and to reducing motor vehicle emissions.

Learning Ladder

Climate change

1 Match each of the activities below with how they reduce greenhouse gas emissions.

Taking public transport	The item can be reused for a longer period of time, rather than being discarded.
Using solar electricity	Greenhouse gas emissions are shared among a larger group of people.
Reforestation projects	Lower transportation needs reduces greenhouse gas emissions.
Eating locally produced food	Electricity production does not result in greenhouse gas emissions.
Repairing an item of clothing instead of replacing it	More trees will take up carbon dioxide from the atmosphere.

2 Describe the effect that not controlling greenhouse gas emissions will have on Earth's climate.

3 Explain one of the impacts on Earth's climate of implementing strategies to reduce greenhouse gas emissions.

4 Analyse the relationship between reforestation projects and a reduction of carbon dioxide in the atmosphere.

5 Assess the effectiveness of the strategy to reduce greenhouse gas emissions by moving all electricity generation to renewable sources such as wind and solar.

Use and influence of science

1 Identify one example of how science can be used to reduce greenhouse gas emissions from:

a energy.
b manufacturing.
c transport.
d agriculture.

2 Describe some examples of how understanding your personal carbon footprint could be interpreted in different ways.

Communicating p. 262

'As I see it, humanity needs to reduce its impact on the Earth urgently and there are three ways to achieve this: we can stop consuming so many resources, we can change our technology, and we can reduce our population. We probably need to do all three.' – Sir David Attenborough, 2009.

1 Imagine you are going to elaborate on the quote above. Identify the three main ideas that you would need to communicate.

2 Suggest one format that you could use to communicate the ideas in Question 1 to your:

a school community.
b parents and their friends.
c local government.

3 Select one of your options from Question 2. Identify a way that you could use digital technologies to enhance your communication of the information.

4 Conduct further research to find scientific evidence to support your answers to Question 1. Use your research to write a short explanation as to how each main idea contributes to reducing humanity's impact on Earth.

5 Use your answers to Questions 1–4 to produce a presentation that justifies Sir David Attenborough's recommendations.

Key idea: Stability and change

'Reduce, reuse and recycle' is a common catchphrase. Predict and explain which of these is the most effective change you could make in your daily life to reduce your greenhouse gas emissions. Evaluate whether this change is realistic.

Success criteria

- I can identify strategies used to reduce greenhouse gas emissions.
- I can suggest some strategies I can use to reduce my carbon footprint.

5·8 ► How can caring for Country support climate change mitigation?

Learning intention

At the end of this lesson, I will be able to explain how ways of caring for Country can support climate change mitigation.

Key terms

carbon farming: where carbon storage in soil and vegetation is maximised

cool burning: a traditional land management practice that involves the controlled use of low-intensity fire

land justice: recognising First Nations Peoples' rights to land

mosaic burning: cool burns conducted in patterns to protect habitat and biodiversity

Key idea: Stability and change

Australia is already feeling the effects of global warming. Efforts to respond to and mitigate the effects of climate change in Australia are increasingly drawing on knowledge and practices developed by First Nations scientists and communities over tens of thousands of years. By recognising First Nations Peoples' rights to land, removing barriers to access and ownership (**land justice**), and centring First Nations methods of caring for Country, we can contribute to a healthy and sustainable future for our planet.

First Nations science and caring for Country

First Nations science has always worked with natural cycles to shape and maintain the landscape. Around 21 000 years ago, during the last Ice Age, First Nations people adapted to the changing climate as parts of the continent became uninhabitable. This long history of adaptation and resilience provides unique expertise and insights into how we can respond to climate change in the present day.

First Nations scientists use a variety of methods to share their results and scientific findings across generations. Stories, songs and dance are ways that science has been effectively practised, taught and learnt for tens of thousands of years. First Nations scientists have a comprehensive understanding of Country, their place in it and how it should be managed.

Figure 5.39: A cool burn in the Kimberley region of north-west Australia. The flames have not burned the tree canopy, and patches have been left to allow animals to escape the flames.

Indigenous rangers work with natural cycles to manage the landscape

One way that caring for Country addresses the impacts of climate change is through Indigenous ranger programs. These programs support the environment by improving biodiversity and soil health and by managing introduced species.

A focus of these programs is how rangers care for Country using fire. While fire does release carbon into the atmosphere, it is still an effective tool in addressing climate change. First Nations Peoples utilise **cool burning** – a traditional land management practice that involves the controlled use of low-intensity fire. Small fires are lit at specific times of the year in carefully chosen locations to reduce fuel load, protect plants and animals, and regenerate habitat. If fuel loads are not managed, uncontrolled bushfires can start which emit large amounts of carbon into the atmosphere.

Cool burns are conducted in patterns, with rangers burning only designated areas at any one time. This allows for the creation of safe zones for birds and animals and ensures that food sources remain available. This is known as **mosaic burning**.

Careful use of fire also protects soil health. Uncontrolled bushfires destroy not just flora and fauna, but also all the microorganisms in the soil and its capacity to continue to hold and absorb carbon. The careful use of fire in land management ensures that when bushfires do inevitably develop, their impact is less, and smaller amounts of carbon are released into the atmosphere.

Figure 5.40: A cool burn in Kakadu National Park, Northern Territory. These small, controlled fires reduce the risk of destructive high-intensity bushfires that release large quantities of carbon into the atmosphere.

Indigenous carbon farming stores carbon in natural carbon sinks

Carbon farming is another way to care for Country and to address climate change. **Carbon farming** involves maximising vegetation and wetland areas, such as mangroves, to absorb carbon naturally. (Carbon sequestration is the storage of carbon in carbon pools such as plants and soil.) An example of this practice is where land that was previously cleared for agriculture or mining is repopulated with native plants. As the plants grow, they absorb carbon from the atmosphere. Over time, a fully grown tree can absorb as much as 40 kilograms of carbon from the atmosphere in just one year. In this way, the revegetation and rehabilitation of Country actively reduces the amount of carbon in the atmosphere.

Land justice and caring for Country are part of Australia's climate change response

To care for Country effectively and to contribute to climate change mitigation, First Nations Traditional Owners need greater control over decisions relating to management of Country. Where non-Indigenous stakeholders control access to and ownership of Country, there can be barriers that prevent First Nations stakeholders from being able to care for Country effectively. First Nations science encompasses wellbeing, reciprocity and ceremony, and rests on an intricate knowledge of natural cycles that has been passed down over millennia. To address climate change in Australia, it will be important to address issues of land justice. Without land justice, First Nations science will be limited in its impact.

Figure 5.41: Eucalyptus trees are a fire-resilient species that often sprout highly nutritious shoots directly from the trunk after being scorched by a fire.

Figure 5.42: Eucalyptus shoots are an important food source for animals returning to their habitat. Very fierce fires kill the trees, which impacts animals and the wider biome.

Learning Ladder

Climate change

1. Identify two ways that caring for Country can reduce and remove carbon from the atmosphere.
2. Propose and describe a trend that might be observed if carbon farming were practised globally.
3. Discuss how the colonisation of Australia has led to more severe bushfires and greater carbon emissions.

Use and influence of science

1. Identify a socio-scientific issue related to cool burning.
2. Describe how First Nations Peoples incorporate song and dance into their scientific approaches.
3. Propose and explain how land justice could support climate change mitigation through the application of First Nations science.
4. Discuss how lighting fires as part of caring for Country can help to reduce carbon emissions, even though it emits carbon initially.

Communicating

p. 262

1. Imagine you are preparing a talk for primary school–aged students about cool burning. List and define four simple vocabulary words you could use to inform students about the topic.
2. Brainstorm and describe three different ways you could communicate your information from Question 1.
3. Select one of the formats from Question 2 and prepare a digital slide show to present the information.
4. Prepare an infographic on mosaic burning techniques that illustrates how they support climate change mitigation through caring for Country. (*Note:* This may require additional research.)
5. Prepare a persuasive essay to justify why land justice and the application of First Nations scientific knowledge could be an important factor in climate change mitigation.

Key idea: Stability and change

Research an example of a climate change mitigation project that is linked to First Nations scientific knowledge, such as the West Arnhem Land Fire Abatement Project or the regenerative grazing project in Lutruwita/Tasmania. Discuss how these initiatives use traditional knowledge to respond to areas affected by climate change.

Success criteria

- I can identify and describe strategies used by First Nations Peoples to care for Country.
- I can explain how First Nations strategies for caring for Country contribute to climate change mitigation.

5·9 ▸ Key idea: Stability and change – the Great Barrier Reef

Learning intention

At the end of this lesson, I will be able to explain how human activity is changing the Great Barrier Reef.

Key terms

symbiotic relationship: an ecological relationship where both parties cannot survive without the other

zooxanthellae: algae that live in coral tissues

Key idea: Stability and change

▲ **Figure 5.43:** The Great Barrier Reef is the world's largest coral reef system.

Coral reefs occur in shallow tropical seas and are among the most diverse ecosystems on Earth. The vast structures built by coral polyps provide homes to thousands of species, many of which are important food sources for humans. The Great Barrier Reef is the world's largest coral reef system. Situated off the coast of Queensland, it extends more than 2000 kilometres from K'gari (Fraser Island) in the south to the Torres Strait in the north. Over recent decades, the Great Barrier Reef has experienced significant damage due to human activity and climate change.

High sea temperatures cause coral bleaching

Coral reefs are calcium carbonate structures built by coral polyps that live in a **symbiotic relationship** with a type of algae called **zooxanthellae**. The algae photosynthesise, providing the coral with glucose in exchange for a place to live. They provide the coral with about 90 per cent of its energy.

Coral bleaching occurs when the coral becomes stressed and expels the algae, leaving the white skeleton behind. The leading cause of coral bleaching is sea temperature rise. Corals grow best at sea temperatures of 23–29 °C, and an increase of about 1 °C over approximately 4 weeks is enough to cause the corals to bleach. If the temperatures quickly return to normal, the coral can recover, but if temperatures stay high for 8 weeks or more, the coral will die.

The Great Barrier Reef first experienced coral bleaching due to sea temperature rise in 1998. It has experienced six bleaching events since – in 2002, 2016, 2017, 2020, 2022 and 2024.

Continuing high temperatures will cause many coral species to disappear from the reef. Algae from bleaching coral reefs is toxic for people who eat fish that have ingested the algae.

Figure 5.44: The Great Barrier Reef is home to thousands of species that depend on the coral for their survival.

▼ **Figure 5.45:** Coral bleaching occurs when coral becomes stressed and expels its symbiotic algae.

Healthy

Algae live inside the coral.

They depend on each other to survive.

Algae give the coral its colour.

Stressed

If the coral gets stressed by rising temperatures or pollution levels, the algae leave the coral.

Bleached

Without the algae, the coral does not receive enough food. It turns white and is more likely to become diseased and to die.

Acidification of oceans is harming coral

Increased levels of carbon dioxide dissolving in the water are making the oceans more acidic. The increased levels of carbon dioxide in the atmosphere are a result of burning fossil fuels.

Coral skeletons are made of calcium carbonate. When carbon dioxide is added to ocean water, carbonate ions bond with excess hydrogen ions instead of with calcium ions. This means that less calcification occurs, and under these conditions corals cannot make their skeletons. If the water becomes too acidic, the skeletons will dissolve.

Figure 5.46: Cyclone Yasi caused extensive damage to the Great Barrier Reef in 2011.

Cyclones damage reefs

Storm surges damage beaches, dunes and urban areas. Floodwaters pick up topsoil and transport it into the oceans.

The intense wind and wave action of a cyclone can kill organisms and damage habitats both on land and in the ocean. Corals are damaged by high-energy waves, and further damage occurs when sediments from the land are washed into the ocean near the reef.

In 2011, Cyclone Yasi – one of the most powerful cyclones to affect Australia in recorded history – caused significant damage to the Great Barrier Reef. Wind gusts of 290 kilometres per hour were reached and about 300 kilometres of the reef was affected to varying degrees. When coral reefs are harmed or destroyed, it can have significant consequences for coastal communities. They become more vulnerable to storms and even to the regular force of waves. Recovery can take decades, and the reef's long-term survival requires ongoing scientific monitoring and conservation measures.

Rising sea levels will cause coral reefs to disappear

Sea levels are important for the distribution of species and habitats in oceans. Many marine organisms have adapted to survive at a certain depth. Coral reefs require shallow water so that enough light can penetrate for photosynthesis to occur. If sea levels rise very slowly, corals will grow towards the surface and the reefs will migrate into shallower water over time.

Since 1991, sea levels have risen about 3 millimetres per year on the Great Barrier Reef. Currently, this small change is not affecting the reefs, as the corals can grow more quickly than this. However, sea levels are predicted to begin to rise much faster and the corals will not be able to keep up. Many reef communities will disappear because they will be unable to reestablish themselves quickly enough.

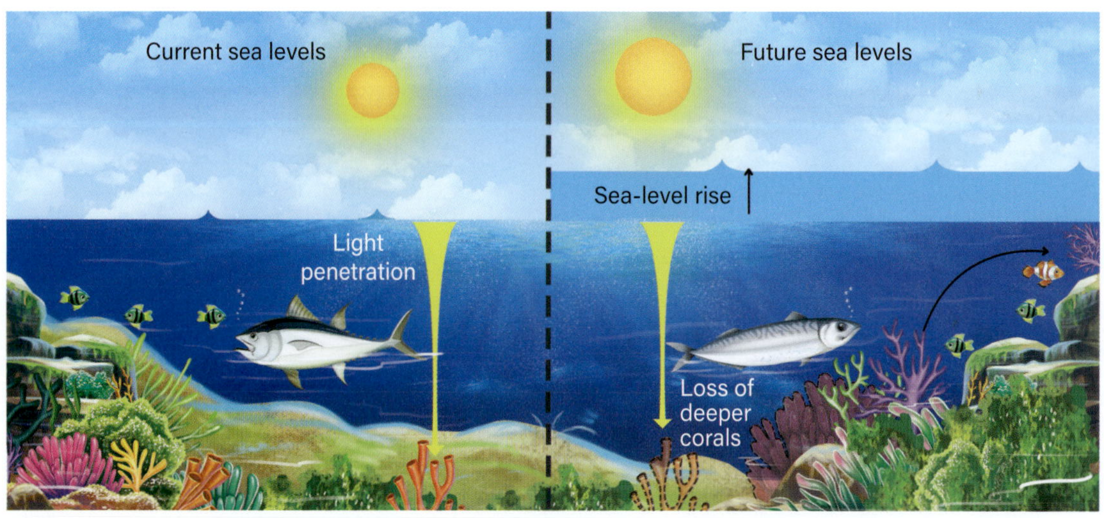

Figure 5.47: When sea levels rise, corals in deeper water cannot survive because they do not receive sufficient sunlight. Corals might establish and grow in shallower waters, but this may not happen quickly enough.

Pollution damages reefs

A growing human population along the Queensland coast is producing more pollution that is damaging the Great Barrier Reef. Development along coastlines is often accompanied by clearing of forests, including mangrove habitats. This clearing has increased the amount of sediment that runs into the ocean during large rain events. Sediments make the water cloudy, blocking sunlight, and settle on and cover the coral. Nutrient run-off from farms encourages algal growth that can smother coral and seagrass beds. Rubbish such as plastic, and fuel from tourists' boats, are also having an impact on the reef.

Figure 5.48: Sediment is carried out of the Burdekin River south of Townsville after a flood event in 2010.

Learning Ladder

Climate change

1. Identify the components of the Great Barrier Reef that can be considered to be part of Earth's:
   a biosphere. **b** hydrosphere.
   c lithosphere. **d** atmosphere.
2. Describe the location and climate where coral reefs can be found on Earth.
3. Explain the cause of the high sea temperatures that have resulted in coral bleaching.
4. Analyse how increased greenhouse gases in the atmosphere will influence the interactions between Earth's systems, leading to a rise in sea levels.
5. Discuss how taking no action to reduce human greenhouse gas emissions would impact the world's coral reefs.

Use and influence of science

1. Marine biologists are undertaking research to develop corals that have a higher heat tolerance. Brainstorm a list of pros and cons for the introduction of such corals on the Great Barrier Reef.
2. Construct a counterargument for at least one statement from your responses for Question 1.
3. Propose why it might be important for marine biologists to research and develop techniques that could be used to protect the Great Barrier Reef.
4. Discuss how public education campaigns can help to support efforts to protect the Great Barrier Reef.
5. 'No one will protect what they don't care about; and no one will care about what they have never experienced.' – Sir David Attenborough. Evaluate this statement.

   In your response, make reference to the key factors that contribute to scientific knowledge being adopted more broadly by society.

Communicating p. 262

Select one of the impacts that humans are having on the Great Barrier Reef. Conduct further research on the impact you selected. You will present your findings to your peers, using appropriate technologies.

1. **a** Use your research to write dot points of key pieces of information.
   b Identify and define any key terms you will need to use in your presentation.
   c Brainstorm some formats that you could use to present your findings.
2. Select one of the formats from Question 1c and develop a draft.
3. Brainstorm how you could use digital technologies to assist you in presenting your findings.
4. Incorporate any appropriate technologies you identified in Question 3 as you complete your final product.
5. Evaluate your final product with regards to its effectiveness in communicating your findings. Can you improve it so that you can better justify the statements you made?

Success criteria

- I can describe how increased sea temperature leads to coral bleaching.
- I can describe how sea-level rise will affect the Great Barrier Reef.
- I can describe the impacts of pollution, cyclones and ocean acidification on the Great Barrier Reef.

Summary

- Solar radiation drives Earth's climate system.
- Incoming solar radiation is balanced with energy that is being reflected, absorbed and radiated through various processes.
- If the amount of solar radiation coming in is higher than that going out, then the planet will be warmer. If the amount of radiation coming in is lower than that going out, then the planet will be cooler.

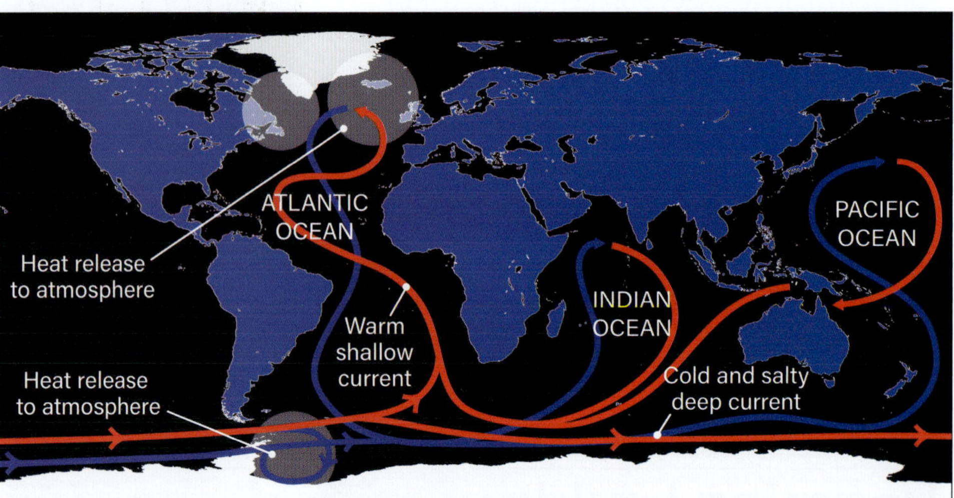

- Ocean currents redistribute heat from the equator around the oceans. This redistribution is driven by thermohaline circulation.
- Circulation in the atmosphere works to balance the difference in temperature between the equator and the poles.

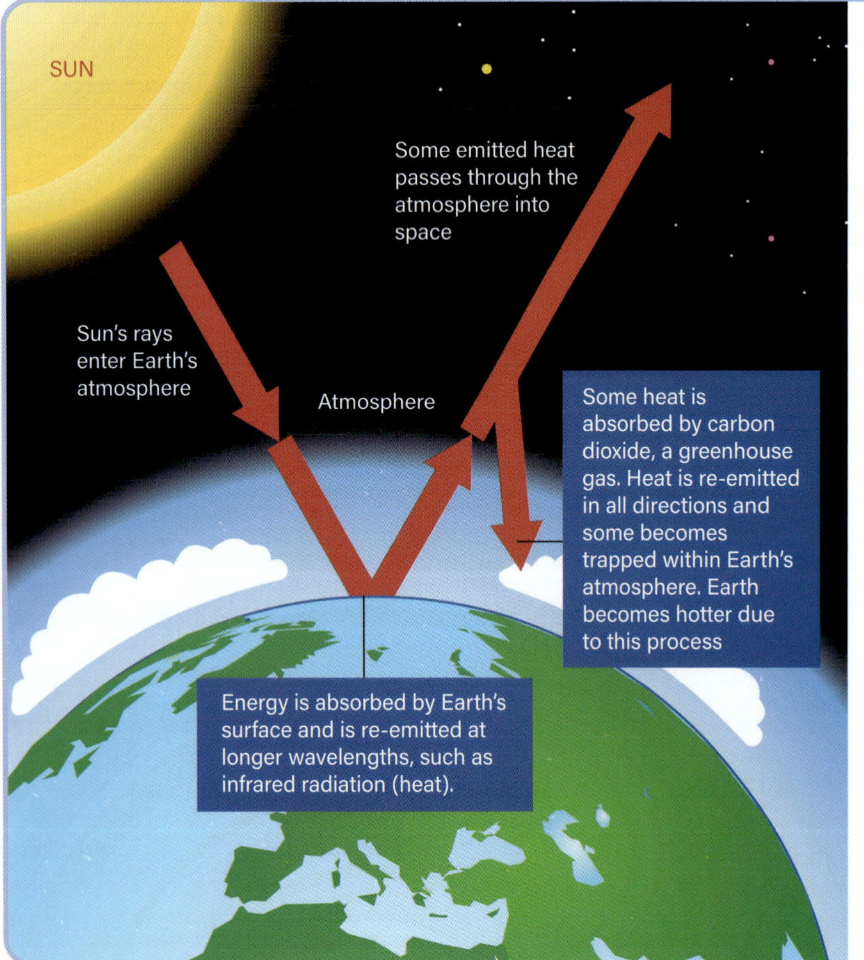

- Greenhouse gases, such as carbon dioxide, water vapour and methane, trap radiation energy produced by the Sun to keep Earth's surface warm. This is called the greenhouse effect.
- Human activity that adds additional greenhouse gases into the atmosphere, such as by the burning of fossil fuels, has enhanced the greenhouse effect, resulting in an increase in atmospheric temperatures.
- Data from ice cores and atmospheric sampling shows that the concentration of carbon dioxide in the atmosphere is increasing rapidly compared to the past.

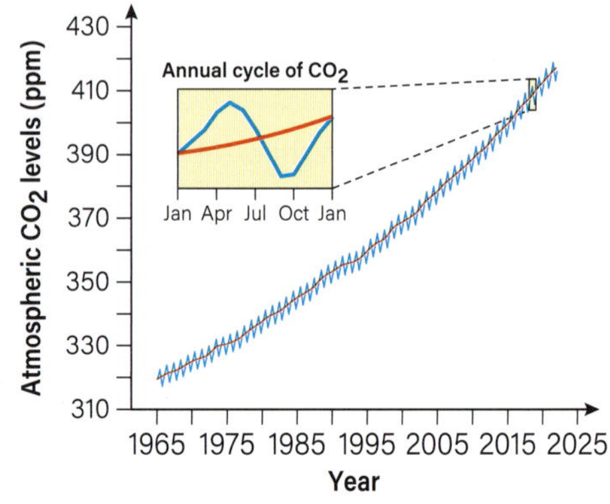

- Strategies to reduce greenhouse gas emissions include:
  - generating electricity from renewable sources
  - switching to more energy-efficient appliances
  - switching to hybrid or electric vehicles
  - changing agricultural practices to reduce methane production by livestock.
- Challenges in reducing greenhouse gas emissions:
  - Fossil fuels are currently the cheapest form of energy.
  - The mining of fossil fuels contributes considerable amounts of money to the economy.
  - Developing new technologies is expensive.
  - Processes for building new technologies can damage the environment.

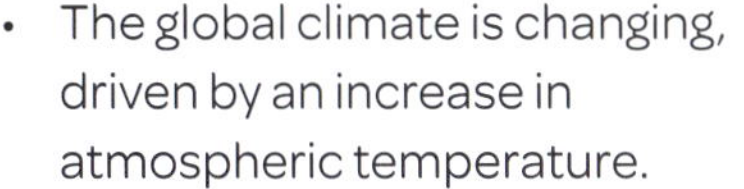

- The global climate is changing, driven by an increase in atmospheric temperature.
- Long-term data shows that Earth's average temperature has increased since 1880.
- This rise in temperature is influencing weather patterns, sea levels, ocean temperatures and the presence of ice at the poles.

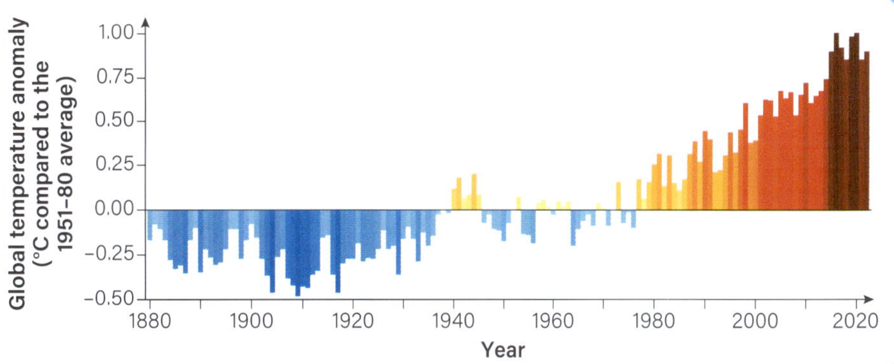

Effects of climate change include:

- increased evaporation and precipitation
- increased acidification of the oceans, resulting in bleaching of coral reefs
- more frequent heatwaves and bushfires
- reduced presence of ice at the poles, contributing to rising sea levels
- the loss of habitats (e.g. from reduced snow cover of sea ice in the Arctic) and reduced biodiversity.

Masterclass

Steps in progression

		1	2
Science understanding	**Climate change**	Define the: • atmosphere. • hydrosphere. • lithosphere. • biosphere.	Describe the trend in Earth's climate as you move from the poles to the equator.
Science as a human endeavour	**Nature and development of science**	Propose an example of a technology that has enabled advances in Antarctic scientific knowledge.	Describe how scientific research in Antarctica has changed over time.
Science as a human endeavour	**Use and influence of science**	Identify one example of scientific knowledge being used to mitigate climate change impacts.	Describe an example of how climate science could be interpreted in different ways.
Science inquiry	**Evaluating** Select one of the investigations in this chapter	Identify any errors present in the investigation.	Select one error you identified in step 1. Describe the impact it would have on the results and propose one way to reduce it.
Science inquiry	**Communicating** Select a topic from this chapter	Brainstorm some ways you could communicate your knowledge to your teacher. *Hint:* Revisit the Learning Ladder at the start of the chapter.	Select one of the formats you identified in step 1 and create a draft.

Figure 5.49: Antarctica and the Southern Ocean play important roles in Earth's climate system.

Researching climate change in Antarctica

Antarctica and the Southern Ocean play key roles in Earth's climate system. The vast Antarctic ice sheet reflects solar energy back into space, keeping the Earth cool. The circulation of the Southern Ocean around Antarctica (the Antarctic Circum Polar Current) helps to drive the global ocean conveyor belt as it mixes warmer surface waters with deep, cold, dense water, while also connecting the Pacific, Indian and Atlantic oceans.

Demonstrate your understanding

3	4	5	
Explain how polar ice contributes to Earth's climate system.	Analyse how an increase in greenhouse gas emissions can impact polar ice caps and ocean currents.	Identify and evaluate one strategy for mitigating climate change. Analyse how it would address the changes observed in Antarctica.	
Explain how the needs of scientific researchers in Antarctica have led to developments in technology and engineering.	Conduct research and discuss the importance of peer review in validating scientific findings on climate change.	Analyse how advances in technologies, such as the development of floating laboratories like RSV *Nuyina*, have enabled advances in climate science.	
Explain how the values and needs of society have influenced climate science research.	Discuss how scientific information and misinformation about climate science can influence personal decision-making.	Identify and analyse at least one factor that has led climate science and knowledge of climate change to be adopted more broadly by society.	
Consider the investigation as a whole. Discuss how the quality of the data, and therefore the validity of the conclusion, can be improved.	Write a discussion for your investigation that includes evidence and scientific knowledge to support your findings.	Use the data you gathered in the investigation to evaluate the validity and reproducibility of the investigation.	Science how-to p. 253
Brainstorm how you could use digital technologies to enhance your presentation.	Refine your draft from step 2, incorporating any digital technologies you identified where appropriate.	Write a short argument to justify the importance of understanding the cause and impacts of climate change.	Science how-to p. 262

In recent years, scientists have observed rapid changes in Antarctica and the Southern Ocean. These changes include:

- warming of surface and deep waters
- faster moving and melting of ice shelves (where glaciers meet the sea)
- reduction of sea ice forming during winter
- declines in populations of organisms such as krill, penguin and seals.

In the past, much of the scientific work conducted in Antarctica was done in the name of exploration. Today, much of the scientific research that happens on the continent is driven by the need to gain a deeper understanding of how the region contributes to Earth's climate system and how a warming Earth may change it.

At the same time, this need for data has driven many technological advances, as well as the adaptation of technologies for scientific purposes. RSV (Research Survey Vessel) *Nuyina*, launched in 2021, provides transportation of people and supplies to Australian research bases in Antarctica. In addition, it contains equipment and laboratories to allow scientists on board to collect samples and gather data to further our understanding of Antarctica and the Southern Ocean, including the properties of ocean water, the ocean floor, the atmosphere, and ecological communities in the water column (layers from surface to seafloor).

Figure 5.50: Scientists conducting research in Antarctica

6.0 The universe

Humans have gazed up at the night sky for millennia. Our ancestors observed that the sky changed with the seasons. They used the stars to help them navigate and linked their observations to stories of creation and mythology. We still look to the stars to help us understand our place in the universe, how it began and evolved, and how our Sun and its solar system formed. Humans have begun to travel further from our planet and its moon. We send spacecraft to the outer reaches of our solar system. As technology improves, scientists unlock and understand more about the mysterious universe and strive to explore it further than ever before.

Learning Ladder

The Learning Ladder for each chapter maps the Science Understanding, Science as a Human Endeavour and Science Inquiry strands that will be covered. Each ladder has five levels of progression, called steps. To climb the ladders, you need to develop fluency at each step. This will help you develop the ability to complete tasks that are more complex.

Steps in progression	Science understanding: Earth and space science: The universe	Science as a human endeavour: Nature and development of science	Science as a human endeavour: Use and influence of science
5	I can evaluate models and techniques used to explain and explore the origin of the universe	I can analyse how advances in technologies enable advances in science	I can analyse the key factors that contribute to scientific knowledge being adopted more broadly by society
4	I can discuss the advantages, disadvantages and impacts of space exploration	I can discuss how scientific knowledge is validated, including the role of publication and peer review	I can discuss how scientific information and misinformation may inform personal and social decision-making
3	I can describe the scientific evidence for the big bang theory, and methods used in space exploration	I can explain how science has contributed to developments in technologies and engineering	I can explain how the values and needs of society influence the focus of scientific research
2	I can sequence key events in the origin and evolution of the universe and space exploration	I can describe how scientific knowledge is refined over time	I can describe scientific knowledge that may be interpreted in different ways
1	I can identify features of the universe and technologies used for space exploration	I can identify technologies that have enabled advances in science	I can identify scientific knowledge that can address socio-scientific issues

Figure 6.1: Located 161 000 light-years from Earth, the Tarantula Nebula is part of the Large Magellanic Cloud, a satellite galaxy of the Milky Way. The nebula is the most active star-forming area in the Local Group of galaxies and gets its name from its hollow appearance, which resembles the silk-lined burrow of a spider.

Questioning and predicting	Communicating	Steps in progression
I can develop explanatory models when investigating scientific questions, problems and claims	I can justify scientific ideas, findings, arguments and proposals for diverse audiences	5
I can discuss what is needed for a question to be investigable or a prediction to be reasoned	I can communicate scientific findings and arguments effectively for specific purposes to specific audiences	4
I can develop a hypothesis that predicts the relationship between investigation variables	I can use digital technologies and/or scientific representations to communicate data and information	3
I can formulate questions to investigate scientific problems	I can prepare a variety of presentation formats to communicate ideas and findings	2
I can make simple predictions based on what I know and observe	I can select appropriate formats, content and vocabulary to communicate scientific ideas and findings	1

Science inquiry

6·1 ► The big bang theory

Figure 6.2: Technological developments have expanded our observations and theories about the origins of the universe.

Learning intention

At the end of this lesson, I will be able to use scientific evidence to outline how the big bang theory explains the origins of the universe and its age.

Key terms

antimatter: particles that have properties opposite to that of normal matter

astrophysicist: a scientist who studies the physics of the universe

big bang theory: the current accepted model of the beginning of the universe, where the universe rapidly expanded from a single dense, extremely hot region

cosmologist: a scientist who studies the origins and structure of the universe

redshift: a change in the wavelength of light towards the red end of the visible spectrum

singularity: an infinitely dense point of matter that existed before the big bang

Investigation 6.1

Modelling the expanding universe, p. 361

Key idea: Scale and measurement

Improvements in technology have led to more accurate and extensive observations and calculations about the universe. This new information has increased our understanding of the universe and how it formed.

The universe began as a singularity

The **big bang theory** states that the universe began about 13.8 billion years ago as a small, dense region called a **singularity**. This singularity expanded rapidly, and the expansion is referred to as the 'big bang'. During the expansion, matter and elements were formed, which then became the basis for the creation of stars and galaxies.

After the big bang, things initially happened very quickly.

- As the universe continued to expand, it started to cool, allowing the formation of particles of matter and **antimatter** (particles that are the opposite of matter).
- At about 0.001 seconds after the big bang, these opposite particles annihilated each other, releasing energy. The matter that we observe today is what is left over from this interaction.
- From three minutes after the big bang, protons and neutrons combined in nuclear reactions to form the nuclei of atoms (about 75 per cent hydrogen, 25 per cent helium and a small amount of lithium). At this stage, all the matter and energy that would *ever* exist in the whole universe was formed – and all in the time it takes to boil a jug of water!
- Between the formation of these nuclei and 380 000 years, the universe was a plasma of hydrogen, helium and lithium nuclei and free electrons (electrons not attached to atoms).
- At about 380 000 years, the universe had cooled down enough for the nuclei to attract electrons and to form atoms, which released light energy.
- The first stars began to form after about 200 million years.
- The first galaxies began to form about 400 million years after the big bang.

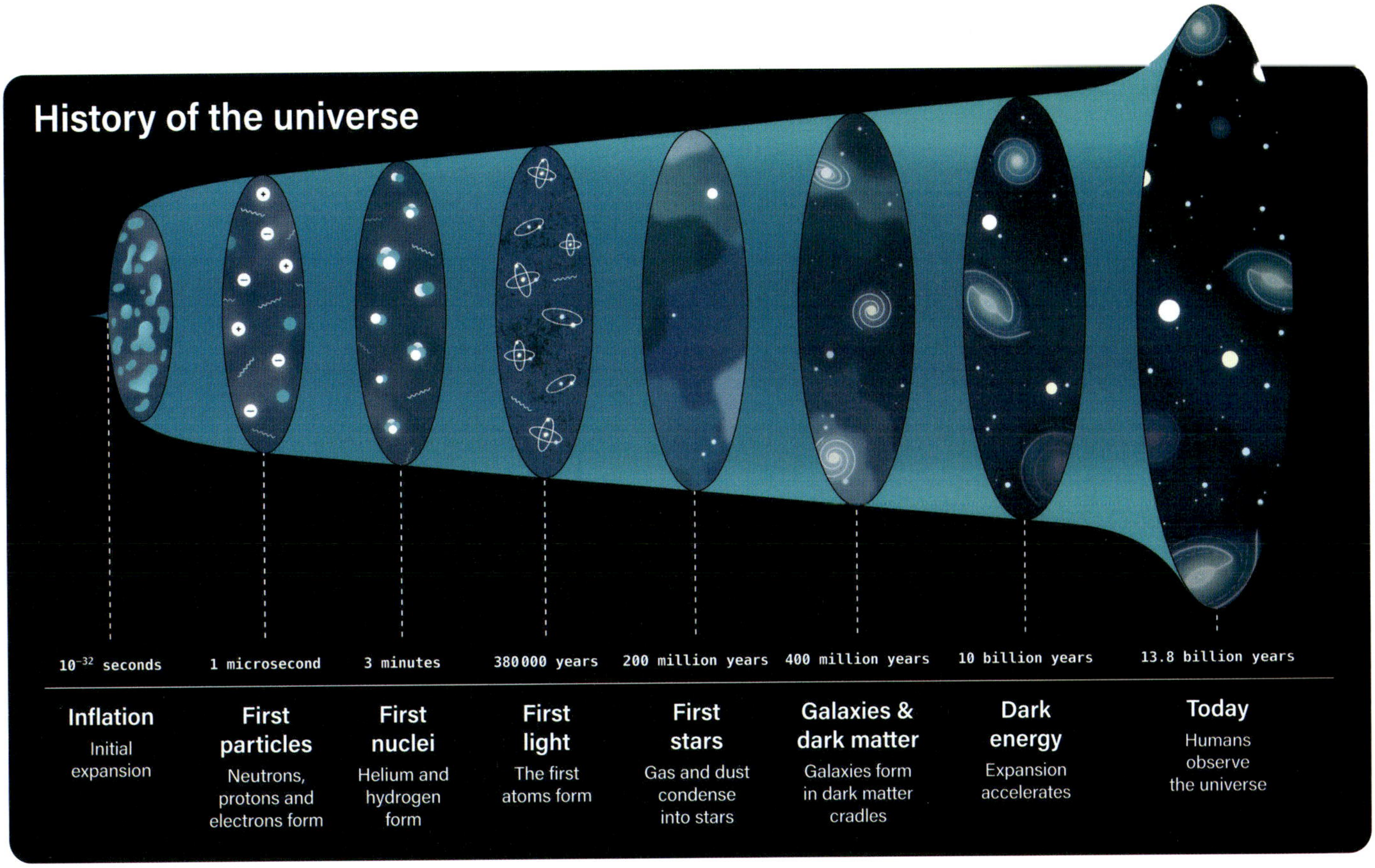

History of the universe

10^{-32} seconds	1 microsecond	3 minutes	380000 years	200 million years	400 million years	10 billion years	13.8 billion years
Inflation	**First particles**	**First nuclei**	**First light**	**First stars**	**Galaxies & dark matter**	**Dark energy**	**Today**
Initial expansion	Neutrons, protons and electrons form	Helium and hydrogen form	The first atoms form	Gas and dust condense into stars	Galaxies form in dark matter cradles	Expansion accelerates	Humans observe the universe

▲
Figure 6.3: The timeline of the big bang, showing events from the beginning of the universe until now

There are three main pieces of evidence for the big bang

The continued expansion of the universe

Edwin Hubble observed that light coming from distant galaxies was stretched into longer wavelengths; that is, it was **redshifted**. This indicates that galaxies are moving away from us and that the universe is expanding. The movements of the galaxies can be traced back to a single point, meaning that the universe must have once been contained in a small region of space.

The abundance of light elements

The universe is about 74 per cent hydrogen and 24 per cent helium by mass, with the other 2 per cent being all the other heavier elements in the periodic table. This abundance supports the big bang theory, because if helium had not been made by the big bang and was made only by fusion in stars, it would account for significantly less than 24 per cent of the universe.

The existence of cosmic microwave background radiation

Cosmic microwave background radiation is leftover heat released about 380 000 years after the big bang, when the universe had cooled enough to allow electromagnetic radiation to pass through it. Because the universe was not of a uniform density when the cosmic microwave background was released, images show 'clumps' of matter. The cosmic microwave background would have originally been released as visible and ultraviolet light, but the expansion of the universe has redshifted it into the microwave band.

Figure 6.4: This image of cosmic microwave background radiation was taken from the Wilkinson Microwave Anisotropy Probe and shows clumps of matter.
▼

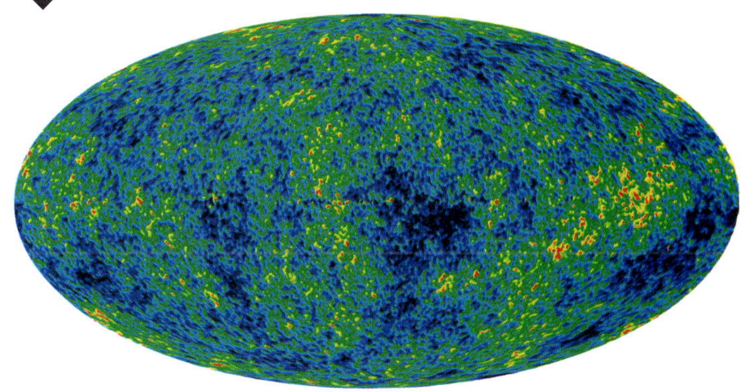

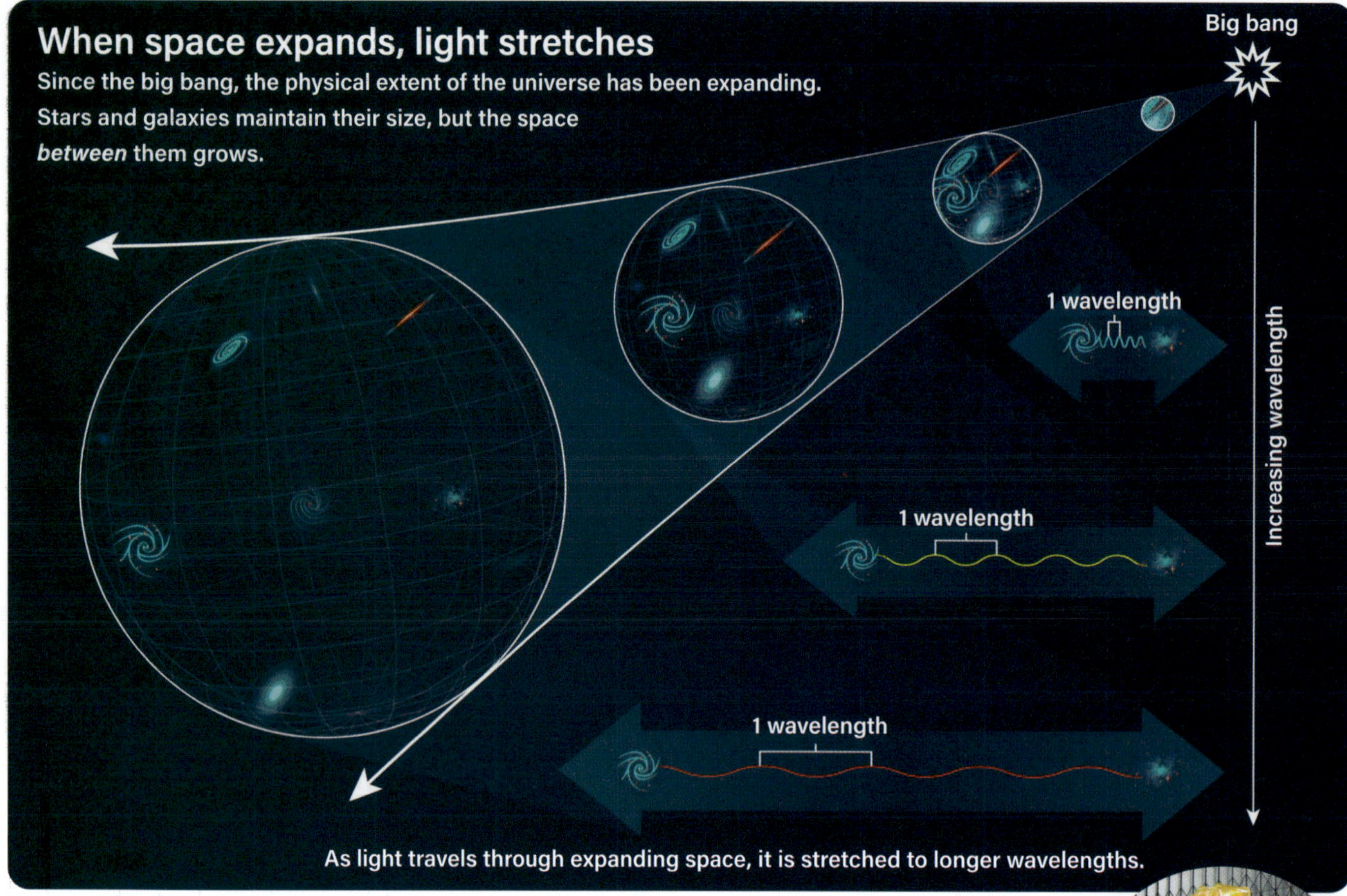

Figure 6.5: Light from distant galaxies is redshifted.

Data from space telescopes helps scientists to calculate the age of the universe

Current observations and calculations put the age of the universe at 13.8 billion years. The European Space Agency's Planck space observatory was used to create a detailed map of the cosmic microwave background radiation. This map allowed **astrophysicists** (scientists who study the physics of the universe) to use Einstein's theory of general relativity to calculate the age of the universe by 'winding the clock backwards'.

Previous estimates were made using the Hubble Space Telescope to measure how far away galaxies were from us and the speed at which they were moving. The telescope was used to calculate the ages of the oldest star clusters – the first stars that would have formed after the big bang. This allowed astrophysicists to calculate how long it took galaxies to reach their current locations and thus to estimate the age of the universe.

Figure 6.6: The Planck space observatory was used to create a detailed map of the cosmic microwave background radiation.

Our universe is expanding at an increasing rate

In 2011, Professor Brian Schmidt from the Australian National University won the Nobel Prize for Physics in conjunction with Adam Riess and Saul Perlmutter. Their research in the late 1990s provided evidence that the universe was expanding. Using new, more highly powerful telescope and computer technology, Schmidt, Riess and their team were working at the same time as Perlmutter and his team on the same problem. They expected to take measurements of supernovae to support the theory that the expansion of the universe was slowing down. Instead, they found (after double checking the calculations, because they thought they had made an error) that the supernovae were accelerating away from their reference point. This finding changed the accepted understanding of the universe and the direction of astrophysics.

It led to the proposal for the existence of dark energy – a force that is causing the acceleration and one that science still does not understand.

Other models of the universe

There have been other models developed of how the universe formed and is evolving. The steady-state theory was proposed in the mid-20th century and was a very popular alternative to the big bang theory. The steady-state theory proposed that the universe was infinitely old, was the same in all directions and had not changed over time. The known expansion of the universe was explained by new matter always being created to fill the space, keeping the density of galaxies in the universe constant. The discovery of the cosmic microwave background radiation helped to disprove this theory.

More recently, cosmologists (scientists who study the origins and structure of the universe) have proposed a multiverse model, where our universe is one of many universes that make up a multiverse. Others have proposed the 'oscillating universe' model, where the universe is in a cycle of expansion and contraction. This theory proposes that eventually the universe will stop expanding and collapse in a 'big crunch', before expanding in a 'big bang' again. Astrophysicists continue to gather evidence to improve our understanding of the universe.

Figure 6.7: Whereas the big bang theory states that the universe is expanding and no new matter is being made, the steady-state theory stated that new matter was constantly being created to fill spaces in the universe.

Big bang theory

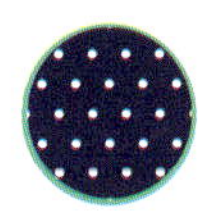

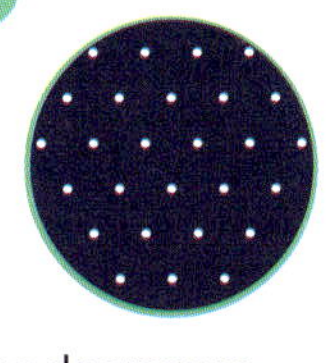

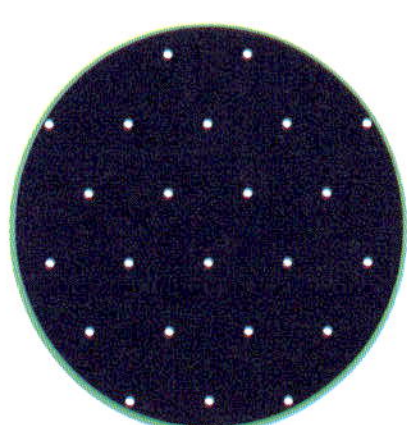

The density of galaxies decreases as the universe expands.

Steady-state model

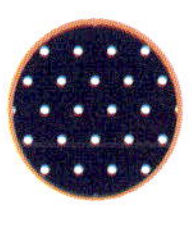

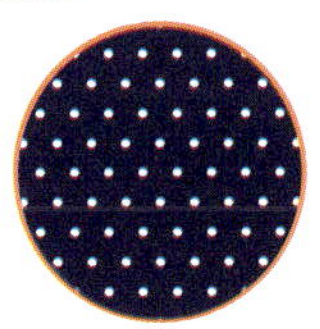

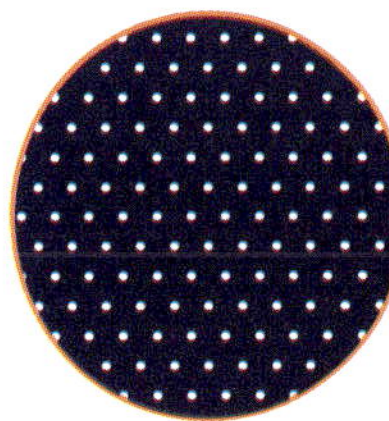

The density of galaxies remains constant as the universe expands. (Spaces are filled by new galaxies.)

Learning Ladder

The universe

1. Identify the correct order of steps for the big bang theory.
   - **A** First particles, singularity, first elements, galaxy formation
   - **B** Singularity, first particles, first elements, galaxy formation
   - **C** Singularity, first elements, first particles, galaxy formation
   - **D** First elements, first particles, galaxy formation, singularity
2. Our Sun is about 4.6 billion years old. Calculate the age of the universe when the Sun first formed.
3. Summarise in your own words how the following provide evidence that supports the big bang theory.
   - **a** Redshifted light from galaxies
   - **b** Cosmic microwave background radiation
   - **c** Abundance of helium and lithium
4. Compare the steady-state theory with the big bang theory of the universe.
5. Evaluate the impact that continued research and exploration has on our understanding of the universe.

Nature and development of science

1. Identify at least two different technologies that have provided evidence to support the big bang theory.
2. Describe how Brian Schmidt's work helped to refine our understanding of the universe.

Communicating

p. 262

Construct an annotated visual timeline of the evolution of the universe that includes the birth of our solar system.

1. **a** Identify the main steps to include in the timeline.
   b Write a dot point that describes what happened for each step.
   c Select or draw a diagram to represent each step.
2. Construct your timeline, incorporating the information from Question 1a–c.

Key idea: Scale and measurement

Conduct further research into Brian Schmidt's team's work on the expanding universe. Write a summary of the observations and measurements they made and how this provided evidence for an expanding universe.

Success criteria

- I can describe the big bang theory.
- I can outline three major pieces of evidence that support the big bang theory.

6·2 ▸ Features of the universe

Figure 6.8: The night sky over Kata Tjuta, in Central Australia, looking towards the centre of the Milky Way galaxy.

Learning intention

At the end of this lesson, I will be able to:

- describe key features of the universe, including solar systems and galaxies
- use appropriate scales to describe sizes and distances between features of the universe.

Key terms

astronomer: a scientist who studies space, stars and celestial objects

astronomical unit (AU): the average distance between Earth and the Sun (about 150 million kilometres)

comet: chunks of rock and ice moving through space

dark energy: a theoretical form of energy causing the expansion of the universe

dark matter: a theoretical form of matter contributing to the mass of the universe

exoplanet: a planet outside our solar system

frost line: a boundary just inside Jupiter's orbit

galaxy: a system of millions or billions of stars

gas giants: large planets with low density that are primarily made up of hydrogen and helium gas

interstellar: the areas of space between stars

light-year (ly): the distance that light travels in one Earth year

Local Group: the group of galaxies that the Milky Way is part of

nebula: a vast region of gas and dust

solar system: a system of planets that orbits around one or more stars, held together by gravity

spiral galaxy: a spiral-shaped galaxy with 'arms' that extend out of its dense, rotating centre

Investigation 6.2

Investigating orbits, p. 363

Key idea: Scale and measurement

The universe is everywhere. It is everything we see and a lot that we cannot see. Humans have always looked out into the universe and asked questions about what it is and our place in it. Astronomers have long observed and studied solar systems, stars, nebulae, galaxies and even black holes, but we still have a lot to understand, including about dark matter and dark energy.

The universe is really big

In the universe, distances are so great that the kilometre is too small a measurement to be useful. **Astronomers** use the light-year to measure distances between stars and galaxies. A **light-year (ly)** is the distance light travels in one Earth year (365.25 days). One light-year is about 9.46 trillion kilometres (9.46×10^{12} km).

Our solar system is in the Milky Way galaxy. The closest galaxy to the Milky Way is Andromeda, which is 2.5 million light-years away. When astronomers look at the light coming from the Andromeda galaxy, it is like looking back in time, because it has taken light from Andromeda 2.5 million years to reach Earth.

The **astronomical unit (AU)** is equivalent to the average distance between Earth and the Sun – about 150 million kilometres! It is a unit of measurement that is used to describe distances within our solar system.

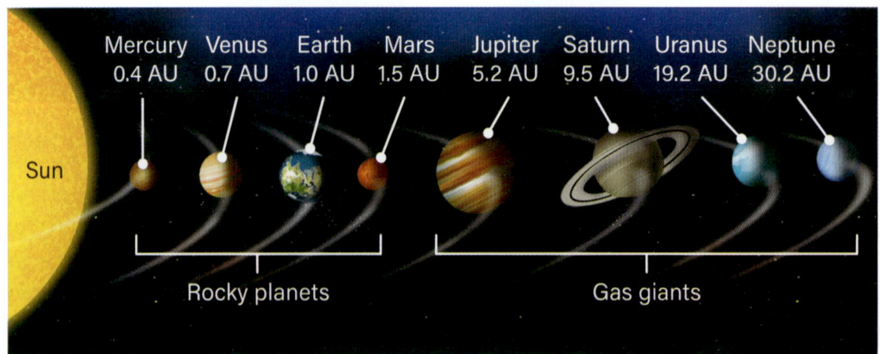

Figure 6.9: The astronomical unit (AU) is the average distance between Earth and the Sun. 1 AU = 150 million kilometres. It is useful for describing distances within our solar system.

Stars are colossal balls of gas

When we look up to the night sky, one of the first things we will notice is the stars. Stars are enormous balls of hot gas, mostly made up of hydrogen. In their centre, nuclear reactions occur in which hydrogen atoms fuse (join) together to form helium atoms. These reactions release a lot of energy, including visible light.

Nebulae are stellar nurseries

Nebulae are **interstellar** regions filled with gas and dust and are where stars are born. A star begins to form when gravity causes a region in the nebula to start to contract. As more gas comes together, the mass starts spinning, forming a hot, dense core called a protostar. The increasing gravity attracts more mass, and eventually the protostar will gather enough mass and a high-enough gravitational force for the hydrogen atoms to start to fuse and form helium, releasing energy and igniting the star.

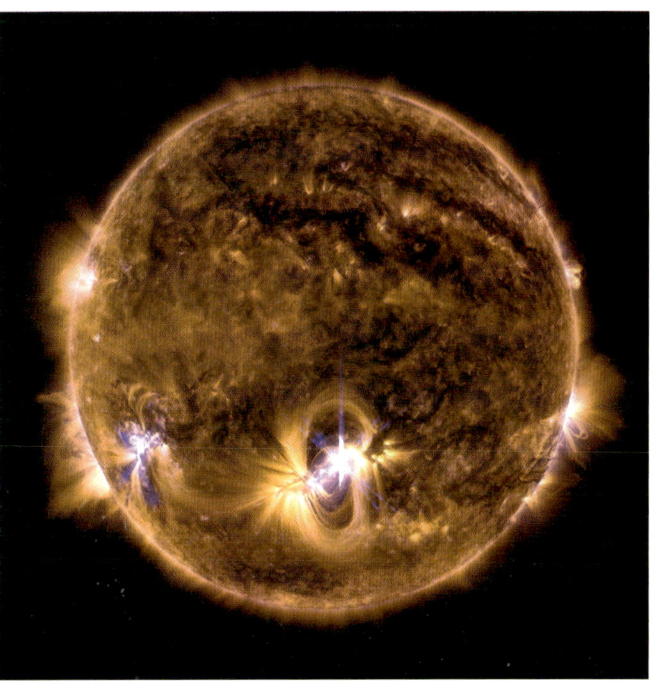

Figure 6.10: Nuclear fusion reactions in stars convert hydrogen into helium, releasing light and heat energy. Our Sun fuses about 600 billion kilograms of hydrogen into helium every second.

Figure 6.11: Nebulae are regions of gas and dust known as stellar nurseries. Our solar system would have formed in a region of a nebula.

Solar systems are systems of planets orbiting stars

A **solar system** is a planetary system that orbits around one or more stars. Observations of much younger stars and nebulae show that solar systems form at the same time as their stars. After the young star has ignited, it will continue to spin. The gas and dust around the spinning star will form a 'protoplanetary disc' that orbits the star, as shown in Figure 6.12. Collisions of matter occur within this disc, and this matter clumps together to form larger and larger objects that can eventually form planets. Solar winds produced by the young star will also blow and clear out any remaining dust and gas from the new solar system. Samples taken from meteorites and asteroids provide evidence that our solar system formed around 4.6 billion years ago.

The rocky planets were formed from heavy elements

The nebula that our solar system formed from was mostly made up of lighter elements such as hydrogen and helium. Heavier elements such as iron, nickel, silicon and aluminium were much rarer. This is why the inner, rocky planets, such as Venus and Earth, are relatively small, because they are mostly made up of these heavier elements.

The inner solar system was too warm for substances with low boiling points (such as water and methane) to condense and exist as liquids. So, only compounds with high melting points, such as metals, were formed, creating the rocky planets. These compounds are quite rare in the universe, so the rocky planets are relatively small.

Many other solar systems have been detected

Astronomers have found many different types of solar systems. Some are similar to ours, while others are very different. Planets that orbit stars other than the Sun are called **exoplanets**. The first exoplanets were observed in 1995 and since then thousands of exoplanets have been discovered. Small rocky planets like Earth, Mars, Venus and Mercury are very common in other solar systems.

Studying other solar systems helps astronomers to build a better picture of how our solar system formed. The main way that astronomers search for exoplanets is to look for a dip in the light coming from the star when the planet passes across it.

Figure 6.13: An artist's impression of K2-18b, an exoplanet located 124 light-years from Earth. In 2019 scientists discovered water vapour on K2-18b, and in 2023 the James Webb Space Telescope discovered methane and carbon dioxide in its atmosphere. K2-18b is the only known exoplanet with both water and suitable temperatures to potentially support life.

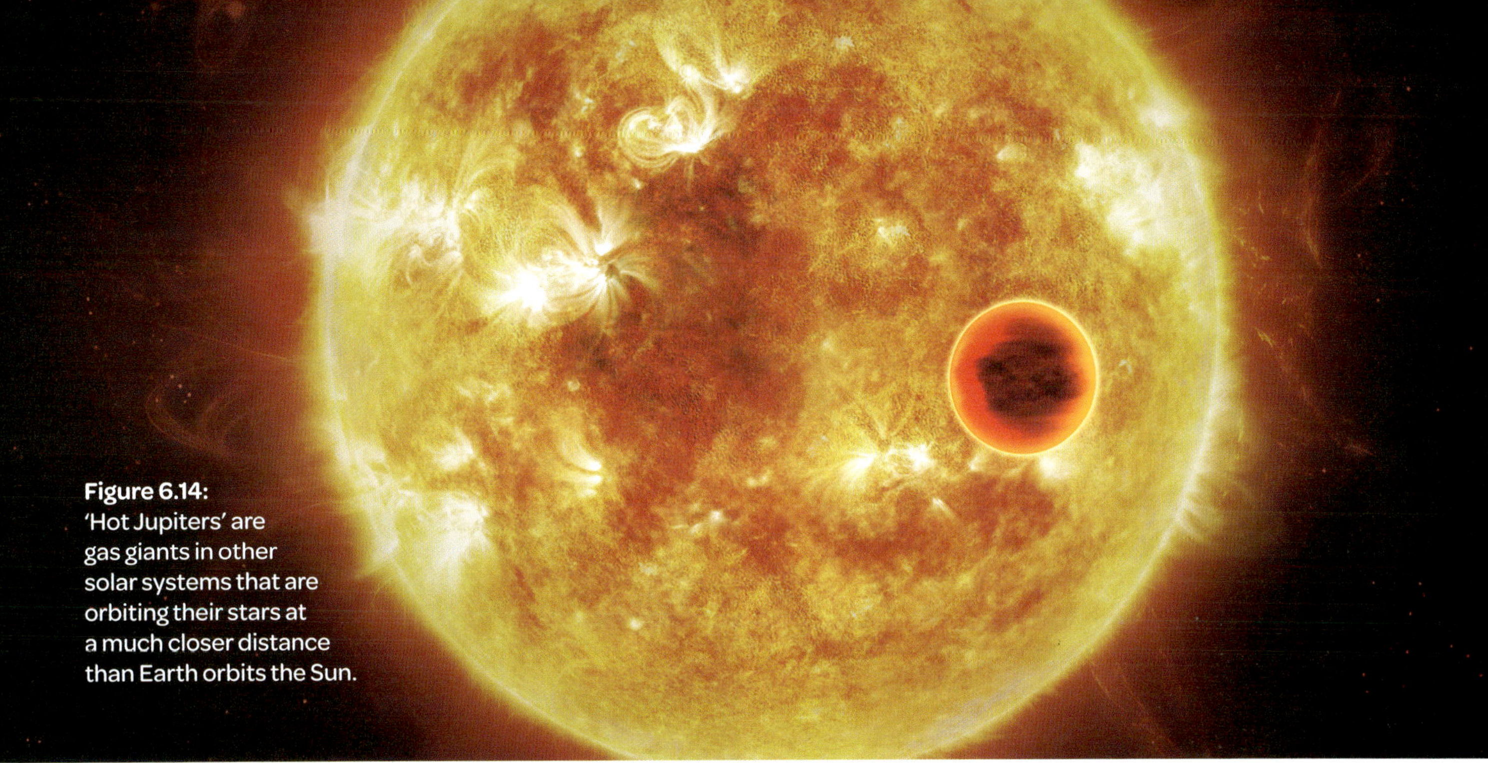

Figure 6.14: 'Hot Jupiters' are gas giants in other solar systems that are orbiting their stars at a much closer distance than Earth orbits the Sun.

The gas giants mostly formed from hydrogen and helium

The **frost line** is found just inside Jupiter's orbit. Outside this boundary, temperatures are so low that hydrogen, helium and other compounds that are gases on Earth (such as methane and carbon dioxide) are able to condense and exist as liquids.

Gas giants have cores made up of rock and ice but are otherwise mostly gases. Jupiter and Saturn, the largest planets in our solar system, are made up of large amounts of hydrogen and helium. Their large cores allowed the planets to attract lots of hydrogen and helium before the solar wind cleared the solar system. Uranus and Neptune are further out, with larger orbits.

Comets come from the Kuiper Belt and the Oort Cloud

Comets are like dirty snowballs; chunks of rock and ice moving through space. The 'tails' that we observe are formed when they melt as they move closer to the Sun.

Short-period comets, which orbit the Sun fairly frequently, originate in the Kuiper Belt. The Kuiper Belt is an area of the outer solar system that extends from the orbit of Neptune to about 55 AU from the Sun. It is a region of icy bodies left over from the formation of the solar system. The Kuiper Belt is home to dwarf planets, such as Pluto.

Long-period comets have very slow orbits; Comet Hale-Bopp orbits the Sun once every 2500 years. These comets come from the Oort Cloud, a spherical cloud of icy bodies also left over from the formation of the solar system. The Oort Cloud lies in the outermost parts of the solar system, 5000–100 000 AU from the Sun.

Figure 6.15: The Kuiper Belt is a ring of asteroid-like bodies orbiting beyond Neptune. The Oort Cloud is a spherical cloud of comets and space debris that orbit the Sun at the edge of our solar system.

Figure 6.16: Comet Hale-Bopp is a large, bright comet that takes 2500 years to complete one orbit of the Sun. In 1997, it passed close enough to Earth to be visible. Its slow orbit and large size meant it could be seen with the naked eye for 18 months!

Figure 6.17: In 2025, NASA released a massive mosaic image of a section of the Andromeda galaxy, created by stitching together more than 600 images taken by the Hubble Space Telescope over more than 1000 orbits. Taking more than 10 years to make, the image includes more than 200 million stars and is so detailed that you would need 600 HD television screens to display the whole thing! The image is the largest and sharpest Hubble has ever taken and provides new insights into Andromeda's history and formation.

Galaxies are massive groups of stars

Galaxies are massive groups of stars, gas, dust and other matter, bound together by forces of gravity. The existence of galaxies beyond our own Milky Way was first proven by Edwin Hubble in the 1920s. Hubble used the Hooker Telescope in California – at the time, at 100 inches in diameter, the largest telescope in the world – to make observations of the night sky. He found that what were previously thought to be nearby spiral-shaped nebulae were actually much further away and were made of stars, rather than gas and dust.

Figure 6.18: Edwin Hubble discovered galaxies and made other important contributions to the understanding of our universe using the 100-inch Hooker Telescope in the 1920s.

Galaxies are classified by their shape

Astronomers classify galaxies into three major groups according to their shape: elliptical, spiral or irregular (Figure 6.19). We cannot look at our galaxy, the Milky Way, from outside it. However, by observing our night sky and comparing it with images of other galaxies, astronomers have determined that the Milky Way is a **spiral galaxy**. It is part of a cluster of galaxies known as the **Local Group** that includes the nearby Andromeda galaxy, the Triangulum galaxy and the Large and Small Magellanic Clouds.

Figure 6.19: Galaxies can be classified into three main categories: spiral, elliptical and irregular.

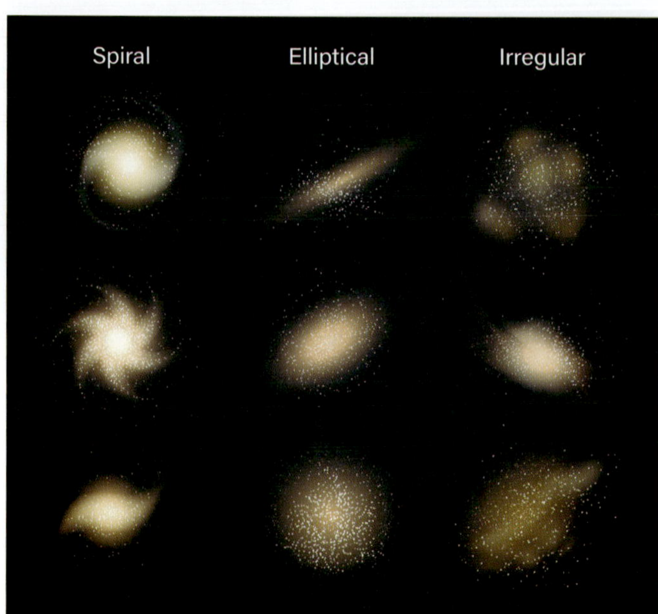

Scientists are still studying how galaxies form

Astronomers do not yet have a good understanding of how the different types of galaxies form, although they know that the forces of gravity cause stars, gas and dust to clump together. One popular theory is that irregular and elliptical galaxies form when galaxies collide and interact with one another, and that spiral galaxies form on their own, with their spiral shape resulting from the spinning motion of the galaxy.

Using telescopes, astronomers have observed that galaxies that are further away appear different from closer galaxies. This is because galaxies further away are much older, but we are observing them at a much younger stage in their life.

Evidence shows that supermassive black holes are in the centres of most galaxies. Their gravitational fields influence the stars, dust and gas in the rest of the galaxy.

Dark matter and dark energy make up a significant portion of the universe

Observations of our universe show that galaxies are rotating at such fast speeds they should fly apart. This means there must be something holding them together. Other observations have indicated there is a force that is causing an acceleration of the expansion of the universe. To try and explain this, astrophysicists have proposed that the visible matter (the matter we can detect using the electromagnetic spectrum) only makes up 5 per cent of the universe, with the rest of it being made up of **dark matter** (27 per cent) and **dark energy** (68 per cent). However, we still do not know what these are. What we *do* know is that, while we are unable to detect them directly, we can observe their effects on visible matter.

Figure 6.20: Experiments conducted by the European Organization for Nuclear Research (CERN) utilising its Large Hadron Collider may help us to learn more about dark matter.

Learning Ladder

The universe

1. Copy and complete the following sentences.
   a A ______________ is a system of planets that orbits a ______________.
   b An ______________ is a planet outside of our solar system.
   c A ______________ is a system of stars.
   d A light-year is the distance that ______________ travels in a ______________.
   e An ______________ is the average distance between ______________ and the Sun.
   f Astrophysicists still do not know much about ______________ ______________ and ______________ ______________.
2. a Outline the key steps that occur in the formation of a solar system.
   b Outline the key steps that occur in the formation of a galaxy.
3. Describe the scientific evidence for dark matter and dark energy.

Nature and development of science

1. Identify the technology that Edwin Hubble used in making observations of galaxies.
2. Describe how advances in telescopes have helped us to refine our understanding of galaxies.
3. Explain how the need to make more detailed astronomical observations would have led to the development of technologies that enable more advanced telescopes.
4. Conduct some research to find out more about Hubble's discovery of galaxies. Write a summary of how his findings were received by the scientific community at the time.
5. Using Hubble's work as an example, analyse how advances in technologies allow advances in science.

Communicating p. 262

Conduct some research, then use digital technologies to create an infographic for display in your classroom on a feature of our universe, such as solar systems or galaxies.

1. a Write a dot point for each of the key pieces of information you will include in your infographic.
   b Draw or select images to support your dot points.
2. Create a draft plan for your infographic, incorporating the information and images from Question 1.
3. Use an app or a program to create your final infographic.

Key idea: Scale and measurement

Conduct research, then draw a diagram representing the Milky Way. Include the location of our Sun and solar system. Label distances with an appropriate unit of measurement.

Success criteria

- I can describe key features of solar systems and galaxies, including how they form.
- I can describe how to measure distances in the universe at different scales.

6·3 ▶ Exploration of the universe

Learning intention

At the end of this lesson, I will be able to describe some examples of how humans have explored the universe using different technologies.

Key terms

lore: traditions and knowledge about a subject

orbit: the curved path of an object as it travels around a planet or moon

payload: the products carried into orbit by the space shuttle

telescope: a device that uses mirrors and lenses to magnify the size of objects

Investigation 6.3

Bottle rockets, p. 365

Key idea: Scale and measurement

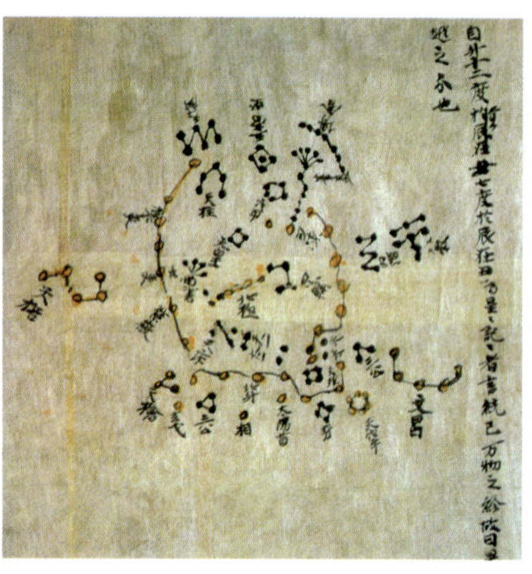

▲ **Figure 6.21:** The Dunhuang star map, from the 7th century CE, is one of the first known graphical representations of stars.

Figure 6.22: ▶ A Babylonian tablet describing Halley's Comet in 164 BCE

Humans have always looked up to the night sky and made observations that have helped them to develop calendars, navigate, follow the seasons, and share culture and **lore** (traditional knowledge). With the invention of the telescope, we were able to learn more about the universe and understand how it began. The invention of rockets that could launch craft into **orbit** around Earth was the next step for exploring space, allowing humans to travel away from our home planet.

Early astronomers made new discoveries

The ancient Babylonians produced the oldest known records of systematic observations of the night sky. They tracked the movements of planets and stars, gave names to stars and constellations, and used mathematics to make predictions.

The first records of Chinese astronomy are from around 3000 BCE. The ancient Chinese used their observations to predict lunar eclipses, record sunspot activity and make the first known star maps.

Telescopes allowed us to see more detail

The first record of a **telescope** comes from the Netherlands in 1608, soon after many people began experimenting with glass lenses to try to improve magnification and make observations of the natural world. In 1609, Galileo Galilei created a refractor telescope that had 30× magnification. Using his device, he made detailed observations of the mountains and craters on our Moon, the phases of Venus and the moons of Jupiter. Galileo published his observations in a book called *Sidereus Nuncius* (Starry Messenger).

A clearer image than was achievable using Galileo's telescopes became possible in 1668, when Isaac Newton created the first reflecting telescope, which used mirrors to collect and concentrate light.

Figure 6.23: Two of Galileo's telescopes ▶

As lens and mirror technologies have improved, so have telescopes. These advances have enabled astronomers to see the universe more clearly. The 100-inch Hooker Telescope, with a 2.5-metre mirror, was the world's most powerful optical telescope between 1917 and 1949, when it was outclassed by the 200-inch Hale Telescope. Edwin Hubble used the Hooker Telescope to discover galaxies and to determine that the universe was expanding.

Figure 6.24: The 100-inch Hooker Telescope was the world's largest optical telescope between 1917 and 1949.

Space-based telescopes are a more recent technology that provide even better images. The Hubble and James Webb space telescopes have been key in making it possible for scientists to explore our universe. The James Webb Space Telescope has very sensitive instruments that can detect both visible and infrared light. This allows it to produce high-resolution images of things that are too faint, old or distant for the Hubble Telescope to photograph, such as the oldest galaxies and stars.

Figure 6.25: In July 2025, the James Webb Space Telescope took this photograph of the Cat's Paw Nebula (NGC 6334), located in the Scorpius constellation – approximately 4000 light-years from Earth. The red–orange circular sections are stellar nurseries, where young stars are forming out of dust and gas.

Figure 6.26: Laika the dog launched on *Sputnik 2* in 1957.

The 'space race' led to new technologies

In the mid-20th century, the United States and the Soviet Union each wanted to prove their superiority in developing rocket technologies that could take people into space and then to the Moon. This rivalry resulted in many scientific and technological developments. The following were some key events.

- *1957:* The Soviet Union launched *Sputnik 1*, the first satellite to orbit Earth.
- *1957:* Laika the dog became the first animal in space, aboard the Soviet craft *Sputnik 2*.
- *1959:* The Soviet craft *Luna 2* became the first spacecraft to get to the Moon.
- *1961:* The Soviet craft *Venera 1* made the first flypast of another planet – Venus; the Soviet cosmonaut Yuri Gagarin became the first human in space when he made a single orbit of Earth aboard *Vostok 1*.
- *1963:* The Soviet cosmonaut Valentina Tereshkova became the first woman in space when she orbited Earth in *Vostok 6* for nearly three days.
- *1964:* The US craft *Mariner 4*, launched by the National Aeronautics and Space Administration (NASA), performed the first successful flyby of Mars.
- *1967:* The crew of NASA's *Apollo 1* were killed in a fire during pre-launch testing.
- *1968:* NASA's *Apollo 8* was the first human space flight to orbit the Moon.
- *1969:* NASA's *Apollo 11* was the first lunar landing, with Neil Armstrong and Buzz Aldrin becoming the first humans to walk on the Moon.
- *1972:* NASA's *Apollo 17* was the last manned lunar landing (to date).

Figure 6.27: *Apollo 11* astronauts Neil Armstrong, Michael Collins and Buzz Aldrin

Figure 6.28: The International Space Station is used to undertake scientific research. Astronaut Jasmin Moghbeli tests microbe samples.

Space shuttles and stations allow humans to spend time in space

The space shuttle was developed as a reusable vehicle that could launch, orbit Earth and return to land. It was designed to launch upright and to land like a glider. There was room for the crew, along with a large **payload** bay. The payload bay was used to transport components of the International Space Station, the Hubble Space Telescope and many satellites. Five space shuttles operated 135 missions between 1981 and 2011. The space shuttle program was not without issues: the shuttle *Challenger* exploded soon after launch in 1986, and the shuttle *Columbia* broke apart on re-entry in 2003, with both disasters resulting in the loss of the crew.

Space stations have enabled humans to learn much about how to survive in space for longer periods of time while undertaking scientific research.

- In April 1971, the Soviet Union launched the first space station – the Salyut.
- Skylab, the first US space station, was launched in 1973.
- Mir was launched in 1986 as a replacement for Salyut. It was larger and allowed for more cosmonauts to have longer stays in space – with some staying a year or more.
- The International Space Station (ISS) was built as a collaboration between 15 countries and has been continuously crewed since November 2000. The ISS was built in stages, with modules being constructed on Earth before being assembled in space using robotic systems and astronauts undertaking space walks.
- A second space station currently in operation is China's Tiangong Space Station, launched in 2021 and crewed since 2022.

Probes explore the solar system

Many different missions have been launched, sending probes to other planets, moons and asteroids to find out more about our solar system. *Voyager 1* and *Voyager 2* were launched in 1977 to fly by Jupiter, Saturn, Uranus and Neptune. They have remained operational and have now passed through the edges of our solar system. *Voyager 1* is currently more than 162 astronomical units from the Sun.

Figure 6.29: *Voyager 1* space probe

To find out more about the surface of Mars, to determine if there has been life on the planet and to prepare for human missions, NASA has sent five rovers to Mars that were designed to move around the surface and take and analyse samples. The first was *Sojourner* in 1997. Twin rovers *Spirit* and *Opportunity* landed in 2004 and lasted for well past their planned 90 days, with last communications being sent in 2010 and 2018. Rovers *Curiosity* (landed 2012) and *Perseverance* (landed 2021) are still operational as of 2025.

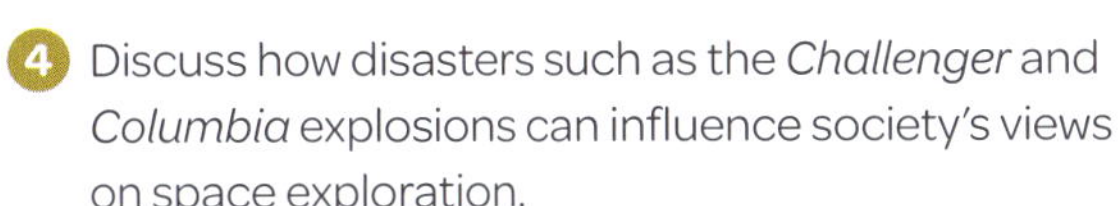

Figure 6.30: NASA's *Curiosity* rover took this selfie in 2019. ▶

Learning Ladder

The universe

1. Identify the technology that enabled:
   - **a** the first observations of the moons around Jupiter.
   - **b** the first humans to land on the Moon.
   - **c** a reusable spacecraft.
   - **d** samples on Mars to be taken and analysed.
2. Using the headings in this section, select 10 different technologies referenced and place them on a timeline.
3. Describe the contributions to explorations of the universe of:
   - **a** Galileo's telescope.
   - **b** *Voyager 1* and *Voyager 2*.
   - **c** the International Space Station.
4. Using examples, discuss the advantages, disadvantages and impacts of space exploration.
5. Evaluate the importance of space exploration for being able to develop a deeper understanding of the origin of our universe.

Use and influence of science

From Velcro and memory foam to camera phones and insulin pumps, space exploration has led to the development of many technologies that have changed how we live on Earth. Undertake some further research to help you answer the questions below.

1. Memory foam was developed to improve seat cushioning and crash protection for astronauts. Identify some examples of how it is used by people on Earth today.
2. Identify and describe a technology developed for space exploration that has been modified for use by people on Earth.
3. Propose and explain a benefit for society of continued research into space travel.
4. Discuss how disasters such as the *Challenger* and *Columbia* explosions can influence society's views on space exploration.
5. Analyse key factors that contribute to scientific knowledge about the universe being adopted more broadly by society.

Questioning and predicting p. 234

Many experiments conducted on board the International Space Station investigate the impacts of microgravity on living things. Think of an investigation that could be conducted on the space station.

1. Describe your investigation and use your understanding of the effects of microgravity to make a prediction of what would happen.
2. Construct a scientific question that could be answered by your investigation.

Key idea: Scale and measurement

Undertake some further research on *Voyager 1* and *Voyager 2* or other space probes.

- **a** Construct a timeline that highlights the main milestones and discoveries of the program.
- **b** Construct a scale diagram that shows how far these probes have travelled.

Success criteria

- I can describe examples of how humans have explored the universe using different technologies.
- I can outline the advantages and challenges of space exploration.

6.4 ▸ Observing the universe

Learning intention

At the end of this lesson, I will be able to:

- describe the type of data gathered by optical, radio and space telescopes
- list some of the discoveries telescopes have been used for.

Key terms

electromagnetic spectrum: all the different electromagnetic waves

infrared light: an electromagnetic wave with longer wavelength than red light

resolution: the ability of a telescope to identify the light of two objects that are close together as separate objects

ultraviolet light: an electromagnetic wave with shorter wavelength than violet light

visible spectrum: the part of the electromagnetic spectrum that humans can see

Investigation 6.4A

Making a telescope, p. 366

Investigation 6.4B

Lens diameter and resolution, p. 367

Key idea: Scale and measurement

Telescopes have enabled us to improve our understanding of the universe. This includes what the universe contains, how it formed and what our place is in it. Technologies have improved greatly from Galileo's simple telescope that allowed him to observe the Moon and planets in our solar system. Telescopes come in many types and are now supported by significant computing power. They can be many metres across in size, set up in arrays across many kilometres and can be launched into space.

Stars emit energy across the electromagnetic spectrum

Stars emit energy as electromagnetic waves, also called electromagnetic radiation (EMR). This energy travels at the speed of light and varies in wavelength and frequency, from very short and frequent gamma rays to very long and low-frequency radio waves. Different parts of the electromagnetic spectrum can provide astronomers with different information about stars, and so different types of telescopes have been developed that specialise in capturing specific parts of the **electromagnetic spectrum**. However, not all wavelengths along the electromagnetic spectrum reach the surface of Earth, due to the protection of the atmosphere. Space telescopes have been developed to enable astrophysicists to collect and analyse energy from parts of the spectrum that are harder to collect on Earth.

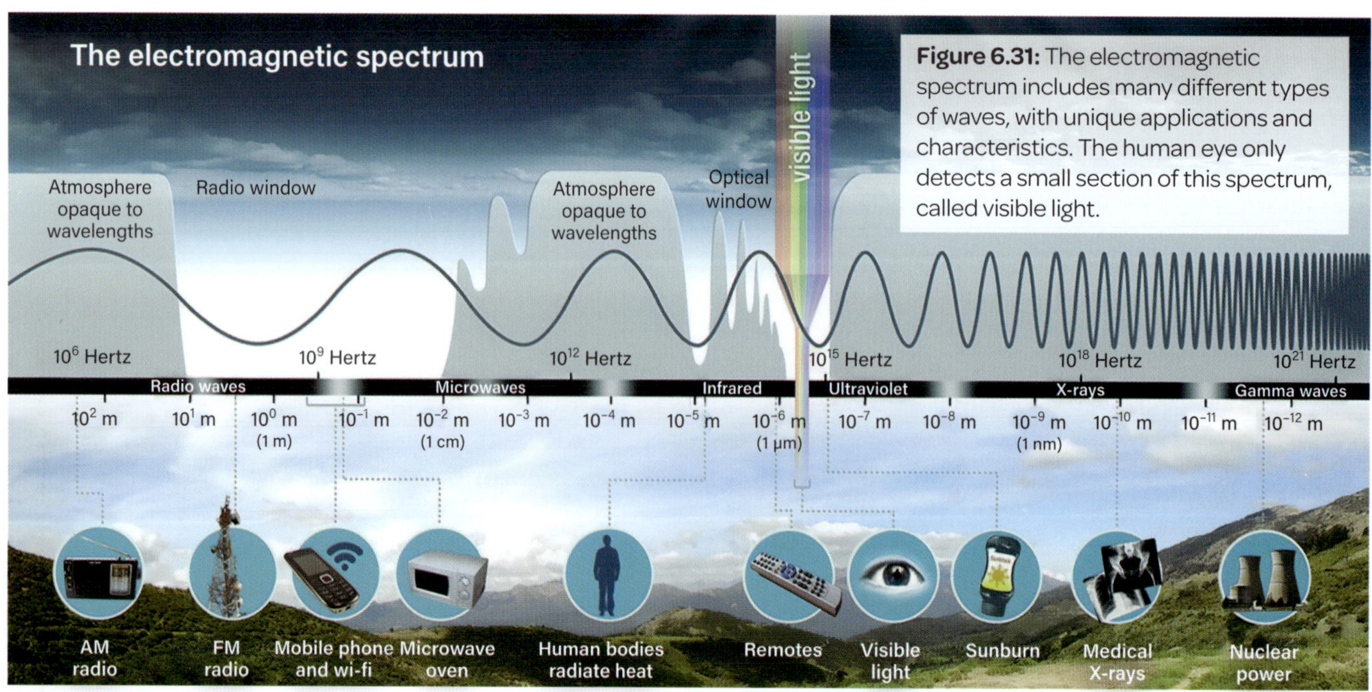

Figure 6.31: The electromagnetic spectrum includes many different types of waves, with unique applications and characteristics. The human eye only detects a small section of this spectrum, called visible light.

Optical telescopes collect and amplify light

An optical telescope works by collecting and amplifying light. Galileo's refracting telescope used two glass lenses – one large lens to collect light, and a smaller eyepiece lens to magnify the image.

Today, large telescopes have mirrors to collect light because large mirrors are easier to make than large lenses. A larger objective lens or mirror will collect more light and therefore allow more details to be observed. Such a telescope has a high resolution. **Resolution** is the ability of a telescope to identify the light of two objects that are close together as separate objects.

The further away an object is, the larger the telescope must be to have good resolution. Most optical telescopes are located at high altitude in dryer climates, to limit distortion from Earth's atmosphere, and well away from populated areas that cause light pollution.

The Extremely Large Telescope (ELT) is currently being built above the Atacama Desert in Chile. When it is completed in 2029, it will be the world's largest optical telescope, with a main mirror 39 metres in diameter. The ELT will be able to collect 100 million times more light than the human eye can detect, and to produce images 15 times sharper than the Hubble Space Telescope. This sensitivity means it will be possible for the ELT to distinguish individual light sources – such as stars – from much further away, allowing scientists to detect stars or planets that have been too distant to see clearly until now.

It is planned that this telescope will be used to investigate extrasolar planets (planets orbiting other stars) and some of the oldest galaxies in the universe.

Figure 6.32: The Extremely Large Telescope, located in Chile, will be the world's largest optical telescope, with a mirror 39 metres in diameter, once it is completed in 2029.

Radio telescopes allow us to see through dust clouds

Hydrogen is the most common element in the universe. When hydrogen atoms have absorbed energy (i.e. when they become 'excited'), they will release radio waves as they return from the excited state. Radio telescopes detect these waves, allowing astronomers to map the shape of galaxies.

Unlike visible light waves, radio waves can travel through dust clouds, so astrophysicists can discover and map out objects that cannot be seen with an optical telescope. Radio astronomy has enabled the discovery and study of pulsars (rapidly spinning neutron stars), quasars (primordial galaxies with supermassive black holes), supernova remnants, and black holes in the centre of galaxies.

Radio telescopes need to be situated away from large populations where there are few radio signals from radio, television and mobile phones. The CSIRO's ASKAP radio telescope is located at Inyarrimanha Ilgari Bundara, the CSIRO Murchison Radio-astronomy Observatory in Western Australia on Wajarri Yamaji Country. The array of 36 antennas, each 12 metres wide, is spread out over a six-square-kilometre area and has a maximum baseline of six kilometres. The set-up, completed in 2012, is supported by a high-speed processing centre that enables the ASKAP to undertake detailed surveys of the whole southern sky, discovering millions of new galaxies in the process. Astrophysicists compare data from the ASKAP taken at different times to analyse how objects in the night sky are changing. The CSIRO is also working with the SKA Observatory to build the SKA–Low telescope at the same location. This will complement the SKA–Mid telescope being built in South Africa. Together they will make up the world's largest and most sensitive radio observatory, enabling us to look back to the very beginnings of our universe, when the first stars and galaxies began to form.

Figure 6.33: CSIRO's ASKAP radio telescope is an array of 36 individual dishes that work together.

Space telescopes allow us to see more

Space telescopes allow astrophysicists to gather data from the electromagnetic spectrum that is blocked by Earth's atmosphere. They also do not experience atmospheric interference and light pollution, like telescopes on Earth can. The first space telescope was known as Stargazer and was launched in 1968.

The Hubble Space Telescope was launched in 1990. At that time, its main mirror, with a diameter of 2.4 metres, was able to produce images about 50 per cent sharper than any optical telescope on Earth. From its orbit 547 kilometres above Earth's surface, it is able to detect in the **ultraviolet**, **visible** and near-**infrared** wavelengths. Its discoveries have significantly advanced our knowledge of the universe; for example, it has verified the existence of black holes, and it has found and measured the movement of distant galaxies.

The James Webb Space Telescope was launched in December 2021 and is the most advanced telescope ever built. Rather than orbiting Earth, it sits at a point 1.5 million kilometres away, orbiting the Sun. This position keeps the telescope in line with the Earth, so the telescope's sunshield can protect it from heat and light from the Sun, Earth and Moon at the same time.

The telescope is able to detect infrared wavelengths with its 6.5-metre-diameter mirror. This allows it to gather very dim, redshifted light from much further away than Hubble. Data from the two telescopes can be combined to build a more complete understanding of our early universe.

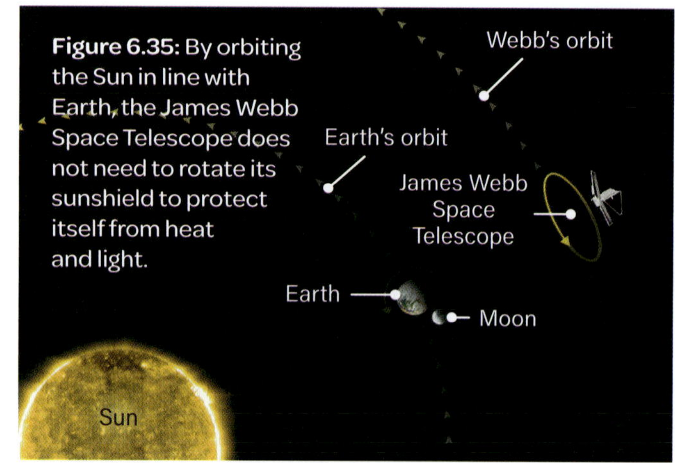

Figure 6.35: By orbiting the Sun in line with Earth, the James Webb Space Telescope does not need to rotate its sunshield to protect itself from heat and light.

Figure 6.34: The James Webb Space Telescope's mirror is almost three times larger than that of the Hubble Space Telescope. While the Hubble captures images at ultraviolet and visible light wavelengths, the James Webb Space Telescope can also capture infrared images.

Figure 6.36: The James Webb Space Telescope provides much more detailed images of the universe than the Hubble Space Telescope.

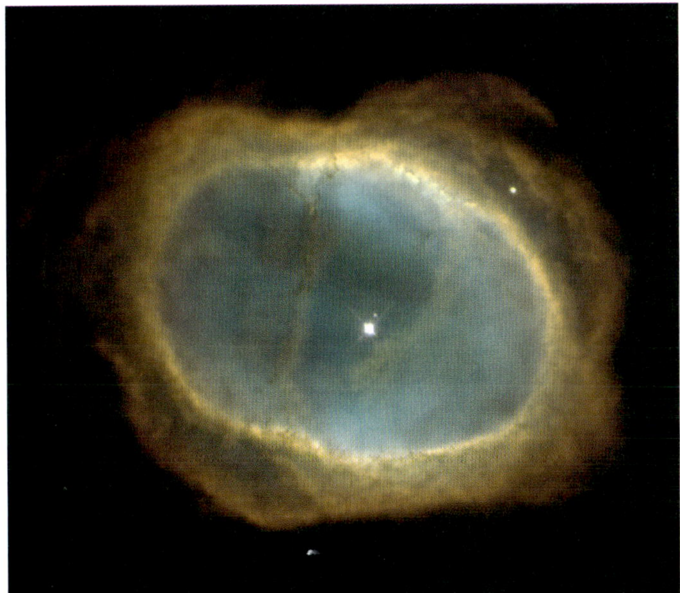

Learning Ladder

The universe

1. Copy and complete the following sentences.
   a ________ telescopes collect and amplify visible light.
   b ________ telescopes have enabled the discovery of pulsars, quasars and black holes.
   c ________ telescopes allow information to be gathered without interference from the atmosphere.
2. Use the information in this section to construct a timeline that includes each of the telescopes mentioned.
3. Explain the advantage of using a space telescope over a ground-based telescope.
4. Some telescopes can focus on an object and then move as Earth rotates. This allows more energy to be gathered, like a long-exposure photograph. Discuss the advantage this would have over a telescope that does not move.
5. Evaluate the need to have different types of telescopes that collect light from different parts of the electromagnetic spectrum.

Nature and development of science

1. Identify the part(s) of the electromagnetic spectrum that can be detected by the:
   a Hubble Space Telescope.
   b James Webb Space Telescope.
   c ASKAP radio telescope.
2. Describe how telescopes have allowed our knowledge of the universe to be refined over time.
3. Suggest how the development of space telescopes has led to improvements in engineering and technology.

Questioning and predicting

p. 234

Read Investigation 6.4B on page 367 before answering the questions below.

1. Make a prediction of how far away from the wall you will be when you can first observe the two lines as being distinct from one another.
2. Construct a scientific question that could be answered by this investigation.
3. a Identify the dependent, independent and controlled variables in this investigation.
   b Develop a hypothesis that predicts the relationship between the dependent and independent variables.
4. Discuss how you would conduct the investigation in order to be able to gather reliable and valid data to address your scientific question.

Key idea: Scale and measurement

Use the information on these pages, as well as additional research, to create and complete a table that includes columns for each of the following: telescope name, wavelength collected, size of the telescope, discoveries made.

Success criteria

- I can describe optical telescopes, radio telescopes and space telescopes.
- I can identify the type of data these different telescopes gather and some of the discoveries they have been used for.

6·5 ▸ Challenges of human space exploration

Learning intention

At the end of this lesson, I will be able to discuss the challenges for humans living in space.

Key terms

cosmic rays: high-energy particles that move through space at the speed of light

magnetic field: a region surrounding Earth where magnetic forces are exerted

microgravity: a condition where gravity appears to be almost absent, making objects and people seem weightless

Key idea: Scale and measurement

Figure 6.37: Spacesuits are made of different layers to protect astronauts from harmful radiation.

There are many challenges that need to be overcome for humans to be able to travel away from Earth, whether it be orbiting in a space station, travelling to the Moon or Mars, or eventually exploring much further away. Space agencies such as NASA have identified five major hazards that need to be overcome to ensure the safety of astronauts. By living and working in space, astronauts are also research participants contributing data to deepen our understanding of the impact of space travel on the human body. The data obtained in these studies is used to better prepare for future space missions and deepen our understanding of human health – both on Earth and in space.

Radiation in space damages cells

Earth's atmosphere and **magnetic field** protect life on our planet's surface from harmful radiation. The atmosphere blocks harmful ultraviolet radiation from the Sun. The magnetic field protects us from **cosmic rays**, high-energy particles that originate from outside of the solar system and move through space at the speed of light, as well as from the charged particles that make up the solar wind. If this radiation comes in contact with living cells, it can damage their cellular structure and DNA. Such damage can lead to cancer or other diseases. Astronauts are protected by spacecraft and spacesuits that are designed to deflect or absorb this harmful radiation.

Changes in gravity impact the human body

The human body has evolved under Earth's gravitational field. Astronauts experience different amounts of gravity when travelling to, from and in space. When leaving and re-entering Earth's atmosphere, astronauts can experience gravitational forces that are around three times the normal forces on Earth. To help them prepare for this change, they undergo training that simulates different amounts of gravity. Each astronaut's health is closely monitored throughout their mission, and they follow individual nutrition and exercise plans to keep them as healthy as possible. Their specially designed spacesuits prevent blood from pooling in their legs and keep it circulating to the brain and other organs, so that they do not lose consciousness.

Your everyday movement on Earth helps to maintain your muscle strength and bone density. If you have ever broken a bone, you may have noticed that the muscles get smaller due to inactivity as the bone heals.

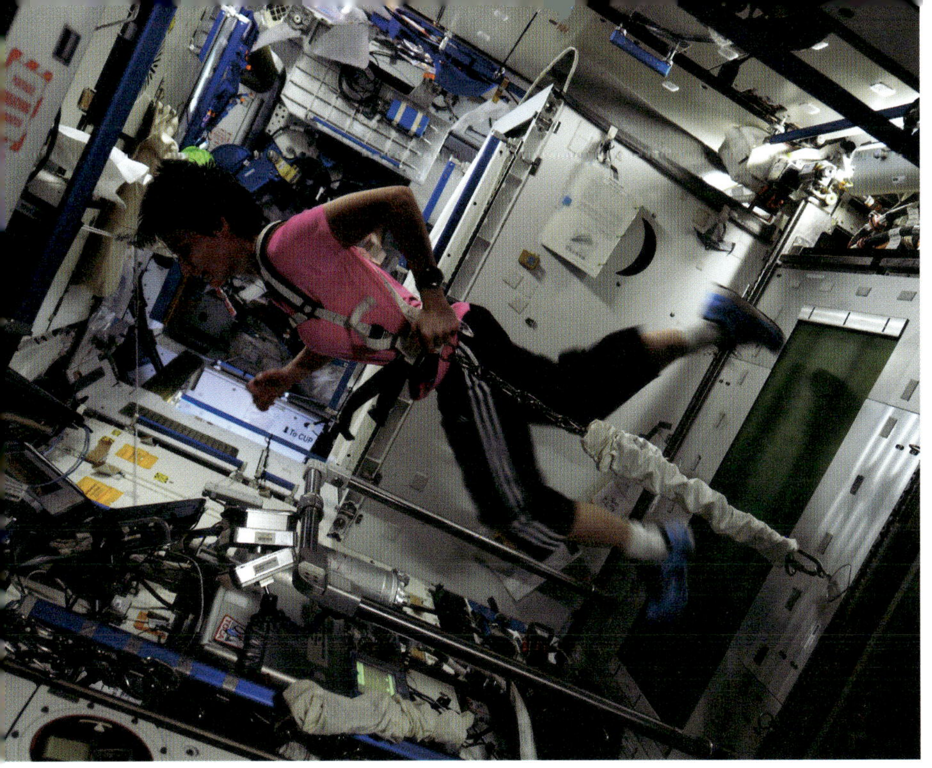

Figure 6.38: Astronauts exercising on a treadmill on the ISS are held in place with elastic straps that simulate gravity pushing them down.

Astronauts who spend a significant amount of time in space lose muscle mass because of the decreased use of their muscles in the lower gravity. Bone density also decreases due to low gravity: NASA's research has suggested that without Earth's gravity, weight-bearing bones lose around 1 per cent of their mineral density each month. To help mitigate this, astronauts on the International Space Station (ISS) have a regular exercise schedule that includes lifting weights and running on a treadmill. Each astronaut contributes data to studies that aim to reduce the impact on the body of being in **microgravity** by constantly refining the exercises.

NASA's Human Research Program (HRP) studies how different systems in the human body respond to time spent in space, including the effects on an astronaut's immune system. Factors such as radiation exposure, stress, microgravity and disrupted sleep–wake cycles can all affect immune function. Research indicates that infection-fighting white blood cells may not work as effectively in space, leaving astronauts more vulnerable to illness. HRP researchers study astronaut urine, saliva and blood samples collected before, during and after spaceflights, and compare these samples to those of a control group on Earth. This allows them to investigate ways to reduce the impact of microgravity on the body.

Isolation and confinement impact wellbeing

Space travel means being isolated from friends and family and spending time with just a small group of people in a confined space for a long period of time. This can negatively impact a person's wellbeing and therefore threaten the mission's objectives. People who are selected to be astronauts have a high resilience, undergo specific training and are well supported to cope with these issues. Space agencies have conducted extensive studies of people's responses to being in an isolated environment. They have found that adequate lighting, good-quality sleep and regular movement are key for maintaining an astronaut's wellbeing.

Astronauts are also encouraged to keep journals as a way to vent their frustrations and process their thoughts and feelings. The astronauts on the ISS are able to communicate with friends and family using email, weekly video calls and voice calls. However, the speed at which the ISS is moving means that call quality can be patchy, and the software they use means that they can make calls but not receive them. Due to the distance between Earth and Mars, astronauts on a Mars mission would not be able to call Earth in real time, as it takes between 4 and 24 minutes for signals to travel through space, depending on the relative locations of the planets.

Figure 6.39: Astronaut Serena Auñón-Chancellor provides a saliva sample, which will be used to investigate changes to her immune system while in space.

Figure 6.40: Pizza night for ISS crew members Christina Koch, Luca Parmitano, Nick Hague, Alexey Ovchinin and Alexander Skvortsov

Missions need to be self-sufficient

The vast distances and time involved in space travel mean that missions must be self-supporting and not rely on resupply. Communication with Earth is delayed, and medical care is very difficult to provide. Careful planning is needed to ensure that astronauts have the right equipment and training to be able to support themselves and each other. The ISS is relatively close to Earth and has regular resupplies; however, a mission to Mars will need to be completely self-sufficient.

Food needs to be nutritious but also able to be stored for a long period of time. Systems will need to be developed to allow astronauts to grow some of their own food using minimal resources. Studies of crew members on the ISS have highlighted the medical conditions astronauts typically face, so that other crews can be better prepared for longer trips. It may be that a future Mars mission would need to include medical staff among the crew.

Enclosed environments encourage the growth of microbes

The environment within a spacecraft needs to be carefully designed and able to be maintained to ensure the health of the astronauts. The temperature, light, air and noise levels all need to replicate those on Earth as closely as possible.

Disease-causing microbes (such as bacteria and viruses) can easily spread through a crew. The design of spacecraft needs to assist regular cleaning and air filtering to minimise the growth and spread of microbes. To reduce risks, more research is needed to understand how normal skin and gut microbes change and behave in space, and how the immune system is impacted by a high-stress environment. On the ISS, astronauts regularly have their blood and urine checked to monitor their health. Surfaces on board are regularly swabbed to test for microbes.

◀ **Figure 6.41:** Astronaut Ricky Arnold processes DNA taken from swabs of surfaces inside the ISS to help identify microbes.

Figure 6.42: Astronauts need to undertake regular maintenance inside the ISS.

Learning Ladder

The universe

1. Identify an example of a technology that would be useful for an astronaut on a mission to the Moon.
2. Propose how an astronaut on a mission to the Moon would experience changes in gravity during their mission.
3. Explain, with reference to an example, why changes in gravity need to be considered for humans exploring space.
4. Space exploration is challenging at an individual level. Discuss the advantages and disadvantages of space exploration from the point of view of an astronaut.

Use and influence of science

1. Identify a benefit to the rest of society of studying the health of astronauts over a long period of time.
2. Cosmic rays can cause damage to living cells. Propose how scientists could use this knowledge to create different solutions that protect astronauts.
3. Explain how developing knowledge on how best to support astronauts in space has benefited humans living on Earth.

Communicating p. 262

Conduct research into NASA's *Artemis* missions. You have been asked to communicate the aims of the mission and the planned milestones to display in your school's science department.

1. **a** Identify the main pieces of information you need to communicate and write a dot-point summary for each.
   b Identify images and diagrams that you could use in your product.
2. Brainstorm ways to present your information to other students.
3. Brainstorm ways you could use digital technologies to present or enhance your presentation.
4. Select the best way to present your information and prepare your presentation.
5. Prepare for a class debate about whether the large amounts of money invested into NASA's *Artemis* missions are justified, by writing dot-point summaries supporting both positions.

Key idea: Scale and measurement

Conduct research to construct a timeline that compares the time spent travelling in space from Earth to the Moon with a trip from Earth to Mars.

Success criteria

- I can identify and discuss some of the challenges for humans living in space.

6·6 ▸ The life cycle of stars

Learning intention

At the end of this lesson, I will be able to:

- describe the life cycle of stars of different mass
- use a Hertzsprung–Russell diagram to identify where a star is in its life cycle.

Key terms

black hole: a body with such a strong gravitational field that it attracts all nearby matter

luminosity: a measure of the brightness of an object

main sequence star: a star that is fusing hydrogen in its core

neutron star: the small, extremely dense core left over from the collapse of a star, when matter is compressed into a small space

protostar: a contracting mass of gas that is in the early stages of forming a star

red giant: a star that has stopped fusing hydrogen in its core

solar mass: a unit of measurement based on the mass of the Sun

supernova: the explosion of a massive star at the end of its life

white dwarf: a small, very dense star formed at the end of a small star's lifetime

Investigation 6.6

Stargazing, p. 368

Key idea: Scale and measurement

When we look up at the night sky on a clear night we can observe thousands of stars. If we use a telescope, we can observe that the stars are different colours – white, blue, red and yellow – and that some appear brighter than others. By looking at many different stars, astrophysicists have been able to determine many things about stars, including their temperature and how they are born, live and die.

The Hertzsprung–Russell diagram shows the relationship between key properties of stars

The Hertzsprung–Russell (H–R) diagram was developed by Ejnar Hertzsprung and Henry Norris Russell. Working independently, they both plotted the relationship between the temperature and brightness (**luminosity**) of stars and came up with very similar diagrams. The H–R diagram shows that most stars are clustered in a band called the main sequence, with brightness increasing with temperature.

Hertzsprung and Russell found two other main clusters of stars.

- **Red giants and supergiants:** Stars that are cool but bright, meaning they must be very large.
- **White dwarfs:** Stars that are very hot but dim, meaning they must be very small.

These clusters on the H–R diagram all relate to a star's stage of life.

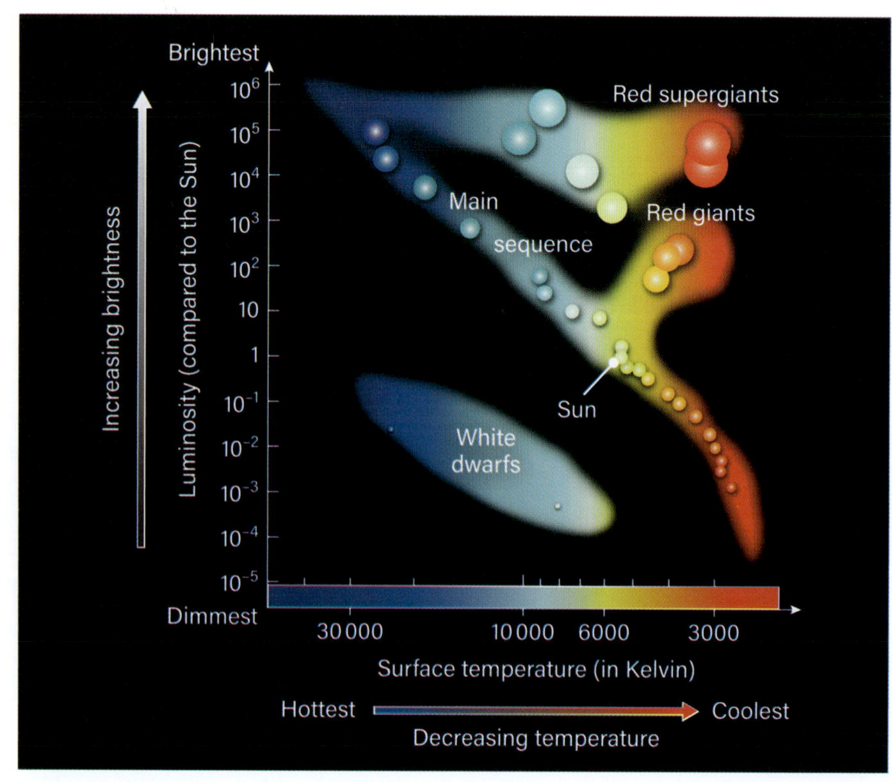

Figure 6.43: Astronomers use the Hertzsprung–Russell (H–R) diagram to classify stars and identify where they are in their life cycle.

Figure 6.44: The Orion nebula is located near the tip of the sword in the constellation of Orion.

The size of a star determines its life cycle

The size of a star determines how long it will live for, as well as the stages in its life cycle, because size affects how quickly all the hydrogen in its core fuses. A star such as our Sun (1 solar mass) has a lifetime of about 10 billion years. Much larger stars (more than 7 solar masses) have much shorter life spans because they fuse hydrogen quickly. Smaller stars of less than 1 solar mass live a lot longer than the Sun because they fuse hydrogen more slowly.

Nebulae are where stars are born

Nebulae are vast regions of gas and dust. Like the rest of the universe, they are mostly made up of hydrogen gas, with a small amount of heavier elements. A star begins to form when a region in the nebula starts to contract and gravity brings the gas together. As more gas comes together, the mass starts to spin, forming a hot dense core called a protostar. The protostar continues to contract until the gas in the centre becomes dense and hot enough that the forces result in the fusion of hydrogen atoms into helium atoms, releasing a lot of energy. When this occurs, the star is born.

Main sequence stars are in hydrostatic equilibrium

A large star will have more gravity than a smaller star like the Sun. Anything in the universe that has mass has gravity – the more mass, the more gravity. In main sequence stars, the release of energy from the fusion of hydrogen into helium pushes outwards from the core. This outward force counteracts the force of gravity due to the mass of the star pushing inwards. This balance of forces is known as hydrostatic equilibrium. Stars will remain in hydrostatic equilibrium while they are fusing hydrogen in their cores. Stars in this phase of their life are in the main sequence on the H–R diagram. Our Sun will spend about 10 billion years in the main sequence.

Eventually, when these forces become unbalanced, the star changes and moves through different stages of its life cycle.

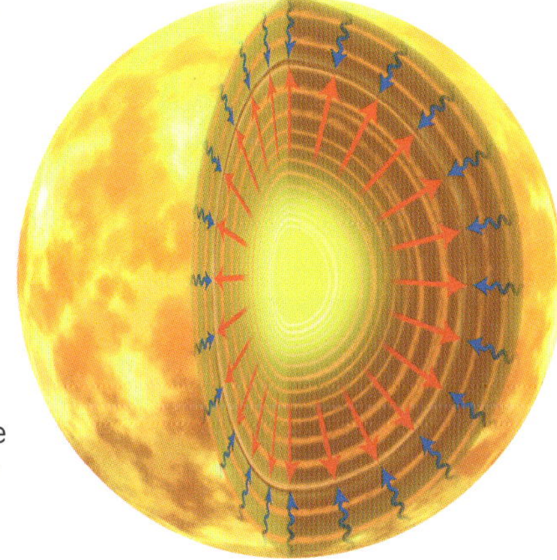

Figure 6.45: Stars ▶ fusing hydrogen into helium in their cores are in hydrostatic equilibrium: the pressure outwards is balanced by the force of gravity pushing inwards.

Stars leave the main sequence to become red giants and supergiants

Once the supply of hydrogen in the core of the star has depleted, the pressure pushing outwards reduces and the helium-rich core begins to contract under gravity. The helium in the core starts to fuse into carbon and oxygen.

At the same time, the hydrogen-rich shell around the core ignites and begins to fuse hydrogen into helium. This increased pressure outwards causes the outer layers of the star to expand to 100–200 times its original size. These layers cool down and turn red. The star is now a red giant. (Really massive stars form red supergiants.) Our Sun will become a red giant in about 5 billion years and could expand out as far as Earth's orbit, destroying the inner planets as it expands.

Sun-like stars will become white dwarfs

After about 1 billion years, the helium in the core of our Sun will have been depleted and it will contract again, forming a small, dense, hot star called a **white dwarf**. At this time the outer layers of the star will become unstable and eject away from the core, forming a planetary nebula. Planetary nebulae are not true nebulae; they were so-named by astronomers because their disc-like shape looked like planets.

Figure 6.46: Ultraviolet image of the planetary nebula NGC 7293, also known as the Helix Nebula. A white dwarf can be seen in the centre.

Massive stars end their lives explosively

Massive stars have a very different ending compared to smaller stars like our Sun. When a red supergiant star runs out of helium in its core, it will contract and become dense and hot enough for carbon to fuse into elements such as neon, magnesium and sodium. As the temperature and pressure keeps rising, oxygen fuses into elements such as silicon and sulfur. Eventually, the atoms will fuse together to produce iron.

At this point, the process changes. Fusing iron requires more energy than it releases. This means gravity pulls inwards much more strongly than the outward pressure. This process triggers the star's core to collapse suddenly, causing a huge explosion called a **supernova**. A supernova is so powerful that atoms collide and fuse, forming elements heavier than iron. The collapsed core of the star is left behind, forming an extremely dense **neutron star**.

Figure 6.47: The life cycle of a star depends on its mass.

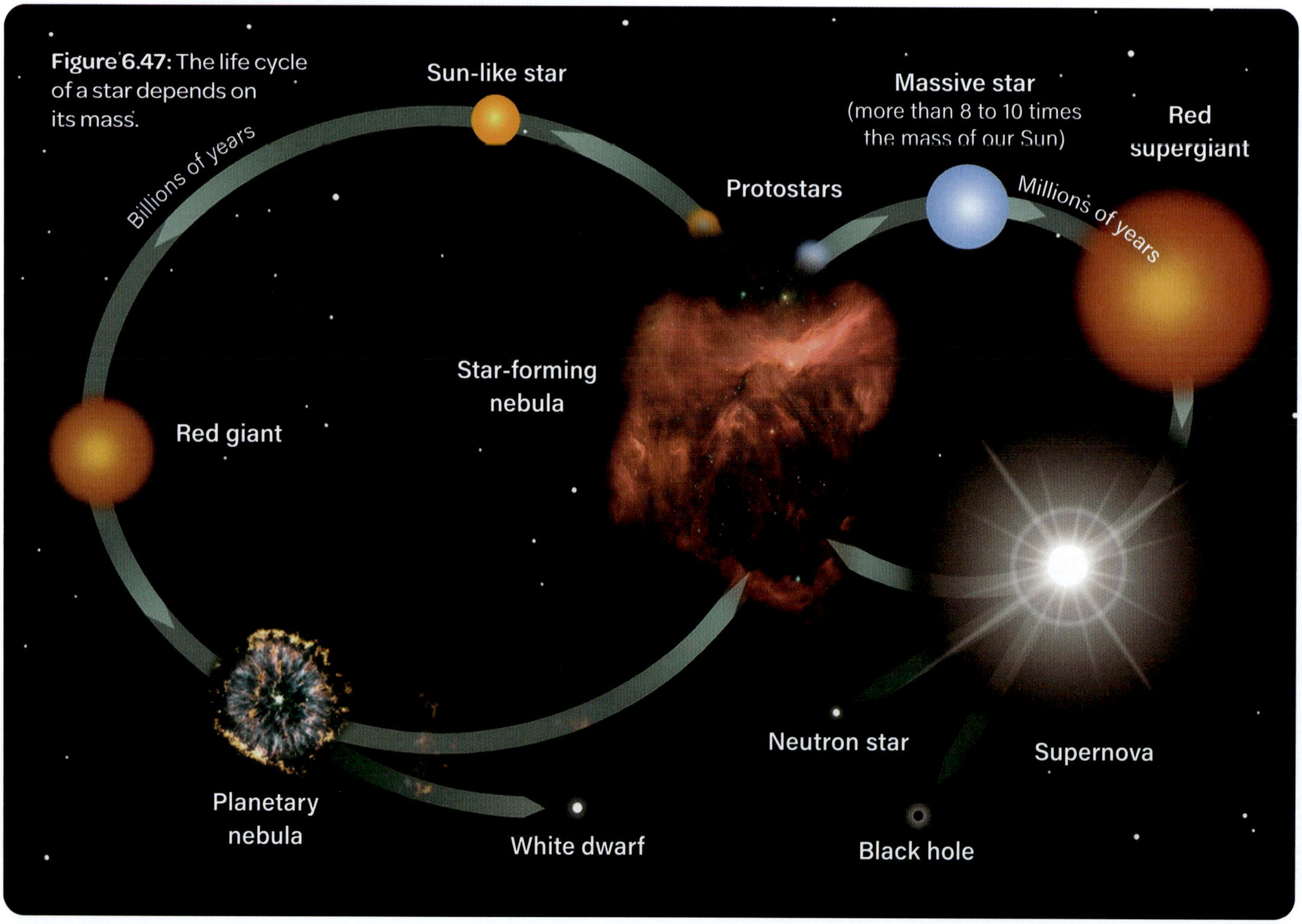

Sun-like star
Billions of years
Massive star
(more than 8 to 10 times the mass of our Sun)
Red supergiant
Protostars
Millions of years
Star-forming nebula
Red giant
Neutron star
Supernova
Planetary nebula
White dwarf
Black hole

Supermassive stars form black holes

Supermassive stars will also explode in a supernova, but instead of forming a neutron star, the core is so massive and dense that gravity causes it to collapse on itself. This forms a **black hole**, a body with such a strong gravitational field that it attracts all nearby matter and even bends light around it.

Black holes were first proposed theoretically, before they were discovered. They do not produce light, making them difficult to detect. Instead, astronomers look for clues that indicate their presence, such as the movement pattern of stars that could be orbiting them, the bending of light around them, and light emitted by disks of matter around them called accretion discs.

Figure 6.48: ▸ A simplified representation of the life cycle of a star.

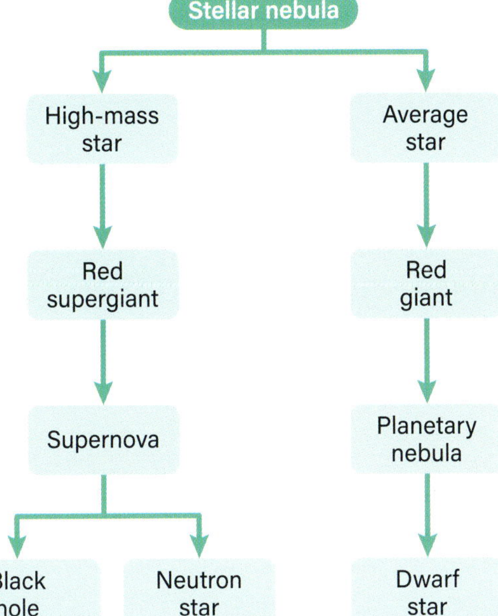

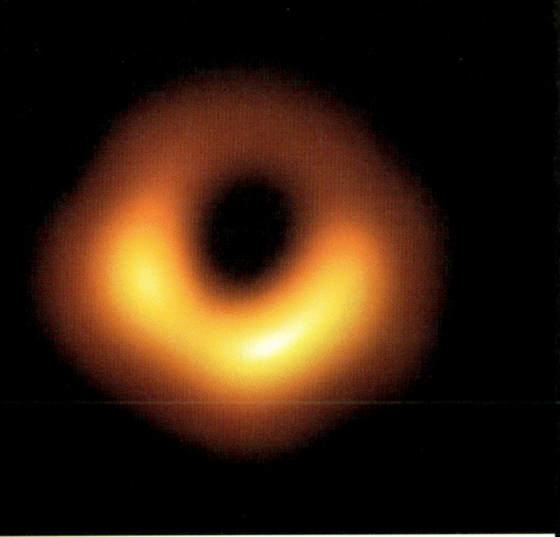

Figure 6.49: The black hole in the centre of galaxy M87 is 6.5 billion solar masses. This image shows the silhouette of the black hole against superheated gas falling into it.

Learning Ladder

The universe

1. **a** Identify the most common element in the universe.
   b Identify the element that is produced in the core of the Sun.
   c Identify the property of a star that determines the stages of life it will go through.
   d Identify what the Sun will become at the end of its life.
2. Construct an annotated diagram that describes the life cycle of a:
   a star similar in size to our Sun.
   b supermassive star.

Nature and development of science

1. Identify the clues astronomers look for to detect black holes.
2. Describe why it might have been difficult for astronomers to differentiate between nebulae and planetary nebulae when they were first observed.

Questioning and predicting p. 234

Use the H–R diagram (shown in Figure 6.43) to answer the questions below.

1. Predict the path our Sun will take over its life.
2. Construct a scientific question that can be answered using the H–R diagram.
3. Describe the relationship between the temperature and brightness of stars in the main sequence.
4. The brightness of stars is also related to their size. The larger the star, the brighter it is. Explain why a white dwarf could appear dimmer than a red supergiant.
5. Neutron stars are very dim and are usually detected using radio telescopes. Conduct some research and explain why neutron stars are not plotted on the H–R diagram.

Key idea: Scale and measurement

Conduct research to allow you to create a scale that compares the size of the Sun with other stars such as Proxima Centauri, Betelgeuse, Sirius and Arcturus. Discuss what this research suggests about how each of these stars will end their lives.

Success criteria

- I can describe the stages of life a star will cycle through, based on its size.
- I can identify the stage of life of a star using the H–R diagram.

6.7 ▸ Key idea: Scale and measurement – stars and the electromagnetic spectrum

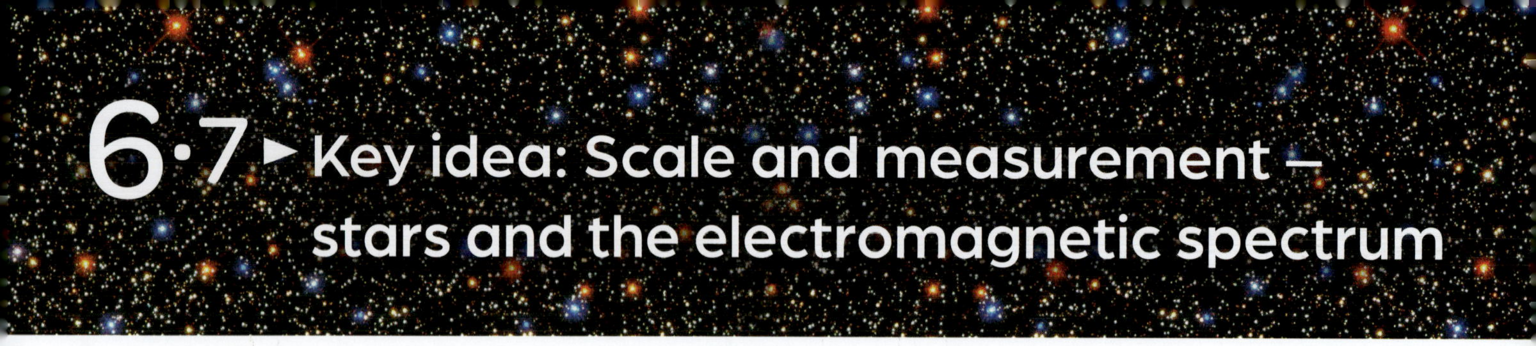

Figure 6.50: The more than 100 000 stars in this image of Omega Centauri are a variety of colours and temperatures.

Learning intention

At the end of this lesson, I will be able to describe why telescopes that can detect different wavelengths of the electromagnetic spectrum are used to take measurements of stars.

Key terms

ionise: to remove an electron from an atom

star: a burning ball of gas, mostly hydrogen and helium, held together by its own gravity

Key idea: Scale and measurement

Table 6.1: The colour and temperature of stars

Colour	Temperature (°C)
Blue	30 000–60 000
Blue-white	10 000–30 000
White	7000–10 000
Yellow-white	6000–7000
Yellow	5000–6000
Orange	3500–5000
Red	<3500

Astrophysicists analyse wavelengths across the electromagnetic spectrum to learn more about stars and the rest of the universe. Different wavelengths can provide different types of information. Telescopes that detect specific wavelengths are used to take measurements that will provide answers to their questions.

The colours of stars reveal their temperature

Stars release most of their energy as visible light. Hotter stars release a higher-energy wavelength towards the blue end of the spectrum (see Table 6.1). Cooler stars release lower-energy wavelengths towards the red end of the spectrum. The visible light image of the stars in Omega Centauri in Figure 6.50 was taken by the Hubble Space Telescope. The variety of colours in the stars indicates their different temperatures. Astrophysicists can plot colour and temperature data on a Herzsprung–Russell diagram (refer to Section 6.6) to help them understand a star's stage of life.

Different wavelengths show different stellar objects

Low-energy *radio waves* and *microwaves* allow us to look through clouds of interstellar dust and to see how cold gas moves in space.

We feel infrared energy as radiated heat. It is able to pass through the cold dust in the universe, allowing us to study warm gas and dust, and cooler stars. *Infrared wavelengths* can also help us to identify molecules in the atmospheres of planets.

Visible light tells us what elements are present, along with the temperature of the star. Helium was first discovered by analysing the visible light being emitted by the Sun.

Ultraviolet wavelengths allow us to see the hottest stars that are not easily visible to the eye. They are also used for viewing the hot glow of star nurseries, where new stars are forming.

X-rays are released from the hottest gas clouds containing atoms, when they are heated to millions of degrees. X-rays are released from superheated material that may be spiralling around a black hole or released from a neutron star (see Section 6.6).

Gamma rays have the highest energy and shortest wavelength. They are released from electrons that have been removed from atoms and accelerated by magnetic fields in exploding stars, or when neutron stars collide, and from supermassive black holes.

The Crab Nebula puts out light across the spectrum

Figure 6.51 shows the Crab Nebula, the expanding remains of an exploding star or supernova. Japanese and Chinese astronomers recorded this star exploding nearly 1000 years ago, in 1054.

The images are produced by different telescopes, including the Very Large Array (radio), Spitzer Space Telescope (infrared), Hubble Space Telescope (visible), XMM-Newton (ultraviolet), Chandra X-ray Observatory and Fermi Gamma-ray Space Telescope.

Each telescope produces different sets of measurements and data that astrophysicists can combine together to build a more detailed understanding. For example:

- In visible light, the Hubble Space Telescope tells us which atoms are present in the gas left over from when the star collapsed. Blue in the filaments in the outer part of the nebula represents oxygen, orange is mostly hydrogen, green is sulfur, and red indicates ionised oxygen (oxygen atoms with some electrons knocked off).
- The X-ray image shows a white dot near the centre of the nebula. This is a rapidly rotating, highly magnetised neutron star, which is creating shock waves that can be seen as a surrounding ring.

Figure 6.51: The combined image of the Crab Nebula overlays all light from across the electromagnetic spectrum.

Learning Ladder

The universe

1. **a** Identify the information that visible light telescopes can provide about stars.
   b Identify the type of star that X-ray telescopes are useful for detecting.
   c Identify the type of telescope that is useful for detecting stellar nurseries.
2. Summarise the information that the combined image in Figure 6.51 provides astronomers.

Nature and development of science

1. Refer back to Figure 6.31 in Section 6.4. Draw a line representing the electromagnetic spectrum and label each section. Add the name of a telescope that can measure light at each part of the spectrum.
2. Describe how the development of telescopes that can view different parts of the electromagnetic spectrum has helped us to understand stars.
3. The James Webb Space Telescope was launched in 2021. Research how engineers provided it with a mirror large enough to make observations of far-away galaxies.

Questioning and predicting

p. 234

1. Predict and compare the temperatures of the two stars in this image of the constellation Orion.
2. Write a scientific question about the colour and temperature of stars that can be answered with the data in Table 6.1.
3. Use the data in Table 6.1 to state the relationship between star colour and temperature.
4. A scientist discovers a new star and claims it is yellow with a temperature of 12 450 °C. Refute or support this claim, based on the data in Table 6.1.
5. **a** Explain how the brightness of stars relates to their distance from Earth and their size.
   b Explain how scientists use this information to work out how far away and how large a star is.
   c Explain the difference between absolute magnitude and apparent magnitude. *Hint:* You may need to conduct some further research.

Success criteria

- I can describe why telescopes that can detect different wavelengths of the electromagnetic spectrum are used to take measurements of stars.

► Summary

- The ancient Babylonians made some of the earliest recorded measurements of the night sky.
- The ancient Chinese made some of the earliest recorded star maps.
- Galileo made important observations of the solar system using an early version of the telescope.
- The space race between the United States and the Soviet Union led to the launch of the first satellites and humans in space before humans first landed on the Moon.
- The space shuttle and space stations have allowed us to learn more about living in space.
- Space probes sent throughout the solar system have allowed us to learn more about our solar system.

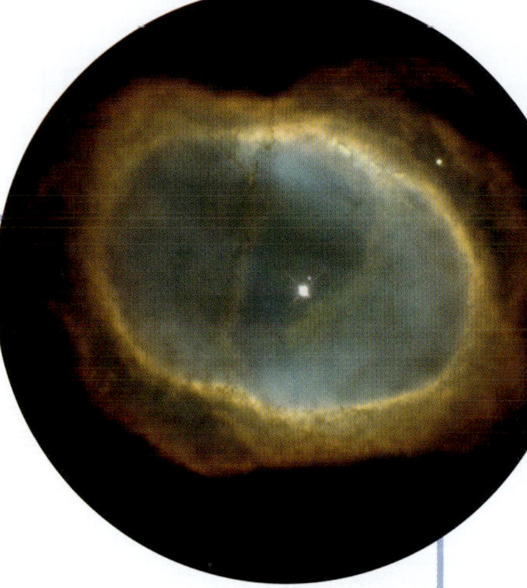

- Improvements in telescopes have led to discoveries that have increased our understanding of the universe.
- Different telescopes are used to collect data at different wavelengths on the electromagnetic spectrum.
- Optical telescopes detect light in the visible part of the spectrum and can provide information about the temperature of a star.
- Radio telescopes detect radio waves and can provide information about the formation of galaxies.
- Space telescopes enable clearer images by avoiding the interference of Earth and its atmosphere.

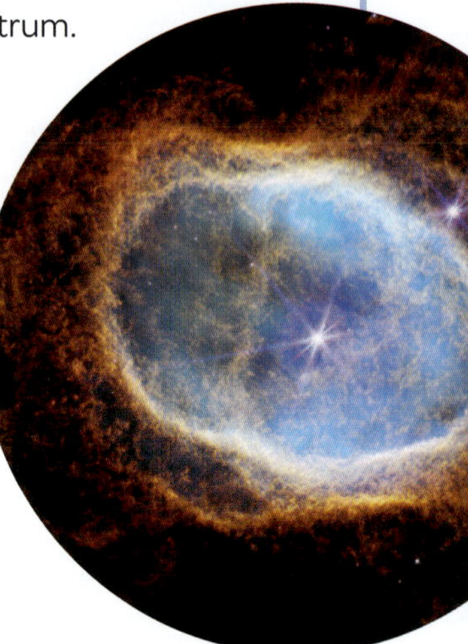

- The mass of a star will determine the stages it will pass through during its life cycle.

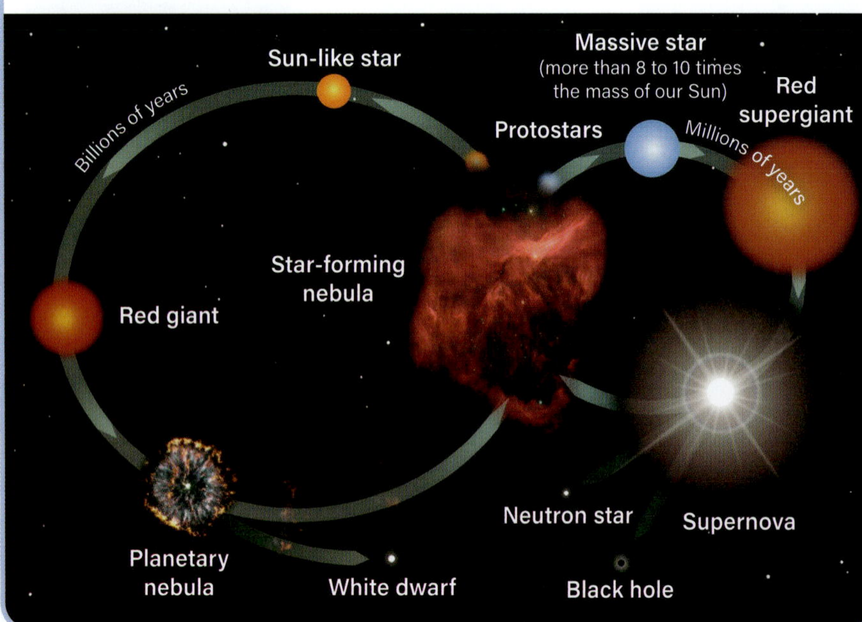

- Galaxies are groups of stars, gas and dust bound together by gravity.
- Solar systems are systems of planets that orbit stars.
- A light-year (ly) is the distance that light travels in one year.
- An astronomical unit (AU) is the average distance between the Sun and Earth.

- The big bang theory is the currently accepted theory explaining the origins of the universe.
- Evidence for the big bang theory includes:
  - the continued expansion of the universe
  - the abundance of light elements
  - the existence of cosmic microwave background radiation.

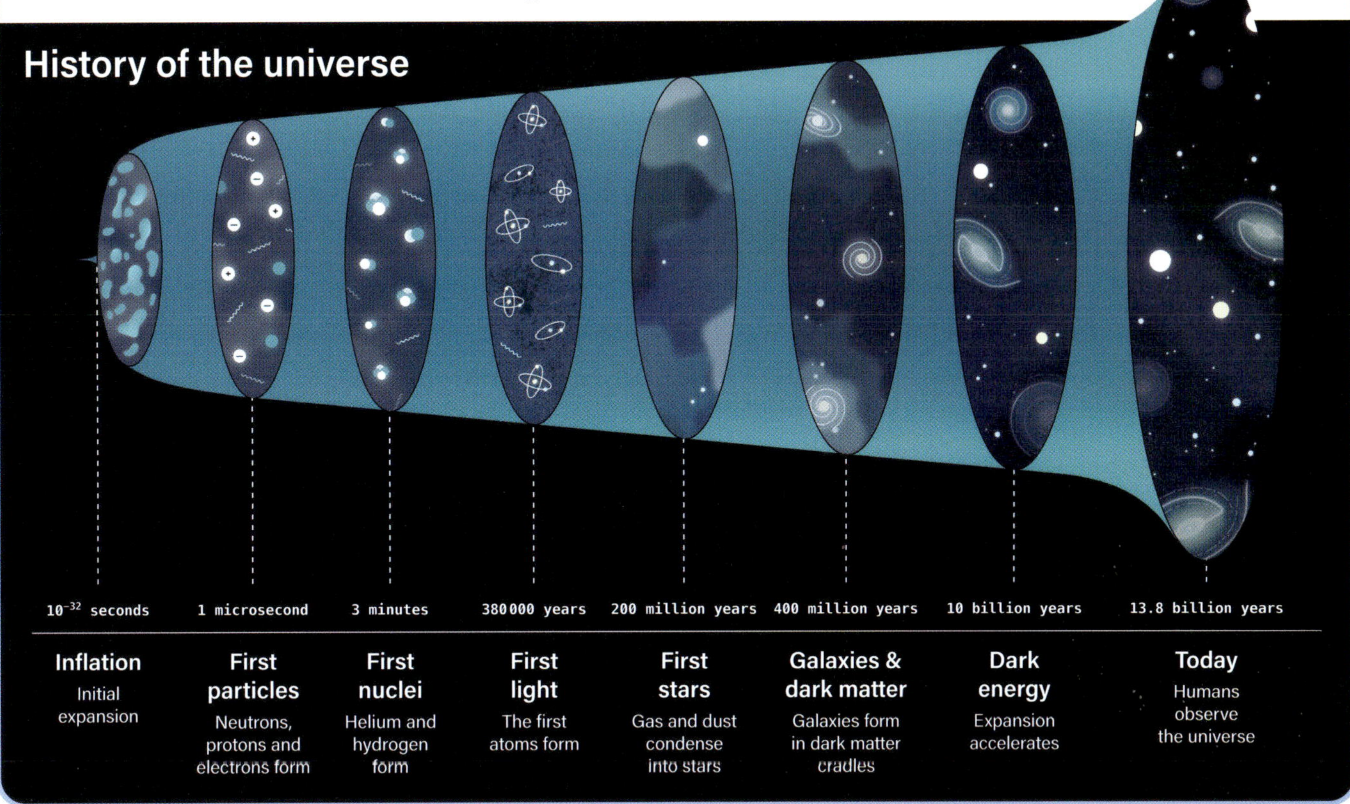

The major challenges of human space exploration are:

- space radiation
- isolation and confinement
- need for self-sufficiency due to distance from Earth
- impacts of changes in gravity on the human body
- health hazards of enclosed environments.

Key idea: Scale and measurement

- Astrophysicists use different wavelengths across the electromagnetic spectrum to learn more about the universe.
- Radio waves and microwaves provide data on how cold gas and dust moves.
- Infrared energy provides data on how warm gas and dust moves.
- Visible light allows us to measure temperature and what elements are present.
- Ultraviolet wavelengths allow us to measure extremely hot objects.
- X-rays are released from superheated material.
- Gamma rays provide data on high-energy events such as supernovas and black holes.

Masterclass

Steps in progression

		1	2
Science understanding	The universe	Identify the key features of the Hubble Ultra Deep Field.	Describe where you would look in the night sky to find the Hubble Ultra Deep Field.
Science as a human endeavour	Nature and development of science	Identify the technologies that could have been used to create the Hubble Ultra Deep Field.	Describe some of the discoveries made by analysing the Hubble Ultra Deep Field.
Science as a human endeavour	Use and influence of science	Identify an example of how space exploration has benefited society.	Describe examples of how the Hubble Ultra Deep Field image can provide different information.
Science inquiry	Questioning and predicting Think of an investigation that could be conducted on the International Space Station.	Make a prediction that describes a possible outcome of the investigation.	Write a scientific question that could be answered by the investigation.
Science inquiry	Communicating Prepare a short, informative presentation on the Hubble Ultra Deep Field.	Identify the key scientific terminology and concepts you could use in this presentation.	Brainstorm some ways you could communicate the information.

The Hubble Ultra Deep Field

The Hubble Ultra Deep Field is one of the most well-known astronomical images ever taken and has had a significant impact on our understanding of the universe.

Taken by the Hubble Space Telescope over a period during 2003 and 2004, the image combines ultraviolet, visible and near-infrared data to show a tiny patch of sky in the constellation Fornax (Figure 6.52) — equivalent in size to a grain of sand held at arm's length. It was not known what the image would show, and astronomers were astounded that it included around 10 000 galaxies. These galaxies span a vast range of distances, including some of the most distant and earliest galaxies ever observed, dating back over 13 billion years. Using this image, astronomers have been able to estimate that there are over 260 billion galaxies in the universe.

The Hubble Ultra Deep Field is a snapshot of the universe when it was just a few hundred million years old. It helps astronomers to understand how galaxies formed and evolved after the big bang. Many of the galaxies in the image appear as faint smudges, some distorted by gravitational interactions, offering clues about the turbulent early universe. The James Webb Space Telescope is being used to take images in the same part of the sky to provide even more detail.

Demonstrate your understanding

3	4	5	
Describe how the Hubble Ultra Deep Field supports the big bang theory.	Discuss the impact that the Hubble Ultra Deep Field has had on our knowledge of the universe.	Evaluate the importance of the use of the James Webb Space Telescope for developing a deeper understanding of our universe.	
Suggest some examples of how space exploration has led to the development of new technologies.	Discuss the role of peer review in the analysis of the Hubble Ultra Deep Field image.	Evaluate this statement using evidence to support your response. *Our understanding of the universe cannot improve unless technology to observe the universe also improves.*	
Explain how exploration of the universe has been driven by society's interest in astronomy.	You have been offered the opportunity to be on a Mars mission. Discuss how scientific knowledge could influence your decision as to whether to join this mission.	Analyse the key factors for the big bang theory becoming accepted by society as the explanation for the origins of the universe.	
Write a hypothesis that addresses your scientific question from step 2.	Discuss how you would design this investigation to ensure it is valid and reliable.	Explain what you would hope to discover by conducting this investigation and how it would build on our scientific understanding.	Science how-to p. 234
Brainstorm how you could use digital technologies to enhance your presentation.	Select one of your ideas from step 2 and create it, incorporating any digital technologies you identified where appropriate.	Construct a short argument to justify the importance of scientific exploration to better understand our universe.	Science how-to p. 262

Figure 6.52: The famous Hubble Ultra Deep Field is an image of a small section of space that contains 10 000 distant galaxies. These galaxies are so far away that the light they emit takes millions of years to reach Earth. The oldest galaxies in this image appear to us as they did when the light started travelling, providing a snapshot of the early universe. The small red galaxies are the oldest, dating back around 13 billion years – 400 to 800 million years after the big bang. This snapshot is the result of 800 exposures, taken across 400 orbits of the Hubble around Earth.

7.0 Motion

We travel constantly, and often at very fast speeds. But it has taken us some time to understand the science of motion. When trains were first invented, some people believed that travelling at just 50 kilometres per hour might cause passengers to suffocate or even to go insane! Now that we have a better understanding of motion, we can travel safely, even at speeds that are faster than the speed of sound (343 metres per second, or 1235 kilometres per hour). Humans have built machines that fly, we have launched rockets and even travelled into space, and we travel on the roads nearly every day. By understanding forces and motion, we can travel safely, such as by creating safety features in cars and on the roads that protect us in the event of an accident.

Learning Ladder

The Learning Ladder for each chapter maps the Science Understanding, Science as a Human Endeavour and Science Inquiry strands that will be covered. Each ladder has five levels of progression, called steps. To climb the ladders, you need to develop fluency at each step. This will help you develop the ability to complete tasks that are more complex.

Steps in progression	Science understanding: Physical science: Motion	Science as a human endeavour: Nature and development of science	Science as a human endeavour: Use and influence of science
5	I can evaluate the impact of Newton's laws on our development of technologies and policies	I can analyse how advances in technologies enable advances in science	I can analyse the key factors that contribute to scientific knowledge being adopted more broadly by society
4	I can apply Newton's laws to analyse and predict the motion of objects in a system	I can discuss how scientific knowledge is validated, including the role of publication and peer review	I can discuss how scientific information and misinformation may inform personal and social decision-making
3	I can explain Newton's laws of motion using examples	I can explain how science has contributed to developments in technologies and engineering	I can explain how the values and needs of society influence the focus of scientific research
2	I can describe the motion of objects using Newton's laws and a range of representations	I can describe how scientific knowledge is refined over time	I can describe scientific knowledge that may be interpreted in different ways
1	I can identify features, properties and laws of motion	I can identify technologies that have enabled advances in science	I can identify scientific knowledge that can address socio-scientific issues

Figure 7.1: Using crash-test dummies to understand how forces and motion operate when a vehicle is involved in an accident has enabled the invention and improvement of car safety features such as seatbelts and airbags.

Steps in progression	Planning and conducting	Processing, modelling and analysing
5	I can evaluate scientific methods regarding safety, ethical and procedural considerations	I can analyse the quality of data using descriptive statistics
4	I can design and conduct reproducible investigations that consider safety, ethical and procedural factors	I can discuss relationships and anomalies that emerge in processed data
3	I can use equipment to generate and record data with precision, to obtain replicable data, using digital tools as appropriate	I can identify and explain trends and/or patterns in a range of dataset representations
2	I can develop and follow risk assessments that consider safety and ethical issues	I can process data by using mathematical relationships and/or constructing graphs
1	I can select appropriate equipment to collect precise data for scientific investigations	I can organise data and information using tables, keys and/or models

Science inquiry

7·1 ▸ Introducing motion

Figure 7.2: Hitting a hockey ball involves a contact force that changes the speed and direction of the ball.

Learning intention

At the end of this lesson, I will be able to:

- describe forces and motion
- explain why forces and motion are important to our everyday activities.

Key terms

contact force: a force that is applied by touching

force: a push or pull between objects that changes the motion speed and/or direction of their motion or their shape

friction: a force that resists motion when two surfaces rub on each other

gravity: a force of nature where two objects with mass attract each other

motion: a change in position of an object over time

non-contact force: a force that is applied without touching

non-uniform motion: a motion where there is a change in speed or direction

uniform motion: a motion with constant speed and direction

Key idea: Scale and measurement

Figure 7.3: Learning to drive requires that you understand the concept of motion.

Driving is a skill that has to be learnt, as being in control of a vehicle is a big responsibility! When learning to drive, you experience forces and motion in practice. Understanding the properties of motion, such as speed and acceleration, while driving can make travelling safer and more comfortable. Knowing about forces and motion may help you to avoid an accident or to save a life on the roads, so increasing your understanding of them can be very useful in everyday life.

Motion is a change in the position of an object

Motion is a change in the position of an object over time. In other words, an object has motion if it moves from one place to another during a period of time. When moving, an object will display a number of properties of motion that we can describe and measure. These properties include speed, velocity, distance, acceleration and direction. We will learn more about them in this chapter.

Motion can be **uniform** – like a constant speed in one direction. A car travelling down a straight road at the same speed is an example of uniform motion. Motion can also be **non-uniform**, such as when there is a change in speed or direction of the motion. A car travelling over a winding road that requires speeding up, slowing down and turning corners is an example of non-uniform motion.

Forces change motion

A **force** is a push or a pull between objects that changes the motion of an object. A force can change the speed, direction or shape of an object. For example, applying the brakes in a car can slow it down. The motor can apply a force that speeds up the car, and turning the steering wheel applies a force that turns the car in a different direction. In a collision, the shape of the car can be changed by the forces acting on it during the crash.

We have already learnt that force can be a **contact force**, such as **friction**, or a **non-contact force**, like **gravity**. Forces are measured in newtons (N).

Understanding motion helps us to analyse everyday life

Do you know how cars measure speed, or why we wear seatbelts? Can you explain how an airbag works? What about how far a car travels when slowing from 100 kilometres an hour to a stop? Is the distance to stop longer when the road is wet? Many of the map applications (apps) we use today on our devices measure and calculate information about motion, like the average speed of a trip and how long it will take to get somewhere. But how does the app calculate these properties? To explain all of these things and to ensure we stay safe on the roads, we will learn about motion.

The number of lives lost on our roads is currently increasing. Table 7.1 shows the lives lost per year in Victoria, including by gender. There are now more cars on the road than ever before, which may partly explain this increase. But cars are also safer than they used to be, with better safety standards and technology to help them avoid accidents. So, this means there must be other possible reasons why the number of lives lost is increasing. These may include:

- an increase in the number of vulnerable road users, such as scooter riders or motorcyclists
- poorer-quality roads, including in regional Victoria
- a higher reliance on technology to save us
- risky behaviours, such as speeding, or texting while driving
- other factors.

As we learn about motion, we will consider how that knowledge is used in sports, road safety and many other contexts.

Table 7.1: Lives lost on Victorian roads, 2020 – June 2025

Gender	2020	2021	2022	2023	2024	2025 (to 30 June)
Female	52	64	61	89	76	36
Male	159	170	179	206	208	111
Total	211	234	241	295	284	147

Learning Ladder

Motion

1. Identify:
   - **a** a contact force.
   - **b** a non-contact force.
   - **c** the unit for measuring force.
2. Describe an example of non-uniform motion in a sporting activity.
3. Explain, using an example, what a force is.

Use and influence of science

1. Identify some properties of motion that, if understood, can help to address the issue of road safety.
2. Describe two different ways the data shown in Table 7.1 may be interpreted by a scientist studying road safety.
2. Explain how the data in Table 7.1 may influence research on improving car safety.

Processing, modelling and analysing *p. 244*

1. Copy and update Table 7.1 to include complete data for 2025 and any further years for which data is available. You will find the data on the Victorian Transport Accident Commission website.
2. Construct a line graph of all the data you have included in Table 7.1.
3. Using your graph from Question 2, identify and explain any overall trends in the data.
4. Using your graph from Question 2, discuss and compare the relationship between the lives lost for males and females.
5. Still using Table 7.1, calculate:
   - **a** the average number of lives lost on Victorian roads per year:
     - **i** in total.
     - **ii** for males.
     - **iii** for females.
   - **b** the overall percentage of male and female lives lost across all years.

Key idea: Scale and measurement

Undertake research to collate a table of data that measures lives lost for all of Australia. Describe and analyse the data to identify whether the trends are similar to or different from Victoria and suggest possible reasons why.

Success criteria

- I can describe forces and motion.
- I can explain why forces and motion are important to everyday activities.

7·2 ▸ Motion of objects

Learning intention

At the end of this lesson, I will be able to:

- analyse and calculate the relationships between distance, time, speed, displacement and velocity
- describe motion using diagrams and graphs.

Key terms

average speed: the speed averaged for an entire journey

displacement: the distance an object moves from its starting position

distance: the total length an object travels

instantaneous speed: the speed at a particular time in a journey

speed: the distance an object travels divided by the time taken to travel that distance

velocity: a measure of how quickly displacement changes

Investigation 7.2

Cars and pedestrians, p. 370

Key idea: Scale and measurement

The world's fastest person is Jamaican sprinter Usain Bolt. Bolt is 1.96 metres tall and holds world records in the 100- and 200-metre sprints and the 4 × 100-metre relay. You can use the distance of his races and the time he ran them in to calculate his speed. You can also determine his velocity from his displacement and time. How fast do you think Usain Bolt can run? What units are you using to estimate this speed?

Speed is distance divided by time

Distance is how far an object travels. **Speed** is the distance an object travels divided by the time it takes to travel that distance. Speed is often measured in metres per second (m s^{-1}) or kilometres per hour (km h^{-1}). The formula for speed is:

$$\text{Speed} = \frac{\text{distance}}{\text{time}}$$

If a car travels 60 kilometres in 1 hour, its speed is 60 km h^{-1}. From this equation, you can see that if Usain Bolt runs a certain distance in less time than other runners, then he has a higher speed.

A formula triangle allows you to calculate speed

Speed is usually measured in metres per second (m s^{-1}), distance is usually measured in metres (m) and time is usually measured in seconds (s). To calculate distance or time, you can use a formula triangle (Figure 7.4).

For example, in 2009 Usain Bolt ran 100 metres in a world-record time of 9.58 seconds. Using the distance and time, we can calculate his speed. We know that:

$$\text{Speed} = \frac{\text{distance}}{\text{time}}$$

$$s = \frac{D}{t}$$

$$= \frac{100}{9.58}$$

$$= 10.24 \text{ m s}^{-1}$$

This is really the average speed, because it would have varied during the run.

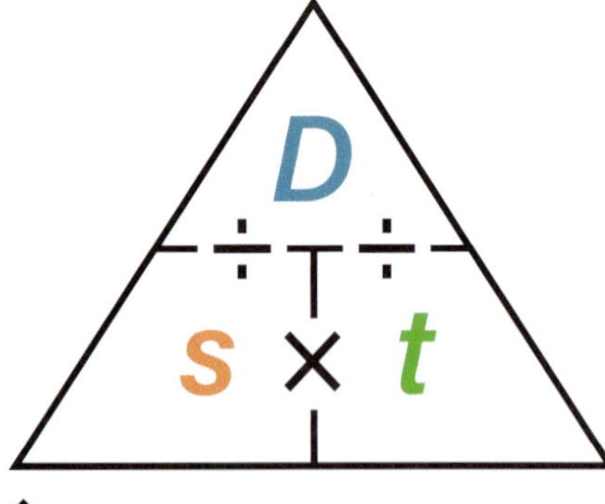

Figure 7.4: A formula triangle for distance (D), speed (s) and time (t)

Average speed: 10.423 m s^{-1}

Time: 19.19 s

Distance ___?___ m

Figure 7.5: Distance travelled is calculated using speed and time. How far did Usain Bolt travel in his gold-medal run in 2016?

If we know speed, we can calculate distance

If one person is jogging and another person is sprinting, and they maintain a constant speed for the entire time, who will travel further in 1 minute? The sprinter is moving at a higher speed , so will cover a greater distance in 1 minute.

If you are riding a bicycle at a constant speed of 4 m s^{-1}, how far would you travel in 3 seconds? From the formula triangle (Figure 7.4), you can see that the formula for distance is $D = t \times s$. So, your distance travelled is 3 s × 4 m s^{-1} = 12 m.

Distance is different from displacement

Distance is how far an object travels, but **displacement** is how far the object is from its initial starting position. If you walk 6 metres north, then 8 metres west, you have travelled a total distance of 6 + 8 = 14 metres. (Try drawing this on a piece of paper.) However, you are only 10 metres from your starting position, so your displacement is 10 metres north-west. Figure 7.6 shows an example.

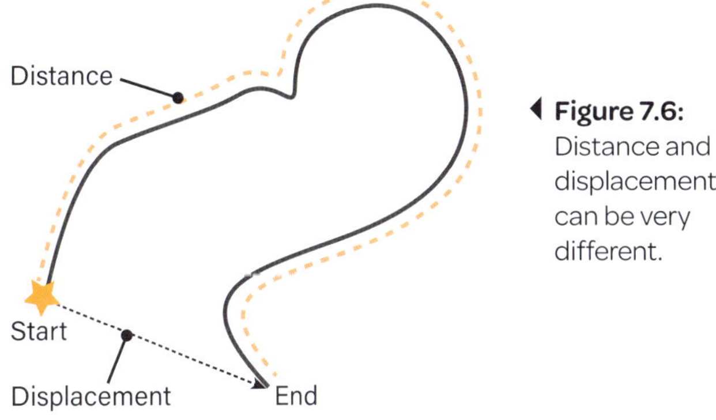

Figure 7.6: Distance and displacement can be very different.

Velocity is a change in speed with direction

Velocity is a measure of the change in displacement in a certain time. In the previous example, if you were walking for 10 seconds, your average speed is distance travelled divided by time taken, so $\frac{10}{14}$ = 1.4 m s^{-1}.

Your average velocity is your displacement divided by the time taken.

The formula for average velocity is:

$$v_{av} = \frac{\Delta s}{\Delta t}$$

where:

Δs is the change in displacement. (Note that '*s*' in this formula is not speed.)

So, your average velocity is:

$v_{av} = \frac{10}{10}$ = 1.0 m s^{-1} north-west

Distance–time graphs represent motion

We can show motion in a distance–time graph, with time on the *x*-axis and distance on the *y*-axis. A distance–time graph tells you the following.

- When distance increases, the object moves away from the starting position.
- The slope or gradient of the graph indicates the speed – the steeper the slope, the higher the speed, because the object is travelling a greater distance in a certain time.
- If the line is flat, the object is stationary (not travelling any distance). From $s = \frac{D}{t}$, if the distance is zero (not changing), then the speed is zero.
- A straight line indicates the speed is constant (not changing).
- A curved line indicates the speed is changing – either increasing (accelerating) or decreasing (decelerating).
- You can work out the distance travelled from a starting point, and whether the object may return to the start.

From the graph in Figure 7.7, you can work out what each journey shows.

- Journey 1 shows a fast constant speed (steep, straight line going up).
- Journey 2 starts with a slower constant speed (straight line going up), then the person stops (flat line) and returns to their starting position (straight line going down).
- Journey 3 shows a person speeding up (line curving upwards), then slowing down (line curving downwards) and stopping (line flat).

To find the speed at any point on the graph, find the gradient (or slope) at any point in time. For example, in Journey 1, from 0 to 5 seconds, the person travelled 100 metres. From $s = \frac{D}{t}$, the speed is $\frac{100 \text{ m}}{5 \text{ s}} = 20 \text{ m s}^{-1}$.

For Journey 3, the line is bending, so the slope is changing and the speed is not constant. We can still calculate the speed:

1. The **average speed** can be calculated for the entire journey by the total distance over the total time until they stop. Using $s = \frac{D}{t}$, average speed is: $\frac{90 \text{ m}}{11 \text{ s}} = 8.2 \text{ m s}^{-1}$.
2. But we know that the speed changes along the journey. We can calculate the **instantaneous speed** (the speed at any point in time) by calculating the slope of the graph at a particular point. For example, at 50 metres, we can calculate the speed from the slope between 6 and 8 seconds using $s = \frac{D}{t}$, $s = \frac{70-30}{8-6} = \frac{40}{2} = 20 \text{ m s}^{-1}$. This is possible because the slope is constant for this section of the graph.
3. If the slope is curving, we can calculate the speed by drawing a tangent to represent the slope at that point. For example, at 4 seconds into the journey, using the tangent in the graph and $s = \frac{D}{t}$, the instantaneous speed is: $\frac{15}{3} = 5 \text{ m s}^{-1}$.

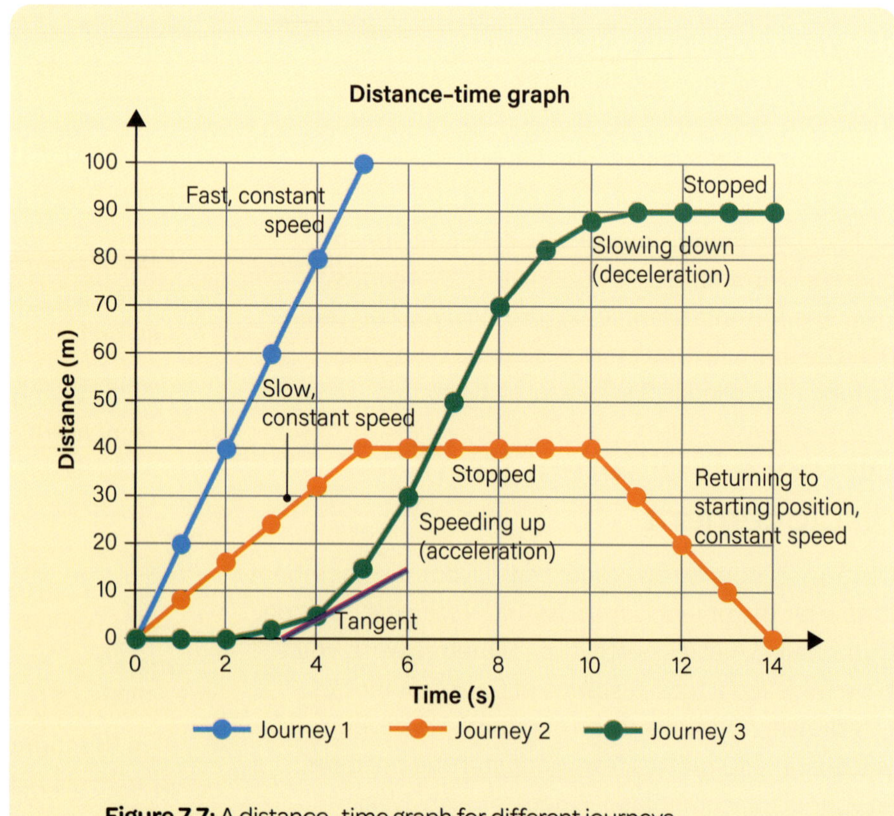

Figure 7.7: A distance–time graph for different journeys

Learning Ladder

Motion

1. Identify the symbols and units for distance, time, speed, displacement and velocity.
2. Describe how distance, time and speed are related, including the formula for this relationship.
3. Explain the difference between speed and velocity and outline an example for each.
4. Analyse the distance–time graph in Figure 7.7 to calculate or identify the:
   a speed for Journey 1 and Journey 2 between 0 and 5 seconds.
   b total distance and displacement travelled for Journey 1.
   c total distance and displacement for Journey 2.
   d instantaneous speed in Journey 3 at 9 seconds.
5. Evaluate, using an example, how our understanding of motion, including speed and distance, has contributed to the improvement of performance in sport.

Use and influence of science

1. Identify and outline an example of how speed and velocity are used in real life.
2. **a** Distinguish between distance and displacement, using an example.
   b A person walks 8 metres north and then 2 metres east. Draw a diagram of their motion, showing the distance and displacement.
   c If it took the person in part b 8 seconds to complete the walk, calculate their speed and velocity.
3. Explain how society's interest in sport and human performance influences research into the performance of athletes.
4. Discuss how data collected about athletes can be used to determine whether a person will remain as an amateur athlete or may be able to compete as a professional.

Processing, modelling and analysing p. 244

1. A student sketched and labelled the following graphs from an investigation. Some of the labels are not correctly matched to the graphs. Copy the graphs and label them correctly.

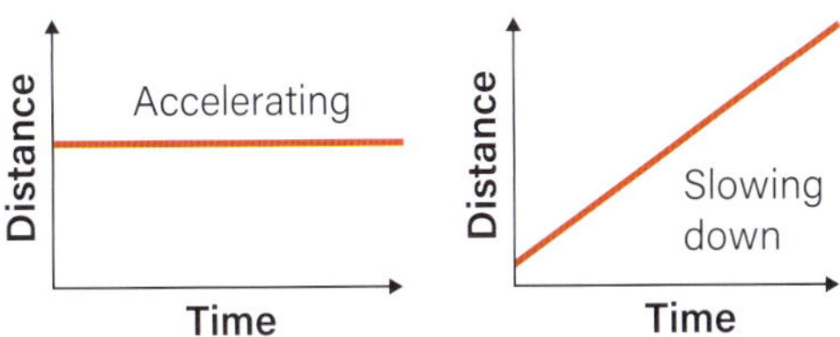

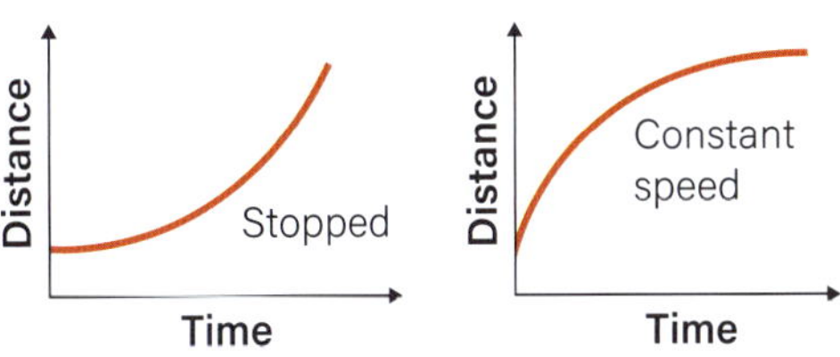

2. The following data was collected about a person's travel. Construct a distance–time graph for this journey.

Time (s)	Distance (m)
0	0
1	4
2	8
3	12
4	12
5	12
6	12
7	20
8	30
9	40
10	50
11	80
12	85
13	90
14	90

3. Using your graph from Question 2, explain the motion of the person at each stage of the journey.
4. Using Figure 7.7, discuss the relationship between the slope (gradient) of the graph and speed of the motion.
5. Using the data from Question 2, calculate the:
   a average distance travelled per second.
   b average speed.

Key idea: Scale and measurement

Usain Bolt is the world's fastest person. Research his incredible 200-metre world-record race in 2009. Using the data measured from his run, calculate his average speed and velocity during the race, using a diagram of the track as well as calculations for speed and velocity. *Hint:* You will need to determine his displacement first.

Success criteria

- I can analyse and calculate the relationships between distance, time, speed, displacement and velocity.
- I can describe and analyse motion using diagrams and distance–time graphs.

7.3 ▸ Speed and acceleration

Learning intention

At the end of this lesson, I will be able to:

- analyse and calculate the relationships between time, speed and acceleration
- describe motion using diagrams and speed–time graphs.

Key terms

acceleration: a change in speed over time

deceleration: a decrease in speed over time

Investigation 7.3

Ticker timers, p. 372

Key idea: Scale and measurement

Figure 7.8: A rollercoaster speeds up, changes direction and slows down – all of which are types of acceleration.

Acceleration, that feeling of speeding up or slowing down, is an everyday part of travelling in a car. It can be very exciting, such as when a rollercoaster drops from the highest point and accelerates so fast you feel as though you have left your stomach behind!

Acceleration is a change in speed

Acceleration is a measure of how quickly an object's speed changes. If you drop a ball, it accelerates towards the ground – that is, its speed increases downwards. If you press the accelerator pedal in a car, the car's speed increases in the direction it is moving.

Acceleration is *any change* in speed or direction over time – speeding up, slowing down or turning.

- If acceleration is in the same direction as an object's motion, the object speeds up.
- If acceleration is in the opposite direction to an object's motion, the object slows down (known as **deceleration**).
- If acceleration is not parallel to the object's motion, the object changes direction.

As long as speed is changing, the object is accelerating. If an object is not moving or is moving at a constant speed, then it is not accelerating. Acceleration (a) is measured in metres per second squared ($m\ s^{-2}$) and equals the change in speed divided by the change in time:

$$a = \frac{\Delta v}{\Delta t}$$

Notice the use of the symbol v for speed. This is velocity, which we learnt is speed with direction.

For example, if a car is travelling at 5 m s^{-1} and increases its speed to 25 m s^{-1} over a period of 10 seconds, what is the acceleration?

1 Change in speed = 20 m s^{-1} (final speed – initial speed).

2 Change in time is the 10 seconds it took to speed up.

3 Applying the formula:

$$a = \frac{\Delta v}{\Delta t}$$
$$= \frac{25-5}{10}$$
$$= \frac{20}{10}$$
$$= 2 \text{ m s}^{-2}$$

Speed–time graphs give us more information

We can show speed, acceleration and distance on a speed–time graph, where time is on the *x*-axis and speed is on the *y*-axis.

- When the graph is flat, the speed is constant.
- Acceleration is shown by the gradient (slope) of the line:
  - When the graph slopes upwards with a positive gradient, the object is accelerating.
  - When the graph slopes downwards with a negative gradient, the object is decelerating.
- The area under the graph gives the distance travelled.

We can use this information to analyse the person's journey shown in Figure 7.10.

1 Initially, the person starts with zero speed and accelerates to 10 m s^{-1} over 4 seconds. We can calculate the gradient of the line to find the acceleration:

$$a = \frac{\Delta v}{\Delta t}$$
$$= \frac{10}{4}$$
$$= 2.5 \text{ m s}^{-2}$$

2 After accelerating, the line is flat, which shows a constant speed of 10 m s^{-1} between 4 and 8 seconds.

3 From 8 to 10 seconds, the person accelerates again, then there is another period of constant speed from 10 to 12 seconds.

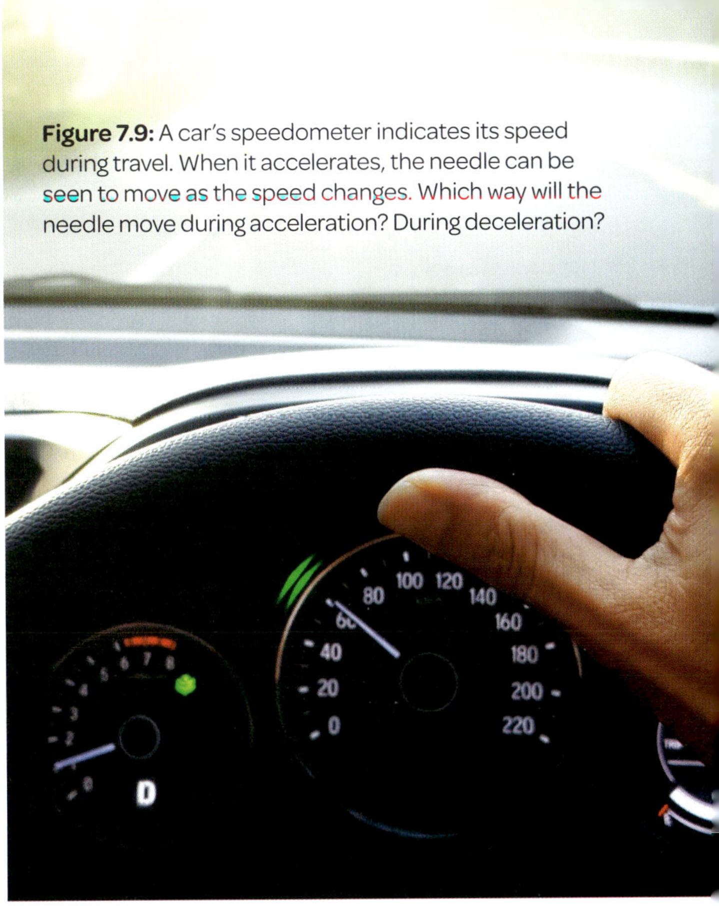

Figure 7.9: A car's speedometer indicates its speed during travel. When it accelerates, the needle can be seen to move as the speed changes. Which way will the needle move during acceleration? During deceleration?

4 Finally, the person decelerates and stops at 16 seconds.

5 We can work out the distance travelled by calculating the area under the graph. For example, between 4 and 6 seconds, the area under the graph is (8 – 4) × (10 – 0) = 4 × 10 = 40 m.

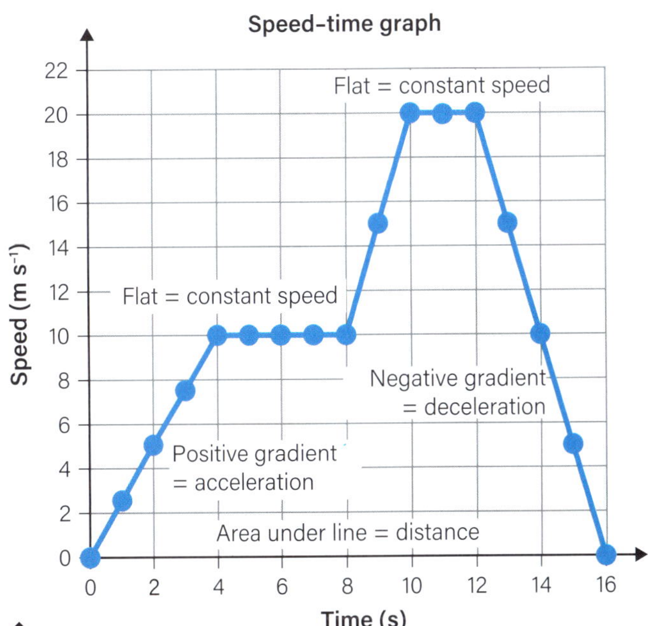

Figure 7.10: A speed–time graph for different stages of a person's journey

Figure 7.11: Different cars accelerate at different rates in a drag race.

Cars accelerate rapidly!

We can compare cars on the basis of how long they take to accelerate from 0 to 100 km h^{-1}. We do this by calculating the average acceleration over a period of time, because the car may accelerate at different rates before it reaches 100 km h^{-1}.

Table 7.2 shows the times for two cars to accelerate from 0 to 100 km h^{-1}. The overall change in speed is 100 – 0 km h^{-1} = 100 km h^{-1}. We need to convert km h^{-1} to m s^{-1}:

$$100 \text{ km h}^{-1} = \frac{100}{3.6} = 27.8 \text{ m s}^{-1}$$

More information about converting between units is available in the Science how-to section on pages 284–85.

Car 1 takes 6.6 seconds to reach 27.8 m s^{-1}. Therefore:

$$a = \frac{\Delta v}{\Delta t} = \frac{27.8}{6.6} = 4.2 \text{ m s}^{-1}$$

Table 7.2: The times for two cars to accelerate from 0 to 100 km h^{-1}

Speed (km h^{-1})	Car 1: Time (s)	Car 2: Time (s)
0	0	0
25	1.6	1.0
50	3.1	2.2
75	5.0	6.1
100	6.6	8.2

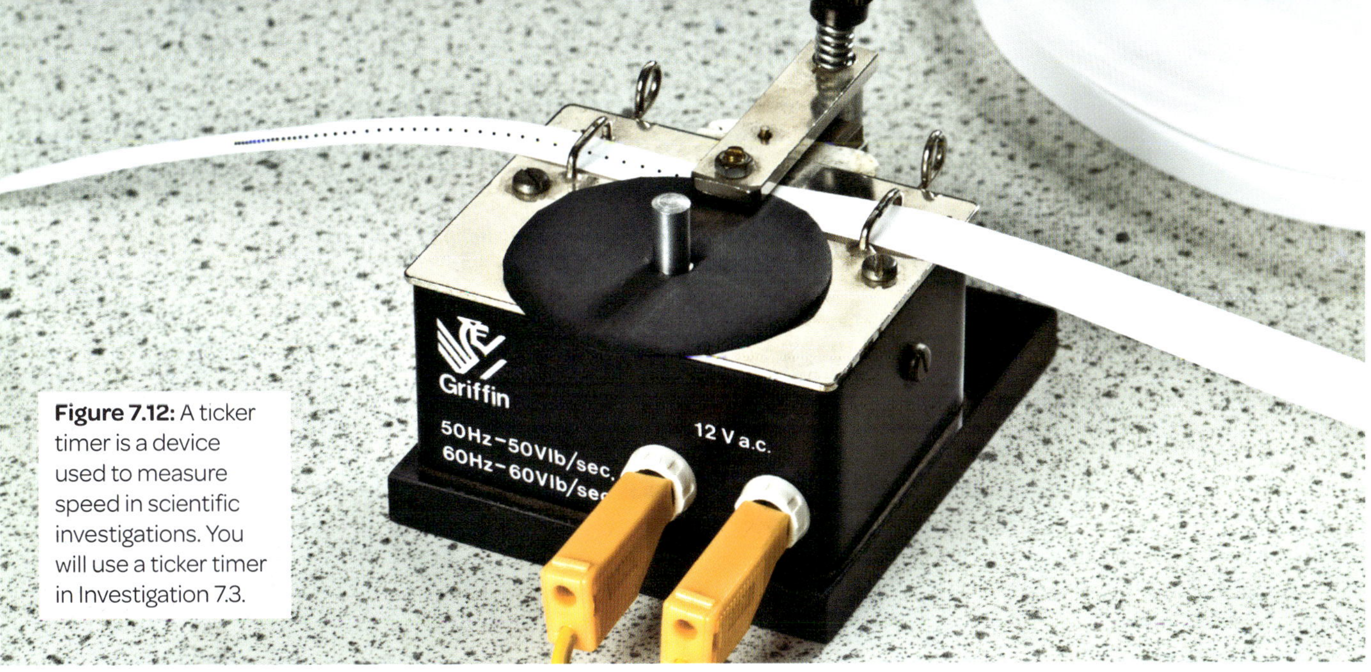

Figure 7.12: A ticker timer is a device used to measure speed in scientific investigations. You will use a ticker timer in Investigation 7.3.

Learning Ladder

Motion

1. Identify the three types of acceleration and describe the features of each.
2. Describe the difference in acceleration in Figure 7.10 from 0 to 4 seconds, and from 8 to 10 seconds. Justify your answer using your knowledge of the gradient.
3. Use the example of motion in the speed–time graph in Figure 7.10 to calculate and explain the:
   - **a** acceleration between 8 and 10 seconds.
   - **b** speed between 10 and 12 seconds.
   - **c** distance travelled between 10 and 12 seconds.
   - **d** acceleration (or deceleration) between 12 and 16 seconds.
   - **e** distance of the entire journey.
4. Analyse the motion of a rollercoaster by answering the following questions.
   - **a** Calculate the change in speed of a rollercoaster that accelerates to the bottom of a drop at 9.8 m s^{-2} for 4 seconds.
   - **b** If the rollercoaster was already moving at 10 m s^{-1}, what is the final speed at the bottom?
   - **c** Convert the final speed to km h^{-1}.
5. Evaluate the importance of speed–time graphs and explain how this understanding may be used to develop a technology in relation to motion.

Use and influence of science

1. Identify an example of how our knowledge of speed and acceleration can be used to address an issue in road safety.
2. Describe the different scientific information that may be interpreted from the movement of the speedometer needle in a car.
3. Explain how our need to make cars safer influences research into acceleration, deceleration and motion.
4. Discuss how your knowledge of acceleration, deceleration and motion may inform your decision about whether to buy an old car or a newer one.

Processing, modelling and analysing p. 244

1. Construct a table of data showing speed and time from the graph in Figure 7.10.
2. Construct a speed–time graph of the data in Table 7.2. You will need a line for each car.
3. Use your graph from Question 2 and Table 7.2 to answer the following questions.
   - **a** Calculate the average acceleration of each car and identify which one accelerated fastest to reach the speed of 100 km h^{-1}.
   - **b** Identify which car accelerated fastest to reach the speed of 50 km h^{-1}. Justify your answer.
4. Using Figure 7.10, discuss the relationship between the slope (gradient) of the graph and the acceleration.

Key idea: Scale and measurement

Investigate measurements for the acceleration of different machines, including rockets, aeroplanes, cars, motorbikes and bicycles. Collect data and construct a table to compare the size of the acceleration of each machine you investigate.

Success criteria

- I can analyse and calculate the relationships between time, speed and acceleration.
- I can describe and analyse motion using speed–time graphs.

7.4 ▸ Newton's laws of motion

Learning intention

At the end of this lesson, I will be able to:

- explain Newton's laws of motion
- apply the laws using examples.

Key terms

balanced forces: forces acting on an object are equal and cancel out, so that motion remains the same

inertia: a property of matter that causes it to resist change in speed or direction (to remain at rest or in a state of uniform motion)

mass: the amount of matter in an object, measured in grams (g) or kilograms (kg)

unbalanced forces: forces acting on an object are different in size and do not cancel out but cause a change in motion

Investigation 7.4A

Car crashes and inertia, p. 374

Investigation 7.4B

Balloon rockets, p. 376

Key idea: Scale and measurement

Isaac Newton was an English scientist who lived in the 1600s. He is so well regarded that the unit of force is named after him. Among his many achievements are his three laws of motion, which describe how objects move in relation to the forces applied to them.

Newton's first law: an object stays at rest or at the same speed unless an unbalanced force acts on it

Newton's first law states that an object at rest remains at rest, and an object in motion remains in motion at a constant speed and in a straight line, unless acted on by an unbalanced force that changes its motion.

An **unbalanced force** is a force that causes a change in motion. This means that a stationary object will remain stationary until a force such as a push or pull starts it moving. It also means that a moving object will keep moving unless a force acts on it, such as pushing the object to change its direction or pushing or pulling it to slow it down or speed it up.

If a car is rolling on the road, it will not keep rolling forever because there is an unbalanced force acting on the car – friction. The car will experience friction from the road and from the air. The friction converts the kinetic energy to heat and sound, and the car slows and stops.

A car travelling at a constant speed also experiences friction. In this case, the friction from the road is equal to the thrust from the engine – the forces are equal in both directions and **balanced** (Figure 7.13a).

If the car accelerates, then thrust is larger than friction and the unbalanced force causes the car to accelerate (Figure 7.13b).

If the car brakes, friction is larger than thrust and the unbalanced force causes the car to slow down (Figure 7.13c).

The property of an object that keeps it doing what it is doing – whether that is to stay in motion or at rest – is **inertia**. Inertia is related to **mass**. An object with more inertia (more mass) requires more force to start or stop it. For example, a truck has more inertia than a car, so the truck is harder to stop because it has more mass.

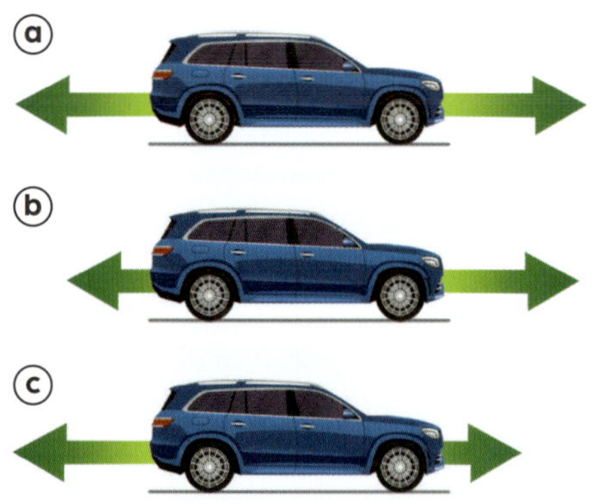

Figure 7.13: Newton's first law: unbalanced and balanced forces on a car. Can you tell which car is accelerating?

Newton's second law: force = mass × acceleration

Newton's second law states that the acceleration of an object is proportional to the force applied and inversely proportional to the mass of the object.

This law is represented by the equation $F = m \times a$, where F is force (N), m is mass (kg) and a is acceleration (m s^{-2}).

The heavier an object is, the more force is required to accelerate it. If the force applied to an object is constant (the same), and the mass is doubled, then the acceleration halves. If mass is halved, then acceleration is doubled. We will learn more about Newton's second law in Section 7.5.

Newton's third law: for every action, there is an equal and opposite reaction

Newton's third law tells us that every force has an equal and opposite reaction force.

This means that if object 1 exerts a force on object 2, then object 2 exerts the same-sized force on object 1, but in the opposite direction. For example, if you push on the wall with both hands, you can feel the wall pushing back on your hands as pressure – the harder you press (the more force you apply), the more you can feel the wall pushing back with equal force.

Riding a skateboard involves pushing backwards to move forwards. These equal and opposite forces are known as action–reaction pairs.

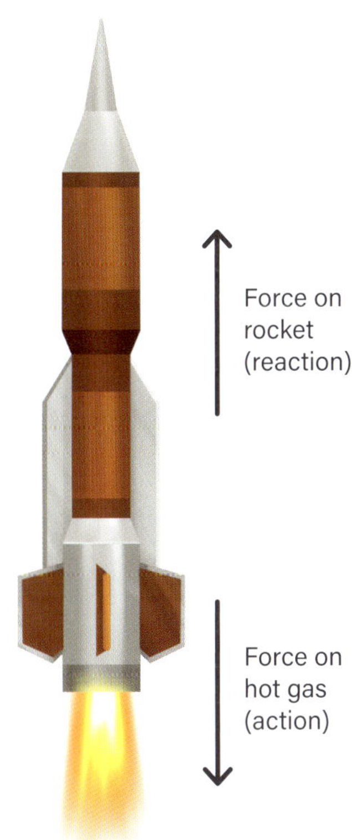

Figure 7.14: Newton's third law: action and reaction forces in a rocket

Learning Ladder

Motion

1. Identify the formula for Newton's second law, including what each variable in the formula represents and their units.
2. Draw a diagram of a real-life example to compare balanced and unbalanced forces on an object.
3. Using Newton's third law, explain how a rocket works.
4. James is standing on a skateboard, holding a large heavy medicine ball. He throws the ball to Jenny.
   - **a** Predict what you would observe when James throws the ball to Jenny.
   - **b** Identify which of Newton's laws this example demonstrates. Justify your answer.

Nature and development of science

1. Suggest an example of technology that could be useful for investigating forces and Newton's laws.
2. Describe how Newton may have tested and developed his laws of motion over time to ensure they were accurate.

Planning and conducting

p. 239

A student performed the following investigation.

- They loaded a toy trolley with some objects to increase its mass. They then applied a gentle force to push the trolley from a start line on a smooth surface and measured the distance the trolley travelled.
- They repeated pushing the trolley with the same mass, but this time they applied more force and pushed harder.
- They added some objects to the trolley to increase its mass. Again, they pushed it gently, and then with more force.

Answer the following questions about this investigation.

1. Construct a list of equipment required for this investigation to collect precise data.
2. Propose some safety measures to consider when developing a risk assessment for undertaking this investigation.
3. Propose how you would use the equipment to generate data, including the sample size required in your data collection.
4. Design a step-by-step procedure and results table for this investigation that will ensure replicable results.
5. Evaluate the student's procedure as to whether it is likely to produce replicable results in a safe manner.

Key idea: Scale and measurement

Using an example of a car being pushed off the road, construct an illustration to show the size of the forces when the car starts moving, moves at a constant speed and then slows down to a stop. Explain which laws and properties of motion are involved in each stage of this example.

Success criteria

- I can explain Newton's laws of motion.
- I can apply Newton's laws of motion to examples.

7·5 ▸ Force, mass and acceleration

Figure 7.15: It requires less force to accelerate a skateboard than to accelerate a heavy car.

Learning intention

At the end of this lesson, I will be able to:

- analyse the relationships between force, mass and acceleration
- apply Newton's second law.

Key terms

inversely proportional: as one quantity increases, the other decreases

proportional: as one quantity increases, so does the other

weight: a force acting downwards on a mass due to gravity

Investigation 7.5

Acceleration and mass, p. 377

Key idea: Scale and measurement

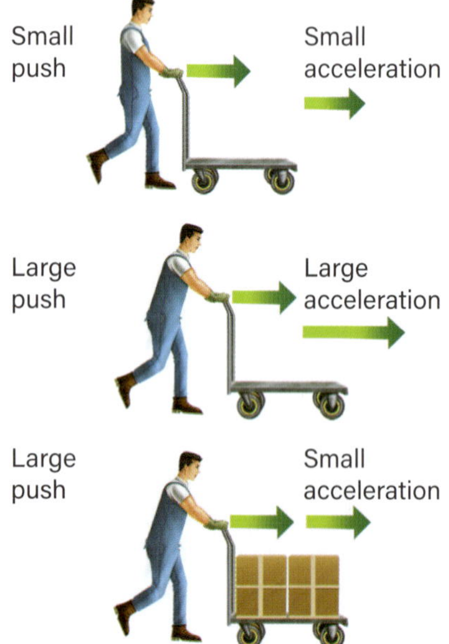

Figure 7.16: Newton's second law: the heavier an object is, the more force is required to accelerate it.

It requires a lot of force to get a car moving. But you can get a skateboard moving with just your foot. The difference is mass: a car is much heavier than a skateboard and so requires a much greater force to make it accelerate.

Acceleration depends on mass and the size of the force being applied

If you use your foot to propel yourself on a skateboard, the force of your foot pushing backwards against the ground causes your acceleration forwards. If someone is standing on the skateboard with you, the total mass increases, so you need to apply more force.

The relationship between force, mass and acceleration is summed up by Newton's second law, which says that the acceleration of an object is **proportional** to the force applied, and **inversely proportional** to the mass of the object:

$$F = m \times a \text{ or } a = \frac{F}{m}$$

As shown in Figure 7.16, if you increase the force on an object of fixed mass, the acceleration increases. If you use the same force but increase the mass, the acceleration decreases.

We can use $F = m \times a$ to calculate force

You can calculate the force required to accelerate an object from the formula $F = m \times a$. To calculate mass or acceleration, you can use the formula triangle in Figure 7.17.

Table 7.3 shows the force required to accelerate three trolleys. Trolleys 1 and 2 are the same mass, but Trolley 2 has more force applied, so it has a higher acceleration. Trolleys 2 and 3 have the same force applied, but Trolley 3 is heavier, so it has a lower acceleration.

Table 7.3: Calculation of forces to accelerate trolleys

Trolley	Mass (kg)	Force (N)	Acceleration ($m\ s^{-2}$)
1	20	50	2.5
2	20	500	25
3	25	500	20

Weight is a measure of force

Mass and weight are not the same. Mass is measured in grams or kilograms and is a measure of the amount of matter in an object, whereas **weight** is the force acting on your body's mass due to gravity.

To calculate the weight force, we can rewrite $F = m \times a$ as $W = m \times g$, where W is weight (N) and g is the acceleration due to gravity (9.8 $m\ s^{-2}$ on Earth).

So, if your mass is 68 kilograms, your weight on Earth is: $W = 68 \times 9.8 = 666.4$ N. Heavier objects with more mass have larger weights.

The acceleration due to gravity on Earth is always 9.8 $m\ s^{-2}$, so two objects that are dropped always accelerate at the same rate and hit the ground at the same time.

Gravity changes as we move away from Earth's surface. Gravity on the Moon and other planets is also different from gravity on Earth. A person would weigh less on the Moon (g = 1.6 $m\ s^{-2}$) and more on Jupiter (g = 24.8 $m\ s^{-2}$).

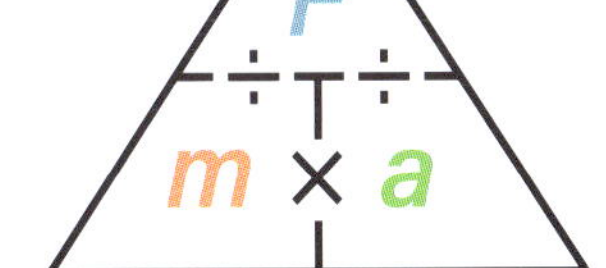

Figure 7.17: A formula triangle for force (F), mass (m) and acceleration (a)

Learning Ladder

Motion

1. Identify:
   - **a** the formula that represents Newton's second law.
   - **b** the units used for each variable when calculating force.
   - **c** the formula and variables used for Newton's second law when calculating weight.
2. Describe the difference in forces needed to accelerate a 22 kg cheetah at 15 $m\ s^{-2}$ and a 15 kg gazelle at 10 $m\ s^{-2}$. *Hint:* You will need to calculate the force in each case to compare.
3. Explain the effect of an object's mass on its weight, using an example.
4. Analyse how force, mass and acceleration are applied in designing trucks of different sizes (and mass) in order to move and stop them.

Use and influence of science

1. Identify which of Newton's laws could be used in designing cars to address the issue of speeding on our roads. Provide a reason for your answer.
2. Explain how the formula $F = m \times a$ can also be used to interpret the weight force of objects.
3. Explain how our need to keep drivers safe influences the focus of research into car safety technology.
4. A student stated: 'You don't need a seatbelt at low speeds – the force is small.' Use this statement to discuss how understanding forces, mass and inertia in an accident may inform people's decision to wear a seatbelt.
5. Analyse how scientific testing of how cars' safety features perform in accidents may contribute to people selecting safer cars.

Planning and conducting p. 239

You are about to complete an investigation, dropping balls of different mass from a balcony and recording the time taken for them to hit the ground.

1. Construct a list of the equipment required for this investigation that, when used, will ensure precise data.
2. Construct a risk assessment for this activity.
3. Outline how you would use the equipment to ensure precise and replicable data for the investigation.
4. Design a step-by-step procedure for this investigation, taking into account safety and procedural factors to ensure that the investigation is reproducible.

Key idea: Scale and measurement

Construct a table showing how the measurement of your weight would change as you move further away from Earth's surface. You will need to research the gravity at different distances from Earth's surface.

Success criteria

- I can analyse the relationships between force, mass and acceleration.
- I can apply Newton's second law of motion in everyday situations, including when calculating weight.

7·6 ▸ Net force

Learning intention

At the end of this lesson, I will be able to use vectors to determine the net force on an object and provide examples of how Newton's laws are applied.

Key terms

force diagram: a simplified diagram showing the direction and size of forces acting on an object

magnitude: the size of a measurement

net force: the sum of all forces acting on an object

normal force: the force of the ground pushing up in opposition to gravity

vector: a quantity that has direction and magnitude

Key idea: Scale and measurement

In a game of tug of war, two teams pull on each end of a rope. They are applying opposing forces. The team that applies the largest force wins by pulling the other team towards them. If we know the force in each direction, we can work out the overall force and who will win the tug of war.

Vectors have a size and direction

We learnt that displacement has a direction as well as a **magnitude** (size) – for example, 6 m s^{-1} east. Quantities that have both a direction and a magnitude are called **vector** quantities. Other vector quantities are acceleration and force, because they act in a direction. We use **force diagrams** to show the forces acting on an object. The arrows represent the size and direction of the forces. Longer arrows represent bigger forces.

In Figure 7.18, we can see two vectors: 2200 N to the left and 2800 N to the right.

Net force is the sum of all forces on an object

The **net force** is all the forces acting on an object added together.

In Figure 7.19, people are shown pushing a trolley. The applied force is represented as a vector by an arrow to show the direction and size. In part (a), a person is pushing the trolley with a force of 500 N to the right. The total or net force is 500 N to the right.

In part (b), two people are pushing in the same direction – to the right. One person pushes with 500 N of force, and the second with 200 N of force. The net force is now 500 + 200 = 700 N to the right.

Figure 7.18: Which team will win in this tug of war, and in what direction will the rope move?

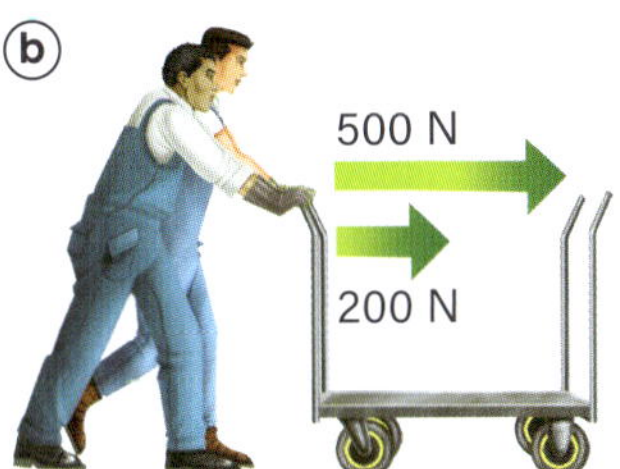

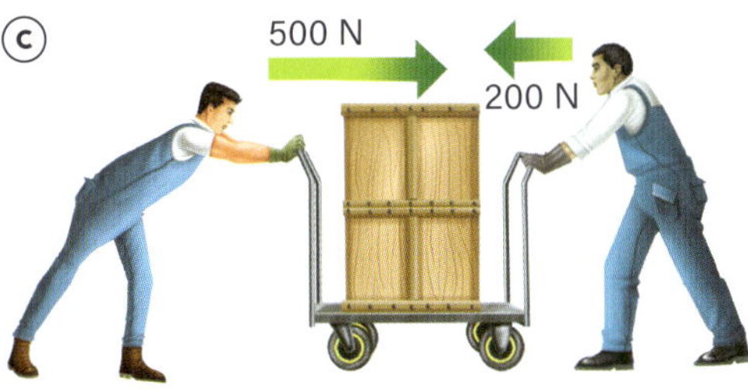

Figure 7.19: Net force is the sum of the different forces acting on an object.

In part (c), one person pushes with 500 N to the right, and the second with 200 N to the left. The forces partially cancel out. To find the net force, we calculate 500 – 200 = 300 N to the right.

Forces can also act up and down. The force due to gravity, or weight, acts downwards in the direction of gravity (Figure 7.20). We know from Newton's third law of equal and opposite forces that the ground pushes back up with an equal force. The force of the ground pushing up is known as the normal force.

Figure 7.21 shows vectors for a person standing still. The net force equals zero (668 – 668 = 0 N). If the person were to jump, they would need to have a net force upwards to move in that direction. In the second diagram, we can see that the vectors will partially cancel, with the net force calculated as 800 – 668 = 132 N upwards, enabling the person to jump.

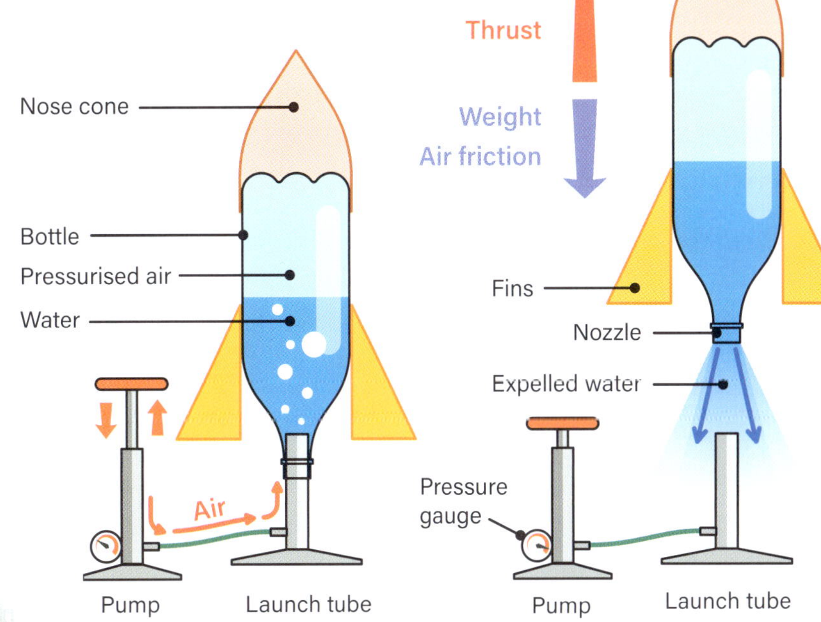

Figure 7.20: For a water rocket to launch, the net force upwards (thrust) must be higher than the force of gravity (weight) and friction downwards.

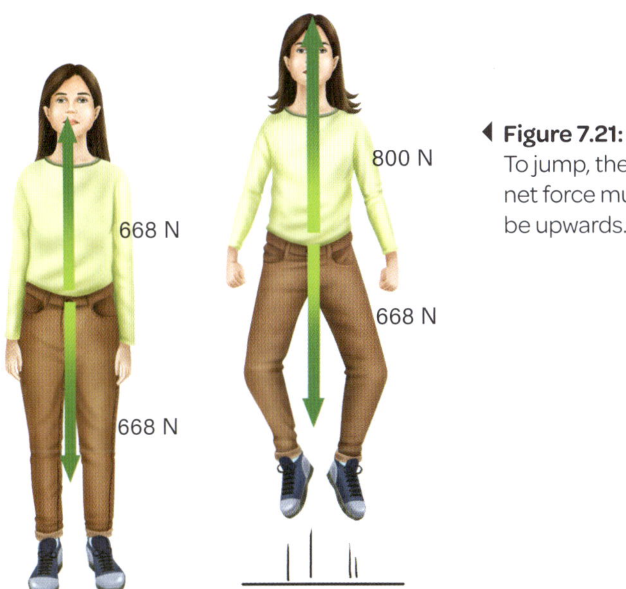

Figure 7.21: To jump, the net force must be upwards.

We can calculate the net force on a rocket

Can you imagine the force required to launch a rocket that has a mass of 1.5 million kilograms? The net force must be upwards, with the thrust being larger than the weight. That is, the net force = thrust – weight.

1 We can calculate the weight using $W = m \times g$:
   $W = 1\,500\,000 \text{ kg} \times 9.8 \text{ m s}^{-2}$
   $= 14\,700\,000$ (i.e. 14.7 million) N downwards.
2 If the thrust is 60.5 million N upwards, we can calculate the net force as follows:
   Net force = thrust – weight, or
   60 500 000 N – 14 700 000 N = 45 800 000 (i.e. 45.8 million) N upwards, and the rocket takes off!

Figure 7.22: The large mass of the rocket means it requires a huge upwards net force to increase its speed.

Forces on an object can act in many directions

Consider the forces acting on a cyclist. The driving force of their feet on the pedals is propelling them forward. Gravity is also acting on them, providing a weight force. Friction between the tyres and the road, as well as air resistance, opposes their motion. Finally, there is the force from the road pushing the bicycle upwards – the normal force.

To calculate the net force acting on the cyclist, using Figure 7.23, add together all the forces acting in the same direction and subtract the forces acting in the opposite direction.

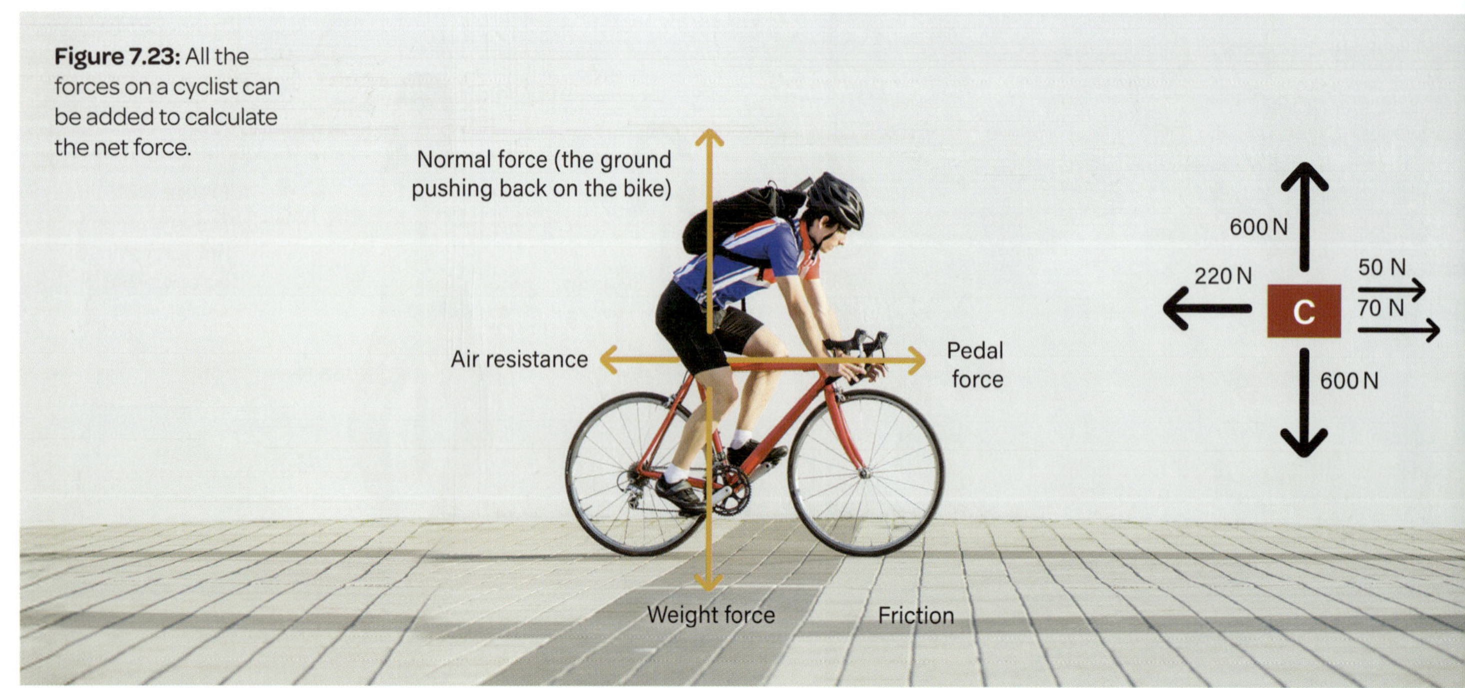

Figure 7.23: All the forces on a cyclist can be added to calculate the net force.

- In the vertical direction, the weight force (600 N) is acting in the opposite direction to the normal force (600 N), so subtract these: 600 N – 600 N = 0 N. In the vertical direction, the sum of the forces is zero, so you can ignore them.
- Air resistance (50 N) and friction (70 N) act in the same direction, so add these together. The total force acting to the right is 50 N + 70 N = 120 N.
- The pedal force (220 N) is to the left, which is the opposite direction to the friction and air resistance, so subtract them.
- Therefore, the net force is 220 N – 120 N = 100 N to the left.

Learning Ladder

Motion

1. Identify the key word that matches each of these definitions.
   a The total of all forces acting on an object
   b A measured quantity with magnitude and direction
   c The force of the ground pushing up in opposition to gravity
2. a Determine the net force from the tug of war shown in Figure 7.18 and identify which team will win.
   b Using the data for the rocket launch, draw a force diagram of a rocket. Add vectors for each force and an overall vector for the net force.
3. Use an example to explain how vectors can demonstrate Newton's laws.
4. Review the cyclist in Figure 7.23.
   a As the cyclist is riding, the wind becomes stronger, and air resistance increases to 120 N. Describe the change to the net force and what happens to the rider.
   b Air resistance continues at 120 N, and the ground becomes rough, increasing friction to 100 N. Describe what happens to the rider.
   c The wind slows and the road returns to normal. The rider picks up a heavy load in their backpack. Describe the effect on the friction force. How can the rider now maintain the same speed?
5. Evaluate how our understanding of net force has allowed us to design machines such as rockets, or cranes that lift and move objects.

Nature and development of science

The Tour de France is a major cycling event, and cyclists look for every advantage to win. Understanding the forces on a cyclist, including air resistance, to reduce their energy use and increase their net force forward and therefore speed is very important.

1. Identify a technology that may be used to study the design of bikes and helmets that reduce air resistance.
2. Describe ways that a scientist researching low-resistance helmets may refine their knowledge over time.

Processing, modelling and analysing p. 244

1. Draw force diagrams showing the vectors for the following and calculate the net force on the objects.
   a A person on a skateboard pushes with a force to the right of 100 N; friction to the left is 75 N.
   b A person jumping on a trampoline lands on the mat. Their downward motion exerts a force of 400 N. Their weight force downwards due to gravity is 700 N. The trampoline mat pushes up with an equal and opposite force.
2. Determine the net force on each of the following boxes.

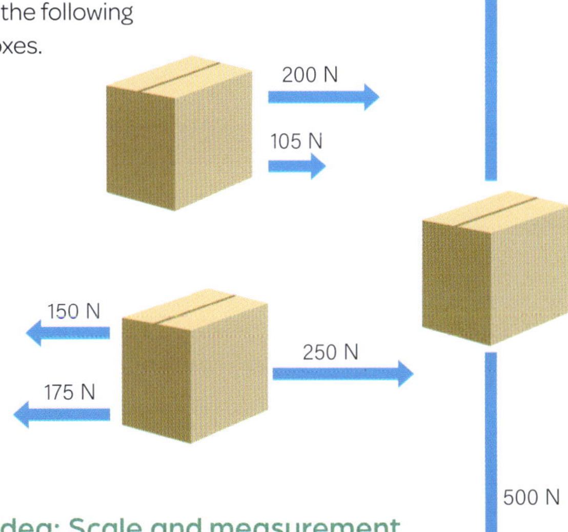

Key idea: Scale and measurement

Investigate the net force on a skydiver during different stages of a jump and produce a report outlining your findings. Use vectors to illustrate the forces at each stage of their fall.

Success criteria

- I can use vectors and force diagrams to determine the net force on an object.
- I can provide examples of how Newton's laws are applied.

7·7 Newton's laws and spearthrowers

Learning intention

At the end of this lesson, I will be able to apply Newton's laws to explain how First Nations Peoples achieved an increase in speed, accuracy and impact force when using spearthrowers.

Key terms

force multiplier: a machine that increases the force applied to an object

impact force: the force applied by an object when it impacts another object

kinetic energy: energy of motion

lever: a simple machine that increases force applied

spearthrower: a device used to increase the velocity of a spear (known as a woomera in some parts of Australia)

Key idea: Scale and measurement

First Nations Peoples have developed many technologies that help with tasks, including hunting as part of their agricultural practice. The **spearthrower**, known as a woomera in some parts of Australia, is a simple machine that is used to throw a spear. Its design increases the speed, accuracy and force of the spear at the point of impact.

The size and design of spearthrowers varies

Spearthrowers can vary in size and design depending on the type of spear they are being used with. A small light spear for fishing may be around 50–100 grams and will require a short spearthrower; whereas a large, heavy spear used for hunting kangaroos could be around 750–1500 grams and will require a longer spearthrower. A long spearthrower will produce a higher speed for the spear and more force when it hits the target.

Most spearthrowers are 60–90 centimetres in length, which enables a balance of leverage for speed and control for accuracy. There is great diversity of spearthrower designs across Australia. For example, in the Kimberley region, woomeras tend to be flat and narrow, while in the Western Desert and Central Australia they are flat and more convex in shape. In south-eastern Australia, including in Victoria, woomeras are rarer, but are usually described as being narrow, similar to a short harpoon.

Figure 7.24: The use of a woomera is demonstrated at Kuranda, Queensland.

The use of a spearthrower demonstrates Newton's first law

When the spear is held in the spearthrower, both the spearthrower and the spear are at rest and the forces acting on them are balanced. The concept of inertia means that the objects will want to remain at rest. The person throwing the spear has to apply a force, using the spearthrower, to overcome the inertia and change the motion. Applying a force means that forces are unbalanced, and the spear accelerates and moves in the direction of the force applied. The spearthrower itself helps to overcome the spear's inertia at rest, as it applies a force over a longer period of time and distance to create higher speed for the spear.

Figure 7.25: Woomeras, or spearthrowers, come in various designs and sizes.

A spearthrower multiplies the force applied

The spearthrower is a **lever**, which is a **force multiplier**. The spearthrower acts like an extension to the thrower's arm. This allows more force to be applied to the spear over a longer distance, and for a longer time, compared to when just the person's arm is used. The result is that the spear is accelerated to a higher velocity.

Figure 7.26: A dog ball thrower is like a spearthrower: it multiplies the force on the ball over a longer period of time and distance.

Figure 7.27: The spearthrower acts like an extension of the thrower's arm. The spear is at maximum velocity when it is released from the spearthrower.

We can understand this by using Newton's second law, using $a = \frac{F}{m}$. If the force is increased and the mass of the spear is the same, then acceleration must increase proportionally to the increase in force. If the force is doubled, so is the acceleration, leading to a significantly higher spear velocity.

Table 7.4 shows some experimental data that estimates how the length of the spearthrower affects the velocity of the spear. It shows that a longer spearthrower produces a higher velocity for the spear. The table also shows that a higher velocity means the spear will travel a greater distance.

Table 7.4: This experimental data shows that the length of the woomera, or spearthrower, is related to the increase in speed and force of the spear when released with a consistent force.

Spearthrower length (cm)	Mass of spear (g)	Distance travelled (m)	Velocity of spear when released ($m\ s^{-1}$)
30	750	7	6.5
60	750	18	11.1
90	750	30	16.0
120	750	38	19.5

A spearthrower increases kinetic energy

With a higher velocity produced from the spearthrower, the spear has more **kinetic energy** (energy of motion). Higher kinetic energy will result in a higher **impact force** when the spear hits the target.

Two factors increase the kinetic energy (and force of impact): the spear's increased velocity, and increased mass. For example:

- A spear with a higher mass of 1000 grams would have more kinetic energy than a spear of lower mass of 750 grams at the same velocity.
- A spear of higher velocity would have more kinetic energy than a spear of lower velocity if both have the same mass.

A spear with higher kinetic energy will have more force at the point of impact. Another way to think of this is using $F = m \times a$. We can see that an increased mass will also increase the impact force, as force and mass are directly proportional. Therefore, hunting larger animals may require a spear with a higher mass, to increase the kinetic energy, and a longer spearthrower, for higher velocity, both of which help to maximise the impact force of the spear.

Spearthrowers demonstrate action/reaction pairs

The spearthrower in action demonstrates Newton's third law with a powerful action and reaction pair of forces. As the thrower pushes on the spear (action), the spear pushes back (reaction) with an equal and opposite force onto the spearthrower and the person holding it.

Figure 7.28: Using a woomera gives the spear more kinetic energy by increasing its velocity. The higher the kinetic energy the spear has, the greater its impact force will be when it lands.

Learning Ladder

Motion

1. For both parts a and b below, identify two factors that will impact:
   a the distance a spear will travel.
   b the velocity of the spear.
2. **a** Construct a force diagram to show the forces on a spearthrower and spear when they are at rest, before being thrown.
   b Identify which of Newton's laws is demonstrated by this force diagram.
3. Use Newton's laws to explain how a spearthrower increases the velocity of the spear.
4. **a** Complete the following table to summarise the properties of motion during the flight of a spear, using the following words: none, low, high, medium, increasing, decreasing. (Ignore the force of gravity and normal force.)

Point in flight	At rest before launch	During launch	In flight	At impact
Force on spear				high
Acceleration	none			
Velocity			high	
Deceleration		none		
Kinetic energy		increasing		

   b Predict how the motion of a spear will change in each of the following scenarios, assuming that only the one variable mentioned is changing.
   i A longer spearthrower is used.
   ii A spear with lighter mass is used.
   iii A higher force is applied during throwing.

Nature and development of science

1. Identify technologies that could be used to investigate the motion of a spear in order to measure and detect the properties of motion.
2. Describe how First Nations Peoples may have refined over time their knowledge of how a spearthrower works.
3. Using an example, explain how our knowledge of levers, and of how they act as force multipliers, has contributed to other technologies.

Processing, modelling and analysing p. 244

1. Construct force diagrams of the forces on a spear during launch and in flight.
2. Construct graphs using the data in Table 7.4 to show the relationships between:
   a length and velocity.
   b length and distance.
3. Identify and explain the trends and patterns in your data and graphs in Question 2.
4. Discuss the relationship between the velocity, mass and kinetic energy of the spear.
5. Use Table 7.4 to calculate the average length of a spearthrower, the average velocity at release and the average distance travelled.

Key idea: Scale and measurement

Select another machine used by First Nations Peoples, such as an axe or bow. Explain how this machine works in relation to Newton's laws and the properties and scale (size) of forces and motion.

Success criteria

- I can explain how the design of a spearthrower impacts the motion of the spear.
- I can analyse the motion of a spearthrower using Newton's laws.

7·8 Key idea: Scale and measurement – road safety

Learning intention

At the end of this lesson, I will be able to apply Newton's laws in the context of road safety and technology that keeps us safe.

Key terms

braking distance: the distance travelled before a car stops after a driver pushes the brake

braking time: the time taken for a car to stop after a driver pushes the brake

momentum: a property of motion that is dependent on the mass and velocity of an object

reaction distance: the distance travelled during the reaction time

reaction time: the time taken to react

stopping distance: the sum of the reaction distance and the braking distance

Investigation 7.8

Car crashes and momentum, p. 379

Key idea: Scale and measurement

There are many properties of forces and motion that are related to understanding safety on the roads. Do you know how the mass of a car, or the velocity, affects how long it takes to stop? What about whether it takes longer when the roads are wet or the tyres are old?

In this section, we will study stopping distances and momentum, including what happens in a crash. We will also explore some of the technologies in cars that keep us safe.

Understanding these properties of motion is a good starting point for anyone who is learning to drive.

Stopping a car takes time

Imagine driving along a road and a dog runs out in front of you. Will you hit it? That depends on many things. The first factor that will determine whether you hit the dog is your **reaction time** – how long it takes you to react and push the brake pedal. During the reaction time, the driver has to do three things:

1. Notice the hazard.
2. Decide what to do or how to respond.
3. React and move their foot to the brake.

During the reaction time, the car will continue to travel at its current velocity until you press the brake. This distance travelled during the reaction time is called the **reaction distance**.

A typical reaction time for a driver is around 1 second in real life, but it is less in testing environments (as low as 0.67 seconds) and higher if, say, the driver is distracted or tired (up to 1.5 seconds or more).

We can calculate the reaction distance as follows:

$$\text{Reaction distance (m)} = \text{velocity (m s}^{-1}\text{)} \times \text{reaction time (s)}$$

$$\text{Reaction distance} = vt$$

For example, if a driver is travelling at 50 km h^{-1} (13.9 m s^{-1}) with a reaction time of 1 second, the reaction distance will be: 13.9 × 1 = 13.9 m. Table 7.5 shows the reaction distance at different speeds. You can see that if the reaction time increases, so does the distance. (If you need help converting kilometres per hour to metres per second, see the Science how-to section on pages 284–85.)

Figure 7.29: Whether the driver will hit the dog depends on how long it takes them to stop.

Table 7.5: Reaction distance calculations

Velocity ($km\ h^{-1}$)	Velocity ($m\ s^{-1}$)	Reaction time (s)	Reaction distance (m)
50	13.9	1.0	13.9 × 1 = 13.9
80	22.2	1.0	22.2 × 1 = 22.2
50	13.9	1.5	13.9 × 1.5 = 20.6
80	22.2	1.5	22.2 × 1.5 = 33.3

After the reaction time, the driver starts to brake and the car then slows and eventually stops. This means we still do not know if the driver will hit the dog. The total time the car will travel before stopping is:

Reaction time + braking time = stopping time

The longer the time, the greater the distance travelled before stopping.

A longer time to stop is a greater distance travelled

Once the driver pushes the brake, it will take a certain distance for the car to stop; this is the **braking distance**. The total **stopping distance** (distance to stop) is the reaction distance plus the braking distance.

Reaction distance + braking distance
= stopping distance

When braking, there are many factors that can affect how quickly you slow (or decelerate). These factors include the quality of the tyres, the condition of the brakes, the type of road surface and the weight of the car. A modern car with good tyres on a dry, good road will slow at about 10.8 $km\ h^{-1}$ (3 $m\ s^{-1}$) every second in normal comfortable braking. However, in an emergency braking situation, it is more like 25.2 $km\ h^{-1}$ (7 $m\ s^{-1}$) every second.

Table 7.6 shows the approximate emergency **braking time** and distance for different speeds when a car slows at 25 $km\ h^{-1}$. The data shows that if you double your speed, the braking distance will be approximately 4 times greater. A gravel or wet road, or an older car, also increases the braking distance.

Remember that to find the total stopping distance, we need to add the reaction distance to the braking distance. Figure 7.30 shows typical stopping distances for a car at different speeds, including the breakdown of reaction distance and braking distance.

Figure 7.30: ▶ Stopping distances for different speeds can be calculated from this representation.

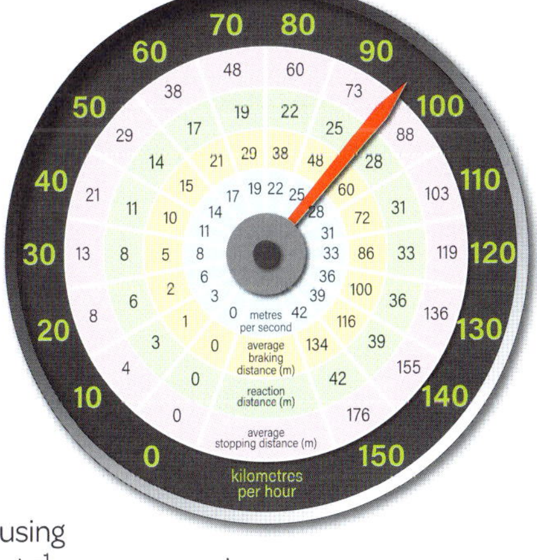

Table 7.6: Approximate emergency braking distances and times using deceleration of 25 $km\ h^{-1}$ every second

Velocity ($km\ h^{-1}$)	Braking time (s)	Braking distance (m)
25	1	3
50	2	13
75	3	31
100	4	55

More force is needed to stop heavier objects in motion

When travelling in a car, you can feel inertia. In Section 7.4, we learnt that inertia makes your body want to remain in its current state of motion. When a car accelerates, your body wants to stay at the same speed and you may feel this as being pushed along by the force of the car. If the car slows down, you may feel your body moving forward as it tries to continue at the same speed even when the car slows. When a car turns, your body tries to continue moving forward in a straight line, and you feel like you are being pushed sideways. A seatbelt is one technology that helps to overcome inertia, especially in accidents. A seatbelt will hold your body in the seat so that when the car stops quickly you do not fly forward into the dashboard or windscreen.

An object with a larger mass will have higher inertia. For example, a truck will have more inertia than a car, and will both be harder to stop and take longer to stop. This is because it will take a larger force to change the motion of a heavier car or truck.

Figure 7.31: The inertia of the balls makes them want to keep moving at their current speed even when the trolley stops. ▼

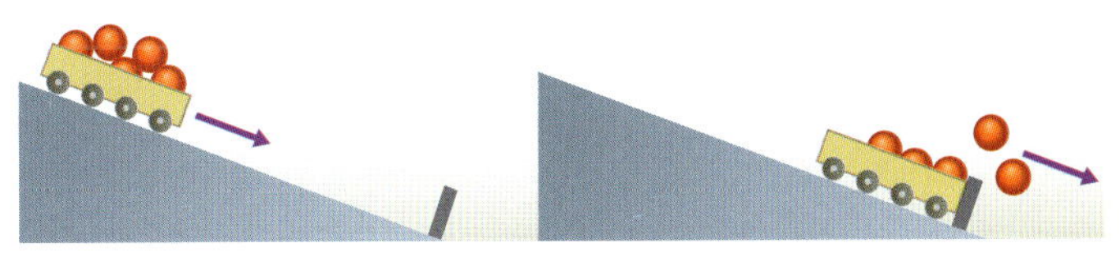

Momentum increases as mass and speed increases

Momentum (symbol P) is a property of motion that is dependent upon the mass (kg) and velocity (m s^{-1}) of the object.

Momentum: $P = m \times v$

This formula shows that if the mass and/or velocity increases, so does the momentum. An object with higher momentum is harder to stop than one with lower momentum. For example, if a truck and car travel at the same speed, the truck will have higher momentum as it has more mass and it will therefore be harder to stop.

A fast-moving car will have more momentum than a slow-moving car if they have the same mass. The fast-moving car will be harder to stop. This also means that in a collision, a car or a truck with higher momentum has the capability to apply more force at impact and to cause more damage. This is one reason why we have speed limits!

Table 7.7: Example of momentum for different cars and trucks. At the same speed, the heavier truck has much more momentum.

Vehicle type	Mass (kg)	Velocity of 50 km h^{-1} (m s^{-1})	Momentum (kg·m s^{-1})
Car	1200	13.9	16 666
SUV	1800	13.9	25 000
Small truck	10 000	13.9	139 000
Large truck	20 000	13.9	277 778

Figure 7.32: Even a car with small mass but high velocity can have enough momentum to cause serious damage.

Technology helps to keep people safe

As cars have developed, new technologies have made them safer, such as airbags, crumple zones, head rests and seatbelts. These technologies are designed to work under the forces of motion that we have explored, in order to reduce damage to passengers, surroundings, other cars and pedestrians. Table 7.8 summarises a number of these technologies and their purpose.

Figure 7.33: This illustration shows an airbag that protects a person, seatbelts to overcome inertia, and a crumple zone that absorbs energy from the crash.

Table 7.8: Some safety technologies in cars that keep us safe

Technology	How it works
Seatbelts	Seatbelts reduce serious injury and death by over 60 per cent in an accident. Seatbelts hold the passenger in place, applying a force on the body and overcoming inertia that keeps the person moving with the car. This stops them being flung into the dashboard, windscreen, or side windows and doors.
Luggage barrier	A luggage barrier in the back of a wagon stops objects that continue to travel forward due to inertia, when a car stops suddenly. This ensures that loose objects do not fly around the car and hit a person. If you are travelling at 60 km h^{-1}, a loose object will continue forward at that speed and can cause serious injury.
Airbags	Most newer cars have airbags in front and back, and on the sides around passengers. These devices fill rapidly with gas in an accident to protect passengers. They are set off by sensors that detect a collision. A chemical reaction produces nitrogen that inflates the bag in around 0.3 seconds. That bag helps to protect your body from objects in the car, and absorbs kinetic energy of your body during the collision to protect you.
Head rests	These protect passengers, especially in rear-end collisions. When a car is hit from behind it is pushed forward, and the person is pushed forward by the seat. Without a head rest, their head is left behind because of inertia. After the initial forward motion the car stops and the person's head is flung back, which can cause serious neck damage. A head rest ensures the head is pushed forward with the body, reducing the chance of whiplash.
Crumple zone	Cars have a 'crumple zone' at the front and rear. These zones collapse in an accident and absorb the energy of the crash. This means less energy is transferred to the passengers while also ensuring the cabin does not crumple and squash the occupants.

▼ **Figure 7.34:** Airbags can create a 'cushion' to protect passengers and absorb energy during an accident.

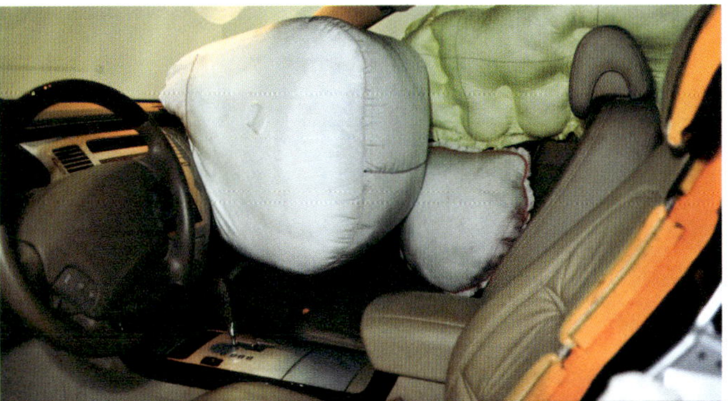

▼ **Figure 7.35:** This test shows that the crumple zone has collapsed but the cabin is still intact.

Learning Ladder

Motion

1. Name the term that describes each of the following properties of motion.
   - **a** A property of motion dependent upon the mass and velocity of the object
   - **b** A property of motion that makes you want to stay in the same state of motion
   - **c** The time taken to respond to an event and push the brake
   - **d** The sum of distance travelled during reaction time and braking time
2. A person has a loose 2-litre bottle on the back window ledge of a car.
   - **a** Describe the expected motion of the bottle when the driver brakes as hard as possible and without changing direction.
   - **b** Describe, using Newton's laws, why this motion occurs.
3. Explain Newton's first law of motion using examples of a person travelling in a car.
4. Answer the following questions using Figure 7.30.
   - **a** If travelling at 80 km h^{-1}, find the speed in m s^{-1}.
   - **b** If a driver is travelling at 70 km h^{-1} and is 60 metres away from a dog that runs into the road, will they hit the dog?
   - **c** What happens to the braking and stopping distance when you double the speed from 40 to 80 km h^{-1}?
   - **d** Two cars of the same mass are side-by-side on a freeway. One is going at 100 km h^{-1} and the second at 60 km h^{-1}. How far apart will they be when each has stopped?
5. Use your knowledge of Newton's laws of motion to evaluate whether lower speed limits should be retained in school zones.

Nature and development of science

1. Identify a technology that has enabled us to advance our understanding of the motion of cars and road safety.
2. Use Figure 7.35 as an example to describe how our knowledge of car safety may be refined over time.
3. Explain how our scientific understanding of Newton's laws has contributed to the development of seatbelts.
4. There is a lot of scientific knowledge about road safety. Explain how a scientist may validate their results from a crash-test facility.
5. Undertake some research about how crash-test dummies have changed over time. Analyse how improvements in crash-test dummies have enabled advances in safety design.

Processing, modelling and analysing

p. 244

1. Construct diagrams to model the following properties of motion.
   - **a** A car on a road, showing reaction distance, braking distance and stopping distance
   - **b** Two cars in motion, showing how each can have a different momentum
2. Construct a graph, using Table 7.6, to show the relationship between velocity and braking distance.
3. The reaction distance is calculated using a reaction time. Study the data in Figure 7.30 to work out the reaction time used for the calculations made in Table 7.6. Explain your answer.
4. Discuss the relationship between velocity, reaction time and reaction distance using the data in Table 7.5 to give examples in your response.

Success criteria

- I can analyse the reaction and stopping times and distances for cars.
- I can describe inertia and momentum and explain how they relate to road safety.
- I can explain what happens in a road accident, and describe technologies that keep us safe.

Summary

- Motion is a change in position of an object over time.
- A force is a push or a pull between objects that changes the motion speed and/or direction of their motion or their shape.
- Understanding forces and motion is useful in many everyday situations, such as improving safety when driving, or in sport.

- Distance is how far an object travels. Speed is the distance an object travels divided by the time it takes to travel that distance. Speed is measured in metres per second ($m\ s^{-1}$) or kilometres per hour ($km\ h^{-1}$).
- Displacement is how far an object is from its initial starting position.

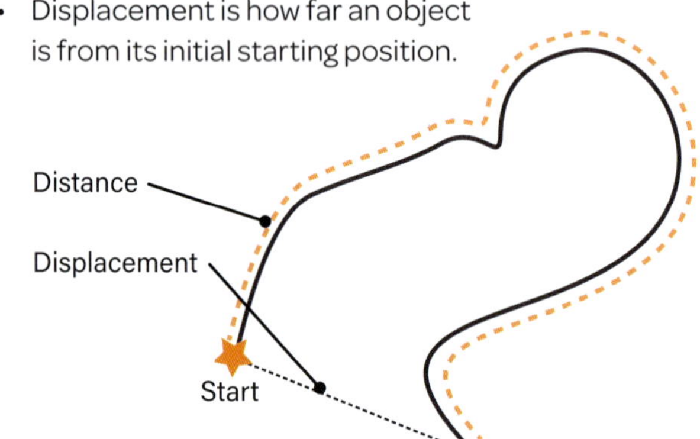

- Distance–time graphs represent motion.

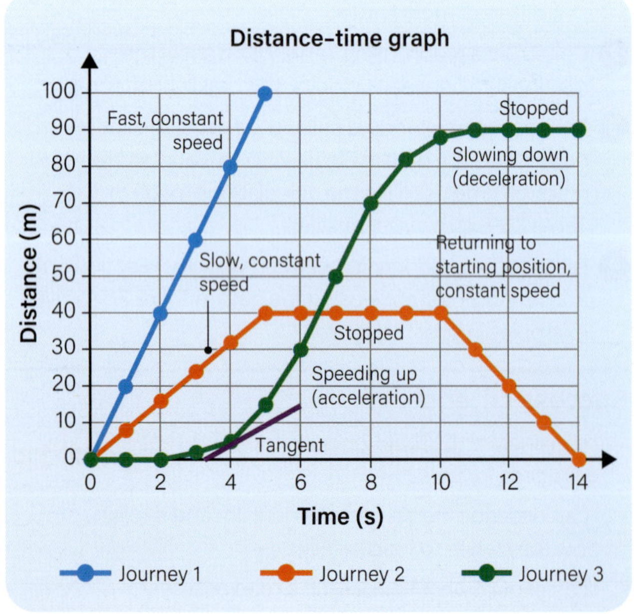

- Acceleration is any change in speed or direction over time (i.e. speeding up, slowing down or turning).
- We can show speed, acceleration and distance on a speed–time graph (with time on the *x*-axis and speed on the *y*-axis).

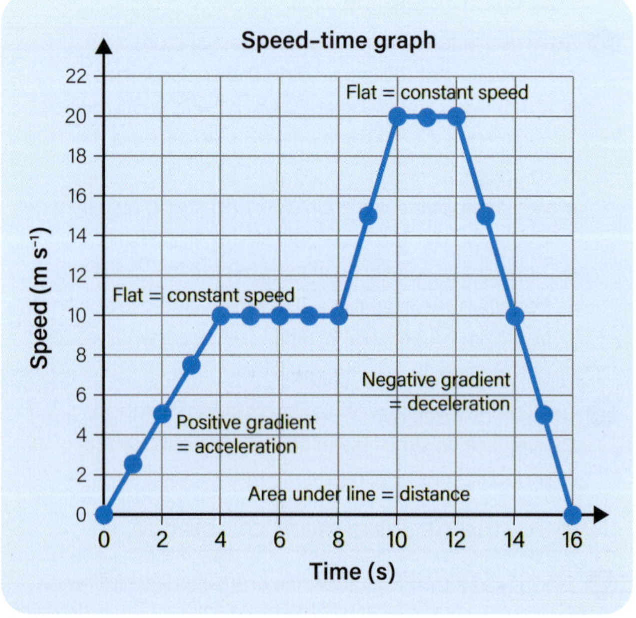

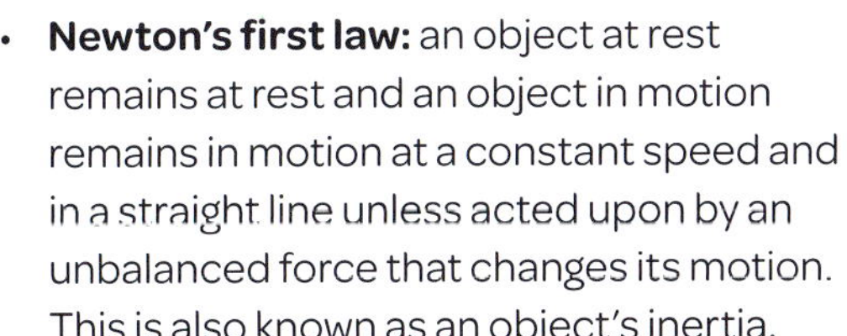

- **Newton's first law:** an object at rest remains at rest and an object in motion remains in motion at a constant speed and in a straight line unless acted upon by an unbalanced force that changes its motion. This is also known as an object's inertia.
- **Newton's second law:** the acceleration of an object is proportional to the force applied and inversely proportional to the mass of the object. $F = m \times a$, where F is force (N), m is mass (kg) and a is acceleration (m s^{-1}).
- **Newton's third law:** for every action, there is an equal and opposite reaction.
- The net force is all the forces acting on an object added together.

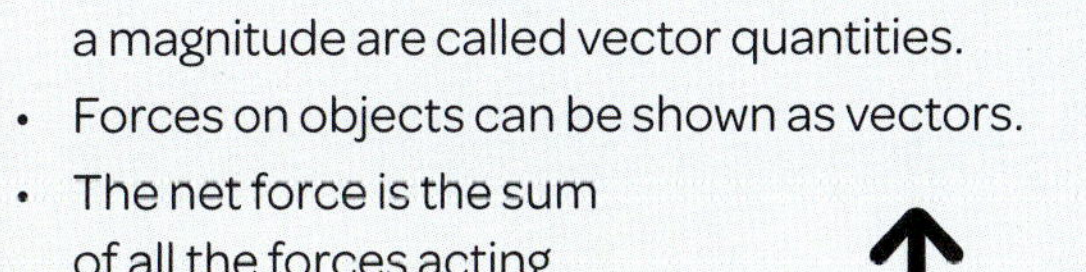

- Quantities that have both a direction and a magnitude are called vector quantities.
- Forces on objects can be shown as vectors.
- The net force is the sum of all the forces acting on an object added together.

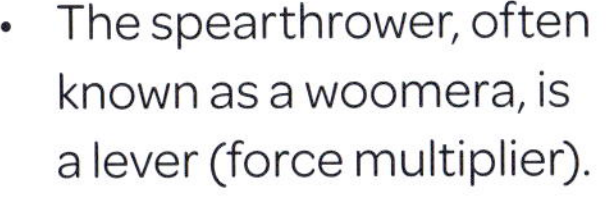

- The spearthrower, often known as a woomera, is a lever (force multiplier).
- The spearthrower acts like an extension to the thrower's arm that applies more force, over a longer distance, for a longer time, causing greater acceleration ($f = m \times a$).

Key idea: Scale and measurement

- Stopping a car takes time.
  - Reaction time is how long it takes you to react.
  - During reaction time, you travel the reaction distance (reaction distance = $v \times t$).
  - The distance travelled while braking is the braking distance.
  - The total stopping distance = reaction distance + braking distance.

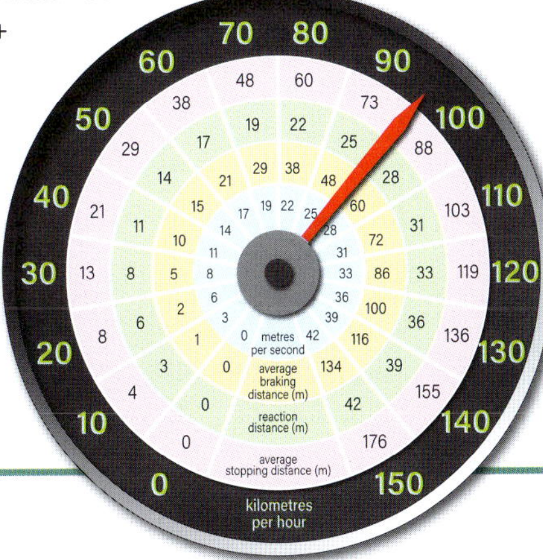

- Larger mass = higher inertia and larger force needed to change the motion.

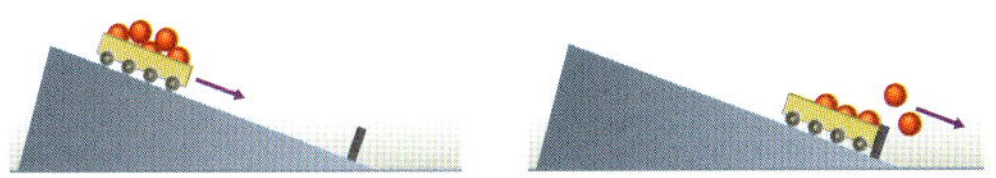

- Momentum: $P = m \times v$ (mass (kg) and velocity (m s^{-1})).
- An object with higher momentum is harder to stop.
- Technologies such as airbags and crumple zones apply Newton's laws to protect occupants of vehicles.

Masterclass

Steps in progression

		1	2
Science understanding	Motion	Identify the point during a swimming race when there are/is: **a** balanced forces. **b** unbalanced forces. **c** maximum acceleration. **d** maximum deceleration.	Construct a diagram to show the size of each force acting on the swimmer when they are: **a** standing on the blocks. **b** pushing off the wall when turning. **c** swimming at a constant speed.
Science as a human endeavour	Nature and development of science	Identify a technology that has enabled better understanding of performance in swimming.	Describe how a scientist studying new swimsuit designs could, over time, refine their knowledge of what works.
	Use and influence of science	Identify the properties of motion that can be used to improve performance in sport.	Describe the differences and similarities between speed and velocity using the 100 m race as an example.
Science inquiry	Planning and conducting	Identify some equipment that would be needed in a swimming pool to accurately measure a swimmer's time.	In an investigation to measure the acceleration of a swimmer off the blocks, outline some safety considerations.
	Processing, modelling and analysing	Copy Table 7.9 and list the swimmers from fastest to slowest.	Add a column to your table from step 1 and calculate the average speed for each swimmer.

100-metre freestyle swimming race

Swimming is a competitive sport worldwide. It also demonstrates many of the concepts of motion and forces we have learnt about in this chapter.

The current world records for a 100-metre freestyle race are as follows:

- *Men:* 46.40 seconds by Pan Zhanle (China), set on 31 July 2024 at the Paris Olympics.
- *Women:* 51.71 seconds by Sarah Sjöström (Sweden), set on 23 July 2017 at the World Championships in Budapest, Hungary.

A swimming race demonstrates many aspects of forces and motion. Consider the following description of the motion during a 100-metre swimming race.

1. The swimmer stands on the starting block (see Figure 7.36).
2. From a stationary position, the swimmer pushes off the block and dives into the water.
3. After the dive the swimmer glides underwater, usually performing a kick to help propel them along. During this time, they maintain a long, thin position until they return to the surface.
4. Once at the surface, the swimmer maintains and manages their speed by kicking and using their arms to pull against the water and move forward.
5. A 100-metre race requires two laps of the pool. At the end of the first lap, the swimmer must flip, or turn, and push off the wall in the opposite direction.
6. Just like at the start after the dive, the swimmer stays streamlined underwater and kicks to return to the surface. At the surface, they again manage their speed by swimming faster or slower.
7. Towards the end of the race, for the last few metres the swimmer increases their rate of kicking and stroke to accelerate to the wall.
8. To end the race, they touch the wall and stop their motion.

Demonstrate your understanding

3	4	5	
Explain Newton's three laws of motion, using examples of each from the swimming race.	**a** Construct a diagram to predict the displacement of the swimmer at the halfway mark and at the end of the race. **b** Sketch a distance–time graph and a speed–time graph for the race.	Evaluate how understanding of Newton's laws has impacted the development of technologies used in swimming.	
Use an example to explain how our scientific understanding of Newton's laws of motion has contributed to improved swimming performance.	Discuss how a scientist may validate their data about new swimsuit technologies. What role would publication in a peer review play in this?	Analyse, using examples, how advances in swimming- related technologies would enable other advances in science.	
Explain, using an example, how interest in sport and performance improvement have influenced the focus of scientific research.	Discuss, using an example, how knowledge of forces and motion may inform your decision-making.	Use an example to analyse how scientific knowledge from studying motion, forces and performance in swimming may be used in other ways in everyday life.	
Describe how you would use digital tools to record a swimmer's race data with precision.	Design a reproducible procedure and data table for the investigation in step 2.	Evaluate your procedure from step 4: identify sources of error and propose how your procedure may be improved.	Science how-to p. 239
Explain any trends in the data table you have created.	Discuss any potential relationship between mass and time in the data table and explain this in terms of $F = m \times a$.	Calculate the average and medium time, and average and medium speed, for the race.	Science how-to p. 244

In swimming there are many technologies that improve performance. One example is modern swimsuits that reduce friction with the water. Less friction means that more of the force applied by the swimmer will go into forward motion, rather than be used up against the friction (or drag), allowing a faster swim. Other technologies include more accurate sensors for measuring speed and time of races.

Table 7.9: Race results for a 100-metre swimming final

Lane number	Participant name	Mass of swimmer (kg)	Time (s)
1	Jasmine	68	51.0
2	Sun	82	62.3
3	Liam	75	48.5
4	David	90	70.1
5	Isabelle	72	53.2
6	Aisha	78	58.0
7	Luca	75	55.5
8	Kimiko	85	66.8

Figure 7.36: Swimmers on the starting blocks. What forces are acting on them?

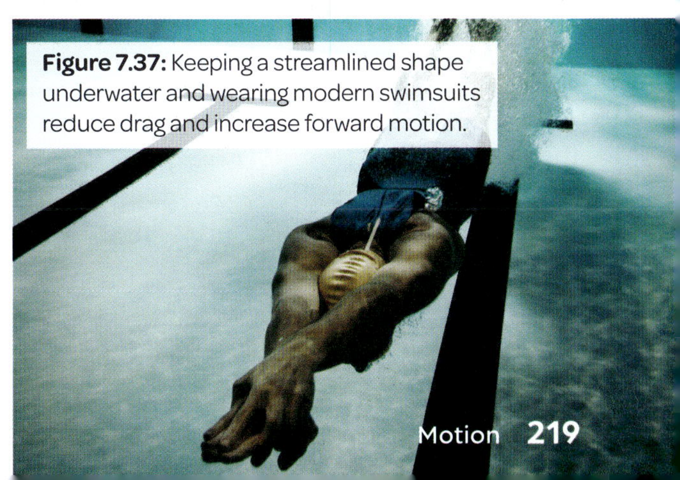
Figure 7.37: Keeping a streamlined shape underwater and wearing modern swimsuits reduce drag and increase forward motion.

Science how-to

Science skills are the tools you can apply when investigating the world around you. Learning how to use these skills will help you to understand scientific processes, ask relevant questions and communicate what you discover.

Command terms and response modelling

Command terms are the instruction words in a question. They tell you the approach to use when you answer it.

The command terms are listed alphabetically over the following pages. We have also included an example of how they would be used in a question and a sample response.

Account for	
Explanation	State reasons for; report on.
Sample question	Account for the increase in heat of a light bulb when it is in use.
Sample response	*Light bulbs transform electrical energy into light energy so that we can see. However, this energy transformation is not 100 per cent efficient, as some of the energy is transformed into heat and sound energy. So, when light bulbs are on, we can feel that they have heated up as a result of the inefficient energy transformations.*

Analyse

Explanation	Identify components/elements and the significance of the relationship between them; draw out and relate implications; determine logic and reasonableness of information.
Sample question	Analyse how the body responds to a drop in temperature to maintain homeostasis.
Sample response	*The human body uses negative feedback loops which allow us to maintain consistent conditions, a state known as homeostasis.* *Thermoreceptors detect a drop in the surrounding temperature. This stimuli is transmitted via neurons to the hypothalamus, which interprets and transmits the message to mechanoreceptors.* *Mechanoreceptors enact responses, including shivering to move muscles and vasoconstriction to decrease temperature loss through the skin.* *These mechanisms, facilitated through a negative feedback loop, allow our bodies to increase their temperature, bringing it back to within a normal range and maintaining homeostasis.*

Apply

Explanation	Use; employ in a particular situation or context.
Sample question	Apply your understanding of chemical reactions to categorise the following reactions: **1** $Mg(s) + 2HCl(aq) \rightarrow MgCl_2(aq) + H_2(g)$ **2** $CH_4(g) + 2O_2(g) \rightarrow CO_2(g) + 2H_2O(g)$
Sample response	*Chemical reactions can be categorised by their reactants and products.* *In reaction 1, an acid and a metal are reacting to form a metal salt and hydrogen gas, making it an acid–metal reaction.* *In reaction 2, methane gas is reacting in the presence of oxygen, producing carbon dioxide and water as steam, making it a complete combustion reaction.*

Assess

Explanation	Make a judgement about, or measure, determine or estimate, the value, quality, outcomes, results, size, significance, nature or extent of something.
Sample question	Assess the effectiveness of using solar energy to reduce household power costs.
Sample response	*Solar energy is energy from sunlight that is converted by panels into usable electrical energy.* *Arguments for:* • *Using solar energy can reduce or remove a household's reliance on obtaining electrical energy from non-renewable power sources.* • *After the initial panel installation, solar energy does not cost anything to obtain, which means households will save money.* • *Households can get credit for any unused solar electricity they send back to the grid, which further reduces their bills.* *Arguments against:* • *Installing solar panels is expensive. To completely remove dependence on the power grid, householders need also to install expensive solar batteries to store the energy.* • *Solar energy requires regular sunny weather to function effectively. Households may have minimal power after a series of cloudy or rainy days.* *Assessment: Even though it costs a lot to set up at the start, solar power is highly effective at reducing household power costs. This is because power bills are much lower when you use solar energy.*

Calculate	
Explanation	Determine from given facts, figures or information; obtain a numerical answer showing the relevant stages in the working; determine or find (e.g. a number, answer) by using mathematical processes.
Sample question	Conduct a first-hand investigation to determine the effect of string length on the time it takes a pendulum to swing. Calculate the average length of time for a 60 cm string length.
Sample response	*The pendulum took 1.5 seconds in trial 1, 1.6 seconds in trial 2 and 1.6 seconds in trial 3.* *This means the average time taken is:* $\frac{(1.5 + 1.6 + 1.6)}{3} = 1.57$ *seconds.*

Compare	
Explanation	Recognise similarities and differences and their significance.
Sample question	Compare the use of renewable and non-renewable resources in energy production.
Sample response	*Similarities:* • *Renewable and non-renewable resources can both be transformed into electrical energy.* • *Renewable and non-renewable energy both rely on the construction of infrastructure (e.g. wind turbines and power stations) in order to produce electrical energy.* *Differences:* • *Renewable energy can be used without being depleted, while there is only a finite amount of non-renewable resources such as coal and oil.* • *Using renewable energy releases minimal pollutants into the atmosphere, while burning non-renewable energy releases pollutants and greenhouse gases such as carbon dioxide.*

Construct	
Explanation	Make, build, create or put together by arranging ideas or items (e.g. an argument, artefact or solution); display information in a diagrammatic or logical form.
Sample question	Construct a poster, diagram or presentation to show your understanding of the electromagnetic spectrum.
Sample response	Different versions of responses to this question are shown in 'Presenting scientific information' on pages 272–77.

Figure S1.1: Solar farms help to reduce the amount of non-renewable resources that are needed for electrical energy production.

Define	
Explanation	Give the precise meaning and identify essential qualities of a word, phrase, concept or physical quantity.
Sample question	Define the term 'pathogen'.
Sample response	*A pathogen is an agent that can cause infectious disease in a host.*

Describe	
Explanation	Provide characteristics, features and qualities of a given concept, opinion, situation, event, process, effect, argument, narrative, text, experiment, artwork, performance piece or other artefact in an accurate way.
Sample question	Describe what occurs to molecules during a chemical reaction.
Sample response	*In a chemical reaction, molecules are changed. The bonds between atoms are broken and rearranged, which forms new substances.*

Discuss	
Explanation	Present a clear, considered and balanced argument or prose that identifies issues and shows the strengths and weaknesses of, or points for and against, one or more arguments, concepts, factors, hypotheses, narratives and/or opinions.
Sample question	Discuss the use of non-renewable energy such as coal for large-scale electrical energy generation.
Sample response	*Coal has historically been the most widely used resource for power generation across Australia. It can provide reliable energy on a large scale, but using it has both advantages and disadvantages.* *Advantages:* • *Infrastructure in place at current power stations is set up to use and burn coal, meaning that it is an accessible and cheap method of providing power to many homes.* • *Most households are set up to receive electrical energy generated from already existing grids and power stations.* *Disadvantages:* • *Burning coal releases large amounts of carbon dioxide into the environment, contributing to global warming. It also produces particles and pollutants, such as sulfur dioxide, which cause acid rain.* • *Coal is a non-renewable fossil fuel, meaning that if we continue to use it to generate energy at a large scale, we will eventually run out of this resource.*

Distinguish	
Explanation	Make clear the differences between two or more arguments, concepts, opinions, narratives, artefacts, data points, trends and/or items.
Sample question	Distinguish between infectious and non-infectious diseases.
Sample response	*Infectious diseases are diseases that are caused by a pathogen and can be transmitted from one host to another. Non-infectious diseases, however, are caused by chemicals, the environment or genetics, and cannot be transmitted from one individual to another.*

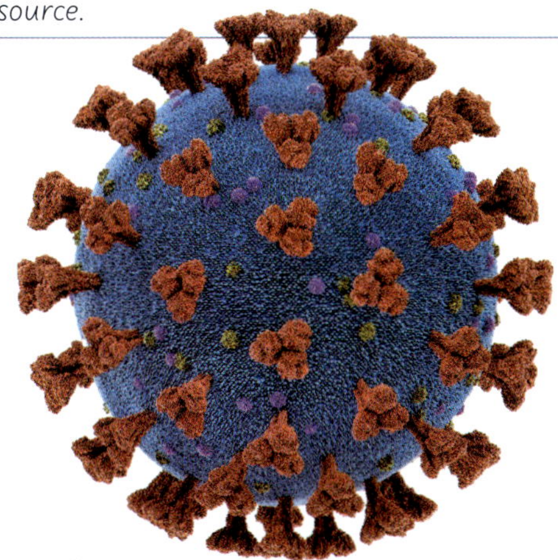

Figure S1.2: Infectious diseases are transmitted by pathogens, such as this COVID-19 virus.

Evaluate	
Explanation	Ascertain the value or amount of; make a judgement using the information supplied, criteria and/or own knowledge and understanding to consider a logical argument and/or supporting evidence for and against different points, arguments, concepts, processes, opinions or other information.
Sample question	Evaluate the use of wind energy as a power source, taking into account levels of pollution and reliability of the energy source.
Sample response	*Wind energy is a renewable form of electricity that uses turbines to transform kinetic energy into electrical energy.* *Arguments and evidence for:* • *Wind energy is renewable, meaning it will not run out and it can be used continually.* • *Wind energy is 'clean' – it does not release unwanted pollutants into the atmosphere.* • *In favourable conditions, a single wind turbine is capable of supplying the power demands of over 300 homes.* *Arguments and evidence against:* • *Current infrastructure does not support the wide-scale use of wind turbines, and constructing them is expensive.* • *Wind turbines rely on environmental conditions – they need to be constructed in windy areas that are obstruction free, so that they can spin and generate electrical energy.* *Judgement: Using wind energy as a power source is highly effective. While upfront costs to set up wind farms are high, the benefits outweigh these costs as they can power multiple homes and reduce the amount of pollution in the atmosphere.*

Figure S1.3: A wind turbine transforms kinetic energy into electrical energy.

Examine	
Explanation	Consider an argument, concept, debate, data point, trend or artefact in a way that identifies assumptions, possibilities and interrelationships.
Sample question	Examine this argument: 'We should gain 100 per cent of our energy needs from renewable sources.'
Sample response	*Renewable energy sources are ways of generating electrical energy that are able to be continuously renewed. These methods of electricity production typically do not produce waste products, making them better for the environment. However, our global energy demands are currently much higher than our ability to use renewable sources for all of our power generation. Instead, we should focus on putting more structures for renewable energy generation in place, so we can switch across from non-renewables without preventing access to electricity around the world.*

Explain	
Explanation	Give a detailed account of why and/or how, with reference to causes, effects, continuity, change, reasons or mechanisms; make the relationships between things evident.
Sample question	Explain how increasing the mass of a car can change its impact force in a collision if its acceleration remains constant.
Sample response	*Newton's second law of motion states that 'the acceleration of an object is proportional to its net force and inversely proportional to its mass'. This means that as the mass of the car gets bigger, its acceleration gets smaller. So, if the acceleration was kept the same, this would mean that the force of the car colliding with another object would increase as its mass increased.*

Extrapolate

Explanation	Infer and/or extend information that may not be clearly stated from a narrative, opinion, graph or image by assuming existing trends will continue.
Sample question	Looking at the graph, extrapolate how much of a radioactive isotope would be left after three half-lives have passed.
Sample response	*The graph shows that the percentage of radioactive isotope decreases over time. After three half-lives, there would be 15 per cent of the original amount of radioactive isotope remaining, according to the trendline.* **Figure S1.4:** A graph showing how much radioactive isotope remains after each half-life has passed Decay of a radioactive substance Substance remaining (%): 0, 10, 20, 30, 40, 50, 60, 70, 80, 90, 100 Number of half-lives: 1, 2, 3, 4, 5, 6, 7

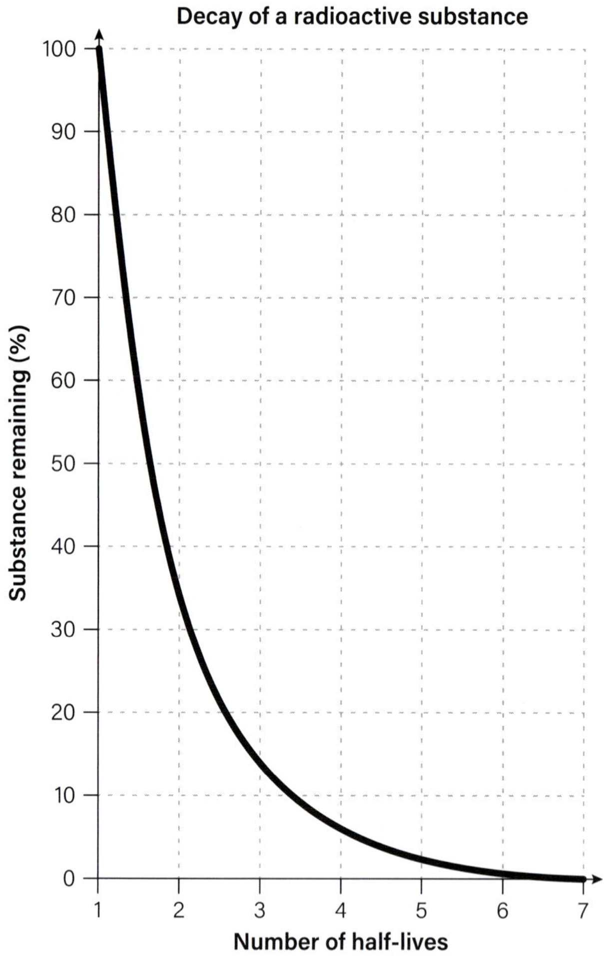

Identify

Explanation	Recognise and name and/or select an event, feature, ingredient, element, speaker and/or part from a list or extended narrative or argument, or within a diagram, structure, artwork or experiment.
Sample question	Identify the type of chemical bond in a molecule of sodium chloride (NaCl).
Sample response	*Ionic bond.*

Investigate

Explanation	Observe, study or carry out an examination in order to establish facts and reach new conclusions.
Sample question	n/a
Sample response	Investigating is covered in the 'Writing investigation reports' section on pages 270–72.

Justify	
Explanation	Show, prove or defend, with reasoning and evidence, an argument, decision and/or point of view using given data and/or other information.
Sample question	Justify the use of energy-efficient appliances.
Sample response	*We should use energy-efficient appliances whenever possible to reduce the amount of electrical energy we consume in our homes on a daily basis. Reducing the amount of energy we use is beneficial, because it reduces the amount of fossil fuels we require in order to produce electrical energy. It also means that we consume fewer kilowatts, which reduces our power bills!*

Figure S1.5: Using energy-efficient appliances helps the environment and saves you money on your power bill!

Outline	
Explanation	Provide an overview or the main features of an argument, point of view, text, narrative, diagram or image.
Sample question	Outline the law of conservation of energy.
Sample response	*The law of conservation of energy states that energy cannot be created or destroyed; it can only be transferred or transformed.*

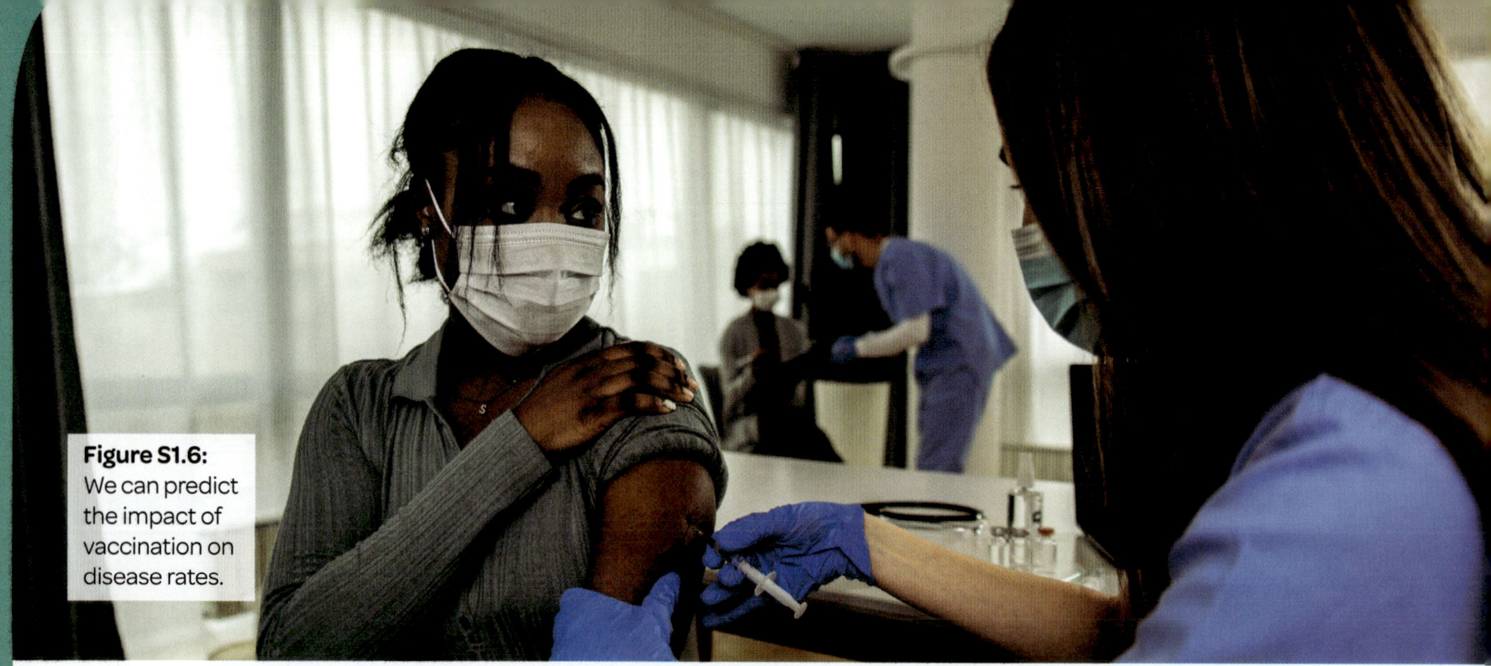
Figure S1.6: We can predict the impact of vaccination on disease rates.

Predict	
Explanation	Give an expected result of an upcoming action or event; suggest what may happen based on available information.
Sample question	Predict the effect of vaccination levels on the occurrence of chickenpox cases.
Sample response	*If the number of vaccinated individuals increases, then the number of chickenpox cases will decrease, because more people will be immune to the disease.*

Propose	
Explanation	Suggest or put forward a point of view, idea, argument, diagram, plan and/or suggestion based on given data or stimulus material for consideration or action.
Sample question	Propose a reason for wearing safety glasses when working with sodium hydroxide.
Sample response	*Sodium hydroxide is a basic solution that is caustic and can cause burns to our skin and eyes. Wearing safety glasses prevents the solution from contacting our eyes and causing damage or blindness.*

Recall	
Explanation	Present remembered ideas, facts and/or experiences.
Sample question	Recall the names of two scientists who have contributed to our understanding of the atom.
Sample response	*One scientist who contributed to our understanding of the atom was J.J. Thomson, who discovered electrons using cathode ray experiments. Another was Niels Bohr, who proposed the idea that electrons exist around the nucleus in orbits, or shells.*

Summarise	
Explanation	Retell concisely the relevant and major details of one or more arguments, text, narratives, methodologies, processes, outcomes and/or sequences of events.
Sample question	Summarise the methods of infectious disease transmission.
Sample response	*Infectious diseases can be transmitted through direct transmission, where the pathogen is passed directly from one person to the next through actions like shaking hands or kissing, or through indirect transmission, where the pathogen can be transmitted by an intermediary such as air, water or animal vectors.*

S2 Science inquiry

Key terms

accuracy: how close a measured value is to the true, exact value; how closely a recorded value matches the expected outcome of an investigation

anomalies: data points or findings that do not follow the normal trend or expected findings

assumption: something accepted as true that may not be tested, which impacts the results of an investigation

controlled variable: variable in an investigation that must be kept the same for all trials. Only the independent variable should change, otherwise it is not a fair test

correlation: a relationship between two factors or variables that shows them changing together, in either the same or opposite way

data: facts and information collected for reference or analysis

data point (datum): a single identified element in a dataset

dataset: a collection of data, often from numerous trials related to a single factor

dependent variable: the thing that is measured in a first-hand investigation

ethical: in science, minimising harm to those involved, and ensuring investigations are conducted honestly and data is collected and recorded accurately

evaluate: judge value based on scientific evidence

fair test: a test where all variables are kept the same, except for the independent variable and the dependent variable

hazard: something that can harm living things, objects or the environment

hypothesis: a suggested explanation or prediction of a scientific problem that can be tested with an investigation

independent variable: the thing that is deliberately changed in a first-hand investigation

pattern (data): when data repeats in a predictable way

plausible: could be reasonably accepted based on available evidence

precision: how close measured values are to each other within a dataset

raw data: data that is collected directly from the investigation

relationship: a link between two factors

repeatable: similar results can be obtained using the same method and equipment

replicable data: data that can be reproduced using the same method and equipment

reproducible: similar results can be obtained by someone else using the same method and equipment

trend (data): when data moves in a general direction, usually up or down

true value: the actual, exact value of a measurement, error-free, which would be obtained if a perfect measurement were made

validity: when an investigation meets its intended purpose

variable: a factor in an investigation or a model that can be changed, measured or controlled

Science has its own skills that scientists use to conduct investigations and test their ideas. Science inquiry allows you to collect information and test your ideas about how different parts of the world work.

▼ **Figure S2.1:** Inquiry is a fundamental part of science.

Unpacking science inquiry

Science inquiry comprises five skills.

1 Questioning and predicting
2 Planning and conducting
3 Processing, modelling and analysing
4 Evaluating
5 Communicating

This section will break down each skill for you – where to begin, how you can apply it as you learn more, and how mastering the skill will elevate your science learning.

The Learning Ladder and science inquiry

The Learning Ladder at the start of each chapter lists two of the five science inquiry skills (two have been selected for each chapter) and the five steps of progress for each of these skills.

Figure S2.2 shows the five science inquiry Learning Ladders, with step 1 being the first level and step 5 the final, most challenging level. You can only climb the ladder by mastering each step. This approach is called developmental learning and puts students in charge of their own learning progress.

Figure S2.2: Start at step 1 and work your way up to mastery.

Learning Ladder

Steps in progression	Questioning and predicting	Planning and conducting	Processing, modelling and analysing
5	I can develop explanatory models when investigating scientific questions, problems and claims	I can evaluate scientific methods regarding safety, ethical and procedural considerations	I can analyse the quality of data using descriptive statistics
4	I can discuss what is needed for a question to be investigable or a prediction to be reasoned	I can design and conduct reproducible investigations that consider safety, ethical and procedural factors	I can discuss relationships and anomalies that emerge in processed data
3	I can develop a hypothesis that predicts the relationship between investigation variables	I can use equipment to generate and record data with precision, to obtain replicable data, using digital tools as appropriate	I can identify and explain trends and/or patterns in a range of dataset representations
2	I can formulate questions to investigate scientific problems	I can develop and follow risk assessments that consider safety and ethical issues	I can process data by using mathematical relationships and/or constructing graphs
1	I can make simple predictions based on what I know and observe	I can select appropriate equipment to collect precise data for scientific investigations	I can organise data and information using tables, keys and/or models

Science inquiry

Each section of this book contains a Learning Ladder question block. The questions relate to both content and inquiry skills. Question 1 corresponds to step 1, Question 2 corresponds to step 2, and so on.

The green headings mean these are science inquiry questions. They will test you on a specific skill, such as evaluating. There is a page reference to the Science how-to section, where you can get more information on what is required at each step.

Learning Ladder

Climate change

1. Identify the feature(s) of the atmosphere that cause the enhanced greenhouse effect.
2. Describe the difference between the natural and enhanced greenhouse effect.
3. Explain a major cause of the enhanced greenhouse effect.
4. Discuss a potential impact of increased concentrations of greenhouse gases in the atmosphere.

Evaluating p. 253

1. Identify an assumption that scientists may have made about the atmospheric data gathered at Mauna Loa and Kennaook/Cape Grim.
2. Describe the impact of the assumption you identified for Question 1 on the atmospheric data.
3. Some students use data loggers to obtain carbon dioxide concentration from the atmosphere at their school in inner Melbourne. Propose ways that they could improve the quality of their data.

Next we will consider each science inquiry skill and break down what is required at each step of the Learning Ladder to develop your skills and achieve process mastery.

Steps in progression	Evaluating	Communicating
5	I can evaluate the validity and reproducibility of investigation methods	I can justify scientific ideas, findings, arguments and proposals for diverse audiences
4	I can construct evidence-based arguments to justify conclusions, address ethical issues or assess claims	I can communicate scientific findings and arguments effectively for specific purposes to specific audiences
3	I can discuss ways to improve the quality of data and validity of conclusions and claims	I can use digital technologies and/or scientific representations to communicate data and information
2	I can describe the impact of assumptions and errors, and propose ways to reduce them	I can prepare a variety of presentation formats to communicate ideas and findings
1	I can identify assumptions and types of errors in an investigation	I can select appropriate formats, content and vocabulary to communicate scientific ideas and findings

Questioning and predicting

Questioning is the basis of all science. By asking questions, we can come up with ideas that we can test scientifically. Predicting scientific outcomes helps us to make better decisions. Predicting weather patterns can help people to prepare for extreme temperatures, while predicting which illnesses might spread rapidly can help health-care workers to adapt their treatments. The development of reasonable and investigable questions and predictions helps scientists to develop appropriate models that explain phenomena.

Step 1 I can make simple predictions based on what I know and observe

When you are writing questions and making predictions, you need to ensure you are proposing things that are **plausible**. To help you predict something plausible, you can think back to experiences you have had or match the investigation to a relevant scientific theory you already know.

For example, when you go out and socialise with a group of people and one of them has a cold, you may have observed that you or other group members become unwell after being in contact with them. You can use this observation to write a prediction, using a specific framework (an *'if ..., then ...'* statement – see step 3).

Figure S2.3: The observation of a hot-air balloon in the sky supports the prediction that heat can make objects rise.

If one person in a group is sick, *then* the others may also become unwell *because* some diseases can be passed between living things.

You can use this experience to predict how being exposed to different diseases will impact other people. Other examples of experiences that can help you to make predictions are:

- slipping on a puddle of water on tiles but not on a similar-sized puddle on concrete
- noticing that someone bigger than you is faster when you roll down a steep hill on your bikes
- seeing a hot-air balloon in the sky and being told that the heat from the flames makes the balloon lift off the ground.

Step 2 I can formulate questions to investigate scientific problems

We call the factors that we change and measure in the scientific question the **variables**. The factor that we change is called the **independent variable**. The factor that we measure is called the **dependent variable**. We want to see if there is a **relationship** between these two variables. In our scientific question structure, this looks like:

How does the *independent* variable affect the *dependent* variable?

For example, we may have the following scientific problem: The type of light bulb used impacts how much money we spend on our electricity bill based on the wasted energy. So, our question is:

How does the *type of light bulb used* affect
how *hot it gets* when used in a circuit?

The independent variable is the type of light bulb. The dependent variable is the temperature it reaches. The additional step is to identify the relationship. In our example, we are testing the relationship between bulb type and temperature.

Figure S2.4: The type of light bulb used affects how hot it gets – but how?

The next step is to use scientific knowledge to help you make predictions. To link your prediction to scientific knowledge, you must match the content of your knowledge to the investigation factors. Your prediction statement will then be followed by a 'because':

If an LED light bulb is used, then it will stay at a lower temperature
because less energy is transformed into wasted heat energy.

The scientific knowledge you use can be general – you do not need to include specific details. However, it must be relevant to the investigation and the predicted effect.

Step 3 I can develop a hypothesis that predicts the relationship between investigation variables

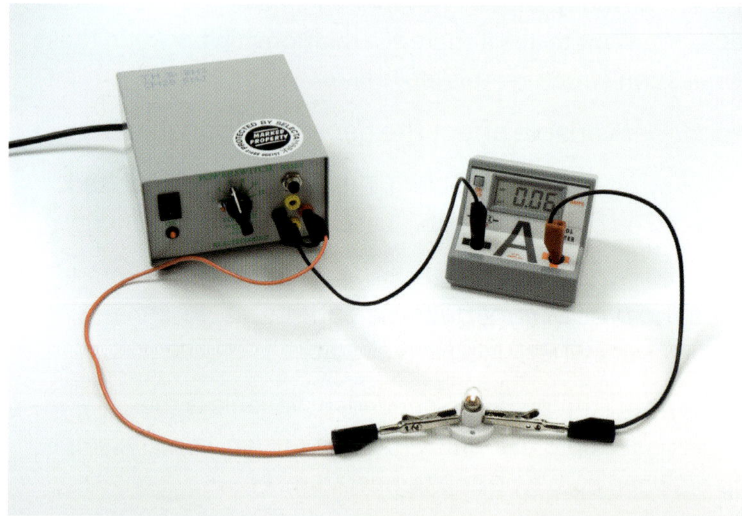

Figure S2.5: There can be a variety of relationships between independent and dependent variables. Scientific knowledge and our own experiences can help us to predict what that relationship will be.

Now that you can write and support a prediction in general, you can update your prediction with the specific variables of the investigation. This means we can also look at the relationship between the variables.

The following method will help you to formulate a scientific **hypothesis**.

- **a** Read the question.
- **b** Identify the independent variable.
- **c** Identify the dependent variable.
- **d** Identify some **controlled variables**.

Step 3 requires more detail on how you will change the independent variable to test your hypothesis, the relationship with the dependent variable and the predicted effect.

Once you have identified and understood your variables, include them in your prediction using the following template:

If [the independent variable and the specific change],
then [the dependent variable and the predicted effect].

So, your hypothesis will become:

If an LED light bulb is used in a circuit,
then it will remain at a lower temperature.

Step 4 I can discuss what is needed for a question to be investigable or a prediction to be reasoned

In the previous three steps, you refined your ability to structure scientific questions and predictions, as well as to support your predictions. In step 4, the focus is on discussing questions and predictions. What makes the question or prediction reasonable and investigable?

This is a higher-order thinking skill that requires you to think critically about questions and predictions.

- Do they include everything they should?
- Do they exclude things that are not appropriate?
- Can they be tested and answered fairly?

When discussing whether a question is investigable and a prediction is reasoned, the following checklists can be applied.

QUESTIONING CHECKLIST

- ☐ It is a question; therefore, it ends with a question mark.
- ☐ It is not a definitive question; i.e. it is not a 'yes' or 'no' question and does not require a particular answer, such as a 'this' or 'that' response.
- ☐ It includes a factor that can be changed (independent variable) and a factor that can be measured (dependent variable).
- ☐ It can be investigated.

PREDICTION CHECKLIST

- ☐ States the predicted effect in a statement that can be structured using the '*If* [independent variable and specific change], *then* [dependent variable and predicted effect] *because* [scientific theory]' template.
- ☐ Includes the independent variable and associated change.
- ☐ Includes the dependent variable and the predicted effect in relation to the change.
- ☐ Includes an explanation of scientific theory that justifies the prediction, based on previous study or knowledge.

If anything from these checklists is missing, you should be able to use your knowledge to modify or adjust the question/prediction so that it meets the required standards. Table S2.1 demonstrates how a discussion about a question and prediction could be formatted, including suggested improvements.

Table S2.1: What to look for when setting up an investigation

Question: How does the *mode of agitation* (stirring) affect the *rate of a chemical reaction*?	
Independent variable	The mode of agitation: swirl by hand, magnetic stirrer
Dependent variable	Rate of reaction
Controlled variables	None
Discussion	This is a good scientific question that changes only one factor; however, it should specify how the rate of reaction will be measured – e.g. volume of gas generated over time. To improve it further, it should also include the controlled variables.
Improvement	This question could be improved by controlling the variables, stating how the dependent variable will be measured. The details of the chemical reaction under investigation should be included. For example, all the factors that affect rate of reactions, besides agitation – temperature, concentration, and surface area of reactants – must be controlled and kept the same.
Prediction: If the angle of incidence of light hitting the solar panel is decreased, then it will generate more electricity because less light is reflected off the surface. This allows more light to be absorbed and converted from solar energy to electrical energy by the photovoltaic cells.	
Independent variable	The angle of incidence of sunlight hitting the solar panel surface
Dependent variable and predicted effect	The amount of electricity generated is expected to increase as the angle decreases
Controlled variables	One – the source of light: the Sun
Scientific understanding: Prediction is justified by reflection/absorption theory	
Discussion	This is an excellent scientific hypothesis that includes one changed and one measured variable, as well as the predicted effect. It also includes how the investigation is controlled and is justified by scientific knowledge. No improvements are required.

Step 5 I can develop explanatory models when investigating scientific questions, problems and claims

An important part of experimental set-up is being able to use previously gained scientific knowledge to develop explanatory models that explain the reason why a specific result has occurred or will occur. This can allow us to decide on the best solutions to problems and help us to determine the **validity** of claims as they are made.

For example, data relating to vaccinations allows us to develop an explanatory model that higher levels of vaccination lead to more of the population being protected from a disease. Claims that are made relating to vaccinations can be assessed against this model, using the following steps.

1 Identify the claim.
2 Identify the data relating to the claim.
3 Check that the data matches what the claim is saying.
4 Describe any benefits or limitations of the data.
5 Make a judgement on whether the data supports or refutes the claim.

For example, the claim 'Vaccines effectively prevent infectious diseases when 50 per cent of the population is vaccinated' can be evaluated using these criteria.

1 Identify the claim.
   The claim states that 50 per cent of a population needs to be vaccinated to prevent susceptibility to a disease.
2 Identify the data relating to the claim.
   Table S2.2 shows data on herd immunity from vaccinations.

Table S2.2: Data on herd immunity from vaccinations

Proportion of population vaccinated (%)	Susceptible individuals (%)	Risk of death from disease
0	100	High
50	50	Moderate
75	15	Low
90	5	Very low
95	<1	Negligible

3 Check that the data matches what the claim is saying.
   This data shows that a 50 per cent vaccination rate means that 50 per cent of the population is still susceptible to the disease and at risk. This does not match the claim.
4 Describe any benefits or limitations relating to the data and claim.
   Benefits:
   - The data shows that 50 per cent of the population will be protected from a disease, and from the risk of death, when 50 per cent of the population is vaccinated against the disease.

   Limitations:
   - Fifty per cent of the population is still susceptible to the disease.
   - The data shows that 95 per cent of the population needs to be vaccinated to effectively prevent susceptibility to a disease.

5 Make a judgement on whether the data supports or does not support the claim.
The claim that 'Vaccines effectively prevent infectious diseases when 50 per cent of the population is vaccinated' is not supported by the data. An evidence-based claim would be 'Vaccines effectively prevent infectious diseases when 95 per cent of the population is vaccinated', as the data indicates that this results in less than 1 per cent of the population being susceptible to the disease.

The same steps can be used when a problem is encountered, as long as the data and predictions have allowed us to develop a model that states the expected outcome of the results.

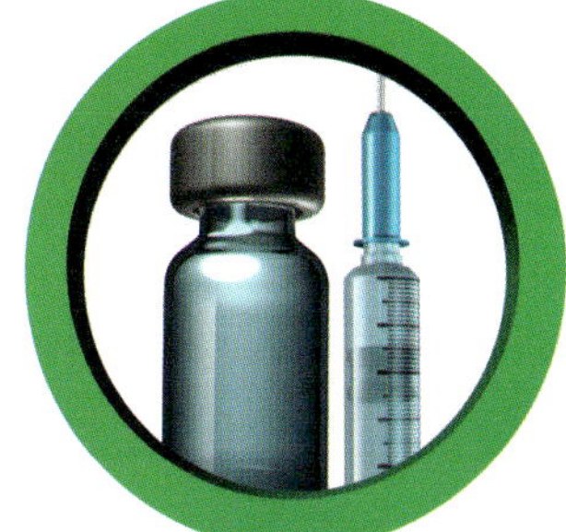

Figure S2.6: Scientific claims must be supported by data.

Planning and conducting

Planning will help you to come up with well-constructed investigations to answer scientific questions. A well-planned investigation should use appropriate scientific equipment, allow you to gather and record **data** precisely, and be **reproducible** by someone else using the same method and equipment as you used. This allows you to safely and **ethically** conduct the investigation by following your plan.

Step 1 I can select appropriate equipment to collect precise data for scientific investigations

Scientific equipment can help us to make precise measurements. It is important, therefore, to know how to select the best piece of equipment that will improve our observations. Selecting the best piece of equipment will improve the precision of our measurements, making our investigation more accurate.

To select the best piece of scientific equipment, ask yourself:

1 What equipment would allow me to measure accurately and precisely?
2 Which version of this equipment will give me the smallest margin of error on my measurements?

Figure S2.7: A digital thermometer provides higher precision of measurement than an alcohol thermometer.

Table S2.3: Selecting the most precise scientific equipment for a measurement is important, as it improves precision and helps to reduce errors in investigations.

Measurement	Scientific equipment	Most precise equipment
Temperature	Thermometer	A digital thermometer: This will help me to measure to within 0.1 °C, rather than to within 1.0 °C using an alcohol thermometer.
pH	Indicator, or pH probe	The pH probe: This will help to measure pH to two decimal places, rather than using whole numbers.

Step 2 I can develop and follow risk assessments that consider safety and ethical issues

A risk assessment allows us to identify potential **hazards** and the likelihood that they will cause harm to those involved in conducting the investigation. A risk assessment can be structured in a table such as Table S2.4. This shows us what we need to do in the investigation to reduce the risk of injury. By following the steps in the table, we are implementing the safety practices we have identified and keeping everyone safe. These practices can also ensure that the investigation is ethical.

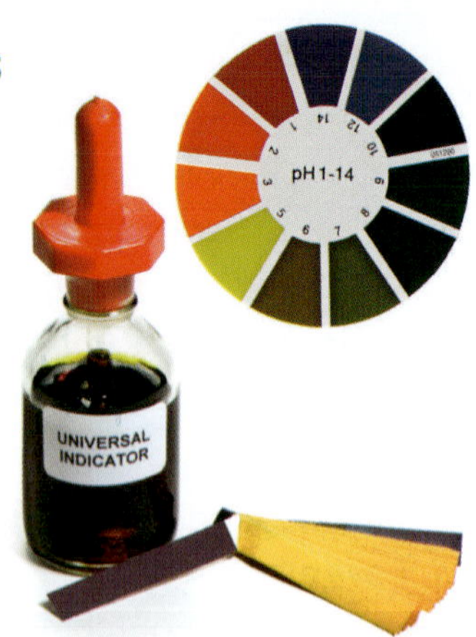

Table S2.4: Reducing risks when working with acids and bases

Hazard	Risk	Strategy
Universal indicator	High likelihood of eye damage if it splashes into the eye	• Use a pipette to transfer indicator from the container to the test tube. • Wear safety glasses to create a barrier between indicator and eyes.
Hydrochloric acid	High likelihood of burns if it splashes onto skin	• Use a pipette to transfer acid from the container to the test tube. • Wear gloves and long sleeves to protect skin from acid.
Glass test tube	Very high likelihood of cuts if glassware breaks	• Never run while carrying glassware. • Place the test tube in a test tube rack away from the edge of the lab bench to prevent it from being knocked onto the floor.

Step 3 I can use equipment to generate and record data with precision, to obtain replicable data, using digital tools as appropriate

As well as being safe, the data you collect in your investigations must also be recorded with **precision**. To check you are being precise in your measurements, you must ensure the following:

- A results table is constructed to record data from the independent and dependent variables of the investigation.
- The equipment used is appropriate for what you are testing (including using digital equipment where necessary).

You can use a checklist to help you confirm that your data is generated and recorded precisely. Your precise data-recording checklist might include the following points.

ACCURATE DATA-RECORDING CHECKLIST

- ☐ I constructed a results table that allows me to record data about the independent and dependent variables.
- ☐ I chose specific scientific equipment that allows me to record data precisely.
- ☐ I checked whether digital equipment was available to measure my variables.

Another important factor to consider when generating data is your sample size. What is the scale of your investigation? Scientific findings need to be able to be **replicated** consistently for data to be considered reliable. This means that 'one-off' results, or results from a small number of tests (or sample), may not be replicated in a larger scale investigation.

For example, if you are measuring the height of 13-year-olds at your school and only measure the height of 10 people in your class when you have 500 students in your year group, that is a small sample size and may not reflect everyone's height. A better study would measure the height of 100 students, giving you a larger sample size that is more likely to reflect the whole year group's height data.

Step 4 I can design and conduct reproducible investigations that consider safety, ethical and procedural factors

Before conducting an investigation, you will need to construct a plan so that you know what you are going to do. Plans are written in the form of scientific methods. They should be written as a series of numbered steps. They should also be written in the *third person*. Do not use *I*, *we* or *you*.

In addition, your investigation should include:

- the *independent and dependent variables* and how you will measure them
- the *controlled variables* and how you will keep them the same
- evidence of *repetition*
- a way to *record the results*.

As an example, imagine we are investigating the following question:

Does the force a toy car is pushed with affect how far it travels across a flat surface?

Our method or procedure would follow the process below.

1 Collect all equipment.
2 Find a flat surface and place a line of tape down to mark the starting point of the car.
3 Place the car level with the starting line.
4 Gently push the car so that it starts to move.
5 Let the car travel along the surface until it comes to a stop.
6 Measure the distance the car travels from the starting line in centimetres, using a tape measure.
7 Record the distance in a results table.
8 Return the car to the starting line.
9 Repeat steps 3 to 8 two more times, applying the same force to the car.
10 Repeat steps 3 to 8 two more times, this time applying medium and then firm force to the car to start its movement.

Having detailed steps in your method will make it reproducible; that is, it can be conducted by other scientists using the same method that you have used.

When conducting your method, you need to complete all parts of each step before moving onto the next one. So, for the first step of your investigation, you would collect all of the equipment you are going to need. While conducting your investigation, it is important that you are safe and your methods are ethical, so that your data can be trusted.

To check that your design is safe and ethical, ensure that:

- you or your teacher has conducted an appropriate risk assessment for the investigation
- the investigation is conducted honestly; no one has tried to skew the test or measurement to obtain a certain result
- the data is recorded exactly as measured.

You can use a checklist to confirm that your investigation is safe and ethical.

RISK ASSESSMENT CHECKLIST

- ☐ I made a risk assessment that identifies at least three hazards and how to prevent them from happening.
- ☐ I conducted the investigation exactly as it was written in the method.
- ☐ I recorded the data I collected exactly as measured in a results table.

Step 5 I can evaluate scientific methods regarding safety, ethical and procedural considerations

Once your plan has been constructed, it is important to evaluate it to ensure it is safe, ethical and follows appropriate procedures. If your constructed method meets all the required criteria, then it can be judged a high-quality method.

Consider the method in step 4. Each of the three factors listed in Table S2.5 can be considered to determine if it is an appropriate method or could be further improved.

Table S2.5: Evaluating safety, ethical and procedural considerations

Investigation safety	**How are investigation hazards managed?**
Benefits	• This method does not include equipment that poses a high risk of injury. • The method provides clear direction on how to use the equipment.
Limitations	There is no evidence of safety equipment (e.g. safety glasses) being used in the method.
Method ethics	**Is this method ethical?**
Benefits	• Specific equipment is referenced for collecting precise measurements. • How to record measurements is explicitly stated.
Limitations	There are no quantitative measures provided for the 'force' on the car.
Procedural considerations	**Does the method contain the features of a good investigation?**
Benefits	• The method includes identified independent and dependent variables. • There is evidence of repetition for each trial. • Precise measurements are included
Limitations	The type of surface is not specified.

Overall judgement: This method is of moderately high quality. It will result in a good investigation. The method could be further improved by explicitly stating the safety considerations for specific pieces of equipment, and the surface to test on, and by providing a precise way to increase the force applied to the car for each of the different trials.

Figure S2.8: We are studying whether the force applied affects how far a toy car will travel along a flat surface.

Processing, modelling and analysing

Information generated during a scientific investigation should be recorded in an organised way so that you can process it or use it to model an idea. Processing data involves presenting information in a way that clearly displays what you have found. This sometimes involves using models to show information in a different way to help make sense of it. Once data is processed or modelled, it should be analysed. This includes looking for trends and patterns that emerge in the visual data, and can also involve using statistics.

Step 1 I can organise data and information using tables, keys and/or models

Table S2.6 is organised in a way that clearly displays information. From the title, we can recognise that the data will show us how the rate of a reaction is affected by the mode of agitation. The headings indicate that measurements were taken in millilitres every 10 seconds for 60 seconds, and the values in the table show the volume of gas evolved from each reaction for each of the three modes of agitation.

▲ **Figure S2.9:** Because the chemical reaction under investigation happens quickly, it makes sense to measure the gas evolved in millilitres (mL) every 10 seconds, as recorded in Table S2.6.

Table S2.6: Effect of mode of agitation (stirring) on rate of reaction

Mode of agitation	Volume (mL) of gas collected every 10 seconds over 1 minute					
	10 s	20 s	30 s	40 s	50 s	60 s
Swirl every 10 seconds	0	5	12	18	21	23
Swirl continuously	8	17	28	36	41	47
Continuous magnetic stir	10	20	35	45	50	53

Being able to organise the data you collect in an investigation is a basic step for processing data and information. Tables can be used and constructed to organise your data, which can later be represented visually in a range of ways.

Follow these instructions to prepare your own table to organise data for a particular investigation.

a **Identify the independent and dependent variables.** Use these to write a descriptive title for the table. The descriptions representing the independent variable are listed in the left column and the dependent variable in the other columns. For example, in Table S2.6 we list the modes of agitation down the left column (independent variable). Then we have a main heading for the measurement that will be taken (the dependent variable), which is divided into six columns for each of the 10-second intervals when measurements are recorded.

b **Determine how many rows and columns are required to construct the table.** The number of **datasets** indicates the number of rows required (plus one or two rows for headings). In Table S2.6, we have three datasets – one for each mode of agitation: (1) swirl every 10 seconds; (2) swirl continuously; and (3) continuous magnetic stir. The number of columns depends on how many times you choose to take measurements. This could be based on time intervals, different items to measure or the number of trials you conduct. The example above uses 10-second time intervals across the first minute of reaction time. Units should be included in the headings when the data is a form of measurement – for example, (cm) or (mL).

c **Construct a table with the appropriate number of rows and columns.** Ensure they are evenly spaced and that there is enough room to write the information clearly.

d **Write headings to represent the variables.** The left column includes the variations of the independent variable: the three modes of agitation. Ensure your headings give the reader a clear picture of the data collected, including the units of measurement used.

The next example shows how to construct a table based on the following investigation question:

How does the type of wheel – bottle cap, compact disc or washer – affect the distance travelled by a model car?

a Identify the independent and dependent variables. Use these to write a descriptive title for the table.

*The independent variable is the type of wheel, because this is the thing that changes. We will measure the distance travelled, so that is the dependent variable. A good title is '**Effect of wheel type on distance travelled**'.*

b Determine how many rows and columns you need.

There are three types of wheels, so there will need to be three rows for recording data, plus heading rows. We can measure distance in centimetres (cm). We will not measure the time it takes. We think it is best to run three trials and then determine an average distance. One column is required for each trial, as well as one for the average, plus one for the wheel types.

c Construct a table with the appropriate number of rows and columns.

d Write headings to represent the variables. Remember that the independent variable belongs in the left column and the dependent variable goes along the top, with its unit of measurement.

Now you are ready to conduct the investigation and to record the data in an organised way, as shown in Table S2.7.

Table S2.7: Effect of wheel type on distance travelled over three trials

Wheel type	**Distance travelled (cm)**			
	Trial 1	**Trial 2**	**Trial 3**	**Average**
Bottle cap	11	9	13	(11 + 9 + 13) ÷ 3 = 11
Compact disc	33	31	35	33
Washer	26	20	32	26

Note that in the cells where we show the averages, we only show the process for calculating the average once. For more information, go to the 'Calculations' section on page 287.

Step 2 I can process data by using mathematical relationships and/or constructing graphs

The next step is to process your data and show it in a graph. Sometimes you will complete calculations, like determining averages as shown in Table S2.7, which are then graphed instead of the **raw data** points. This allows you to show mathematical relationships in the data. Graphs are useful for displaying data, as they can be easier to understand and interpret. The three graphs we will use most often in science are the column graph, the line graph and the scatter plot, shown in Figure S2.10.

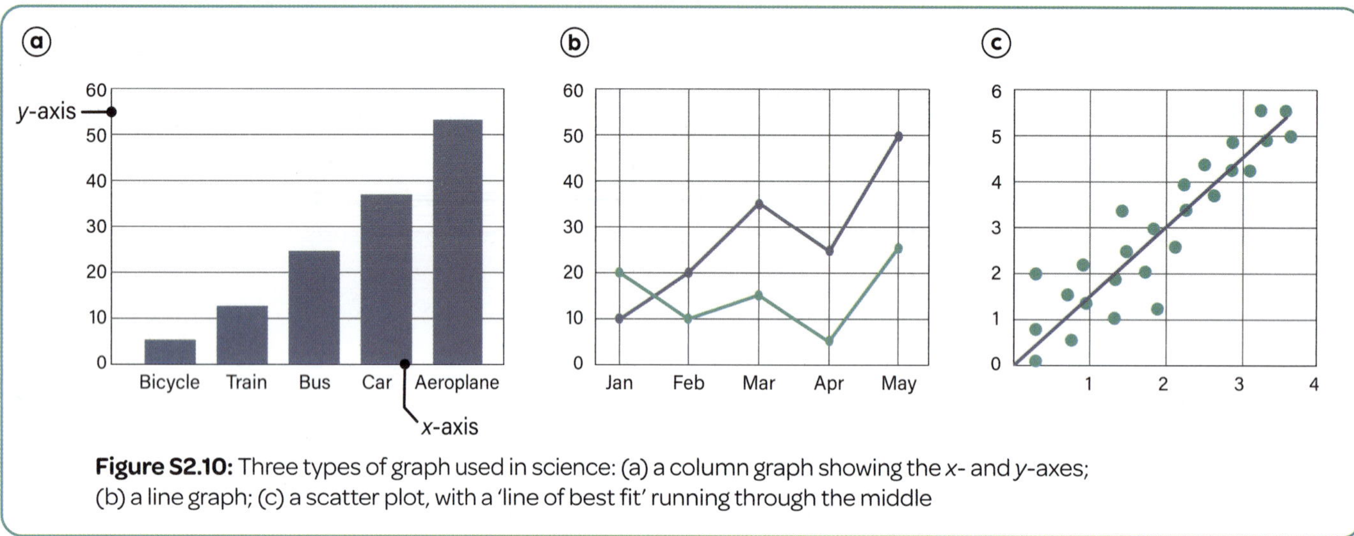

Figure S2.10: Three types of graph used in science: (a) a column graph showing the *x*- and *y*-axes; (b) a line graph; (c) a scatter plot, with a 'line of best fit' running through the middle

Figure S2.11 is a column graph of the results of the model car investigation. The annotation describes how to construct the graph.

ⓐ Determine the range of the axes based on the data you have collected.

Write the dependent variable on the *y*-axis (vertical axis) and the independent variable on the *x*-axis (the horizontal axis).

Plot the measurements along the *y*-axis and the wheel type along the *x*-axis. The lowest data value from our averages is 11. The range of the *y*-axis must begin below this value. We could choose 10, 5 or 0, but it is good to start at 0 for simplicity. The largest data value is 33, so the highest value on the *y*-axis must be higher than that, such as 35.

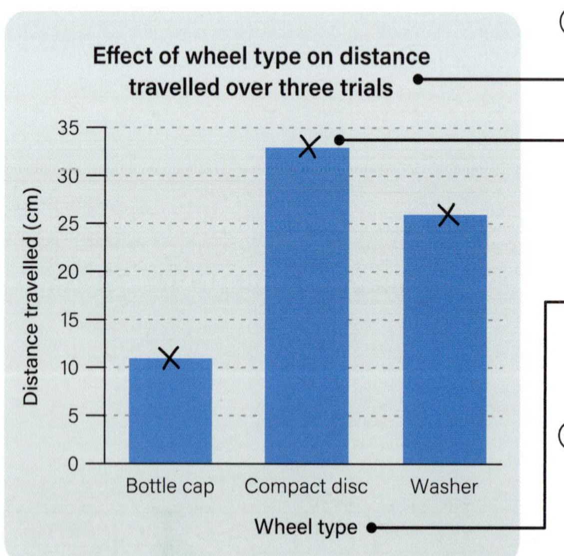

Figure S2.11: A column graph of the results of a model car investigation

ⓑ Use a ruler to draw the axes and select the space between values. Since the *y*-axis will be spread from 0 to 35, it makes sense to count by fives up the axis. Avoid writing every number between 1 and 35, as this will make the graph too cluttered.

ⓒ Write the title, label the axes and plot your points on the graph – the graph title can be the same as the title of your table. Be sure to include the independent and dependent variables, not only in the title but also on the axes labels. Include the unit of measurement, if needed, on each axis label.

ⓓ Draw your graph – a column graph requires you to draw and shade the columns. (A line graph means you connect the points in a line, while a scatter plot graph requires you to draw a 'line of best fit' that represents an average of your points.)

Categories of factors like the wheel type are best represented in a column graph, as shown in Figure S2.12. If measuring continuous time is involved, a line graph with connected points is best. If both the independent and dependent variable values are numbers, a scatter plot is best. A scatter plot includes a 'line of best fit', to show the trend.

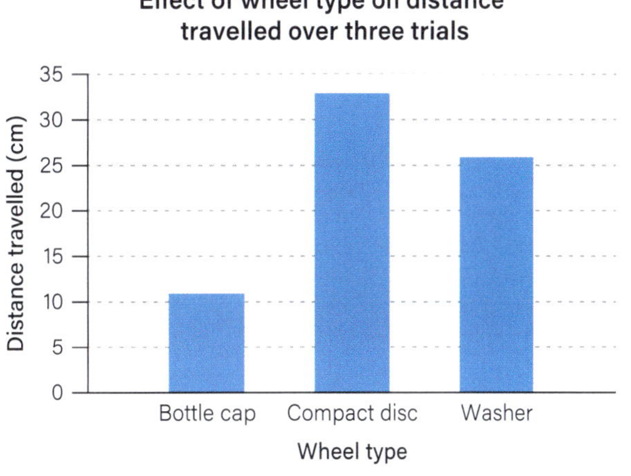

Figure S2.12: An example of a column graph using data from Table S2.7

Step 3 I can identify and explain trends and/or patterns in a range of dataset representations

In order to describe a trend, we must first identify it. A **trend** is shown by the line or bars in your graph moving in a general direction, usually up or down, as shown in Figure S2.13. A **pattern** is when the data repeats in predictable ways. Sometimes your data has no clear trend, so your visualisation will show random values all over, or there may be a sideways trend (a horizontal line). A sideways trend means the value may trend up or down in small amounts but there is no clear upward or downward trend.

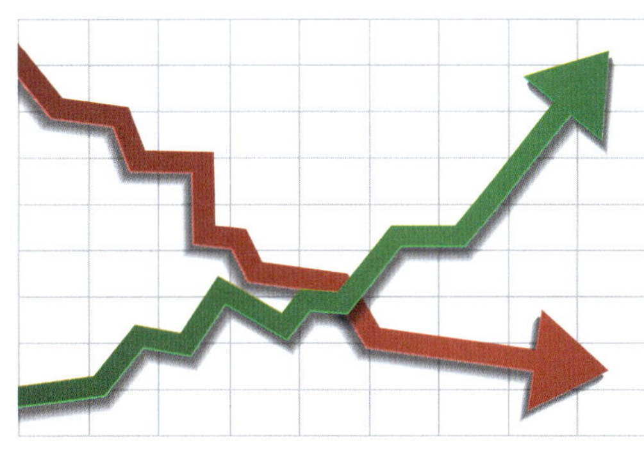

Figure S2.13: The green line is trending up, while the red line is trending down.

Figure S2.14 is a graph of snake sightings. It does not show a trend across the whole year. It shows a trend downwards from January to June and a trend upwards from July to December.

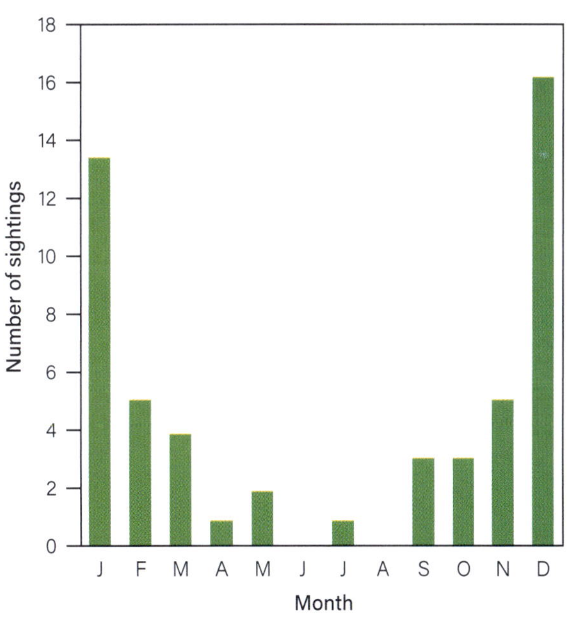

Figure S2.14: Snake sightings, average per month

To recognise patterns, look for recurring values or repeating shapes. For example, Figure S2.14 does not show a year-wide trend, so the next thing you look for is a pattern. It is clear that there are more snake sightings in the warmer months, and not many at all in winter. This is a pattern that we would expect to see year after year.

It is clear that the pattern in Figure S2.14 is that snake sightings trend downwards in the first half of the year and upwards in the second half. The trends alternate in a repeating pattern each year.

Figure S2.15: The number of snake sightings increases with average temperature across the year.

Next, we have to link any trends to the variables of the investigation. The relationship between the variables in an investigation is called the **correlation**. Identifying the correlation enables you to then describe the relationship between the dependent and independent variables.

- *Positive correlation:* the relationship between variables is direct. This means that if the value of one variable increases, the value of the other variable also increases, and vice versa.

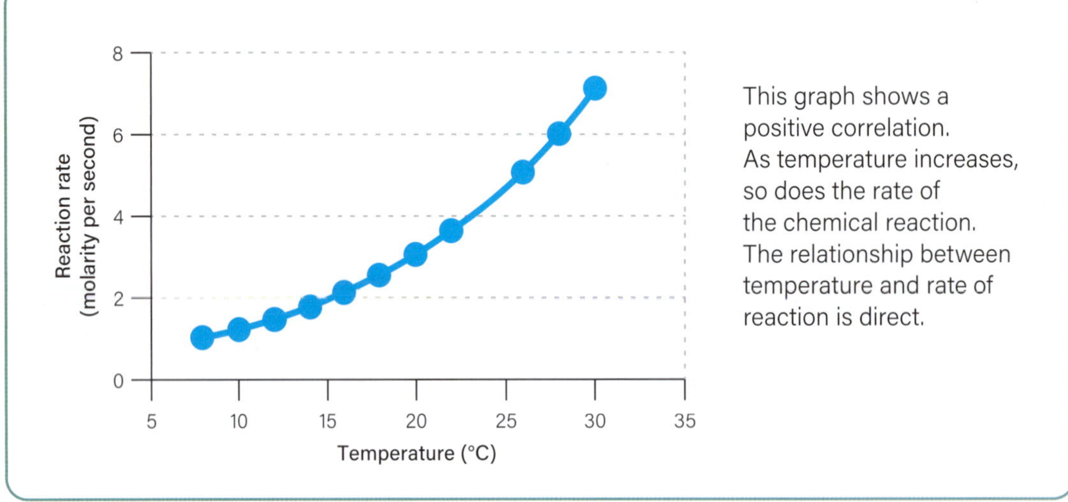

Figure S2.16: ▶ A line graph of reaction rates in an investigation, showing a positive correlation

- *Negative correlation:* the relationship between variables is indirect. This means that if the value of one variable changes in one direction, the value of the other variable does the opposite.

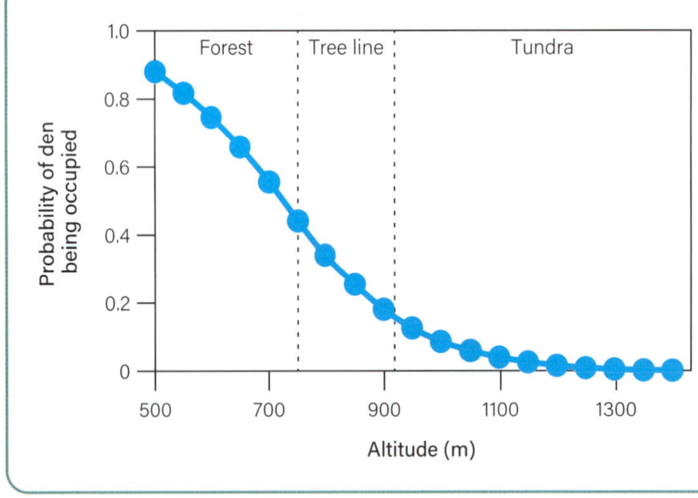

This graph shows a negative correlation. The likelihood of the fox den being occupied decreases as the altitude of the habitat increases. If one goes up, the other goes down. This displays an indirect relationship between whether a den is occupied and its altitude.

Figure S2.17: ▶ A line graph of fox den occupation rates, showing a negative correlation

- *No correlation:* there is no apparent relationship between the variables.

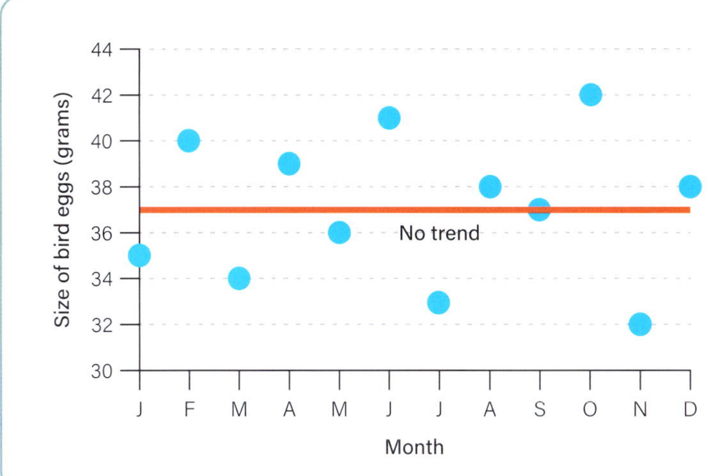

There does not appear to be a correlation in the data of this graph. This indicates there is no relationship between the size of bird eggs and the month of the year. We cannot predict whether bird eggs will be bigger in a particular month because the data does not show a pattern or trend.

Figure S2.18: A graph of the size of bird eggs and month laid

After identifying trends and/or patterns in the data, you should draw on your scientific knowledge and theoretical understandings to explain them. Once the relationship between variables is established, the next step is to ask: *Why is one variable affected by another variable in this way?*

For example, in Figure S2.17, the negative correlation between the fox den being occupied and altitude is linked to habitat theory and the effect of altitude on living conditions for the fox. As altitude increases, temperature and access to food decreases. These are reasons for why the den is less likely to be occupied at high altitudes, especially in winter.

Step 4 I can discuss relationships and anomalies that emerge in processed data

As scientists, we can identify scientific findings based on the data that emerges from an investigation or study by:

a stating the relationship between the specific variables, based on the data.

This finding can then be generalised by:

b replacing the terms specific to the investigation with general terminology, so it can be applied in a variety of contexts.

For example, if you investigate the effects of friction on the time it takes an object to reach the bottom of a ramp, you could use a toy car as the object, and ramp surfaces with varying amounts of friction, such as glass and wood. The results would show that the car reaches the bottom more quickly when rolled down the glass ramp than the wooden ramp. You can then use the points above to identify a finding.

a State the relationship between the specific variables, based on the data.
When the car rolls down the wooden ramp with more friction, it stops sooner than when it rolls down the glass ramp with less friction.

b Replace the terms specific to the investigation with general terminology so the statement can be applied in a variety of contexts.
When an object slides across a surface with more friction, it will come to a stop sooner than when it slides over a surface with less friction.

Figure S2.19: Investigating the effects of friction

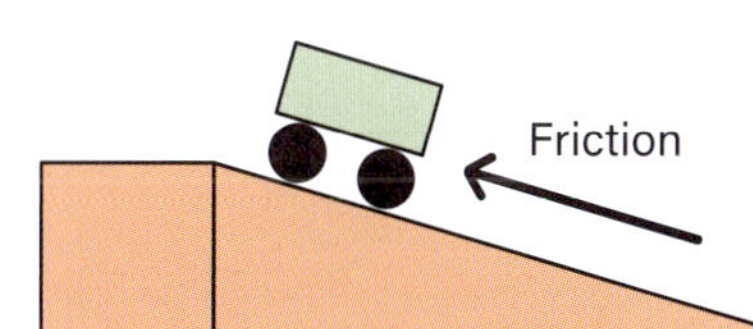

From the results, we analysed the trends and relationships in order to identify the finding and state the conclusion of the investigation. The term *car* was replaced with *object*, and instead of stating the specific surface used in the experiment, we generalised to the amount of friction – more or less. Table S2.8 provides some more examples of drawing conclusions.

Table S2.8: Converting investigation results into broader conclusions

Scientific finding	Conclusion
When carbon dioxide gas from a chemical reaction is captured in a balloon and then heated in a hot-water bath, the size of the balloon increases.	When the temperature of gas particles increases, the volume of the substance expands.
As the number of trees planted in eucalypt forests increases, the population of koalas also increases.	Revegetation efforts increase native animal populations by restoring natural habitats.
The more water a barrel contains, the greater the number of people required to push it 2 metres.	The greater the mass of an object, the more force is required to move it from a stationary position.

This does not mean that all our conclusions are correct. It is simply an opportunity to say what we know at that time and to investigate further.

Sometimes **data points** and conclusions do not follow the observed trend in a dataset or the expected outcome of an investigation. These are called outliers or **anomalies** and must be further analysed to determine whether the deviation is due to either an error in how the data was collected or to genuinely new discoveries. If they are mistakes or errors, they should not be averaged into the results. If it is unclear why there are anomalies, further investigations are required.

Step 5 I can analyse the quality of data using descriptive statistics

Using descriptive statistics to process and represent data can help to inform how we establish the quality of the data. The quality of data is influenced by errors/assumptions, biases and areas of uncertainty, all of which impact the results and can be revealed when the data is processed and displayed visually. Including statistics in a graph helps us to understand the data better. Statistics gives us numbers that show the mean, median and range, as well as trends and differences. For help with calculating averages, see the 'Calculations' section on page 287. When these things are displayed visually, investigators can make generalisations from the results as well as determine the general quality of the processed data.

Consider the data from Table S2.7. The graph of this data in Figure S2.21 does not include any information about the individual trials, which could be very important when analysing it. To reach step 5 of the Learning Ladder, you must consider the spread of data collected across all trials and thereby indicate the quality of the data collected. This can be done by displaying a bar for each trial, as well as the average, as displayed in Figure S2.20. However, this approach to displaying data can appear cluttered and confusing for the audience.

To maintain simplicity in the visual data, error bars can be used to avoid clutter while still showing the spread of trial data (see Figure S2.21). The following are examples of what it might look like to incorporate descriptive statistics into a graph.

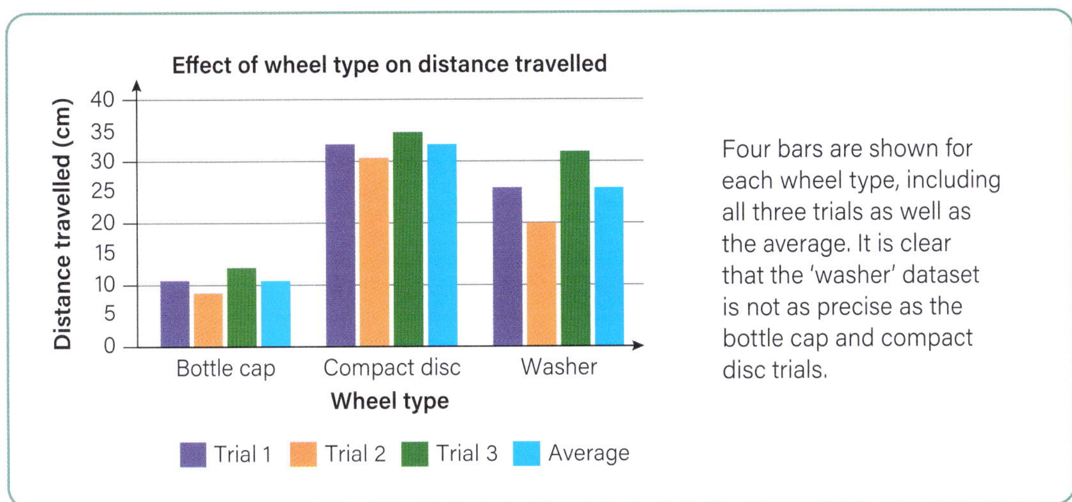

Four bars are shown for each wheel type, including all three trials as well as the average. It is clear that the 'washer' dataset is not as precise as the bottle cap and compact disc trials.

Figure S2.20: Effect of wheel type on distance travelled

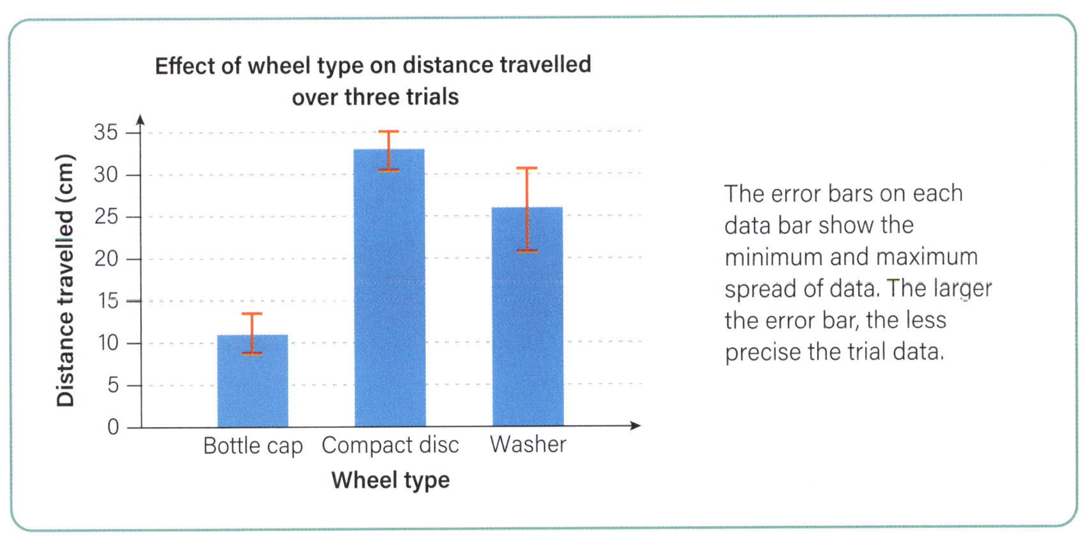

The error bars on each data bar show the minimum and maximum spread of data. The larger the error bar, the less precise the trial data.

Figure S2.21: Effect of wheel type on average distance travelled over three trials

The error bars show a range of values that includes the true value somewhere within that range. This means we cannot be sure of the exact true value, based on the data collected. To determine the error bar range, the half range must be determined, which is the highest data value in a set, minus the lowest value, divided by 2:

$$\text{Half range} = \frac{(\text{highest data value} - \text{lowest data value})}{2}$$

For example, in Table S2.7, the highest value for the bottle cap was 13, while the lowest was 9. The half-range data is therefore (13–9)/2 = 2. This means we can assume that the true value is greater than 2 or less than 2 from the calculated average.

The half-range data value can be used to show the certainty of the results and is written as 11 cm ± 2 cm. For details about the certainty of experimental data values, see the 'Calculations' section on page 297.

Error bars across multiple datasets can be used to analyse the overall data to evaluate its quality, and therefore the quality of the overall investigation. Here are some things to consider when analysing error bars in visual data.

- If the error bars overlap from one dataset to the next, the true value for each dataset could be the same. This is not ideal when you are trying to answer a question scientifically.
- If the error bars across datasets do not overlap, the quality of the data is good. The smaller the individual error bars, the higher the quality of the data.
- If error bars overlap, the quality is not good. The more overlap that exists across the error bars, the lower the quality of the data.

Consider Figure S2.22. Due to the overlap in error bars across four of the five datasets, this data would be considered very low in quality. The only valid finding that can be drawn from this data is that 'B' is higher than the others. Since error bars give a range of where the true value might be, there is no way to know with certainty how A, C, D and E compare to each other, since all of those error bars overlap.

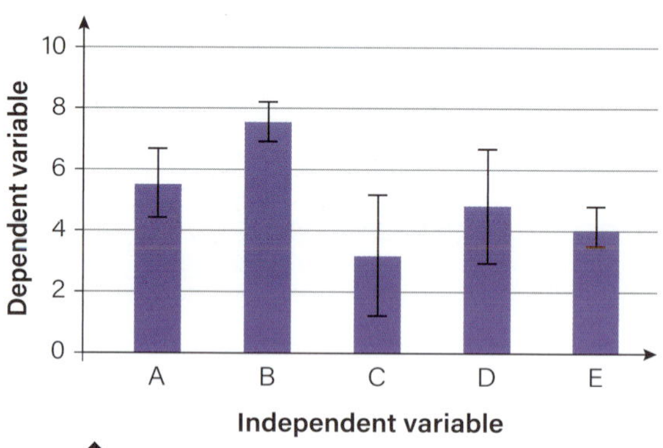

Figure S2.22: Error bars represent a range of values. The true value is somewhere within that range.

Evaluating

Evaluating an investigation means looking closely at how it was carried out and deciding how reliable the results are. This involves checking whether the data was collected carefully, if the results are accurate and whether the investigation is reproducible. It is also important to consider the quality of the results, whether the conclusions are supported by the evidence and how the investigation could be improved next time. Work your way up the Learning Ladder to approach this skill effectively.

Step 1 I can identify assumptions and types of errors in an investigation

In any investigation, our aim is to collect data that is accurate and valid. In other words, we want our results to find the **true value** that answers the experimental question by conducting a **fair test**. It is up to us, the investigators, to identify **assumptions** and errors in the method that lower the **accuracy** and validity of the results.

Let's consider the investigation question:

How does the type of wheel (bottle cap, CD, washer) affect the distance travelled by a model car?

Simple questions shown in Table S2.9 should be asked to help you recognise sources of error in the method. Answering 'YES' to any of these questions can lead to the identification of assumptions and errors in the method.

Table S2.9: Identifying the sources of errors

Questions to reveal sources of error	Describing the scenario	Source of error
1 Did your experimental set-up or process differ from the intended procedure?	*The method required the ramp to be propped up on a printer box, but there was only one box. To save time, different things were used to prop up the ramp for each car. This method went against the intended procedure.*	*Different ramp angles were used for each car.*
2 Did you record any qualitative observations related to difficulties or assumptions in the method?	*The cars did not always move in a straight line, which caused variations in how the distance was recorded. If it went straight, the total path of the car was measured, but if it curved, the distance from the starting line to the stopping point was measured directly. This shows that assumptions were made about the path and distance the cars travelled.*	*The way the distance was measured was not consistent for all three cars.*
3 Were there any new variables identified during the investigation that were not controlled?	*The type of wheel was the independent variable for this investigation, but it is not clear whether the material type or the size of the wheel is responsible for the results.*	*The material and size of the wheels differed, which means there is more than one independent variable.*
4 Did anything unexpected occur during the investigation, including strange results?	*During the third trial for the car with washer wheels, a gust of wind came through the classroom. Interestingly, the car in this trial went the furthest by far, as compared to the other two trials.*	*The surrounding environment was not kept constant for all trials, (e.g the gust of wind that passed through the classroom in one trial).*
	For some trials, the student pushed the car down the ramp instead of letting it roll on its own as the method indicated.	*The force applied to the cars was not consistent across all three trials.*

Figure S2.23: We can see that different-sized wheels were used in this experiment; therefore, this is a source of error in the results.

There are different types of errors that you may encounter in an investigation. Being able to identify and describe the types of errors present will help you to understand how they have impacted on your data. Table S2.10 summarises the different types of errors you will be expected to recognise.

Table S2.10: Types of errors

Type of error	Description
Random	Unpredictable differences in how measurements are taken, resulting in a 'random' variation of values.
Systematic	A consistent difference in measurement from the true value. When a particular value is measured repeatedly, the error is the same.
Personal	Mistakes that can be avoided if the method is followed correctly. Rather than discussing personal errors in a report, the method should be repeated to eliminate mistakes.

Table S2.11 considers the sources of error identified in step 1 and determines their type.

Table S2.11: Types of sources of error identified in step 1

Source of error	Type of error	Supporting information
Different ramp angles were used for each car.	Systematic	*If two ramps had the standard angle but the ramp for the CD car had a higher angle, the CD car would experience more force down the ramp, leading to a greater distance travelled than if it had gone down the standard ramp. The additional distance travelled would be relatively the same for every trial for that car.*
The way the distance was measured was not consistent for all three cars.	Random	*The method did not specify how to measure the distance travelled; therefore, it may have been done differently across all cars and trials. There is no way to know how much or in what direction the values vary from the true values.*
The surrounding environment was not kept constant for all trials (e.g. the gust of wind that passed through the classroom in one trial).	Random	*External factors may have influenced the cars' performance down the ramp for better or worse, but there is no way to know the extent of the impact.*
The force applied to the cars was not consistent across all three trials due to some students pushing the car at the start.	Personal	*This mistake can be avoided if the method is followed correctly. If it is not possible to redo the investigation, this would be considered a random error as you have no way to know the impact of the mistake. Ideally, the trials where the cars were pushed should be eliminated from the data and redone.*

Step 2 I can describe the impact of assumptions and errors, and propose ways to reduce them

Once you understand the difference between various types of errors and their sources, the next step is to discuss how the errors affect your data. It is not enough simply to say that an error will reduce the accuracy or validity of results. This is assumed in step 1 of the Learning Ladder. At step 2, you are expected to discuss specifically *how* the error affects accuracy and/or validity. Does it increase the experimental value? Why does it reduce validity?

Random and systematic errors each have a known impact on the data, as Table S2.12 illustrates.

Table S2.12: Known impacts on data of random and systematic errors

Error type and impact	Elaboration
Random errors reduce precision *Values are spread around the centre true value	*Each data value is not accurate; however, the average of all values is often close to the true value. Therefore, random errors do not necessarily affect the accuracy of an experiment, as long as there are enough trials to average out the error.*
Systematic errors reduce accuracy *Values are close to each other but not to the true value	*Data values are precise but not accurate, because the collection of data is affected in a predictable and consistent way. For example, an uncalibrated scale used to measure mass might add 1 gram to all measurements.*
Personal errors can reduce precision and/or accuracy	*Mistakes can be random errors or systematic errors, depending on whether the impact on the data varies or is consistently predictable. Refer to step 2 for more information.*

To effectively support your findings with evidence-based explanations about the quality of your data, you must first determine the type of error present and its general impact, as outlined above. This is the first item on the checklist below.

CHECKLIST DESCRIBING THE IMPACT OF ASSUMPTIONS AND ERRORS

- ☐ Determines the type of error and its impact on the data; reduces precision and/or reduces accuracy.
- ☐ Compares experimental values to the true values, where possible, referenced in scientific literature (see the 'Conducting scientific research' section on pages 277–80).
- ☐ Uses descriptive words to describe the impact on data (higher, lower, heavier, stronger, shorter, etc.).

Use the checklist to ensure your explanation includes everything necessary for you to display your skills at step 2 of the Learning Ladder.

Let us now revisit a couple of the errors from the previous step to see how they can be explained with reference to the quality of data (see Table S2.13).

Table S2.13: Describing the impact of assumptions and errors

Source of error	Description of impacts
Different ramp angles were used for each car (systematic).	The distance travelled for the CD car (higher angle ramp) is greater than it would have been on a standard ramp. This increases all values recorded for the CD car by the same amount, reducing the accuracy of the average of the trials for this car. This, in turn, reduces the validity of the results as we cannot be sure if the CD would have gone further than the other cars had it been done on a standard ramp.
The way the distance was measured was not consistent for all three cars (random).	As a result of inconsistent measuring, the precision of data values is reduced. There is no way to know how much or in what direction each value differs from the true values. Because there were three trials for each car, the average of all three trials is closer to the true value than a single data value on its own. Averaging trial values improves the quality of the data and, as a result, increases the accuracy of the results.

*The text colour coincides with the checklist items above.

Please keep in mind that there may be several correct ways to describe how errors impact results. The more practice you have, and the more feedback you get from your teacher, the more comfortable you will become with using this skill.

The second part of this skill is proposing ways to modify the method to reduce the impact of errors. Table S2.14 revisits the errors described in Table S2.11 and puts forward ideas for how to improve the method to reduce the impact of each error.

Table S2.14: Proposed improvements to minimise the impact of errors

Source of error	Proposed improvement
Different ramp angles were used for each car.	*Ensure there are enough materials to conduct each trial under the same conditions at the same time. Specifically, there should be three printer boxes so that all three ramps are propped to the same height. If this is not possible, allow more time to complete all the trials using the same ramp.*
The way the distance was measured was inconsistent between cars.	*It should be clear in the method how the measurement should be taken. Measure the total distance that the car travels, even if it does not go in a straight line. This may require a length of string so that curves can be measured in full. Use the string to follow the exact path of the car. Then use a metre stick to measure the string. Conducting more trials will also minimise the impact of this error.*
The material and size of the wheels were changed, which means there is more than one independent variable.	*Instead of testing 'types' of wheels, the investigation question should be more specific about what will change between wheels. It could be split into multiple investigations, each testing a different feature of the wheel.* *For example:* • *How does the* ***size of the wheel*** *affect the distance travelled by a model car?* • *How does the* ***material of the wheel*** *affect the distance travelled by a model car?* *This will ensure that only one thing is being changed.*
Surrounding environment was not kept constant for all trials (e.g. breeze through classroom).	*Ensure that laboratory conditions are kept constant throughout all the trials of the investigation. For instance, windows and doors should be shut to prevent uncontrolled airflow. Conducting more trials will also minimise the impact of this error.*
Force applied to the cars was not consistent across all trials due to some students pushing the car at the start.	*Follow the method correctly by releasing the car at the top of the ramp, allowing gravity to send the car into motion naturally without applying extra force.*

Step 3 I can discuss ways to improve the quality of data and validity of conclusions and claims

We know that sources of error decrease the quality of results. Now that you have proposed ways to reduce errors, the next step is to discuss how the proposed modifications can improve the quality of data and, in turn, the validity of the investigation findings. This skill requires a deeper understanding of the impact of proposed modifications and how they increase the quality of data to better address the investigation question with valid results.

If an investigation is valid, it means the method measures what it is meant to measure. Validity allows us to make generalisations from our conclusion and to apply our learnings to other situations. It considers both accuracy of measurements and the control of variables.

In order for results to be valid, we must design a fair test that tests one variable at a time, while controlling all other variables that could affect the results. This is how we can determine a cause-and-effect relationship. If more than one variable changes, we cannot be sure what caused the effect, which means the test would not be considered fair and the results would not be valid. By considering the accuracy of measurements, the precision of values within each dataset, and the method and materials used to isolate and control variables, you are able to propose and discuss ways to improve the quality of data and the validity of results.

Use this checklist to help you.

CHECKLIST FOR DISCUSSING HOW MODIFICATION CAN IMPROVE THE INVESTIGATION

- ☐ Consider the errors identified in the method.
- ☐ Suggest a modification for each source of error that will minimise its impact.
- ☐ Provide specific details for how the modification reduces error, including scientific theory where appropriate.
- ☐ Make links between modifications and the overall quality of the investigation.

First, we will identify some errors from the wheel-type investigation. We can then suggest modifications and consider how they minimise the errors, with reference to the quality of the investigation. (The text colours in the 'Discussion' column in Table S2.15 coincide with the checklist above.)

Table S2.15: Sources of error in the investigation

Source of error	Explanation of how the error affects the quality of data	Discussion of how modifications can improve the investigation
Inconsistent ramp angle for each car (systematic error)	The distance travelled for the compact disc (CD) car (higher angle ramp) is greater than it would have been on a standard ramp. This increases all values recorded for the CD car by the same amount, reducing the accuracy of the average of the trials for this car. This reduces the validity of the results, as we cannot be sure if the CD would have gone further than the other cars had the test been done on a standard ramp.	*There must be three identical boxes so that all three ramps are propped to the same height. The ramp angle determines how much gravitational force the car will experience down the ramp, impacting the distance it will go. This must be consistent across all trials to remove the ramp angle as a source of error. The ramp angle is a variable that should be controlled. Requiring identical boxes to prop the ramp will ensure a fair test.*
The way the distance was measured was inconsistent between cars (random error)	As a result of inconsistent measuring, there is no way to know how much or in what direction the values vary from the true values. Because there were three trials for each car, the average of all three trials is closer to the true value than a single data value on its own. Averaging trial values improves the quality of the data and, as a result, increases the accuracy of the results.	Measure the total distance that the car travels, even if it does not go in a straight line. Otherwise, the measured value will be shorter than the actual distance travelled, like cutting corners. Since the car going straight or curved was unpredictable, measuring the exact patch of the car reduces the random measurement errors, increasing the precision of the values collected. This improves the quality of data, making the key finding more valid, which improves the overall investigation.

Students often recommend repeated measurements, or *more trials*, to improve the method. This works for random errors, but *not* for systematic errors, as summarised in Table S2.16.

Table S2.16: The impact of repeating measurements on random versus systematic errors

Type of error	Impact of repeating measurements	Why
Random	Increasing the number of trials *does* reduce the impact of random errors on the average results.	The more repeat measurements there are, the closer the average will be to the true value, which means the investigation questions can be answered more accurately.
Systematic	Increasing the number of trials *does not* reduce the impact of systematic errors on the average results.	The error is consistent and predictable for every trial, so more trials will not 'average out' the error. The values are already precise, just not accurate or reflective of the true value. Therefore, the only way to minimise the impact of a systematic error is to address the problem in the method.

Step 4 I can construct evidence-based arguments to justify conclusions, address ethical issues or assess claims

Once you are comfortable discussing the validity of conclusions with reference to errors and assumptions, the next step is to construct evidence-based arguments to discuss a variety of factors involved in first- and second-hand investigations. This includes things related to the quality of data collected, as outlined in step 3, but also whether ethical factors have been considered.

For help with writing an evidence-based scientific argument, see the 'Scientific writing' section on pages 280–81. Now, combine your explanations from step 3 and/or your analysis of whether ethical considerations have been made appropriately, and construct your argument.

Step 5 I can evaluate the validity and reproducibility of investigation methods

Investigation results should be valid and reproducible. In other words, the results of an investigation should be of a high quality and give similar results when repeated by other researchers using the same method.

An investigation that is **repeatable** means that we can trust that if we repeated the investigation, we could produce similar results. An investigation that is *reproducible* means that similar results are obtained, even when carried out by different researchers, using the same method and equipment. When an investigation is both repeatable and reproducible, it indicates that the method used produces consistent and reliable results. If an investigation is not reproducible, it may not be using a valid method.

Use the checklist on the next page to evaluate the validity and reproducibility of an investigation.

CHECKLIST FOR EVALUATING THE VALIDITY AND REPRODUCIBILITY OF INVESTIGATION METHODS

- ☐ One variable was changed.
- ☐ One variable was measured.
- ☐ All other variables were properly controlled, without any assumptions.
- ☐ The quality of data is high (see step 5 of 'Processing, modelling and analysing' on page 250).
- ☐ Biases and areas of uncertainty are acknowledged in the investigation method or report.
- ☐ The investigation gives similar results when repeated by other researchers using the same method and equipment.

Ticking each of the items in the checklist does not guarantee that the conclusions or claims are valid and reproducible. Nor do all items need to be ticked in order for a claim to be true. The checklist is just a guide to help us evaluate a variety of scientific investigations by asking standard questions as a starting point. More questions can always be asked and further studies carried out to help us continue to contribute to the scientific body of knowledge in safe, thorough and ethical ways.

Figure S2.24: Which of these containers has a volume closest to 50 mL? Measurements are always subject to uncertainties.

Communicating

Communicating is essential to science – scientists need to share the ideas they have and the findings they have discovered. When we communicate, we need to consider what we are trying to say and who we are trying to say it to. This allows us to present our ideas and findings using specific methods and language targeted to our particular audience.

Step 1 I can select appropriate formats, content and vocabulary to communicate scientific ideas and findings

You can communicate your scientific findings in many different ways. To decide on a suitable method of communication, ask the following questions:

- What information do I have?
- What do I want to communicate to the audience?
- What format will display this information most clearly?

Table S2.17 shows some examples of how to apply these questions to make your decision.

Table S2.17: Choosing formats for scientific communication

Information	Communication goal	Format
Data from a first-hand investigation	What I found out while investigating a scientific question	Scientific table or graph
A complete first-hand investigation	How I did the investigation and what the results of the investigation were	Scientific report
What causes an eclipse	How the movement of the Sun, Earth and Moon cause an eclipse	Scientific poster with text and diagrams
Carbon movement	How carbon moves through the different stages of the carbon cycle	Scientific diagram with labels
The structure of DNA	How two strands of nucleotides form the double-helix structure of DNA	Scientific model

Figure S2.25: A scientific model of DNA shows the components that make up its structure.

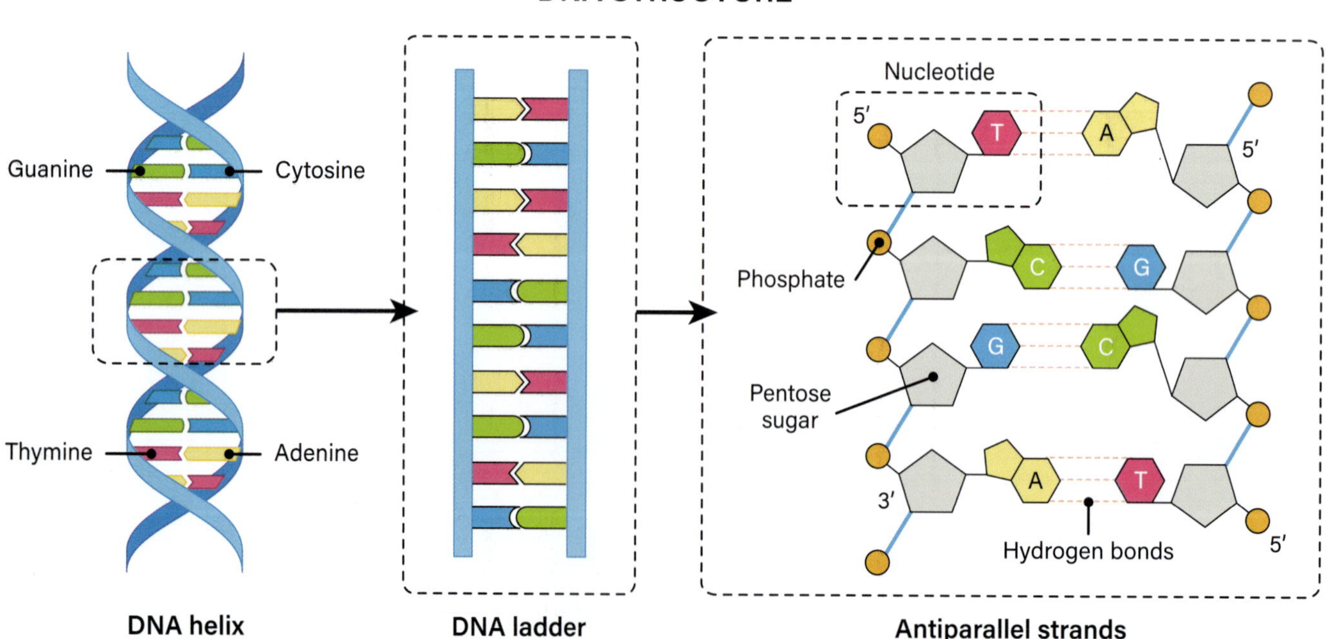

Step 2 I can prepare a variety of presentation formats to communicate ideas and findings

After selecting a method of communication and a format to present it, you can then look at how to construct a presentation. When considering how to present scientific information in a chosen format, you can ask the following questions:

- What is the main point I want to get across?
- What information do I need to include in this format?
- What is the best way to present this information in my chosen format?

Once you have answered these questions, you can start to put your information into your selected format. Make sure you are using correct scientific terminology and information relating to your topic. More detailed examples of how to construct each type of presentation can be found in the 'Scientific writing' section on page 272.

Ideas and findings can be represented in numerous visual formats as well. You should choose an appropriate representation of your information, just like picking the right tool for a job. Figures S2.26 to S2.30 show some ways you can visually communicate the information.

Figure S2.26: An infographic showing the results of the model car investigation

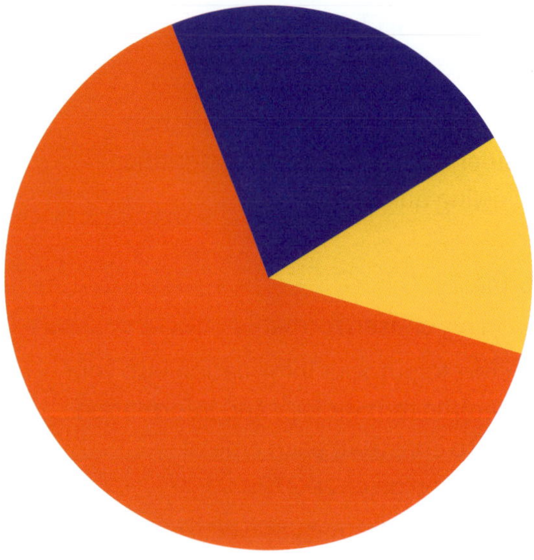

Figure S2.27: A pie chart is best for looking at the parts that make up a whole.

Figure S2.28: A Venn diagram is used to compare similarities and differences.

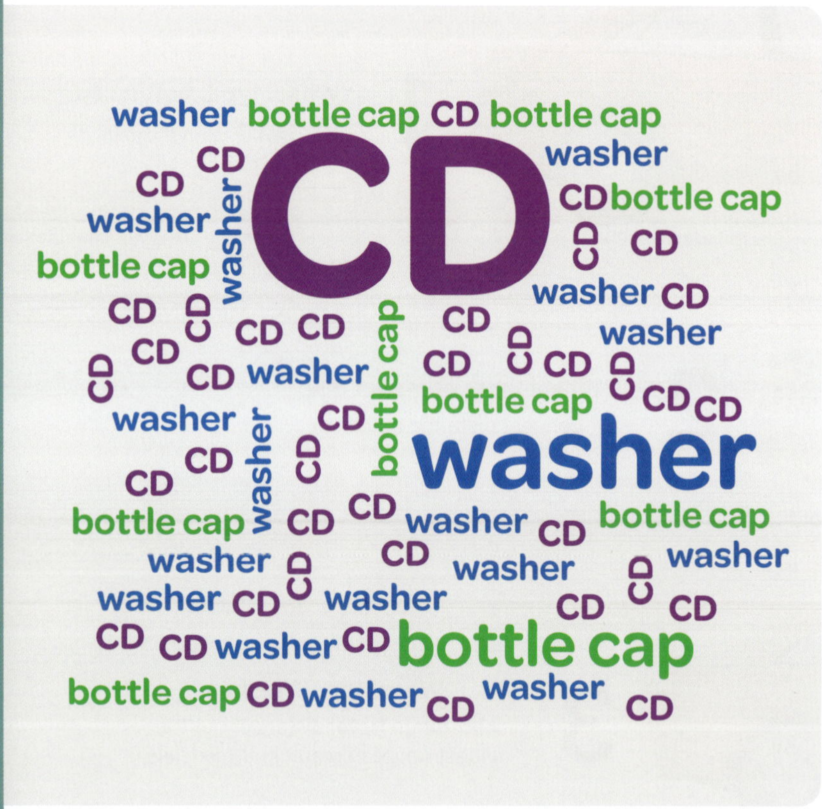

Figure S2.29: A word cloud shows the magnitude of results of the model car investigation.

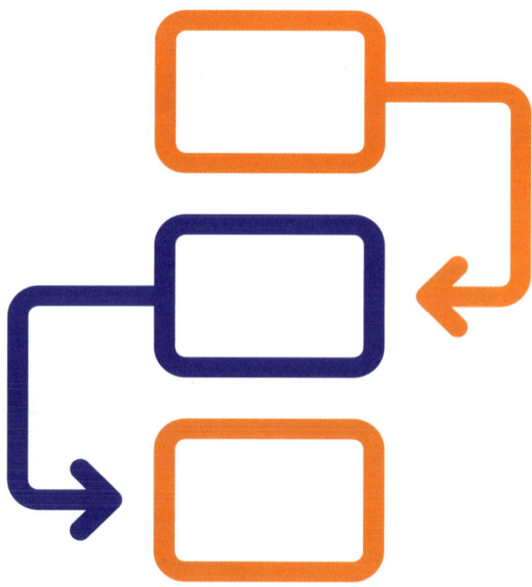

Figure S2.30: A flow diagram can be used to show how one thing moves or transforms to another.

Step 3 I can use digital technologies and/or scientific representations to communicate data and information

Once you have selected an appropriate way to communicate your information, it is important to be able to represent any data you have digitally. Using digital technologies means that information can be presented neatly and succinctly, so that it is easy for audiences to understand. Table S2.18 provides a range of examples. Some examples of these representations can be seen in the 'Scientific writing' section on page 272.

Table S2.18: Digital presentation formats

Information	Presentation format	Example communication presentation method
Data from a first-hand investigation	Scientific table or graph	Google Sheets or Microsoft Excel
A complete first-hand investigation	Scientific report	Google document or Microsoft Word document, with typed information
Infectious disease transmission	Scientific poster with text and diagrams	Canva or slides to visually present information
Chemical molecule structures	Scientific diagrams with labels	Canva, slides or Lucidchart software to create a diagram with labels
Cell structure	Scientific model	Canva or Lucidchart software to create two-dimensional models

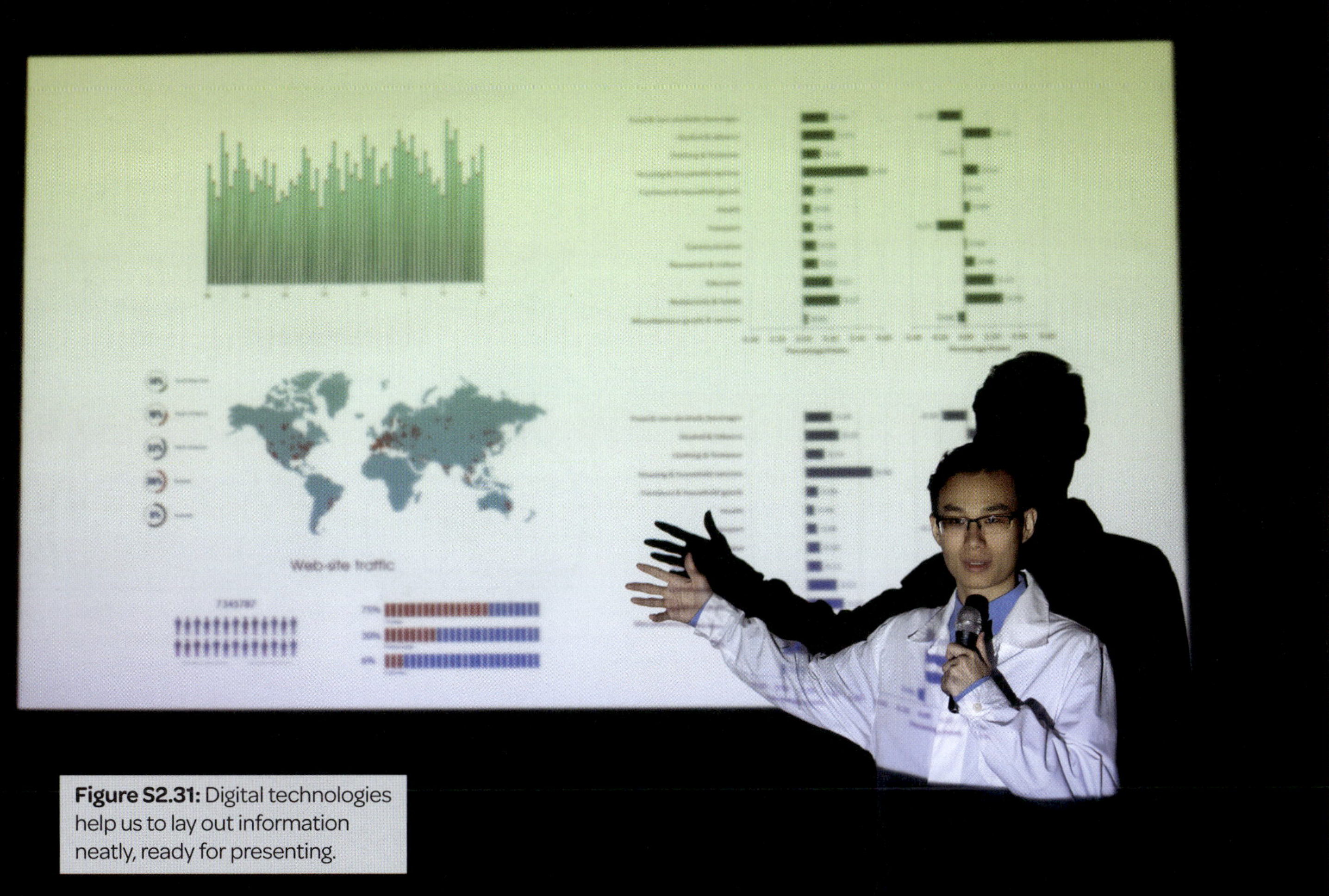

Figure S2.31: Digital technologies help us to lay out information neatly, ready for presenting.

Step 4 I can communicate scientific findings and arguments effectively for specific purposes to specific audiences

Now that you can determine appropriate communication formats, you should select information and text that matches the audience you are trying to communicate with. To help you figure out *how* to tailor your communication, consider the:

- age of your audience
- knowledge level of your audience
- idea or scientific principle you want to communicate
- best way to tell your audience about this principle.

Table S2.19 shows how the same information can be presented to two different audiences.

Table S2.19: Tailoring your message for different audiences

Age	Knowledge level	Idea	Communication
Eight-year-old children	Very limited	Water as a resource	Summarise the water cycle and encourage household water-saving measures on a poster.
Adults	High – environmental scientists	Water as a resource	Present research on ideas to prevent surface evaporation from lakes and dams as an article or oral presentation.

We can then create a communication format for each audience. For the eight-year-olds, we could include the following text on a simple poster.

In the water cycle, rain falls and then runs off into storage areas like dams and lakes. We use some of this water in our homes. To help us use less water, we can take shorter showers or shallow baths, and turn off taps while cleaning our teeth. These actions mean every household uses less water overall, and this is also known as using water sustainably.

The adult scientists would expect more sophisticated language, and the information could be displayed using specific diagrams and figures.

When water that has precipitated runs off into storage areas such as dams and lakes, it is subject to surface evaporation, reducing the amount available for household consumption. Placing products such as evaporation-reducing shade balls on the surfaces of these areas will assist in minimising surface evaporation, reducing water loss and increasing availability for household usage, even in times of drought.

Figure S2.32: A poster showing sustainable water usage in the home would be a suitable communication method to share ideas with younger audiences who have little scientific knowledge.

Step 5 I can justify scientific ideas, findings, arguments and proposals for diverse audiences

In addition to communicating findings, it is important also to be able to justify them to a diverse range of audiences. This means providing your audience with the reason 'why' the information you are providing them is correct, using appropriate data.

To justify what you are communicating, you should consider the following:

- the age of the audience you are communicating the information to
- the scientific principle you wish to communicate
- relevant evidence relating to the concept
- how the principle and evidence can be effectively linked.

For example, the risk of transmission of COVID-19 can be communicated to adults using the following information.

COVID-19 is an infectious viral disease that can be transmitted through direct and indirect contact. The virus is able to remain in the air and to survive on secondary surfaces for 4–5 hours, meaning that others may still be at risk of contracting the disease even if an infectious individual has left a space. To minimise the risk of transmission, individuals can wear well-fitting face masks, improve indoor air quality, wash hands after being in contact with public surfaces, and maintain a distance of 1.5 m between themselves and others when in public or when visiting friends.

The information above communicates how to reduce transmission, but it does not provide any evidence. Evidence would be as follows:

In the scientific article 'A review on COVID-19 transmission, epidemiological features, prevention and vaccination', Zhang et al. (2022) conducted a meta-analysis of different prevention strategies to help minimise the transmission of COVID-19. The researchers found that interventions helped to reduce the rate at which viral particles were successfully transferred, reducing the incidence of the disease.

Figure S2.33: COVID-19 communications required evidence so that they would be accepted by the general public.

When the information is combined with evidence, it provides readers not just with advice on what to do to prevent becoming infected, but also a description of the impact of prevention strategies and why they are important.

> *COVID-19 is an infectious viral disease that can be transmitted through direct and indirect contact. The virus is able to remain in the air and to survive on secondary surfaces for 4–5 hours, meaning that others may still be at risk of contracting the disease even if an infectious individual has left a space. Studies like 'A review on COVID-19 transmission, epidemiological features, prevention and vaccination' (Zhang et al., 2022) have shown that COVID-19 can be transmitted in this way.*
>
> *These studies also show that measures such as wearing well-fitting face masks, improving indoor air quality, maintaining social distancing, and washing or sanitising hands after contacting public surfaces can reduce transmission. This is because these measures reduce the ability of the virus to enter an individual's body, meaning they do not have to fight off the disease as they are not exposed to the virus in the first place. Maintaining these important strategies will help to consistently reduce case numbers.*

Figure S2.34: Supporting evidence helps the general public to understand why certain public health measures might be necessary.

COVID-19 PREVENTION

COVID-19 is caused by a virus that can remain in the air and survive on surfaces for hours after an infectious person has left a space. Prevention strategies and vaccination reduce the risk of contracting and spreading the disease.

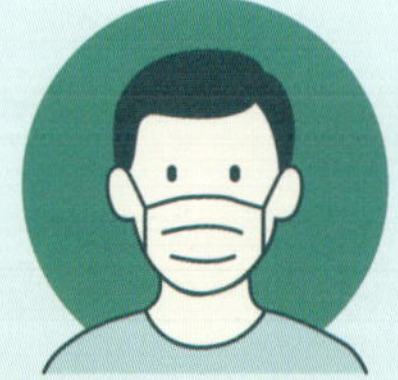

Wear a well-fitting face mask in crowded spaces

Stay home if unwell

Maintain social distancing

Wash or sanitise hands after touching public surfaces

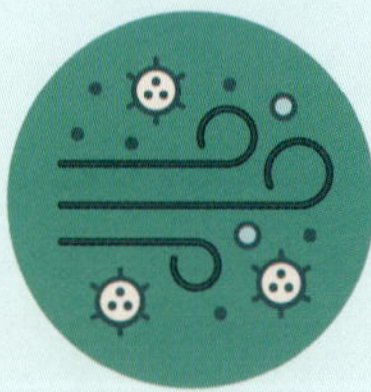

Improve indoor air quality

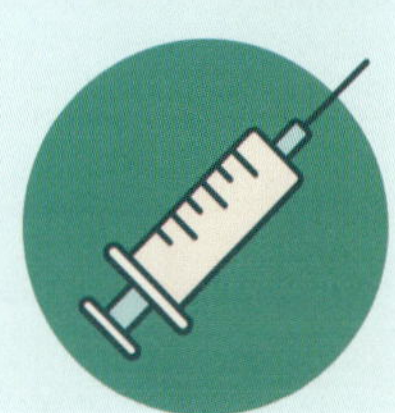

Keep up to date with vaccines

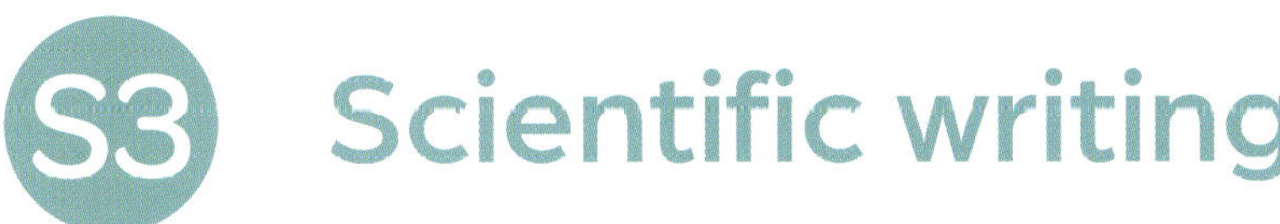

S3 Scientific writing

'Scientific writing' means applying your writing skills to your science studies to help you develop your understanding and communicate what you have learnt.

Key terms

citation: a way of giving credit to a source, usually in the same text where the information appears

plagiarise: to copy someone else's work and present it as your own

Figure S3.1: Scientific writing is fundamental to science communication.

Writing investigation reports

Writing a clear and concise report after your investigation is complete will help other people to understand your work. Table S3.1 provides an overview of the sections that are often included in a standard investigation report, as well as the science inquiry skill used in each section.

Your investigation reports should typically have a similar structure to the one shown below.

Table S3.1: Investigation report sections matched to their working scientifically processes

Section	Process
Title	Questioning and predicting
Introduction • Background • Aim • Hypothesis	Questioning and predicting Planning investigations
Risk assessment	Planning investigations
Materials	Planning investigations
Method	Conducting investigations
Results	Processing data and information
Discussion and conclusion	Analysing data and information
References	Communicating

The title is clear and uses plain language. Many scientists write their title as a research question.

To what extent do different modes of agitation affect the rate of a chemical reaction?

Set the scene. If there is scientific information or context for the investigation, summarise it here.

Introduction

Chemical reactions can occur at different rates, depending on the proportion of successful collisions between reacting particles in a given amount of time. Many factors can influence the frequency of successful collisions, such as temperature, concentration, surface area and the presence of a catalyst. Agitation of, or stirring, a reaction mixture is another factor that could have an impact on colliding particles. It would be interesting to compare how different ways of stirring a reaction mixture impact on the rate of a chemical reaction. This could inform what types of equipment are purchased for the laboratory and help in making methodical decisions for investigations in relation to efficiency.

The background often leads into the objective of the investigation. Use your research question to write the aim starting with a verb, such as: 'To investigate ...'

Aim

To explore how different types of agitation – swirling once every 10 seconds, swirling continuously and magnetic stirring – affect the rate of a chemical reaction.

If your investigation is an experiment, you should predict what you think will happen based on scientific theory. A good hypothesis will include the independent variable and the predicted effect on the dependent variable.

Hypothesis

If a chemical reaction is agitated to a greater extent and continuously during a reaction (magnetic stir), the reaction will occur at a higher rate than if it is only agitated manually and periodically throughout the reaction. This is because agitation increases the amount of collisions that occur between reactants in a given amount of time; therefore, more agitation will result in a higher proportion of successful collisions, increasing the rate of reaction.

List all materials and equipment with amounts and sizes as simple bullet points.

Materials

- *3 × 100 mL conical flasks with stopper and delivery tube*
- *3 × 20 mL 0.5 M HCl*
- *3 × 2 g clean magnesium strip*
- *water trough and inverted 100 mL measuring cylinder filled with water*
- *magnetic stirrer*
- *timer*

Method

1 *Measured 20 mL of 0.5 M HCl into each conical flask.*

2 *Inserted delivery tube under water and into inverted measuring cylinder.*

3 *Dropped one magnesium strip into the first conical flask and quickly replaced the stopper with tube already positioned under water. Started timer for 1 minute.*

4 *Every 10 seconds, recorded the total volume of gas collected in the inverted measuring cylinder, while swirling the flask for 2 seconds (starting at 10 seconds).*

5 *Repeated steps 2 to 4 for the other modes of agitation: swirling continuously and using a continuous magnetic stirrer.*

The method shows how the investigation was conducted. Number the steps, and write in the past tense and in the third person.

Methods should be written like a recipe – simple, clear and detailed.

Results

Table 1: Effect of mode of agitation on rate of reaction

Include recorded data such as the results table. Include any visual elements, such as photos or graphs, here.

Mode of agitation	Volume (mL) of gas collected every 10 seconds over 1 minute					
	10 s	20 s	30 s	40 s	50 s	60 s
Swirl every 10 seconds	0	5	12	18	21	23
Swirl continuously	8	17	28	36	41	47
Continuous magnetic stir	10	20	35	45	50	53

Discussion

As the results in Table 1 show, the mode of agitation that generated gas the fastest was the magnetic stir, with 53 mL of gas collected in 1 minute. Swirling continuously was the next-fastest mode, with 47 mL of gas collected, followed by swirling every 10 seconds, which collected 23 mL in 1 minute.

The results show a trend similar to the intensity and proportion of swirl time for each mode of agitation. Using the same mode of swirling, but for only 20 per cent of the time as compared to continuous swirling, saw very different results (a 24 mL difference), indicating that the duration of agitation makes a significant difference in the rate of the reaction. Continuous stirring, whether swirling or magnetic, had similar results (a 6 mL difference), suggesting that continuous stirring, rather than intensity of stirring, has a more significant impact on the rate of reaction.

Use the discussion to analyse the results and identify a key finding. Evaluate the method to identify the quality and limitations of the data.

Start by summarising the data, using amounts in your table. Identify any trends or patterns that could lead to your key finding. Link your finding to your scientific understandings.

One source of error is that the surface area of the magnesium strips was observed to be inconsistent. Although they were 2 g for each mode, the strip for the magnetic stirrer was longer than the others. This means it had more surface area, which would have led to a higher reaction rate independent of the agitation because, according to collision theory, increased surface area increases reaction rate. This would have caused the amount of gas that was collected to be higher than it would have been had the magnesium's surface area matched the others.

Identify any potential errors here and suggest improvements to try to control them.

The method could be improved by replacing magnesium with a reactant that has a surface area that is easier to control, such as a powdered substance. Carrying out multiple trials with magnesium would also minimise this error by averaging out the inconsistencies in surface area.

Conclusion

The results of this investigation show that the mode of agitation does impact the rate of a chemical reaction, with the duration of agitation playing a larger

Conclude the report by responding to the aim. Mention whether the results supported or refuted your hypothesis. Do not present any new information.

The conclusion can also be the closing paragraph of the discussion.

role in reaction rate than the intensity of agitation. The investigation supported the hypothesis that if this reaction is stirred continuously using a magnetic stirrer, it will react at a higher rate in the first minute than if it is only swirled once every 10 seconds. This is due to a higher proportion of successful collisions occurring in a given amount of time.

References show the source of any information you used that was not your own, including for the background.

References

Constant A, Cashmere R, Fricker A, Herron C and Saffin P, 2025, Good Science 9, 2nd ed. Matilda Education.

Presenting scientific information

The ability to choose and create an appropriate presentation format allows you to transform complex ideas into engaging concepts and to share your discoveries in ways that will best reach your audience. Presenting scientific information effectively is a powerful tool that can help to shape the future of science.

Posters

The purpose of a scientific poster is to communicate information on a particular topic or to display the findings of an investigation in a quick and engaging way. When information is shared on a poster, it should be easy for people to see the main idea as they are walking past. A display of scientific posters might include dozens of posters lined up, one after another. This is why it is important for your main finding to be the central focus.

A poster includes the same sections as an investigation report. However, each section includes only the most important information. Anyone who wants to learn more about the investigation details can read the full report. Figure S3.2 shows the basic framework for a scientific poster.

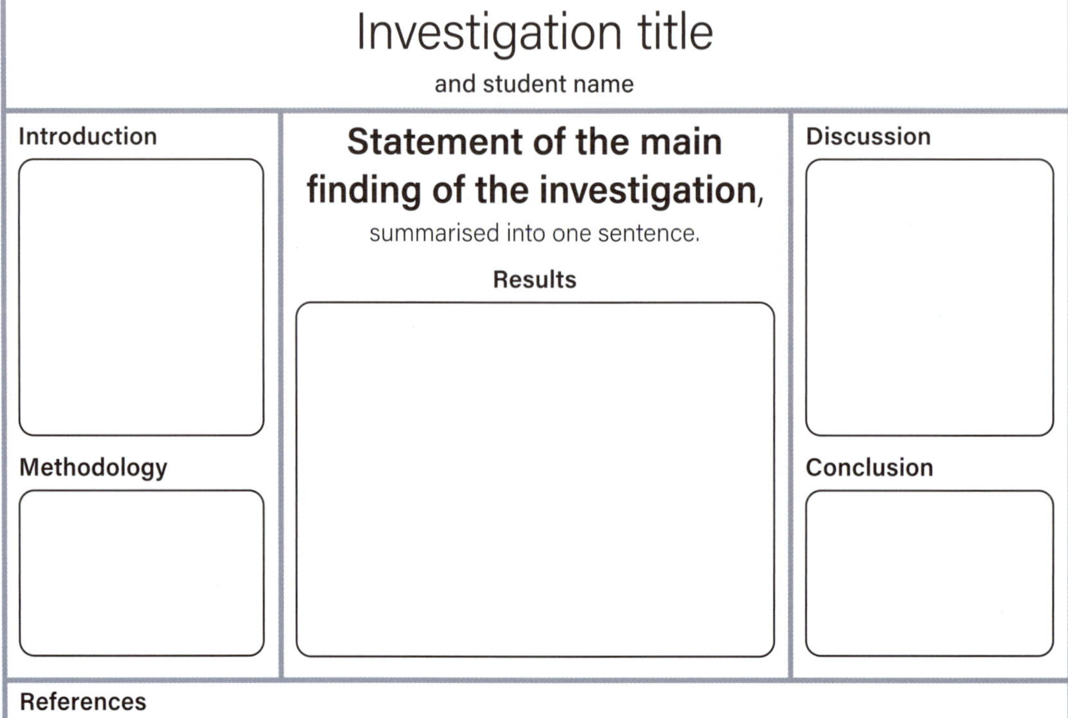

Figure S3.2: A simple framework for the structure of a scientific poster

Figure S3.3 gives an example of how we might present, in poster form, the findings from our investigation report on how different modes of agitation affect the rate of a chemical reaction. We could add visual interest by adding sketches of flasks.

To what extent do different modes of agitation affect the rate of a chemical reaction?

Introduction

Chemical reactions can occur at different rates, depending on the proportion of successful collisions between reacting particles in a given amount of time. Many factors can influence the frequency of successful collisions, such as temperature, concentration, surface area and the presence of a catalyst. Agitation of, or stirring, a reaction mixture is another factor that could have an impact on colliding particles. It would be interesting to compare how different ways of stirring a reaction mixture impact on the rate of a chemical reaction. This could inform what types of equipment are purchased for the laboratory and help in making methodical decisions for investigations in relation to efficiency.

Methodology

Within this investigation, 20 mL of 0.5 M HCl was placed into three conical flasks. Strips of magnesium metal were placed into each flask, and the contents were then agitated at different intervals: once every 10 seconds, hand swirled continuously and agitated quickly using a magnetic stirrer. All flasks contained the same amount of magnesium and acid. The rate of gas production was measured from each flask in an inverted measuring cylinder.

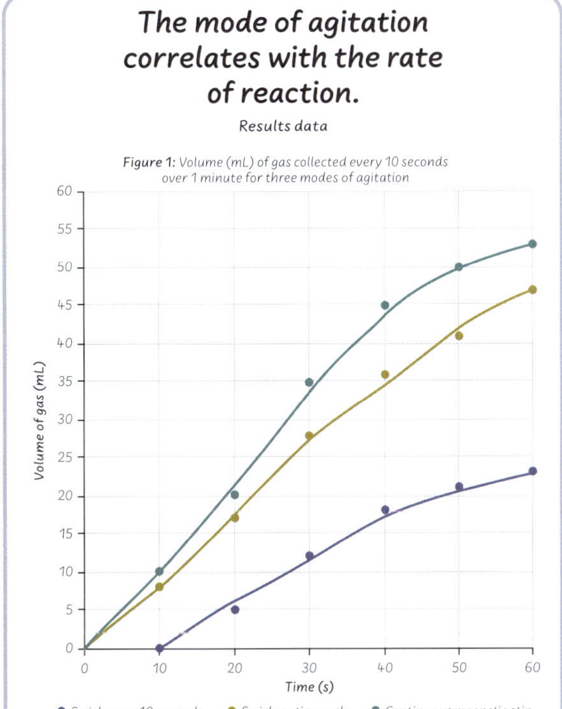

Discussion

As the results in Figure 1 show, the mode of agitation that generated gas the fastest was the magnetic stir, with 53 mL of gas collected in 1 minute. Swirling continuously was the next-fastest mode, with 47 mL of gas collected, followed by swirling every 10 seconds, which collected 23 mL in 1 minute.

One source of error is that the surface area of the magnesium strips was observed to be inconsistent. Although they were 2 g for each mode, the strip for the magnetic stir was longer than the others. This means it had more surface area, which would have led to a higher reaction rate independent of the agitation because, according to collision theory, increased surface area increases reaction rate. This would have caused the amount of gas that was collected to be higher than it would have been had the magnesium's surface area matched the others.

The method could be improved by replacing magnesium with a reactant that has a surface area that is easier to control, such as a powdered substance. Carrying out multiple trials with magnesium would also minimise this error by averaging out the inconsistencies in surface area.

Conclusion

The results of this investigation show that the mode of agitation does impact the rate of a chemical reaction, with the duration of agitation playing a larger role in reaction rate than the intensity of agitation. The investigation supported the hypothesis that if this reaction is stirred continuously using a magnetic stirrer, it will react at a higher rate in the first minute than if it is swirled only once every 10 seconds. This is due to a higher proportion of successful collisions occurring in a given amount of time.

Reference: *Constant A, Cashmere R, Fricker A, Herron C and Saffin P, 2025, Good Science 9, 2nd ed. Matilda Education.*

Figure S3.3: A simple scientific poster

Articles

The purpose of a scientific article is to communicate the findings of an investigation or to summarise research into secondary sources. A scientific article is divided into specific sections, each with a purpose to communicate specific information relating to the scientific concept being studied.

1 *Heading:* a relevant title that is a statement of what the article is about
2 *Author:* the name of the person writing the article
3 *Abstract:* a summary of what was done or what was found in the study
4 *Introduction:* some research on and background to the topic, linked to a hypothesis
5 *Method:* the methods scientists used to investigate the topic
6 *Results:* the results found about the topic
7 *Discussion:* a detailed analysis of the results, linked to the background
8 *References:* a list of your sources. See the 'Referencing' section on pages 279–80 for more information.

Figure S3.4: High-quality scientific articles can be reviewed by other scientists and published in scientific journals.

Figure S3.5 provides an example framework for a science article.

Figure S3.5: The structure of a science article on the effect of agitation on the rate of a chemical reaction

Heading → **The effect of agitation on the rate of a chemical reaction**

Author → Bailey Chang

Abstract → **Abstract**

A review of an investigation into chemical reaction rates was conducted. The investigation and method were studied and analysed to determine the effect of agitation on reaction time. The study found that increased agitation of the chemical solution correlated with increased rate of reaction – 53 mL over 60 seconds with continuous agitation, compared to 23 mL with agitation every 10 seconds. Further investigation into similar studies could assist in supporting this hypothesis.

Introduction → **Introduction**

Chemical reactions can occur at different rates, depending on the proportion of successful collisions between reacting particles in a given amount of time. Many factors can influence the frequency of successful collisions, such as temperature, concentration, surface area and the presence of a catalyst. Agitation of, or stirring, a reaction mixture is a factor that directly impacts the rate of colliding particles in a reaction. By agitating particles, the amount of successful collisions between particles is increased; therefore, the overall rate of reaction is increased. This could inform what types of equipment are purchased for the laboratory and help in making methodical decisions for investigations in relation to efficiency.

Method → **Method**

Within this investigation, 20 mL of 0.5 M HCl was placed into three conical flasks. Strips of magnesium metal were placed into each flask, and the contents were then agitated at different intervals: once every 10 seconds, hand swirled continuously, and agitated consistently using a magnetic stirrer. All flasks contained the same amount of magnesium and acid. The rate of gas production was measured from each flask in an inverted measuring cylinder.

Results

As the results in Table 1 and Figure 1 show, the mode of agitation correlates with the rate of reaction. More consistent agitation led to a greater rate of reaction.

Results

Table 1: Effect of mode of agitation on rate of reaction

Mode of agitation	Volume (mL) of gas collected every 10 seconds over 1 minute					
	10 s	20 s	30 s	40 s	50 s	60 s
Swirl every 10 seconds	0	5	12	18	21	23
Swirl continuously	8	17	28	36	41	47
Continuous magnetic stir	10	20	35	45	50	53

Figure 1: Volume (mL) of gas collected every 10 seconds over 1 minute for three modes of agitation

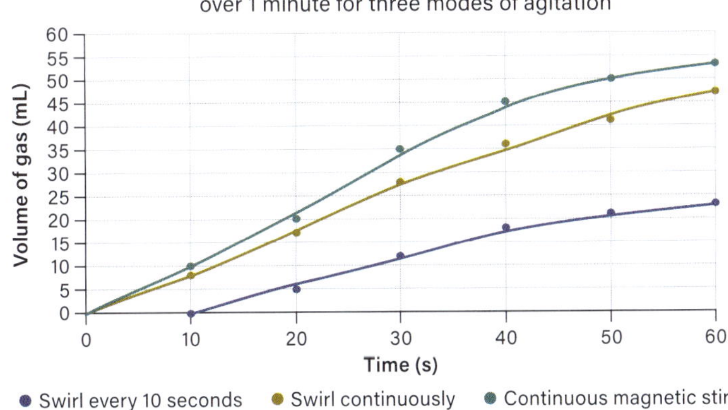

Discussion

As the results in Table 1 show, the mode of agitation that generated gas the fastest was the magnetic stir, with 53 mL of gas collected in 1 minute. Swirling continuously was the next-fastest mode, with 47 mL of gas collected, followed by swirling every 10 seconds, which collected 23 mL in 1 minute.

Discussion

The results show a trend similar to the intensity and proportion of swirl time for each mode of agitation. Using the same mode of swirling, but for only 20 per cent of the time as compared to continuous swirling, saw very different results (a 24 mL difference), indicating that the duration of agitation makes a significant difference in the rate of the reaction. Continuous stirring, whether swirling or magnetic, had similar results (a 6 mL difference), suggesting that continuous stirring, rather than intensity of stirring, has a more significant impact on the rate of reaction.

One source of error is that the surface area of the magnesium strips was observed to be inconsistent. Although they were 2 g for each mode, the strip for the magnetic stir was longer than the others. This means it had more surface area, which would have led to a higher reaction rate independent of the agitation because, according to collision theory, increased surface area increases reaction rate. This would have caused the amount of gas that was collected to be higher than it would have been had the magnesium's surface area matched the others.

The method could be improved by replacing magnesium with a reactant that has a surface area that is easier to control, such as a powdered substance. Carrying out multiple trials with magnesium would also minimise this error by averaging out the inconsistencies in surface area.

References

Constant A, Cashmere R, Fricker A, Herron C and Saffin P, 2025, *Good Science 9*, 2nd ed. Matilda Education.

References

Presentations

The purpose of a scientific presentation is to communicate the findings of an investigation or article quickly and concisely. You need to summarise the key methods and findings clearly, as your audience may view several presentations in one sitting.

Include the same sections that you would in a scientific article to provide your audience with an overview of the concepts. Those who are interested can then read the full report.

To create a clear oral presentation, use the following tips.

- Pick a theme and stick with it.
- Use carefully selected images that help you to make your point.
- Ensure the summary on each slide is large enough to be read from the back of the room.
- If you use slide transitions, only use one typeface for the whole presentation as too many can be distracting.

Introduction

Chemical reactions can occur at different rates, depending on the proportion of successful collisions between reacting particles in a given amount of time. Many factors can influence the frequency of successful collisions: temperature, concentration, surface area and the presence of a catalyst. Agitation of, or stirring, a reaction mixture is another factor that has an impact on colliding particles.

Figure S3.6: An example of a slide layout for a scientific presentation.

Models

The purpose of a scientific model is to provide a representation of a scientific concept that could be difficult to see or understand. Scientific models can show microscopic or large-scale ideas that are impossible to view with the naked eye, or they can help us to visualise the parts and functions of something – for example, a model of how the human eye works.

To construct a good scientific model, use the following criteria.

- Does this model contain all of the relevant components of the concept it is showing?
- Is each section of the model clearly labelled?
- Can the model be easily interacted with (visually or physically) by the audience?

Figure S3.7a aims to inform the audience about the structure of a specific bacteria. However, it is missing specific labels. To improve the information it communicates and make it a better scientific model, labels should be added. Figure S3.7b is appropriate, as it also includes labels of the bacteria and its structures and components.

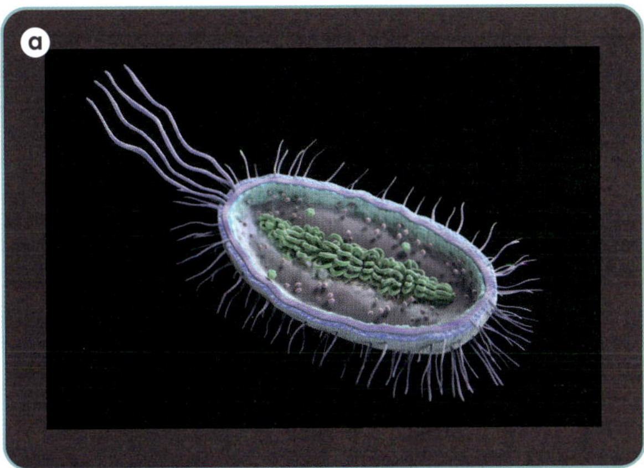

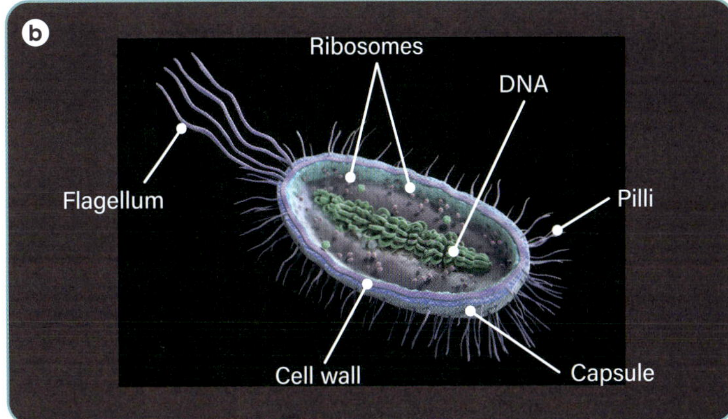

Figure S3.7: (a) A model of a bacteria that is missing labels; (b) an improved model of a bacteria, including labels

Conducting scientific research

CRAAP

When conducting scientific research, we need to confirm that the information is valid. Even if a source looks credible on the surface, it may not have valid information. One way that we can check this is by using the CRAAP test (Figure S3.8).

Currency: Is the information on the source up to date? Was it recently published (or published in a time relevant to the information)?

Relevance: Does the information answer the question I am asking? Is it written in a way that I can understand?

Authority: Does the author of the text have experience in this branch of study? Do they have appropriate qualifications to be giving out information on the topic?

Accuracy: Where is the information from? Can I find similar information across multiple sources? Is there evidence supporting the information?

Purpose: Why is this particular author or company publishing this information? Are they unbiased or are they trying to sell a particular viewpoint? Is the information fact or opinion?

Figure S3.8: The CRAAP test

Source: Meriam Library, California State University, Chico, California

A secondary source of information must meet all of these criteria to be considered a valid source. If the source meets the criteria, use it to support your scientific writing. For example, when looking at the question 'How is water used in Australia?', we can analyse two sources of information about water set out in Table S3.2. The table shows us that the first source meets all of the CRAAP criteria, making it valid, while the second source only meets two criteria and should be disregarded.

Table S3.2: Evaluating research sources for the topic of water

Source	Source 1: 'Water Use in Australia, 2024–25'	Pass?	Source 2: 'My Opinion on Water Use in New Zealand'	Pass?
Currency	Updated recently, with the date of last update clearly stated e.g. *last updated: 7 June 2025* – current	☑	Updated recently, with the date of last update clearly stated e.g. *last updated: 13 June 2025* – current	☑
Relevance	Provides information on the distribution and use of water in Australia – highly relevant	☑	Provides information on water use, but not in Australia – related, but not relevant for our research question	☑
Authority	Reputable scientific organisations or research bodies that state clearly their qualifications and methods e.g. *Geoscience Australia – a government department focused on water research* – good authority	☑	No listed qualifications, or qualifications are from organisations that don't exist. The author might be an expert in a different field but isn't qualified to be considered an expert on water	☒
Accuracy	Similar information can be found across multiple other reputable sources and is supported by evidence	☑	Very few other sources include similar information; claims are not supported by evidence or are rejected by other experts in the field	☒
Purpose	To educate Australians on where water is and how it is used, using current peer-reviewed scientific research	☑	To share the opinion on water use of one person who is not an expert in the field	☒

Types of information

When conducting research for investigations and articles, it is important to remember to include a variety of primary and secondary sources.

Primary sources are original, or first-hand, material and can include:

- letters or diaries
- photos
- research data
- laboratory notes
- data collected during first-hand investigations.

Secondary sources are materials that evaluate primary sources. These include:

- scientific reports
- journal articles.

There are also *tertiary sources* of information, which include both primary and secondary sources. Tertiary sources include:

- textbooks and encyclopaedias
- general-knowledge websites (make sure the authors are trusted and the information is current).

Some sources of information are considered *non-scientific*. You should not include these in your reports or articles. These include:

- your personal stories – back up your information with evidence!
- blogs or articles that are not based in fact
- personal opinions.

A well-constructed scientific report or article will include a combination of sources. This shows that you have researched widely to ensure that you include valid and trusted information to inform your audience.

Figure S3.9: There are many places to access valid sources of information, such as libraries and Google Scholar. All sources of information should be referenced and cited in your written work.

Researching information

When researching a scientific topic, sources include:

- textbooks, books, magazines and journals available from a library or online collection
- search engines such as Google; however, do not just use the information that pops up initially – visit the sites listed and ensure they fulfil the CRAAP test
- specific search areas such as Google Scholar, which will show scientific journal articles relating to your topic.

When searching for information, remember:

1. Go to sources that contain valid information – check each one against the CRAAP test.
2. Choose three or four key words about your topic and use them when entering search terms into Google or Google Scholar; this should increase the number of relevant sites and articles that appear.
3. Save a copy of the URL (web address) for each website you use, and note down the details or take photos of every textbook or physical source you use. You will need this information when referencing your work at the end of your writing.

Referencing

When researching and writing scientific material, you must acknowledge the sources of information you use. This is known as referencing. Referencing, or adding a **citation**, shows the audience that your information is supported by research and that you have been careful not to **plagiarise**, or copy, the work of others. At the end of your report or article, you collect the details of all your sources in a reference list.

A reference list should be in alphabetical order and follow a specific format. The most commonly used format in Australia is 'author–date'. Write each entry as follows:

1. surname of author(s) and first initial (or an organisation's name)
2. year the book or article was published
3. book title in *italics* or <u>underlined</u>; or article title in quote marks, with details of the journal

4 the title of the webpage or the name of the publisher of the book
5 for digital content, the date you accessed it and the URL.

For example, below are references for a book and a journal article.

Constant A, Cashmere R, Fricker A, Herron C and Saffin P, 2025, *Good Science 9*, 2nd ed. Matilda Education.

Schmitz G, 2005, 'What Is a Reaction Rate?,' *Journal of Chemical Education*, Volume 82, Issue 7, accessed 28 September 2024, https://pubs.acs.org/doi/10.1021/ed082p1091.

Writing evidence-based essays

As part of your science studies, you will be required to write essays. Your essays must be supported by evidence. This type of scientific writing is usually informative or argumentative. In either case, you will need to structure your essay. One way to do this is to write an essay with five or more paragraphs or sections, consisting of:

- an introduction paragraph
- three or more body paragraphs
- a concluding paragraph.

Each body paragraph should follow the TEEL approach. The TEEL acronym reminds you of what to include in each paragraph of your essay.

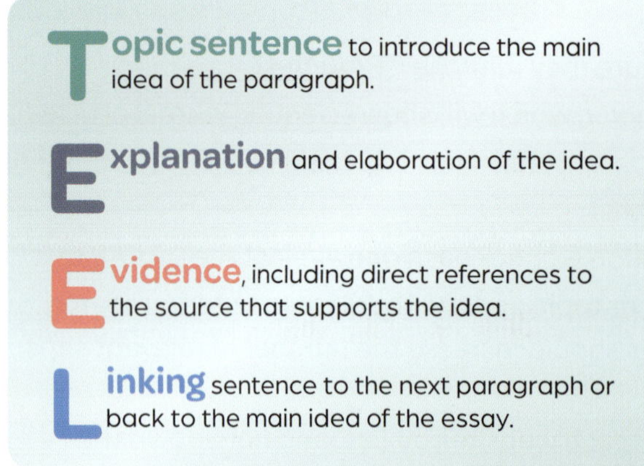

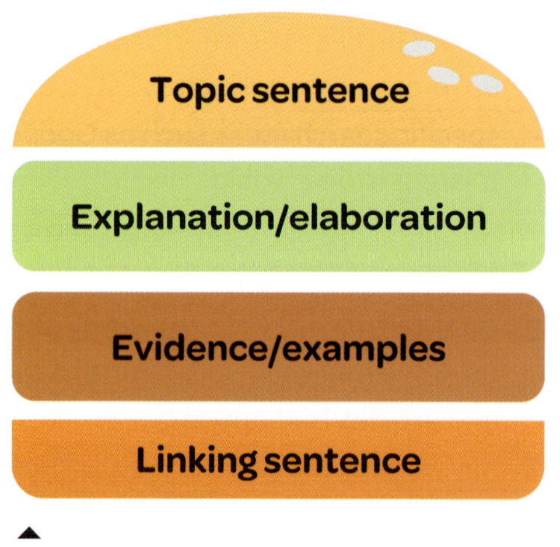

Figure S3.10: Each TEEL paragraph should build on the topic and link to the next idea or the essay as a whole.

Writing an informative essay

Informative scientific essays communicate scientific facts and ideas with a neutral tone, and cover all aspects of a topic. These essays are sometimes called research papers because they summarise multiple sources of research available on a particular topic, without including any opinions or personal narrative. There are many ways to structure an informative essay, such as the TEEL approach.

Writing a scientific argument

A scientific argument presents a position on a particular topic. Instead of communicating all aspects of the research evenly, the writer evaluates and uses the research to form a position.

Before you begin writing your argument, you will need to develop an evidence base by conducting research on the topic and evaluating each source. This will help you to choose your position.

For example, if the topic is wind power, you should find multiple sources that discuss positive and negative aspects of wind energy. Read each research piece carefully and critically. Review and evaluate each source by asking the following questions:

- Is this piece of evidence relevant?
- Is this piece of evidence written by a reputable person or group? (If not, it still may be useful as you could use it to criticise the opposing position.)
- What position does it take?
- Is it valid in taking this position?
- If it is not valid, what are the problems with it?

Use the answers to these questions to decide which position you intend to argue scientifically.

Finally, once you have established your position, you need to write your argument. You can follow the standard essay structure, using five or more paragraphs and the TEEL approach for each paragraph. Usually, each of your three body paragraphs should provide a new reason why your position is valid.

Figure S3.11: The Waubra Wind Farm, located near Ballarat in western Victoria, uses wind turbines to generate electricity.

S4 Maths in science

Key terms

chemical formula: an expression of the elements that make up a chemical compound, usually presented as a ratio using letters and numbers – for example, H_2O

conversion factor: a number used to change one unit of measurement to another

density: how heavy something is for its size; mass divided by volume

equation: a mathematical statement which shows that two things are equal; for example, $2x + 6 = 14$ is an equation that needs to be solved so that $2x + 6$ does actually equal 14

error bar: used on graphs to show uncertainty in repeated measurements

exponent: the superscript value that says how many times to use a number in a multiplication; for example, when we write 10^3, '3' is the exponent. It means we need to multiply 10 by itself 3 times

half-range: shows how much a set of experimental values might vary

mathematical formula: a rule or principle that helps you to find the answer to a question or understand the relationship between variables

mean: a measure of centre (an average) calculated by adding all the numbers together and dividing by how many numbers there are

median: the middle number in a set of numbers when they are arranged in order

mode: the number that appears most frequently in a set of numbers

percentage error: measures how close the experimental value of an investigation is to the theoretical or true value

power: the result of multiplying a quantity by itself one or more times; for example, 2 to the *power* of 3 is $2 \times 2 \times 2$, which is 8, so the *power* of 2^3 is 8

range: in a set of numbers, a measure of spread between the highest number and the lowest number

ratio: a way of comparing like quantities without units

scale factor: the ratio between corresponding measurements of an object and a copy of that object

scientific notation: a way to write very large or very small numbers in a simple form

significant figures (or **sig figs**)**:** digits in a number that indicate the precision of a measurement

subscript: a letter or number written slightly below and to one side of another – for example, '2' in H_2O

uncertainty: a measure of the extent to which data is known (i.e. its variability)

Understanding and applying basic mathematical concepts is essential for engaging with science effectively. It involves using mathematics to interpret and analyse scientific information, enhancing your comprehension of the scientific ideas you encounter in your studies and daily life.

Figure S4.1: In September 1999, NASA's Mars Climate Orbiter spacecraft was destroyed as it entered Mars's atmosphere. A conversion error meant the computer software controlling the probe used SI units but was reading data in non-SI units. Instead of orbiting at 226 kilometres above the planet's surface, the probe descended too low, to a height of just 57 kilometres!

Units of measurement

When we come across numbers in science, we need to identify:

- what the number is measuring
- the symbol used to represent the measurement
- the unit the number is measured in.

For example, you read that a ball takes 10 seconds to roll down a ramp, and you identify that:

- the measurement used is *time*
- the symbol used is *t*
- the unit of measurement used is *seconds*.

In science, we use an international standard system of units for all types of measurements. These are called 'SI units'. We use some of these units every day, such as using metres to measure distance or minutes to measure time, but some others are only used in advanced science and maths. For example, in this book we use the common unit degrees Celsius (°C) for temperature, rather than the SI unit of kelvin (K). (Note that 1 °C is equivalent to +274.15 K.)

Some numbers do not have units. If you swim two laps of a pool, the length can be measured in metres, but the number of laps will just be '2' with no units because it is a general quantity.

Table S4.1 shows measurements you will come across in this text, their symbol and their unit.

Table S4.1: Some common measurements, symbols and units

Measurement	Symbol*	Commonly used unit at this level
Length or distance	*D*	metre (m)*
Time	*t*	second (s)*
Speed or velocity	*v*	metres per second (m s^{-1})*
Mass	*m*	gram (g), kilogram (kg)*
Volume	*V*	millilitre (mL), cubic metre (m^3)*
Density	*d*	grams per millilitre (g mL^{-1})
Temperature	*T*	degrees Celsius (°C)
Force	*F*	newton (N)
Acceleration	*a*	metres per second squared (m s^{-2})*
Energy	*E*	joule (J)
Number of ...	*N*	no units

*SI unit.

Tip 1: **Capitalisation matters**. If you mix up the lower and upper cases, it can change the meaning. But keep in mind that these can vary by year level and course. For example, '*d*' is sometimes used to represent distance.

Tip 2: **Symbols are not the same as units**. Notice that '*m*' in *italics* is the symbol for 'mass', while 'm' in regular text is the unit for metres. Symbols are used to represent variables in an equation, while the unit tells us *how* the variable is measured. You should be able to distinguish between symbols and units.

Converting between units

Sometimes you will need to change measurements from one kind of unit to another. You can do this if you know the **conversion factor**. For example, if you are converting minutes into seconds, the conversion factor is 60 seconds (60 s), which is the number of seconds in a minute. At other times, you may need to look up the conversion factor. For example, on Earth a kilogram is about the same as 9.8 newtons, so the conversion factor from kg to N is 9.8. Once you know the conversion factor, you then need to multiply or divide.

If the new unit you are converting to is larger, you must divide by the conversion factor. If the new unit is smaller, then you multiply by the conversion factor.

For example, converting minutes to seconds is changing from a larger unit to a smaller unit, so you multiply by 60. Converting from seconds to minutes is changing from a smaller to a larger unit, so divide by 60. Worked example 1 shows more examples of conversions.

Worked example 1: Converting between units

1 Convert 2.5 kilometres to metres.

Answer: 2.5 × 1000 = 2500 **m**

- There are 1000 metres in a kilometre, so the conversion factor is 1000.
- This conversion is from a big unit (**km**) to a small unit (**m**), so you need to multiply (**×**).
- Remember to include the unit you are converting to – in this case, metres (**m**).
- If converting in the opposite direction, from m to km, you divide by 1000.

2 Convert 150 grams to kilograms.

Answer: 150 ÷ 1000 = 0.15 **kg**

- There are 1000 grams in a kilogram, so the conversion factor is 1000.
- This conversion is from a small unit (**g**) to a big unit (**kg**), so you need to divide (**÷**).
- Remember to include the unit you are converting to – in this case, kilograms (**kg**).
- If converting in the opposite direction, from kg to g, you multiply by 1000.

3 Convert 4 hours to seconds.

1 hour = 60 minutes

1 minute = 60 seconds

1 (hour) × 60 (minutes) × 60 (seconds) = 3600 **s**

Answer: 4 × 3600 = 14 400 **s**

- You need to do two conversions here: hours to minutes, and then minutes to seconds. Start with calculating 1 hour in seconds.
- These conversions are both from big units to small units, so you need to *multiply* (**×**). The conversion factors are both 60.
- Remember that there are 4 hours, so we need to multiply by 4 for the total number of seconds. Remember to include the unit you are converting to – in this case, seconds (**s**).

However, if you want to keep it simple, you can use conversion charts as shown in Figure S4.2, which provides the conversion factor and shows whether to multiply or divide, depending on which direction you are converting to. Worked example 2 has some examples.

Figure S4.2: Conversion charts for length, time, volume and mass

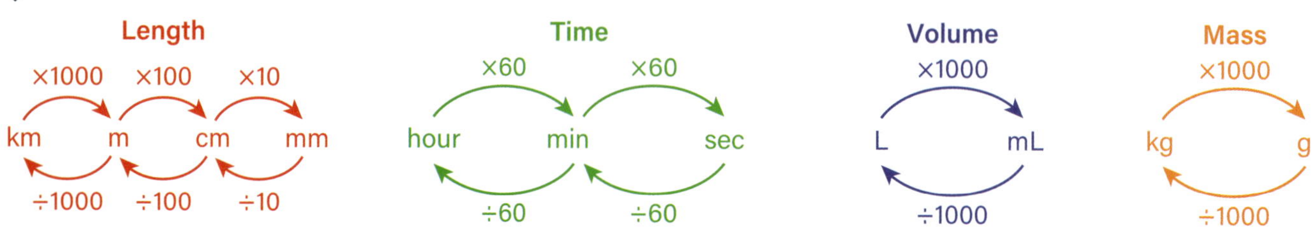

Worked example 2: Converting between units using conversion charts

1. Convert 250 metres to kilometres.
   Hint: You are converting to a *larger unit*; therefore, you need to *divide*.
   The conversion from the chart is to divide by 1000:
   250 ÷ 1000 = 0.25 km
2. Convert 40 kilograms to grams.
   Hint: You are converting to a *smaller unit*; therefore, you need to *multiply*.
   The conversion from the chart is to multiply by 1000:
   40 × 1000 = 40 000 g
3. Convert 7200 seconds to hours.
   First, convert seconds to minutes by dividing by 60, then divide by 60 again to get hours.
   7200 ÷ 60 = 120 minutes ÷ 60 = 2 hours

Figure S4.3: The colours of Labrador puppies in a litter can be expressed as ratios.

Understanding ratios

Ratios are a way of comparing like quantities without units. For example, if a pancake recipe asks for one part milk and two parts flour, that is a ratio of 1 to 2. We can write this as 1:2. Ratios do not require units – no matter how much milk you use, you must always use twice as much flour (× 2).

To identify a ratio, first count the total of each category and write the two totals (e.g. '10 to 20'), then replace the word 'to' with a colon, so '10:20'. Next, simplify the ratio by dividing both sides by a common factor. So, divide by 10 to get a ratio of 1:2.

For example, a black Labrador has eight puppies. Six are black and two are chocolate brown. Including the mother, there are nine dogs in total. Table S4.2 helps us to identify the different ratios for this example.

Table S4.2: Identifying ratios

Ratio	Initial count	Simplify	Answer
Black to chocolate puppies	Black = 6 Chocolate = 2 The ratio is 6:2.	Divide both numbers by the common factor of 2. 6 ÷ 2 = 3 and 2 ÷ 2 = 1 The simplified ratio is 3:1.	The ratio of black to chocolate puppies is 3:1.
Chocolate to black puppies	Switch the numbers around, as the question now asks about chocolate puppies first. The ratio is 2:6.	Divide by 2. 2:6 = 1:3	The ratio of chocolate to black puppies is 1:3.
Black to total puppies	There are 6 black puppies and 8 puppies in total. The ratio of black puppies to total puppies is 6:8.	Divide by 2. 6:8 = 3:4	The ratio of black to total puppies is 3:4.
Total to chocolate puppies	There are 8 puppies in total, and 2 chocolate puppies. The ratio is 8:2.	Divide by 2. 8:2 = 4:1	The ratio of total to chocolate puppies is 4:1.
Chocolate puppies to black dogs	Black dogs = 7 (including the mother). The ratio is 2:7.	This cannot be simplified any further.	The ratio of chocolate puppies to black dogs is 2:7.

Ratios in science

We see ratios all the time in science. A chemical compound, for example, can be represented using a **chemical formula** based on the ratio of elements. We can use the chemical formula and the ratio to solve mathematical problems. The **subscript** to the right of each element is part of the ratio of atoms in the substance.

Water is a substance with the chemical formula H_2O. 'H' is the symbol for the element hydrogen and 'O' is the symbol for the element oxygen. A molecule of water contains 2 hydrogen atoms and 1 oxygen atom. This can be written as the ratio 2:1.

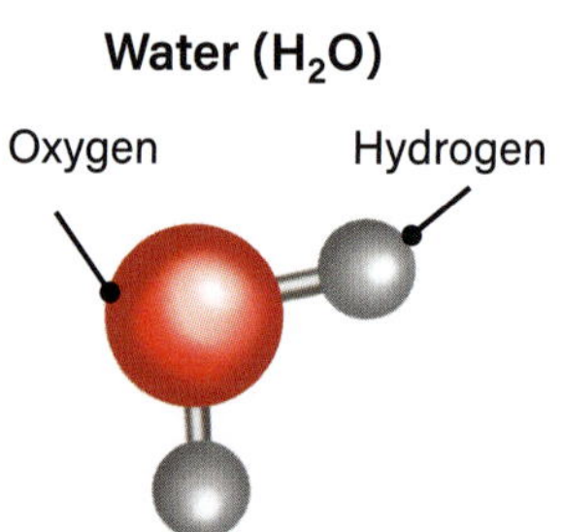

Figure S4.4: This model of water clearly shows the 2:1 ratio of hydrogen to oxygen.

Worked example 3: Ratios in a chemical mixture

How many carbon atoms will there be if there are 40 hydrogen atoms?

1 C_3H_8.
The ratio of carbon atoms to hydrogen atoms is 3:8.

- This means there are 3 carbon atoms and 8 hydrogen atoms in a molecule of propane.
- The coloured subscript numbers indicate how many atoms there are in the molecule.

2 Turn 3:8 into a fraction.

$$\frac{\text{carbon atoms}}{\text{hydrogen atoms}} = \frac{3}{8}$$

- Ratios can be written as fractions.

3 Work out how many carbon atoms we will have if we have 40 hydrogen atoms given the ratio is 3:8.
$40 \div 8 = 5$

- Divide 40 by 8 to find the **scale factor.**
- The scale factor is 5.

4 Multiply the *numerator* x 5 to get the number of carbon atoms.
$3 \times 5 = 15$

- The numerator is 3.
- Multiply 3 by 5 (because 5 is the scale factor).

5 Answer: There will be 15 carbon atoms if there are 40 hydrogen atoms.

We also see ratios in balanced chemical equations. Consider the balanced chemical equation that represents the corrosion of iron. The coefficients indicate the ratio between substances in the reaction.

$$4Fe(s) + 3O_2(aq) + 6H_2O(l) \rightarrow 4Fe(OH)_3(S)$$

iron	:	oxygen	:	water	:	iron(III) hydroxide
4	:	3	:	6	:	4

This also shows us that when iron corrodes in this way, oxygen and water molecules are required in a 1:2 ratio. This is based on the fact that the coefficients for O_2 and H_2O are 3 and 6, indicating a 3:6 ratio. This can be simplified to 1:2 when we are looking at these two substances independently from iron and iron(III) hydroxide.

Worked example 4: Using coefficients in chemical equations to determine the ratio of substances in a chemical reaction

If plenty of iron (Fe) and water (H_2O) are available, how many units of iron(III) hydroxide ($Fe(OH)_3$) will form if only 18 units of oxygen are available to react?

Table S4.3: Ratio of substances in a chemical reaction

Step	Thinking	Working
1	According to the coefficients in the balanced chemical equation, 3 units of oxygen will react with 4 and 6 units of iron and water, respectively. Therefore, to determine how many 'sets' of 3 oxygen units there are in 18 units of oxygen, we divide 18 by 3.	18 units / 3 units = 6 sets
2	The number of 'sets' indicates the factor that can be applied to all coefficients when determining the amount of each present. Since there are 6 sets of oxygen available to react, 6 sets of iron will also react. Multiply 6 sets by the coefficient of iron (Fe), 4, to determine how many total units of iron will react.	6 sets * 4 units = 24 units
3	Since iron (Fe) and iron(III) hydroxide both have coefficients of 4, they react in a one-to-one ratio; 1:1. This means the amount (in units) of iron that reacts will be equal to the amount of iron(III) hydroxide formed.	If 24 units of iron (Fe) react, then 24 units of iron(III) hydroxide will form

Calculations

Mean, *median*, *mode* and *range* are terms you will often hear when working with data. Mean, median and mode are different ways of measuring the centre of the data. Each is used in different situations for different purposes, depending on what you are trying to understand. Range is used to understand the spread of numbers in a dataset. We will go through each of them in turn.

Mean

The **mean** (often called the average) can be used when many measurements are taken of the same thing and you want a value near the centre of all the measurements. This value is often required in scientific investigations when you have conducted multiple trials. The mean is often used in visual representations of the data.

The mean equals the sum of all values divided by the total number of values.

Here is a set of values: 5, 8, 9, 11, 12.

To arrive at the mean or average of the set of values, add them together: 5 + 8 + 9 + 11 + 12 = 45. Then divide 45 by the number of values (in this case, 5 values) to calculate the mean.

The mean is 45 ÷ 5 = 9.

See Worked example 5 on the next page.

Worked example 5: Calculating the mean

Geelong recorded the following temperatures over four consecutive days in January: 29 °C, 39 °C, 35 °C and 33 °C. Calculate the mean of this data.

1 Mean = sum of all the values (add the values up) divided by the total number of data values

- Identify the formula for calculating the mean.

2 $\frac{29+39+35+33}{4}$

- Substitute the known values into the formula.
- There are four temperature readings, so the total number of data values is 4.

3 $\frac{136}{4} = 34$

- Solve for the mean.

4 Answer: the mean is 34 °C.

- Remember to include any units of measurement (in this case, °C) in your answer.

Median

The **median** is the middle value in a group of values. In the set of values 5, 8, 9, 11, 12, the middle value is 9. Therefore, the median is 9.

Mode

The **mode** is the most frequent value in a dataset. If you are asked to identify the mode in a group of values, look for the value that occurs the most often. For example, in the group of values 2, 4, 5, 2, 6, 2, the mode is 2 because it occurs 3 times.

Range

The **range** is the difference between the smallest and largest values in a dataset. In the group of values 5, 8, 9, 11, 12, the smallest value is 5 and the largest value is 12. The range is the difference between them: 12 – 5 = 7. Therefore, 7 is the range of this dataset.

Half-range

The **half-range** is used to show how much a set of experimental values might vary. It is half the difference between the smallest and largest values.

$$\text{Half-range} = \frac{(\text{highest data value} - \text{lowest data value})}{2}$$

This is helpful when drawing **error bars** on graphs to show uncertainty in repeated measurements. (See Step 5 of the 'Processing, modelling and analysing' section on page 250.) For example, if a measurement is repeated and the values are 10, 12 and 14, the smallest value is 10 and the largest is 14.

The range is: 14 – 10 = 4

The half-range is: 4 ÷ 2 = ±2

We would use this value along with the average (12) to show how much the results might vary. The values are spread around the average by the half-range, which is 2. So, an error bar of ±2 would be added to the graph to show the possible variation in the results.

Percentages

Percentages tell you what part of the whole you have. One hundred per cent (100%) is the whole thing, so percentages should always add up to 100. Percentages are another way to represent a fraction or decimal, as shown in Table S4.4. A percentage is converted to a decimal by dividing by 100.

Table S4.4: Percentage can be represented in fraction or decimal form

Percentages	0%	10%	20%	30%	40%	50%	60%	70%	80%	90%	100%
Decimals	0	0.1	0.2	0.3	0.4	0.5	0.6	0.7	0.8	0.9	1
Fractions	$\frac{0}{10}$	$\frac{1}{10}$	$\frac{2}{10}$	$\frac{3}{10}$	$\frac{4}{10}$	$\frac{5}{10}$	$\frac{6}{10}$	$\frac{7}{10}$	$\frac{8}{10}$	$\frac{9}{10}$	$\frac{10}{10}$

Use the following basic formula to calculate percentage:

$$\text{Percentage} = \frac{\text{the part of something you want the percentage of}}{\text{the whole thing}} \times 100$$

$$\% = \frac{\text{part}}{\text{whole}} \times 100$$

If someone wants to eat three pieces of this pizza, what percentage of the pizza will they eat?

1 3 out of 10 pieces were eaten.

2 Percentage = $\frac{\text{part}}{\text{whole}} \times 100$

3 Percentage = $\frac{3\ [\text{pieces}]}{10\ [\text{total pieces}]} \times 100$

4 Percentage = 0.3 × 100 = 30%

Table S4.4 also shows you that

$30\% = \frac{3}{10} = 0.3.$

The person will eat 30 per cent of the pizza if they eat 3 slices out of 10.

Figure S4.5: If a pizza is cut into 10 equal pieces, each piece represents 10 per cent.

Worked example 6: Calculating percentages

There is 41 500 000 km^3 of fresh water on Earth, but only 6 225 000 km^3 is available to use. What percentage of fresh water on Earth is available?

1 Percentage = $\frac{\text{part}}{\text{whole}} \times 100$ — Identify the formula for calculating percentage.

2 The part = available fresh water = 6 225 000 km^3
The whole = total fresh water = 41 500 000 km^3 — Identify and label the values given in the question.

3 Percentage = $\frac{6\,225\,000}{41\,500\,000} \times 100$ — Substitute the given values into the formula. You might need to use your calculator.

Percentage = 0.15 x 100 = 15%

4 Answer: Available fresh water is 15% — The percentage symbol (%) must always accompany a percentage value.

This formula can be adapted to suit a variety of situations, as shown in Worked example 7.

Worked example 7: Energy efficiency and percentages

Your hair dryer uses 28 000 joules of energy but only puts out 15 960 joules of blowing energy. What is the percentage efficiency of your hair dryer?

1 Percentage = $\frac{\text{part}}{\text{whole}} \times 100$ — Identify the formula for calculating percentage.

2 Efficiency = $\frac{\text{part}}{\text{whole}} \times 100$

Efficiency = $\frac{\text{blowing energy}}{\text{total}} \times 100\%$ — Adapt the formula to suit the problem. We are trying to work out the *efficiency* of the hair dryer.

3 The part = blowing energy = 15 960 J

The whole = total energy = 28 000 J — Identify and label the values given in the question.

4 Efficiency = $\frac{15960}{28000} \times 100\%$ — Substitute the given values into the formula. You might need to use your calculator to work this out.

Efficiency = 0.57 x 100 = 57%

5 Answer:

The efficiency of the hairdryer is 57% — The percentage symbol (%) must always accompany a percentage value.

Percentage error

Percentage error tells you how close the experimental value of an investigation is to the theoretical or true value. In this case, the 'part' that you are using to determine the percentage is the difference between the experimental value and the theoretical value. The 'whole' is the true value.

$$\%\ \text{error} = \frac{\text{experimental value} - \text{true value}}{\text{true value}} \times 100$$

The closer your percentage is to zero, the more accurate your results.

Worked example 8: Calculating percentage error

The volume of carbon dioxide collected during an experiment was measured to be 56.5 mL; however, the calculations predicted that 62.3 mL should form. Calculate the percentage error of these experimental results.

1 Identify the formula for calculating percentage error.

$\%\ \text{error} = \frac{\text{experimental value} - \text{true value}}{\text{true value}} \times 100$

2 Substitute the known values into the formula.

$\%\ \text{error} = \frac{56.5 - 62.3}{62.3} \times 100$

3 Solve for the percentage error.

$\%\ \text{error} = \frac{-5.8}{62.3} \times 100 = -9.3\%$

Note: The percentage error is negative, which indicates the experimental value is less than the true value. Sometimes the negative is ignored, so percentage error is always positive.

Normally, you will not see percentages that are higher than 100, but since percentage error is a comparison, it is possible for the percentage error to go above 100. However, this may indicate very low-quality results. For example, if 150 mL of carbon dioxide was collected in Worked example 8, the percentage error would be 141. This tells us not only that more than double the amount of gas was collected than expected, but also that major sources of error are likely to be present in the method.

Scientific notation

Scientific notation, sometimes called standard form, helps us to write very large or very small numbers in a simpler way. Every number can be written in scientific notation as the product (product = multiply, or times) of two numbers that are:

- a decimal greater than or equal to 1 and less than 10
- a **power** of 10 written as an **exponent**. (An example of an exponent is 10^3, which is the same as 10 × 10 × 10. The '3' means that the 10 is multiplied by itself 3 times.)

2.56×10^3 is an example of scientific notation. It is a different way of writing the number 2560. It is the same as writing 2.56 × 10 × 10 × 10.

Scientific notation becomes really useful when you work with very big and very small numbers. For example, 2.56×10^7 is the same as 2.56 × 10 × 10 × 10 × 10 × 10 × 10 × 10, which equals 25 600 000. It is faster and easier to write 2.56×10^7. The bigger or smaller the number, the more useful scientific notation becomes.

In Worked examples 9 and 10, we will explore writing large and small numbers in scientific notation.

Worked example 9: Scientific notation for large numbers

1 Write 65 000 000 in scientific notation.

- 65 million is a very large number.

2 Turn 65 000 000 into a decimal >1 and <10.

6.5 0 0 0 0 0 0.

It becomes 6.5000000.

- To write a number in scientific notation, you need two things: a decimal greater than or equal to 1 and less than 10, and a power of 10. To create your decimal, you need to move the decimal point at the end of 65 000 000 left 7 times (think of 65 000 000 as 65 000 000.0).

3 Answer: 65 000 000 in scientific notation is 6.5×10^7.

- The number of times you move the decimal point to the left becomes the power you need to write your number as in scientific notation. You moved the decimal point 7 times, so the power is 10^7. It is a **positive** power.
- *Remember:* When you are expressing a very large number in scientific notation, move the decimal point *left* to create your decimal. Your power will be *positive*.

Worked example 10: Scientific notation for small numbers

1 Write 0.0000098 in scientific notation.

- 0.0000098 is a very small number less than 1.

2 Turn 0.0000098 into a decimal ≥1 and <10. It becomes:

0.0.0.0.0.0.9.8

- To write this in scientific notation, you must have the same two things: a decimal greater than or equal to 1 and less than 10, and a power of 10.

It becomes 9.8.

3 Answer: 0.0000098 in scientific notation is 9.8×10^{-6}.

- The number of times you move the decimal point to the right becomes the power you need to write your number as in scientific notation. But because your number is less than 1, your power is **negative**. You moved the decimal point right 6 times, so the power is 10^{-6}.
- *Remember:* When you are expressing a very small number in scientific notation, move the decimal point *right* to create your decimal. Your power will be *negative*.

If you want to convert your numbers back from scientific notation into decimal notation, you reverse the method. If the power is positive, move the decimal point to the right to make the number larger; and if the power is negative, move the decimal point to the left to make the number smaller.

3.2×10^4 = move the decimal point 4 places to the right = 32 000

3.2×10^{-4} = move the decimal point 4 places to the left = 0.00032

Table S4.5 compares several examples of decimal notation to scientific notation.

Table S4.5: Comparing decimal and scientific notation

Decimal notation	Scientific notation
6	6×10^0 (10 to the power of zero is 1)
500	5×10^2
4323.7	4.3237×10^3
0.0043237	4.3237×10^{-3}
−23 000	-2.3×10^4
5 830 000 000	5.83×10^9
0.9	9×10^{-1}
0.0000000351	3.51×10^{-8}

Significant figures

Significant figures (or **sig figs**) are the digits in a number that indicate the precision of a measurement. They are particularly important when collecting experimental values and carrying out calculations. To determine how many significant figures are in a number, we must apply a set of rules to each digit in a number.

Here are the basic rules for identifying significant figures.

1 All non-zero digits (1–9) are significant; e.g. 25.844 has 5 significant figures.
2 Zeros between non-zero digits are significant; e.g. the zero in 205 is significant.
3 Leading zeros (zeros before the first non-zero digit) are *not* significant; e.g. the zeros in 0.0045 are not significant.
4 Trailing zeros (zeros at the end of a number) on the right of the decimal point are significant; e.g. the zeros in 2.00 are significant.
5 Trailing zeros at the end of a whole number could be either significant or not, but more information is required.

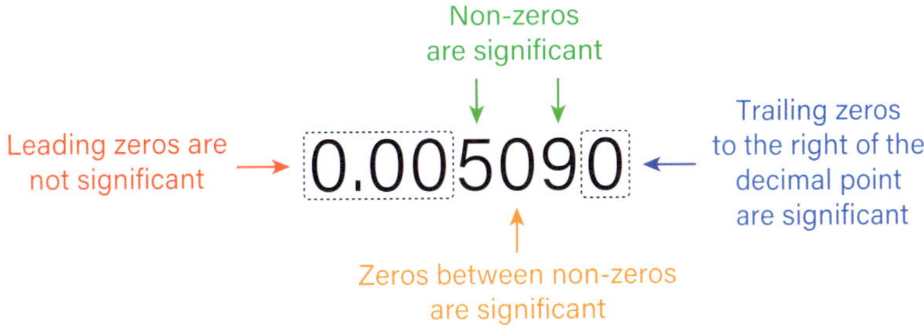

Figure S4.6: This example summarises the rules for determining which digits in a number are significant. 0.005090 has four significant figures.

As stated in rule 5, whole numbers with trailing zeros can be unclear. For example, 5000 could have 1, 2, 3 or 4 significant figures. For clarity, the number should be written in scientific notation. Using scientific notation removes the ambiguity.

- 5×10^3 has **1 significant figure**
- 5.0×10^3 has **2 significant figures**
- 5.00×10^3 has **3 significant figures**

This way, you can clearly show how precise the number really is.

Significant figures in calculations

When doing calculations with measurements, it is important to round your final answer to reflect the correct precision, using the correct number of significant figures. The rules for significant figures depend on the type of calculation.

Addition and subtraction

Your final answer should be rounded to the same number of decimal places as the measurement with the *fewest decimal places*.

- Example:
  12.3 (1 decimal place) + 4.56 (2 decimal places)
  = 16.86 → round to **1 decimal place** → **16.9**

Multiplication and division

Your final answer should be rounded to have the same number of significant figures as the value with the *fewest significant figures*.

- Example:

  3.2 (2 sig figs) × 2.45 (3 sig figs) = 7.84 → round to **2 sig figs → 7.8**

Tip: Always round only at the end of your calculation to avoid rounding errors.

Worked example 11: Significant figures in a motion calculation

You want to calculate the speed of a cyclist using the formula $s = \frac{D}{t}$.

You measure:

- Distance travelled, D = 142.0 m (4 significant figures)
- Time taken, t = 12.4 s (3 significant figures)

1 Write the formula

$s = \frac{D}{t}$

2 Substitute the values

$s = 142.0/12.4 = 11.4516129 \text{ m s}^{-1}$

3 Apply the multiplication/division rule

- Distance (142.0 m) has 4 significant figures.
- Time (12.4 s) has 3 significant figures.

→ Final answer must be given to 3 significant figures because it is the lowest number of sig figs in the values used.

4 Round the answer to 3 sig figs

$s = 11.5 \text{ m s}^{-1}$

Graphing

Graphing data is an excellent way to visually represent the quantitative information you have gained from a first-hand investigation. The type of data will determine the type of graph to make. In first-hand investigation reports, this is usually a scatter plot or a column graph.

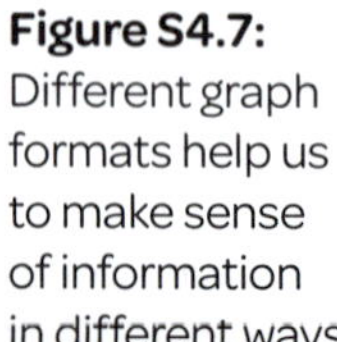

Figure S4.7: Different graph formats help us to make sense of information in different ways.

Scatter plots

Scatter plots are used when two pieces of data are directly related and show trends, such as a change over time. When looking at trends, a line graph is useful as it can show increases and decreases very clearly. For example, in a particular city, the percentage of new cars sold that had airbags was recorded each year. The results were 18 per cent in 2018, 24 per cent in 2019, 32 per cent in 2020, 45 per cent in 2021 and 60 per cent in 2022. This data could be displayed in a line graph.

Remember that the independent variable (changed factor) should go on the horizontal axis, the x-axis. In this case, that would be the year the car was sold. The dependent variable (measured factor) should go on the vertical axis, the y-axis. This would be the percentage of cars with airbags. Each axis should have a label and an even scale.

Plotting points

Once you have drawn an even scale, you can plot the points on the graph.

1 Locate the relevant data on the *x*-axis (e.g. year 2018).
2 Locate the relevant data on the *y*-axis (e.g. percentage 18 per cent).
3 Trace on the graph to where the two points meet.
4 Where the points intersect, plot the point by drawing an 'x' on the spot.
5 Continue plotting points for each piece of data.

Line of best fit

Once all of the data points are on your graph, you can add a trend line, known as the 'line of best fit'. This line will go through the middle of most of the points but will not necessarily connect the dots between each. A line of best fit can be straight (ruled) or curved, depending on the data. Use the following instructions to draw a line of best fit.

1 Look at all the points on the graph. Do they make a straight or a curved line?
2 For a straight line, place a ruler on the graph, through the points. Move it until most of the points are close to or touching the line. Carefully rule the line on the graph, continuing past the final point to the end of the *x*-axis.
3 For a curved line, carefully sketch a single line curve through the data points and continue it to the very top of the graph or the edge of the *x*-axis.

Your finished graph should look similar to Figure S4.8.

Figure S4.8 shows a scatter plot for the data given in the example. From this graph, we can see a *trend*. A trend describes what the data is showing. In this case, the percentage of cars sold with airbags is increasing over time. This is a positive trend.

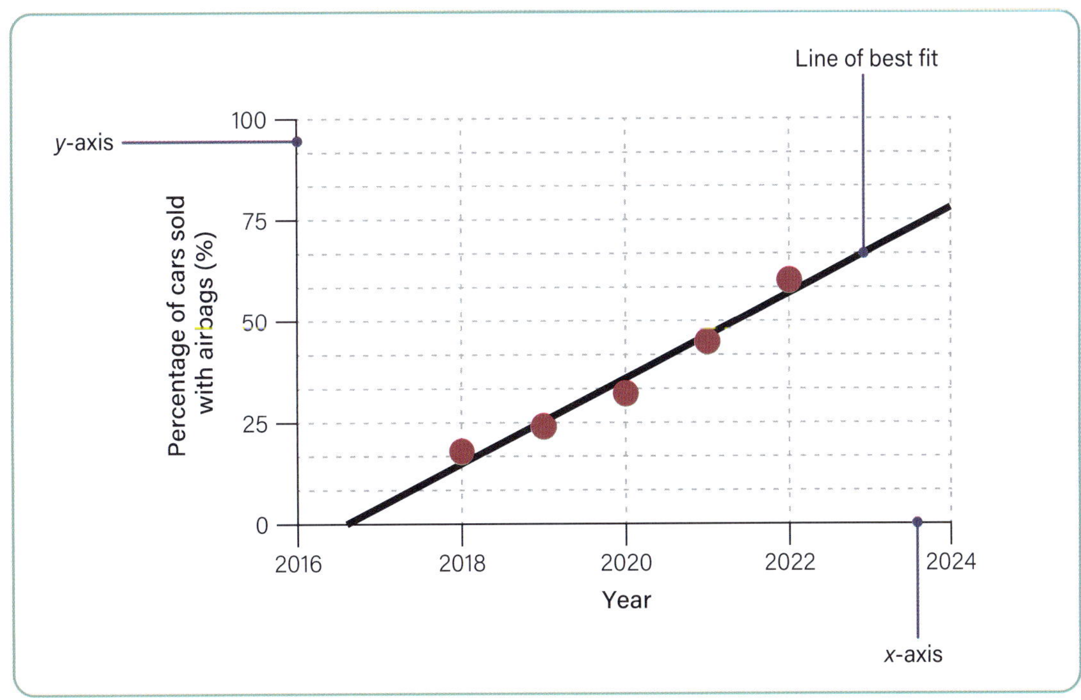

Figure S4.8: Percentage of cars sold with airbags (%) per year, 2016–22

Column graphs

Column graphs are used to compare separate categories of the same type of data. This means there is a distinction between the independent variables – each is its own separate category. However, they are still measured against the same criterion.

For example, in City X the average daily maximum temperature for the month was recorded for every month of a year (2024). The results were 33 °C in January, 30 °C in February, 25 °C in March, 19 °C in April, 17 °C in May, 12 °C in June, 8 °C in July, 10 °C in August, 13 °C in September, 20 °C in October, 27 °C in November and 31 °C in December. This data could be displayed in a column graph.

Remember that the independent variable (changed factor) should go on the horizontal *x*-axis. In this case, it is the month of the year. The dependent variable (measured factor) should go on the vertical *y*-axis. This is the temperature in degrees Celsius (°C). Each axis should have a label and an even scale.

Adding columns

Once you have drawn an even scale, add the columns to the graph.

1 Locate the relevant data on the *x*-axis, such as the month of January.
2 Locate the relevant data on the *y*-axis, such as the temperature of 33 °C.
3 Trace on the graph to where the two points meet.
4 Where the points intersect, put a small pencil mark.
5 Using a ruler, extend the point out on each side of the mark and then down to the *x*-axis to make a column.
6 Continue for each piece of data on the *x*- and *y*-axes.

A completed column graph from the data above should look similar to Figure S4.9.

Note: If you prefer, the data represented as a column graph here could also be represented as a line graph.

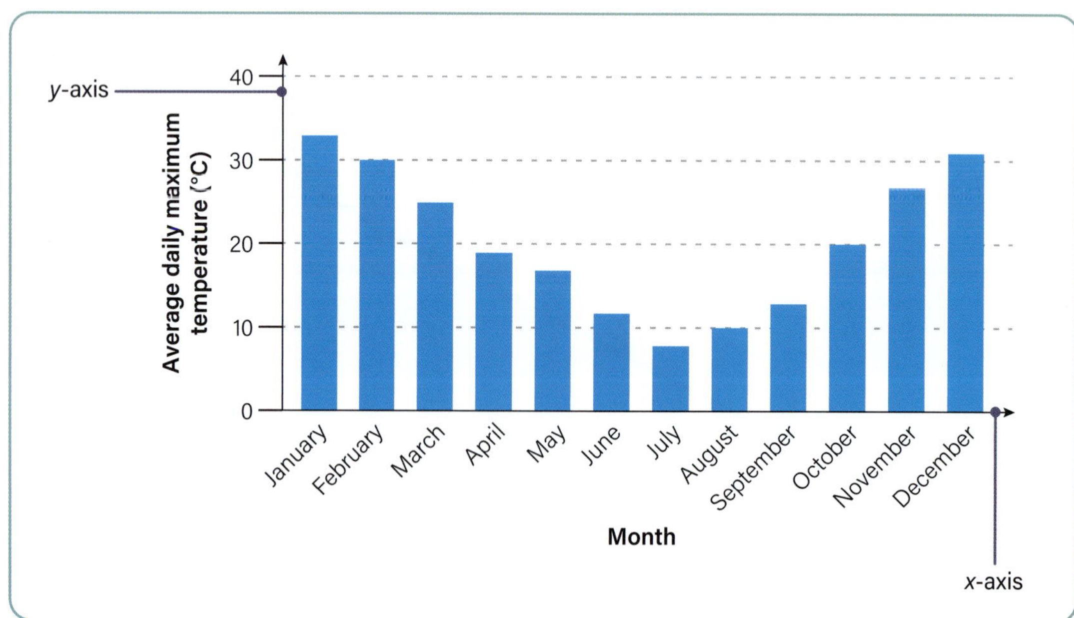

Figure S4.9: ▶ Average daily maximum temperature across month of the year in City X, 2024

Graphing uncertainties

When showing results on a graph, it is important to include any **uncertainty** in your measurements. This is done using error bars.

If you have already calculated the half-range of your repeated measurements (see page 288), you can use that value to show uncertainty. Simply draw a vertical error bar above and below each average point on the graph, extending by the amount of the half-range. This helps to make your graph more accurate by showing variations in the data.

Figure S4.10 shows how a half-range value can be applied to a column graph. The same can be done with points on a scatter plot and other graphs (Figure S4.11).

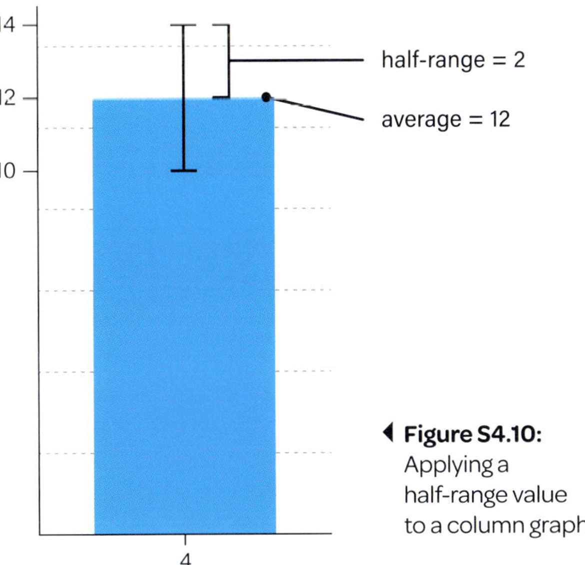

Figure S4.10: Applying a half-range value to a column graph

Figure S4.11: Other examples of applying half-values

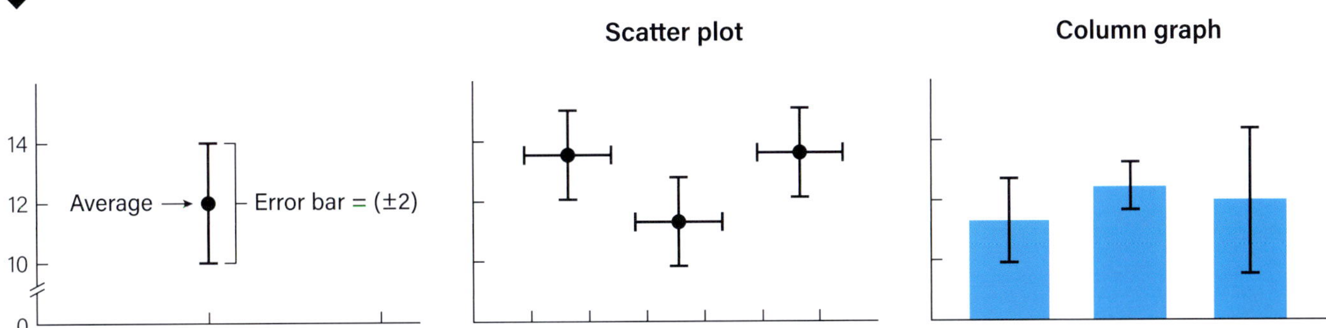

Constructing pie charts

A pie chart is a visual representation of parts of a whole. The 'pie' is a circle where each 'piece' represents a fraction or a percentage. Therefore, the whole circle represents the whole thing: 100 per cent.

To make a pie chart, the size of each 'piece' should reflect the percentage it represents. So, if one of the parts is 25 per cent, that piece should take up a quarter of the pie, which is 25 per cent.

When there are lots of parts, it can be difficult to estimate the size of each piece by hand. Ideally, you should use a computer program such as Microsoft Excel, or an app for making pie charts, to ensure your pieces are the right size for each part they represent, but if you are creating your pie chart by hand, you can estimate. A helpful way to approach this is to divide your circle into eight equal pieces. Each piece is 12.5 per cent of the circle, which will help you to make your estimates.

Figure S4.12: Pie charts are used to visually display the percentage of parts that make up a whole, such as this graphic representation of the percentages of different fruits consumed weekly by a household.

Draw a circle.

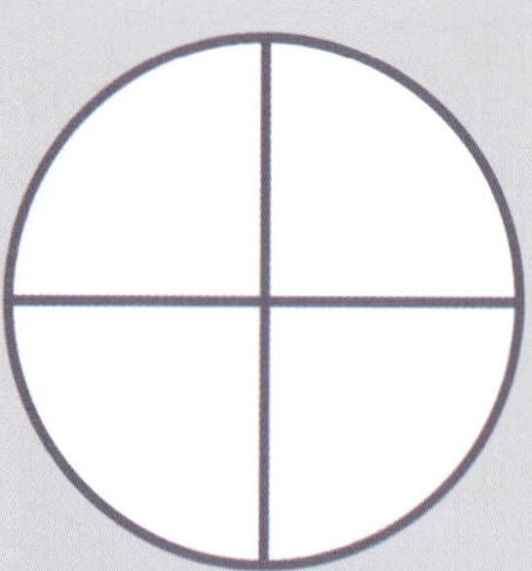

Then, draw two straight lines through the circle to divide it into 4 equal parts.

These are 25% each.

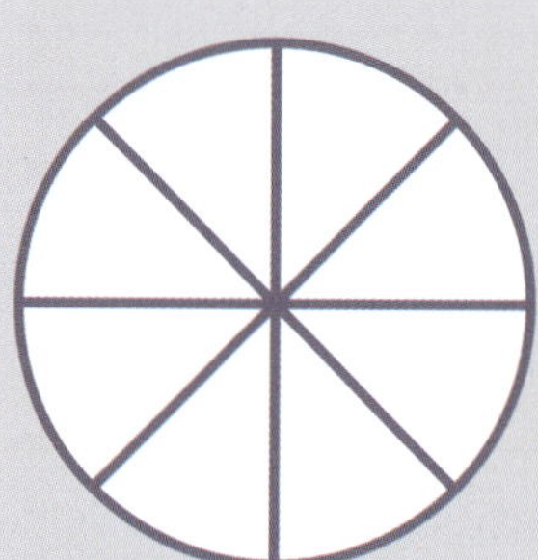

Next, draw two more diagonal lines that divide each of the 4 pieces into 2 equal parts, to make 8 equal pieces in total.

These are 12.5% each.

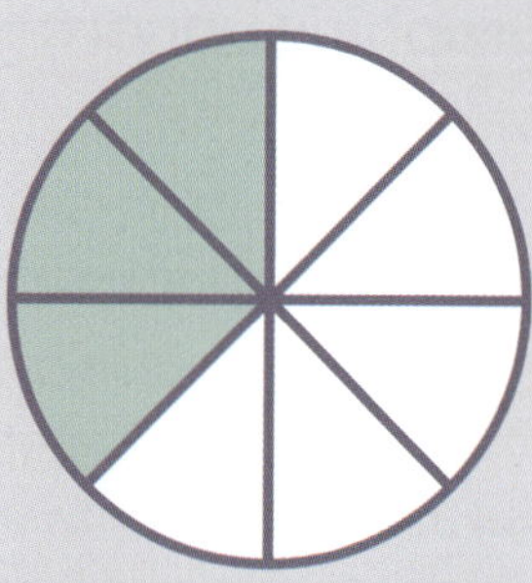

Next, shade the pieces as needed to estimate the size each piece should be for your pie chart.

For example, a pie chart with two parts – 37.5% (shaded) and 62.5% (unshaded) – would look like the circle shown.

Figure S4.13: How to create a simple pie chart using estimates

Figure S4.14: Representing three amounts in a pie chart using estimates

Let's say the percentages you want to represent in a pie chart are 25, 12 and 63.

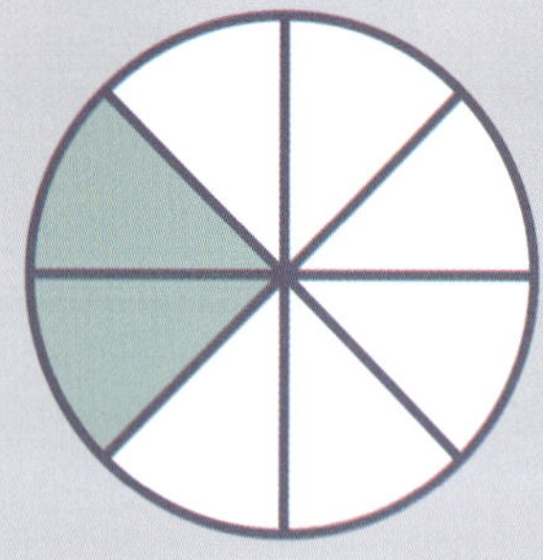

- Start with the pie divided into 8 pieces.
- If any of the parts are exactly 12.5% or multiples of 12.5%, shade those first.
- 25% is 12.5 × 2, so this is a good part to shade first in your pie chart, using 2 pieces to represent 25%.

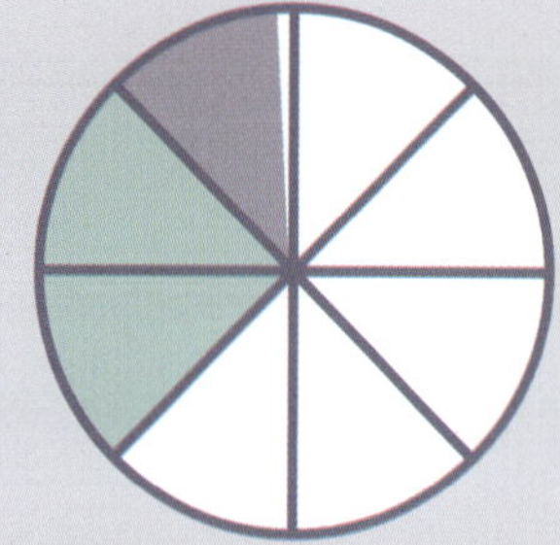

- The next value to represent in your pie chart is 12%.
- Since each piece is 12.5%, you will estimate the part to be slightly smaller than one piece of the pie.
- Shade this in a different colour to the 25% part.

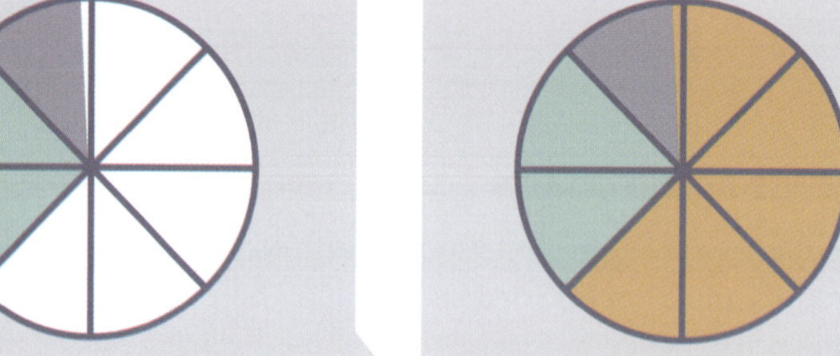

- The remaining part should be 63% of the circle. This is about 5 pieces (5 × 12.5% = 62.5). So, shade 5 more pieces, plus the tiny bit of the piece left over from the 12% part. Use a third colour for the third part of the pie.
- You have created this by hand, so it may not be exact, but using the 12.5% pieces to estimate will help your pie chart be as accurate as possible when digital tools are not available.

This is one way to estimate the size of the pieces of a pie chart when drawing by hand. Explore other ways to draw pie charts by hand, such as by using a tape measure, a protractor, or a circle with 10 equal pieces that are 10 per cent each. However, when you need to formally present your information, it is best to create the pie chart digitally, to make the pieces of your pie chart as exact as possible.

Worked example 12: Drawing pie charts

To create each sector, first add up the total distance. In our investigation results, Table S2.7 on page 245, the total distance all three wheels travelled is 70 cm. Using a calculator, divide each distance by the total distance, then multiply it by 360 degrees to find the angle for each sector, as shown in Table S4.6.

Table S4.6: Average distance and sector angles

Wheel type	Average distance (cm)	Sector angle
Bottle cap	11	11 ÷ 70 × 360 = 56 degrees
Compact disc	33	33 ÷ 70 × 360 = 170 degrees
Washer	26	26 ÷ 70 × 360 = 134 degrees

Draw a circle, locate the centre and draw a radius from the top to the centre. Measure the largest angle from that radius line to make the first sector (in this case, 170 degrees). Starting from the new line, measure the next angle and repeat until finished. If your data is given as percentages, then find that percentage of 360 degrees to determine the angles.

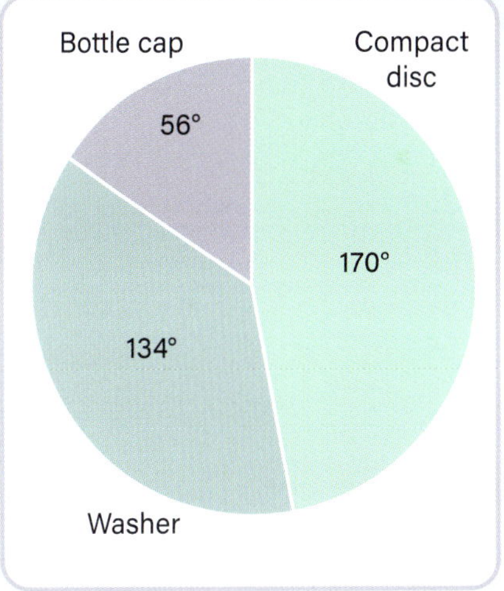

Figure S4.15: The finished chart clearly shows each part of the whole distance.

Using formulas to determine an unknown value

Mathematical formulas are used to show the relationship between variables. Often, you will be given all the values of variables in a particular **equation** except for one (the unknown). You can use the formula to solve the equation to find the unknown. However, before you apply a formula, you need to determine which formula will be best to use.

Follow these steps to identify and use scientific formulas to determine an unknown value.

1. Read the question carefully. Identify and label the values that are given in the question and the variable you are trying to determine.
2. Consider the variables you have listed. Identify the formula that includes those variables.
3. Rearrange the formula, as needed, based on your unknown variable, and substitute in known values.
4. Solve for the unknown, keeping the same units as indicated by the given values, or convert units if required by the question.

To practise these steps, we will look more closely at a specific formula. **Density** is a measure of how heavy something is compared to how much space it takes up. This relationship can be shown mathematically.

$$\text{Density} = \frac{\text{mass}}{\text{volume}}$$

When we replace the words with their symbols (in *italics*), it is now a formula equation.

$$d = \frac{m}{V}$$

Worked example 13 will help you to understand how to use this formula to determine an unknown value.

Worked example 13: Choosing a formula to determine an unknown value

You pour a cup of tea and measure a volume of 30 mL of honey into it. Your tea now weighs (has a mass of) 42 g more than before the honey was added. What is the density of the honey?

1 mass, $m = 42$ g
volume, $V = 30$ mL
density, $d = ?$

- Collect the information you have been given:
You know the volume of honey and the mass of the tea. You don't know the density of the honey. That is what you are trying to work out.

2 $d = \frac{m}{V}$

- Choose the formula that allows you to use the information you have collected. The formula is for density.

3 $d = \frac{42}{30}$

- Calculate the density of the honey by substituting the known values into the formula. (You can use your calculator.)

4 Answer: $d = 1.4\ \text{g mL}^{-1}$

- 42 ÷ 30 = 1.4
- Remember to use the same units as the given values in your answer.

The density formula can be rearranged to determine mass or volume if you know the other two variables. For example, if you are solving for mass, the formula is rearranged to $m = d \times V$. If you know the mass and density, the formula to solve for volume is $V = \frac{m}{d}$.

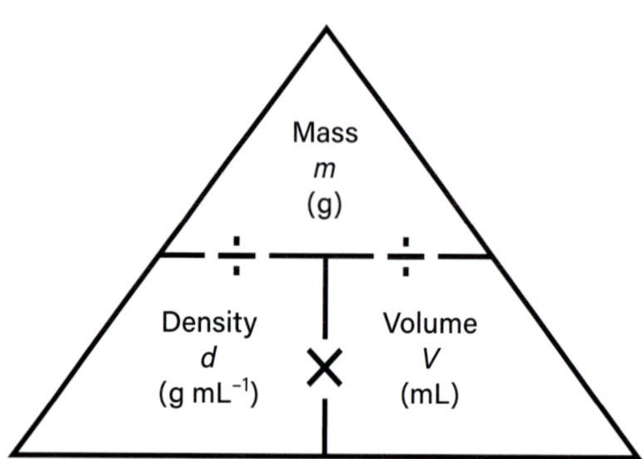

Using scientific equipment

Good science involves curiosity, a desire to understand the world, and a passion for improvement and innovation. Good science also means conducting investigations safely and precisely, which includes knowing your equipment and how to use it. In this section, we will familiarise ourselves with common equipment.

Key terms

concave: curved or rounded inwards

meniscus: the curve seen at the top of a liquid in its container

parallax error: the apparent shift in something's position when it is viewed from different angles

surface tension: where the molecules at the surface of a liquid are more attracted to each other than to the air above the liquid

Safety

Conducting investigations in a safe manner is key to good science. Follow laboratory safety rules to ensure you minimise risk. Use the acronym PIECE to remember these safety rules.

Figure S5.1: Working with chemicals requires following safety information and wearing personal protective equipment (PPE).

Protective equipment: Wear safety glasses, gloves and a laboratory coat to protect you from any chemical splashes or spills.

Instructions: Listen to and read all instructions carefully before you start your investigation.

Equipment: Inspect all equipment to make sure you have the correct items for your specific investigation. Ensure equipment is intact and working properly.

Consumption: Food and drink can become contaminated. Never bring or consume food or drink in the laboratory.

Energy: Manage your energy levels so that you are walking sensibly and holding equipment securely while moving around the lab.

Pictograms

Pictograms are a part of science safety. They are used to label chemicals with their known hazards. Figure S5.2 shows some pictograms you are likely to see in the laboratory.

Pictograms are used across many industries, so you may recognise them from the cleaning products used in your home. Pay close attention to the pictograms on the chemicals you use in the laboratory and follow any warnings. Some chemicals may not be hazardous, so you will not see any pictograms. Others may have several hazards and contain several pictograms for a single substance.

There are many more pictograms than the ones shown here, so be sure to ask your teacher if you come across an unknown pictogram.

Figure S5.2: Common warning pictograms

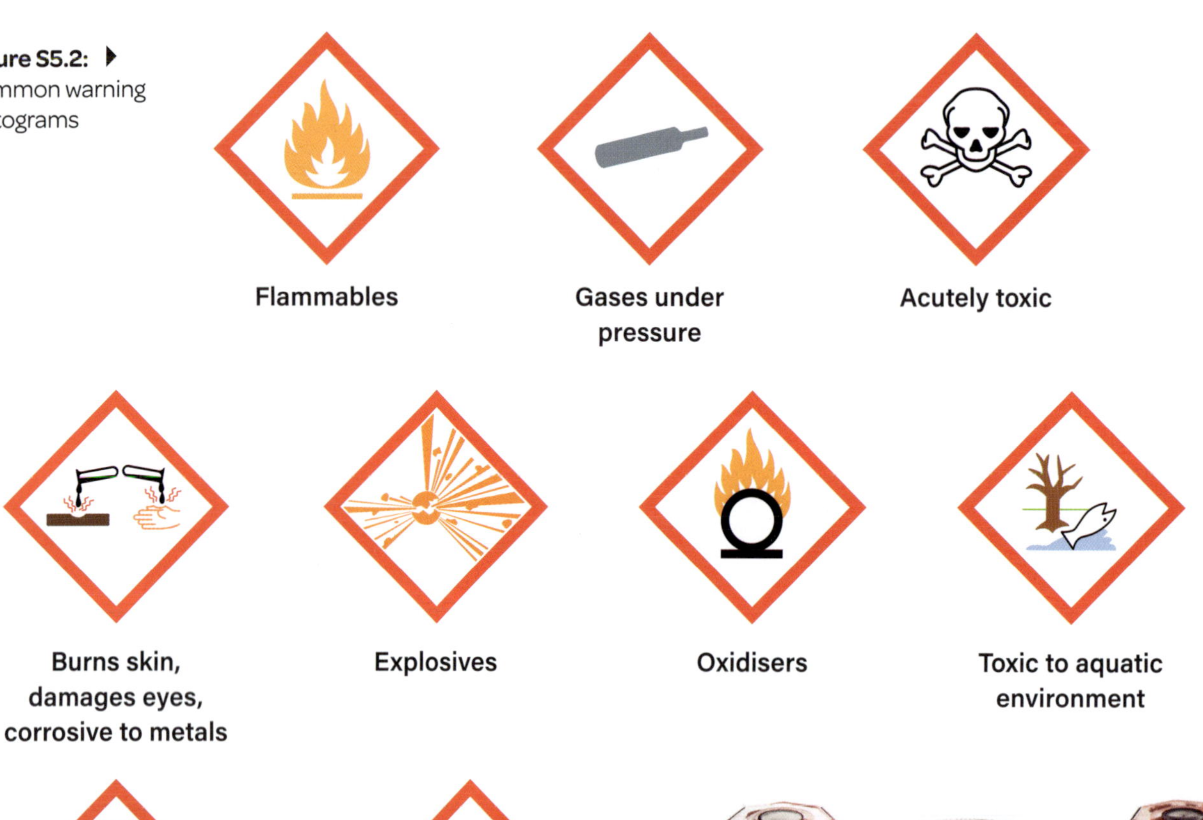

Health hazard
(Chronic health hazards, denoted by the health hazard pictogram, include carcinogens, reproductive toxins, mutagens, specific target organ toxicants and aspiration toxicants.)

Acutely toxic (harmful)
(Other health hazards, denoted by the exclamation mark pictogram, include skin, eye and respiratory irritation, allergic skin reactions, drowsiness and dizziness.)

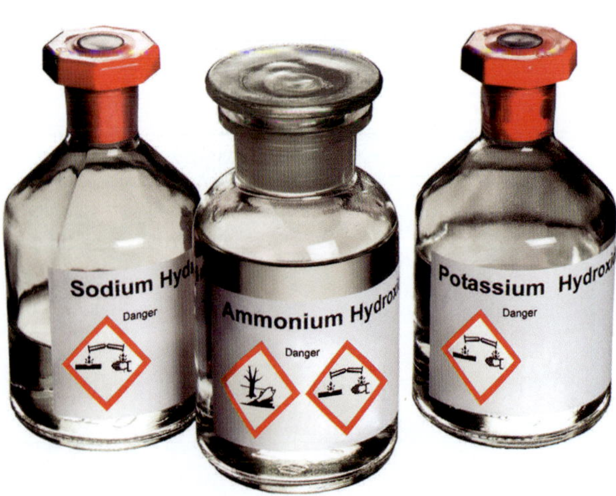

Figure S5.3: Pictograms show the chemicals in your laboratory that are hazardous.

Laboratory equipment

Get to know common laboratory equipment

By learning the names and uses of laboratory equipment, you can select and use the correct equipment for any investigation. Some of the most common items of equipment are shown here.

Figure S5.4: Laboratory equipment

Crucible and lid

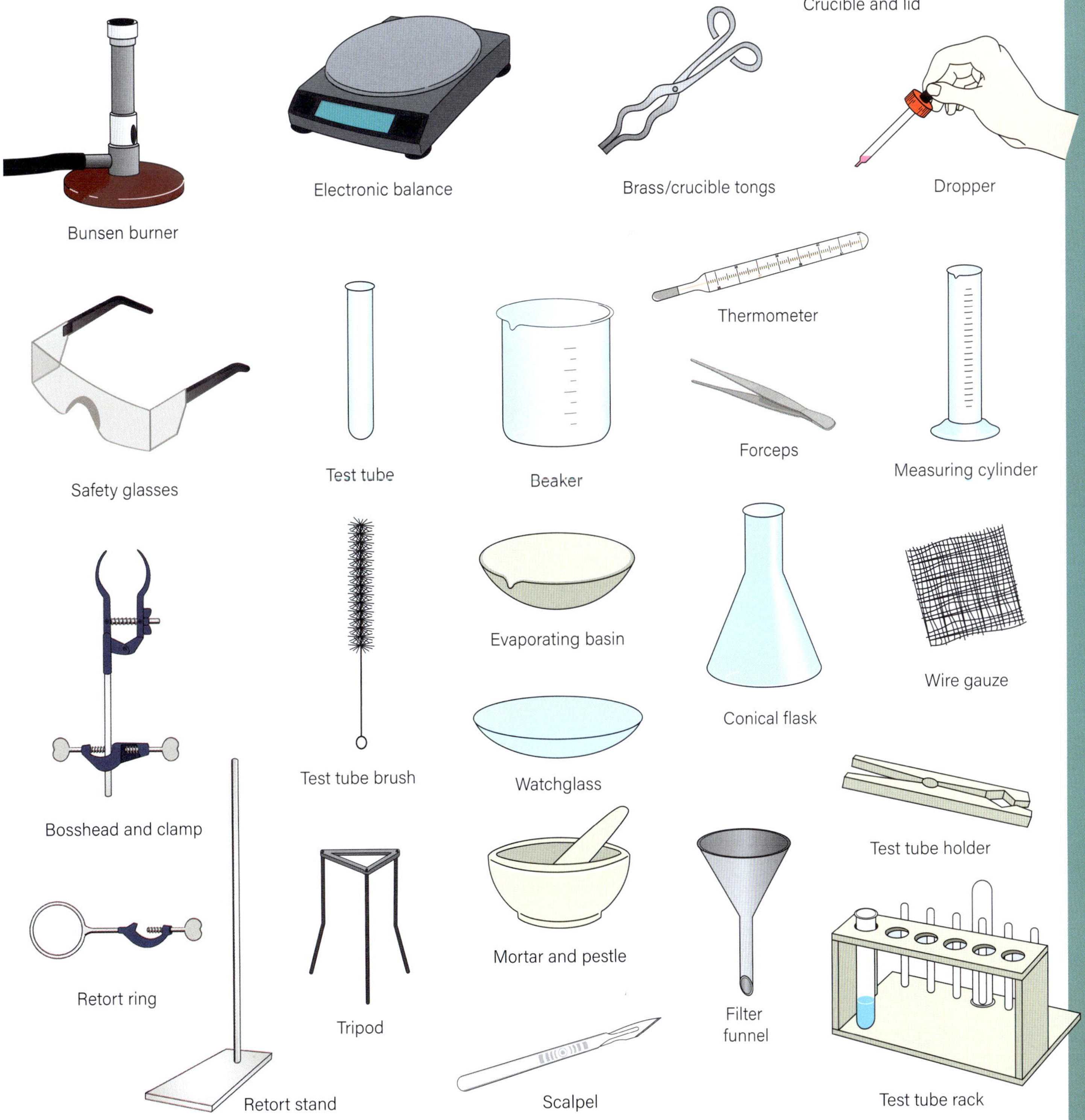

Bunsen burner

Electronic balance

Brass/crucible tongs

Dropper

Thermometer

Safety glasses

Test tube

Beaker

Forceps

Measuring cylinder

Evaporating basin

Wire gauze

Conical flask

Bosshead and clamp

Test tube brush

Watchglass

Test tube holder

Mortar and pestle

Retort ring

Tripod

Filter funnel

Retort stand

Scalpel

Test tube rack

Measurements in science

As discussed earlier, scientists use specialised equipment to measure temperature and volume. In the school laboratory, we measure temperature in degrees Celsius (°C) using thermometers. We also measure volume in litres (L) and millilitres (mL) using measuring cups and beakers. By measuring things carefully, scientists can report their findings reliably.

Figure S5.5: When measuring 8.0 mL, use the smallest measuring cylinder that can hold the required amount.

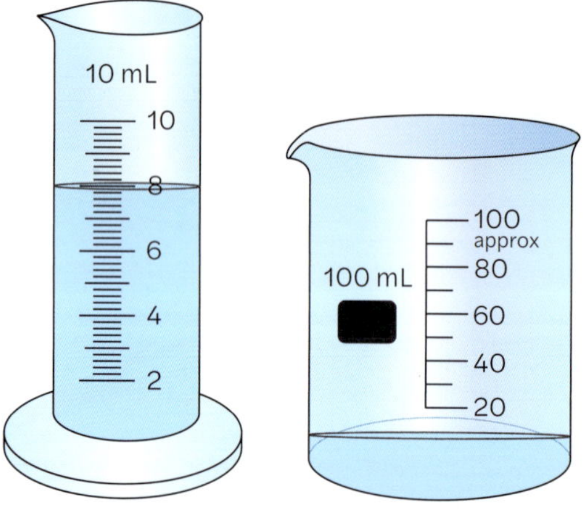

Volume measurements

Reading volume measurements is not as easy as it might seem. It requires an understanding of how liquids can curve and the ability to take this into account. You need to use specific equipment, which will help you to measure accurately.

Equipment for measuring volume

Beakers and measuring cylinders are the most common pieces of equipment you will use to measure volume. Beakers are typically used when volumes are approximate, while measuring cylinders are used when precision (the exact amount) matters.

When selecting a measuring cylinder, always use the smallest one available that can measure the required amount. For example, if you need to measure 8.0 mL of something, it is best to use a 10 mL measuring cylinder, rather than a 100 mL one, to get the most accurate result.

Observing the meniscus

Because of **surface tension**, liquids do not form a flat surface in a relatively small container. Instead, a curve forms, which is called the **meniscus**. You should take volume readings at eye level, at the very bottom of a **concave** meniscus, as shown in Figure S5.6. It is obvious that the bottom of the meniscus is between the 36 mL and 37 mL increment lines, which is recorded with certainty. The place value after the increment marks – in this case, the tenth place (.0) – should be estimated. Therefore, the measurement for this volume is 36.5 mL, where the .5 is estimated and is therefore uncertain.

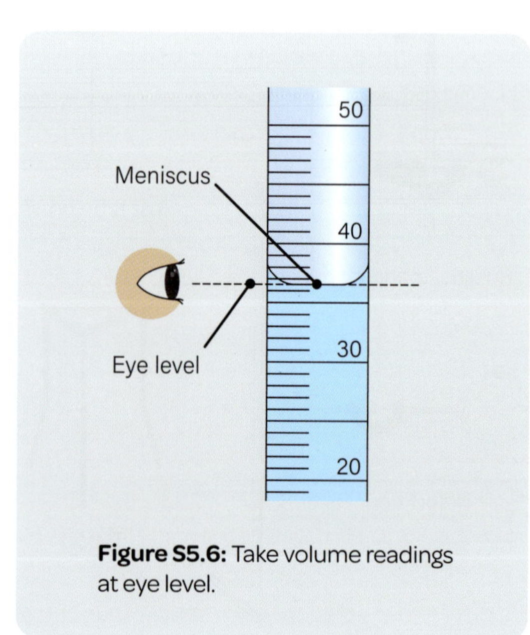

Figure S5.6: Take volume readings at eye level.

If you do not take the reading at eye level, **parallax error** will occur. If this happens repeatedly, you will have a systematic error in your measurement, because the error will be the same for all readings. Figure S5.7 shows how reading the meniscus above eye level results in a low reading as compared to the true value, and reading the meniscus from below gives a higher value.

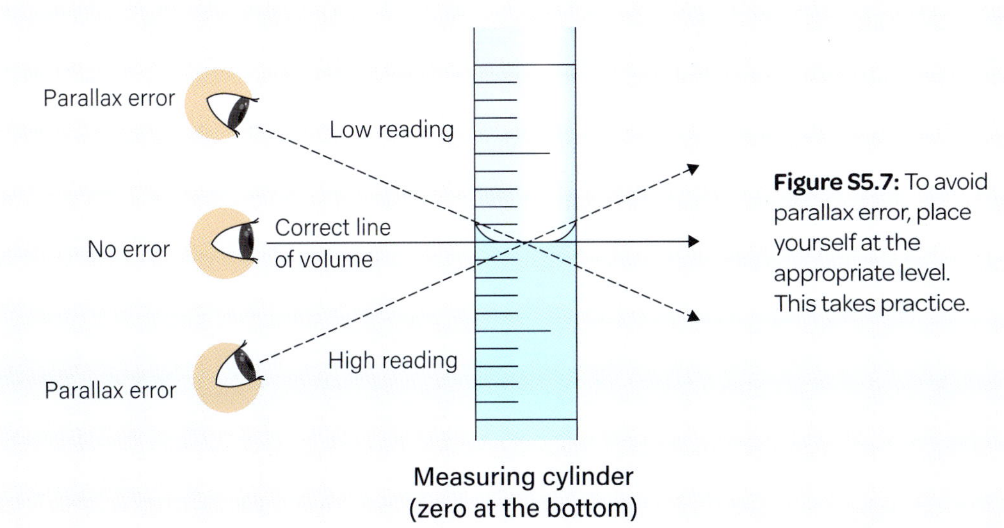

Figure S5.7: To avoid parallax error, place yourself at the appropriate level. This takes practice.

Measurement uncertainties

All scientific equipment has some level of uncertainty, no matter how carefully it is used. This is because every instrument has limits to its precision. For example, a ruler marked in millimetres is more precise than one marked only in centimetres. The last digit of a manual measurement is always estimated by the investigator but will have a given uncertainty range. The uncertainty of a manual or analogue measurement is usually *half the smallest scale division* on the equipment. If the smallest increment is 0.1 cm, then the uncertainty is 0.1 / 2 = 0.05 cm (see Figure S5.8).

Figure S5.8: When the investigator determines the length of the item to be 12.65 cm, the last decimal has been estimated with an uncertainty of +/−0.05, so the true value could be anywhere between 12.60 cm and 12.70 cm.

This is different from half-range uncertainty, which comes from repeated trial measurements in an investigation that vary. While equipment uncertainty comes from the tool itself, half-range uncertainty comes from variation in results.

When drawing error bars on a graph, you need to use the *larger* of the two uncertainties; either the half-range from your data or the uncertainty of your measuring equipment. This gives a fair and accurate representation of the possible variation in your results.

Light microscope

The most common type of microscope in the school science laboratory is a light microscope. It allows you to study samples by shining a bright light through an extremely thin slice of material. The image is magnified by the microscope's lenses, which you look through.

The eyepiece of a light microscope already magnifies samples by 10 times. The objective lens, which is lower down and usually rotates, then magnifies samples by a further amount. To identify the total magnification, multiply the eyepiece magnification (10) by the objective lens magnification.

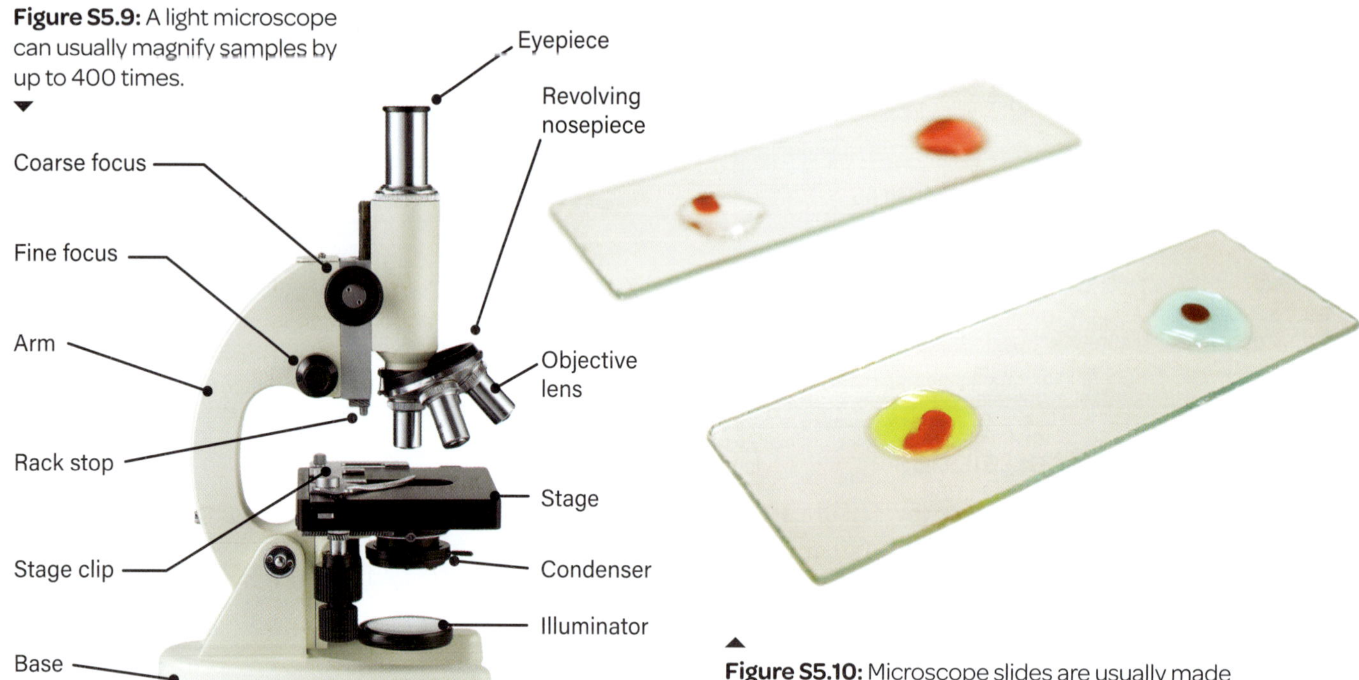

Figure S5.9: A light microscope can usually magnify samples by up to 400 times.

Figure S5.10: Microscope slides are usually made of glass and are very fragile. Treat them carefully.

Setting up and using a light microscope

1 Place your microscope on the bench or table, making sure that it is not too close to the edge and that the arm is facing you.
2 Plug your microscope in and turn it on. The illuminator (light) will come on.
3 Lower the stage as far as it can go, using the coarse focus knob.
4 Consider each of the rotating objective lenses. Often they are different colours and have the magnification written on them.
5 Start with the lowest magnification – this is usually 4× (four times). Your microscope eyepiece already has a 10× magnification on its own, so when coupled with the 4× eyepiece, what you are looking at will be 40 times its actual size.
6 Carefully insert the microscope slide onto the top of the stage and hold it firmly under the stage clip. Position it so the object you need to see is in the middle of the stage.
7 Look through the eyepiece and slowly bring the stage upwards towards you, using the coarse focus knob. This can take some time and everything will look bright and fuzzy until you get a glimpse of the slide as it comes into focus.
8 When the slide is roughly in focus, use the fine focus to turn it into a clear image.
9 Increase the magnification by changing the objective lens to 10×, 20× or 40×.
10 40× is usually the highest available magnification and can be tricky to find and focus on. It can also bring the lens extremely close to the slide, enough to crack and break it. Monitor this carefully.

Investigations

Investigation number	Investigation title	Inquiry skill	Inquiry skill focus
Chapter 1 Genetics		**Processing, modelling and analysing, Communicating**	
1.1	Extracting DNA from strawberries	Communicating	Presenting scientific findings
1.6	Genetic trait survey	Communicating	Using digital technologies to organise and present findings
1.9	Investigating proportions of sprout colours	Processing, modelling and analysing	Processing data
1.10	DNA lolly model	Communicating	Selecting communication formats
Chapter 2 Evolution		**Questioning and predicting, Evaluating**	
2.2	Natural selection in practice	Questioning and predicting	Developing a hypothesis
2.3A	Dating fossils	Evaluating	Identifying investigation errors and assumptions
2.3B	Comparative anatomy	Evaluating	Constructing evidence-based arguments
Chapter 3 The periodic table		**Processing, modelling and analysing, Evaluating**	
3.2	Investigating the reactivity of metals	Processing, modelling and analysing	Organising data
3.3	Student-designed investigation: properties of common substances	Evaluating	Evaluating the validity and reproducibility of methods
3.5	Modelling atomic structure alongside the periodic table	Evaluating	Constructing evidence-based arguments
Chapter 4 Chemical reactions		**Questioning and predicting, Planning and conducting**	
4.2A	Synthesis of iron oxide	Planning and conducting	Describing ways to minimise risk
4.2B	Decomposition of copper(II) carbonate	Planning and conducting	Generating and recording data
4.3A	Growing metallic crystals: metal displacement reactions	Questioning and predicting	Making predictions
4.3B	Precipitation reactions	Questioning and predicting	Developing a hypothesis

Investigation number	Investigation title	Inquiry skill	Inquiry skill focus
4.4A	Neutralising hydrochloric acid	Planning and conducting	Generating and recording data
4.4B	The effects of indicators on acids and bases	Planning and conducting	Selecting scientific equipment
4.4C	Reaction of acids with metals	Planning and conducting	Developing a risk assessment
4.4D	Reactions of acids with carbonates	Planning and conducting	Obtaining replicable data
4.5A	The effect of concentration on the rate of reaction	Questioning and predicting	Constructing scientific questions
4.5B	The effect of temperature on the rate of reaction	Questioning and predicting	Developing a hypothesis
4.6A	The effect of surface area on the rate of reaction	Planning and conducting	Developing a risk assessment
4.6B	The effect of a catalyst on the rate of reaction	Planning and conducting	Obtaining replicable data
4.6C	Student-designed investigation: investigating the rate of chemical reactions	Planning and conducting	Designing scientific investigations
4.7	Predicting energy changes for exothermic and endothermic reactions	Questioning and predicting	Developing a hypothesis
Chapter 5 Climate change		**Evaluating, Communicating**	
5.2	The albedo effect	Evaluating	Describing the impact of assumptions and errors
5.3	Observing the weather	Communicating	Using digital technologies to organise and present findings
5.4	The enhanced greenhouse effect	Evaluating	Proposing ways to improve the quality of data
5.5A	Modelling thermal expansion	Evaluating	Describing types of errors
5.5B	Melting ice and sea-level rise	Evaluating	Constructing evidence-based arguments
5.6	Modelling ocean acidification	Evaluating	Evaluating the validity and reproducibility of methods

Investigation number	Investigation title	Inquiry skill	Inquiry skill focus
Chapter 6 The universe		**Questioning and predicting, Communicating**	
6.1	Modelling the expanding universe	Communicating	Using digital technologies to organise and present findings
6.2	Investigating orbits	Communicating	Preparing representations to communicate findings
6.3	Bottle rockets	Questioning and predicting	Developing a hypothesis
6.4A	Making a telescope	Questioning and predicting	Constructing scientific questions
6.4B	Lens diameter and resolution	Questioning and predicting	Making predictions
6.6	Stargazing	Questioning and predicting	Constructing scientific questions
Chapter 7 Motion		**Planning and conducting, Processing, modelling and analysing**	
7.2	Cars and pedestrians	Processing, modelling and analysing	Processing data
7.3	Ticker timers	Processing, modelling and analysing	Constructing graphs
7.4A	Car crashes and inertia	Planning and conducting	Generating and recording data
7.4B	Balloon rockets	Planning and conducting	Describing ways to minimise risk
7.5	Acceleration and mass	Planning and conducting	Obtaining replicable data
7.8	Car crashes and momentum	Processing, modelling and analysing	Analysing scientific models

Investigation **1.1**

Extracting DNA from strawberries

Inquiry skill: Communicating

Inquiry skill focus: Presenting scientific findings

Scientists can use a variety of representations to simplify and explain ideas. Visual communications make things easier to understand, which helps us to interpret scientific ideas and findings. As you complete this investigation, think about how the process will help you to visualise the DNA of living things.

Hint 1: What equipment are you using, and how could you represent this when constructing diagrams of your set-up and results?

Hint 2: Refer to 'Communicating' in the Science how-to section on page 262.

Aim

To investigate the process of extracting DNA from strawberries

Materials

For the class:

- liquid dishwashing detergent
- salt
- electronic balance
- tap water
- 50–100 mL measuring cylinder
- 10 mL measuring cylinder
- stirrer

For each group:

- plastic spoon
- strawberries
- tap water
- zip-lock bag
- 10 mL extraction buffer
- large test tube
- test tube rack
- funnel
- 10 cm square of Chux cloth
- 30 cm ruler
- plastic pipette
- very cold ethanol (kept on ice or in the freezer)
- wooden skewer

Method

1. Copy the results table from the next page into your notebook, adding a title and rows as needed.
2. Make the DNA extraction buffer (this can also be pre-prepared by a lab assistant). Add 5 mL of detergent, 0.75 g of salt (weighed on an electronic balance) and 45 mL of water to the larger measuring cylinder. Stir to mix. Divide into five 10 mL samples of extraction buffer.
3. Within your group, remove the green stem of a strawberry and wash the strawberry carefully with water.
4. Place the strawberry in a zip-lock bag with 10 mL of the extraction buffer and remove as much air as possible before sealing the bag tightly.
5. On the bench, crush and massage the strawberry inside the bag for a minute.
6. Set up the test tube in a rack on the bench.
7. Place the funnel in the test tube and then line the funnel with the Chux cloth.
8. Pour the mixture from the zip-lock bag into the funnel.
9. When the mixture has filtered (this will take some time), squeeze the Chux cloth to get all the liquid into the test tube. Then discard the cloth and the strawberry pulp, keeping only what was filtered into the test tube.
10. Measure the amount of liquid in the test tube very roughly with a ruler.
11. Use the plastic pipette to add the same amount of ice-cold ethanol as the amount of liquid already in the test tube.
12. Observe what happens in the test tube. You should be able to see changes straight away.
13. Use the wooden skewer to gentle remove some DNA from the top of the test tube.

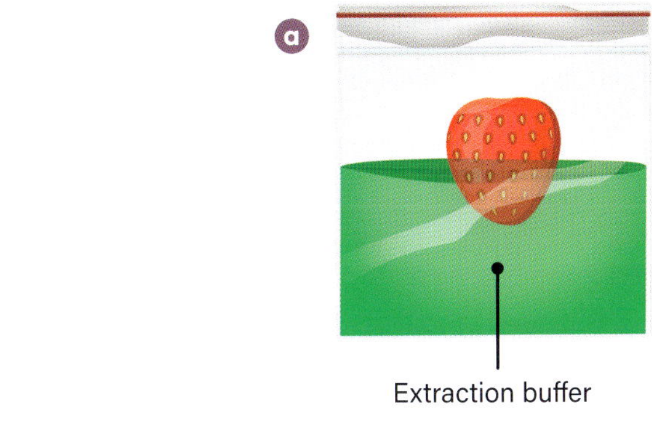

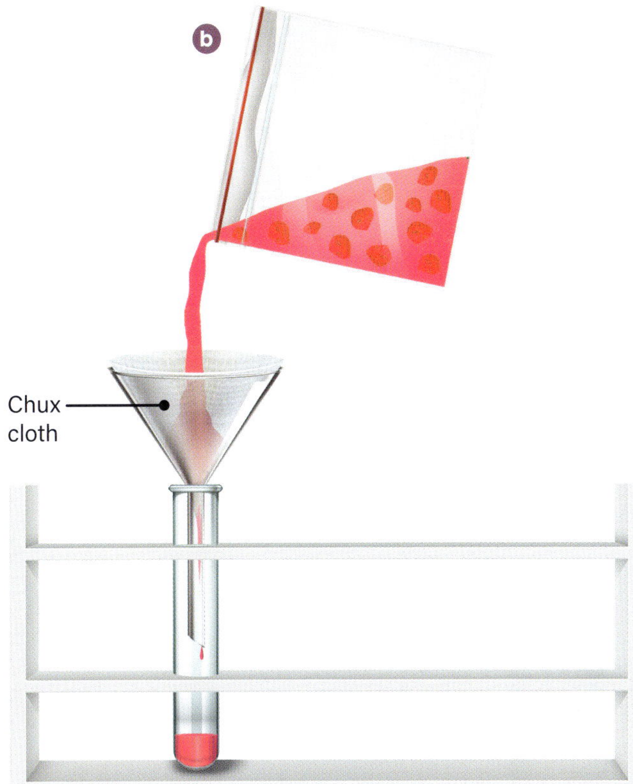

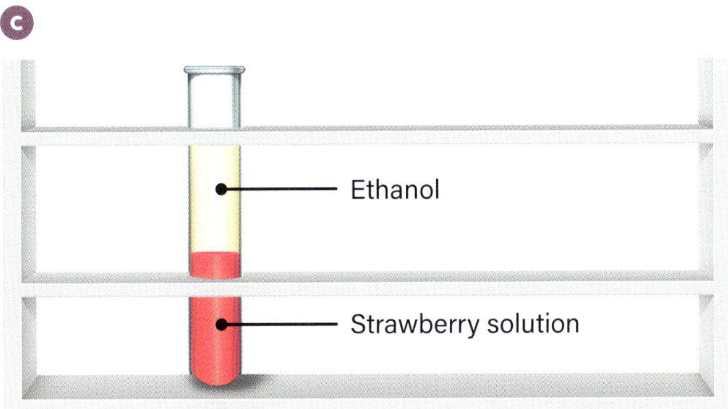

Figure INV1.1: Thoroughly crush the strawberry before filtering it.

Questions

1 Describe the appearance of the strawberry DNA.

2 How does this compare to the description of DNA in Section 1.1?

3 Propose why it was important to crush the strawberry before extracting the DNA.

4 Construct a diagrammatic flowchart, showing the processes involved in the extraction, that could be used by another student to complete the investigation.

Conclusion

Copy and complete:
'The results show that: *(respond to the aim)*.'

Table INV1.1: Results

Observations	Photo or diagram

Investigation 1.6

Genetic trait survey

Inquiry skill: Communicating

Inquiry skill focus: Using digital technologies to organise and present findings

Digital technologies allow us to present scientific findings in many different ways. They can be used to construct tables and graphs, as well as to record images and videos. Think about how you could use digital technologies to organise and present your findings for this investigation.

Hint 1: *Think about the observations you will be making and the results you will be gathering.*

Hint 2: *Think about how digital technologies could help you to display your observations and results. For example, you could construct a digital results table on a computer.*

Aim

To investigate the different genetic traits that exist in a population

Materials

- genetic trait survey
- at least 10 participants
- laptop

Method

1. Construct a digital version of the results table on your laptop or device.
2. In the second column of the table, record whether you have the genetic trait listed in the first column.
3. Share your results with the rest of the class and construct a class tally.
4. Copy the class tally into your digital table.

Questions

1. Digitally graph your class results for each trait. (More information about graphs is available in the Science how-to section on pages 294–99.)
2. Identify the most and least prevalent traits in your class.
3. Propose two more genetic traits you could have checked for.
4. Propose why some traits are more common than others.
5. Considering the audience for this information, propose and construct an appropriate digital presentation to show the outcome of the survey.

Conclusion

Copy and complete:
'The results show that: *(respond to the aim)*.'

Table INV1.6: Results

Trait	My response (yes or no)	Total number in class
Curly hair		
Brown eyes		
Attached earlobe		
Cleft chin		
Hitchhiker's thumb		
Ability to roll tongue		
Left-handedness		
Widow's peak hairline		

Initial set-up

6 weeks for plant growth

Investigating proportions of sprout colours

Inquiry skill: Processing, modelling and analysing

Inquiry skill focus: Processing data

Data can be processed in different ways. One way is by collecting raw measurements and then using mathematical relationships to calculate a new value. This investigation requires you to place your results into a table, and an appropriate graph, and to convert the total numbers into the simplest possible ratio.

Hint 1: See the Science how-to section on page 285 for information about ratios.

Hint 2: See the Science how-to section on page 294 for information about constructing graphs.

Aim

To investigate the ratio of monogenic plants grown from seeds

Materials

- gloves
- multi-tray plastic plant container (or 3–4 small plastic pots)
- vegetable soil
- pea seeds (*Pisum sativum*)
- tap water
- 50 mL measuring cylinder

BIOLOGICAL MATERIALS SUCH AS FERTILISER CAN BE HARMFUL IF CONSUMED. MAKE SURE TO WEAR GLOVES WHILE HANDLING SOIL.

Method

Part A:

1. Select the trait you want to study (tall versus short, round versus wrinkled seed, etc.).
2. Fill your tray or pots with vegetable soil, measuring out the same amount of soil in each.
3. Place one pea seed into each section, or 2–3 seeds in each pot.
4. Water each container with the same amount of water, using the measuring cylinder for precision (15 mL each in plastic tray, 50 mL each in plastic pot).
5. Place seeds on a windowsill where they will all be exposed to the same sunlight and temperature conditions.

Part B:

1. Regularly water all the seeds every 2–3 days.
2. Monitor and record plant growth.
3. After 6 weeks (or when each plant has a pea pod), examine your plants for the trait you selected.
4. Record the number of plants with each trait in a results table. (Table INV1.9 provides a sample layout for recording your results for this investigation. Make sure to include space for every plant you grow!)

Table INV1.9: Results

Observed trait	Pea plant number					Total plants with trait
	1	2	3	4	5	

continues ▶

Questions

1 Identify the dominant trait, based on the total numbers you have calculated.

2 Convert your total numbers to the simplest possible ratio comparing the occurrence of the two traits.

3 Compare your ratio to Mendel's calculated ratio. Is it similar? Propose reasons for any differences in your data. (*Hint:* Think about the number of plants studied.)

4 Convert the data from your table into an appropriate scientific graph.

5 If two pea plants that were heterozygous for your selected trait were crossbred, what percentage would you expect to have the recessive trait? Use a Punnett square to support your answer.

Conclusion

Copy and complete:
'The results show that: '*(respond to the aim)*.'

Investigation 1.10

DNA lolly model

Inquiry skill: Communicating

Inquiry skill focus: Selecting communication formats

An important part of scientific investigations is communicating information and findings in a way that is effective and suitable for the type of investigation being conducted. To decide on a suitable format of communication, ask the following questions: *What information do I have, What do I want to communicate? What format will display this information clearly?*

Hint 1: For an overview of different formats, refer to 'Communicating' on page 262 of the Science how-to section.

Hint 2: Think about the size and location of the DNA molecules.

Aim

To investigate the structure of DNA by making a model out of lollies

Materials

- 2 long red liquorice straps
- 4 of the same type of lollies in 4 different colours (e.g. mini-marshmallows, jelly beans, gummy bears)
- toothpicks
- clean surface or cutting board
- small pieces of paper

Method

1. Assign lolly colours to each of the four DNA bases: adenine, cytosine, guanine and thymine.
2. Use two pieces of red liquorice as the DNA backbone.
3. Pair your lolly bases: guanine with cytosine and adenine with thymine.
4. Thread the matching base pairs onto toothpicks.
5. Insert the ends of the toothpicks into the red liquorice DNA backbone.

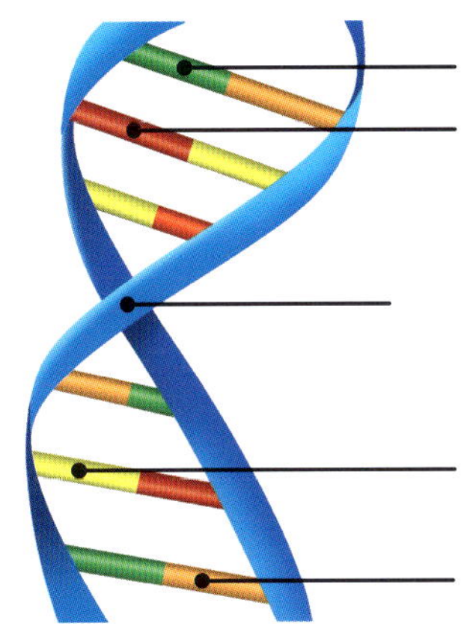

Figure INV1.10: Remember to organise your four colours into two pairs and always put them next to one another between the two liquorice ladders.

6. Twist the red liquorice DNA backbone as you go, creating the DNA twisted double helix model.
7. Use the paper and toothpicks to label each of the four bases and the sugar–phosphate backbone of your DNA model.
8. Draw a diagram or take a photo of your finished DNA model.

Questions

1. Propose why the use of models in science is important.
2. Outline why this model of DNA is useful for presenting to others when explaining the structure of DNA.
3. Explain how your model demonstrates the features and arrangement of a molecule of DNA inside a cell.
4. Discuss one way you could improve this model to make it more accurate. (*Hint:* Think about the backbone.)

Conclusion

Copy and complete:
'The results show that: *(respond to the aim)*.'

Investigation 2.2

Natural selection in practice

Inquiry skill: Questioning and predicting

Inquiry skill focus: Developing a hypothesis

Part of a scientific investigation is developing a hypothesis, which is a proposed prediction of the outcome of an investigation that is intended to address a problem or answer a question. To develop a hypothesis, you must first identify the independent, dependent and controlled variables. The independent variable is the one thing that you purposefully change in the investigation, the dependent variable is what you measure, and the controlled variables are all the things you keep the same throughout the investigation. Develop a hypothesis for this investigation, using the format below.

If ..., then ..., because ...

***Hint 1:** What factor are you changing in this investigation, and what are you measuring?*

***Hint 2:** Refer to the Science how-to section on page 236 to help you.*

Aim

To investigate the effectiveness of different 'beaks' in collecting different types of food

Materials

- marbles
- uncooked rice
- string (cut into 3 cm pieces)
- popcorn kernels
- 4 small plastic tubs
- forceps
- crucible tongs
- kitchen tongs
- blunt pliers
- stopwatch
- 4 paper cups

Method

1. Copy the results table from the next page into your notebook, adding a title.
2. Place the marbles, rice, pieces of string and popcorn into each of the plastic tubs (i.e. marbles in one tub, rice in another tub, etc.).
3. Designate a different 'beak' (forceps, crucible tongs, kitchen tongs or blunt pliers) to each person in the group.
4. Choose a person to use the stopwatch to time 30 seconds.
5. Each student uses their tool to pick up as many objects in one of the tubs as possible in the 30 seconds. For example, the student with the forceps attempts to pick up items from the marble container, the student with the crucible tongs attempts to pick up items from the rice container. Participants place the objects they collect with their tool into their paper cup.
6. After 30 seconds, stop and count how many of each object is in each person's paper cup.
7. Record the totals in your results table.
8. Share your data so that each person can complete their results table.
9. Each student returns their items to the plastic tub.
10. Repeat steps 4–9 three more times. For each turn, students move to the next plastic tub and try to pick up a different item using the same instrument.

Questions

1. Define 'natural selection'.
2. Explain how this investigation demonstrates the principles of natural selection.
3. Identify the 'beak shape' that was the best for picking up each type of food and propose a reason for why this was the case.
4. Identify whether the data you collected supports your hypothesis. Explain why you think this is the case.

Conclusion

Copy and complete:

'The results show that: *(respond to the aim)*.'

Table INV2.2: Results

Tool	Beak type	Number of pieces of string	Number of grains of rice	Number of popcorn kernels	Number of marbles
Forceps	Long, thin, pointy				
Crucible tongs	Long, medium width, pointy				
Kitchen tongs	Long, wide				
Blunt pliers	Short, thick, wide				

Figure INV2.2: ▶ Different beak types are better suited to picking up different food items. Those birds whose beaks are best suited to obtaining the available food source are best placed to survive and reproduce – a process known as natural selection.

Investigation 2.3A

Dating fossils

Inquiry skill: Evaluating

Inquiry skill focus: Identifying investigation errors and assumptions

Errors or assumptions in an investigation are something that will cause you to record a value or observation that is higher or lower than the true value. Errors can be caused by numerous things, such as faulty equipment, an experimental flaw or changing environmental conditions. Unidentified errors can lead to assumptions about the investigation results. Scientists aim to minimise errors, but for errors that cannot be eliminated, it is important that they are identified so we can understand how they may impact the results.

Hint 1: *Refer to step 1 of 'Evaluating' in the Science how-to section on page 253.*

Hint 2: *Carefully read through the materials and method for the investigation. Are there any steps where an error could be present?*

Hint 3: *Can the method be modified to minimise the errors you have identified? Could you use different equipment or a different technique?*

Aim

To investigate a model for absolute dating

Materials

- 100 × 5- or 10-cent coins in a container

Method

1 Each coin represents an atom in the radioactive element carbon-14. You will be investigating how long the half-life of carbon-14 is. Scientists can calculate the age of fossils by identifying the half-lives of atoms surrounding the fossil.

2 Copy the results table into your notebook, adding a title.

3 Shake the container and carefully empty the coins onto the table. Spread them out so that none overlap.

4 Remove all the coins showing tails. These represent atoms that are no longer radioactive and have decayed.

Table INV2.3A: Results

Shake number	Number of coins showing heads (radioactive atoms)
1	
2	
3	
4	
5	
6	
7	
8	
9	
10	
11	
12	

5 Record in your results table the number of coins showing heads. These represent atoms that are still radioactive.

6 Put the coins showing heads back in the container. Shake the container, and carefully empty the coins onto the table. Repeat this process until all the coins are gone or you have completed 12 trials.

Questions

1 Describe how this investigation models the process used to date fossils.

2 Identify any errors in this investigation. Discuss how these errors may lead to assumptions in your results.

3 a Describe how you could minimise the errors you identified in the previous question.

b Discuss how these modifications would impact your data.

4 The half-life of carbon-14 is 5730 years. If each coin shake represents one half-life, calculate the number of years it would take for carbon-14 to completely decay.

5 Construct an appropriate graph showing the decay rate of your coins.

Conclusion

Copy and complete:
'The results show that: *(respond to the aim)*.'

Comparative anatomy

Inquiry skill: Evaluating

Inquiry skill focus: Constructing evidence-based arguments

The results of an investigation can be used as evidence to make claims and construct arguments about scientific concepts and skills. How might the results of this investigation be used to make recommendations about the theory of evolution by natural selection?

Hint 1: How does the limb structure impact your analysis of this evidence of evolution?

Hint 2: Refer to the 'Writing evidence-based essays' section on pages 280–81 of the Science how-to.

Aim

To investigate the comparative anatomy of humans, cats, whales and bats

Materials

- images of human, cat, whale and bat appendages (or full skeletons/models if possible)
- coloured pencils

Method

1. Consider the diagrams of the homologous structures of a human, a cat, a whale and a bat in Figure INV2.3Ba.
2. Copy the diagram into your notebook or obtain a photocopy from your teacher.
3. Colour each ulna the same colour on each of the four appendages.
4. Continue to colour the same bones on each appendage in matching colours. Label the bones.
5. In your results section, create a Venn diagram, like the one in Figure INV2.3Bb, that shows the similarities and differences between each species appendage, based on your observations.

Questions

1. Identify bones that are common to all four species.
2. Describe the function of each of the structures shown in the diagram. What do they tell you about the organisms they come from?
3. Based on your analysis of the structures, construct an evidence-based argument for the statement: 'Comparative anatomy provides evidence showing that species must have diverged from a common ancestor.'
4. Analyse how comparative anatomy is used to provide evidence supporting evolution.

Conclusion

Copy and complete:

'The results show that: *(respond to the aim)*.'

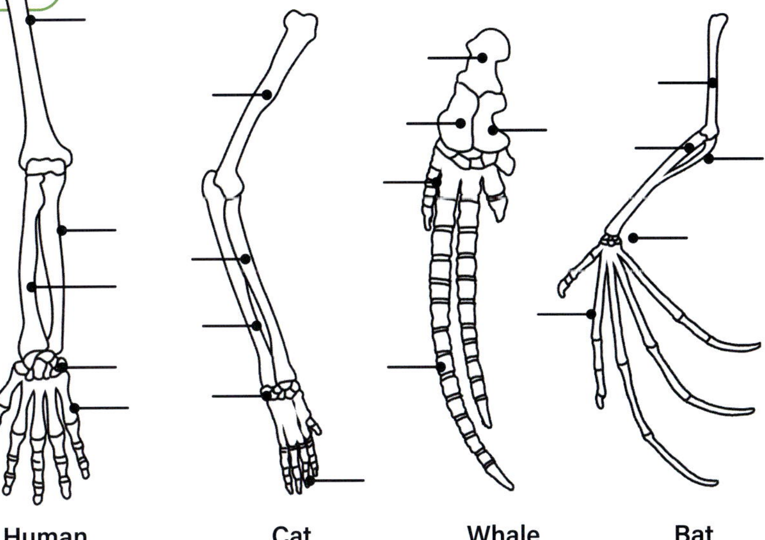

Figure INV2.3Ba: The structure of a human arm, cat leg, whale fin and bat wing

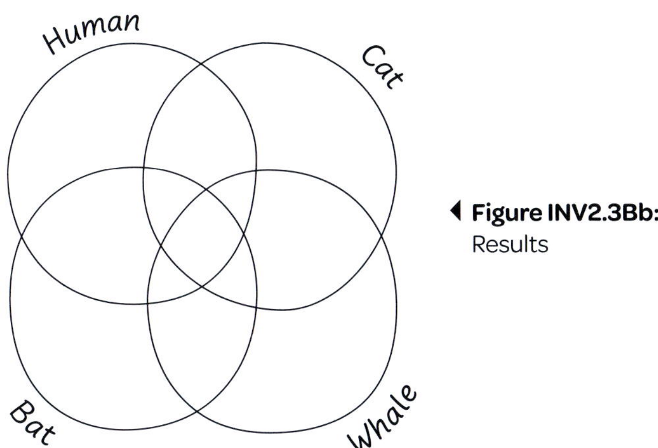

Figure INV2.3Bb: Results

Investigation 3.2

Investigating the reactivity of metals

HYDROCHLORIC ACID IS A HAZARD. WEAR SAFETY GLASSES AND GLOVES DURING THIS INVESTIGATION TO PREVENT ACID FROM BURNING YOUR SKIN OR SPLASHING IN YOUR EYES. DISPOSE OF ALL LIQUIDS AS DIRECTED BY YOUR TEACHER.

Inquiry skill: Processing, modelling and analysing

Inquiry skill focus: Organising data

When recording data and observations that are generated during a first-hand investigation, it is important to use a suitable table to keep the information organised. Before you start your investigation, design a table with appropriate columns and rows to record your data and observations.

Hint 1: Use a ruler to draw your table so that it is neat, clear and easy to read.

Hint 2: Label the rows and columns with headings, including units for quantitative measurements.

Hint 3: Refer to 'Processing, modelling and analysing' in the Science how-to section on pages 244–45.

Background information

In this investigation, we will be testing the reactivity of different metals in hydrochloric acid. If a reaction occurs, hydrogen gas will be released from the hydrochloric acid, and the chlorine from the acid will combine with the metal to form a metal chloride salt. Below is the word equation for a reaction that may or may not occur between aluminium and hydrochloric acid:

aluminium + hydrochloric acid → hydrogen gas + aluminium chloride

Aim

To investigate the reactivity of different metals – aluminium (Al), magnesium (Mg), zinc (Zn) and iron (Fe) – and to consider the positions of these elements in the periodic table

Materials

- safety glasses
- gloves
- 4 test tubes
- 10 mL measuring cylinder
- sandpaper
- test tube rack
- 40 mL of 1 M hydrochloric acid
- a variety of samples of metals (aluminium, magnesium, zinc, iron)

Method

1 Read the entire method, then construct a results table in your notebook to collect the data from this investigation. Use the checklist (Figure INV3.2) to help you. Make sure you include a column to identify the reactivity of each metal.

Figure INV3.2: Checklist for constructing a scientific table

Scientific table checklist

- ☐ The table has a title.
- ☐ Each column in the table has a heading.
- ☐ Units of measurement are included.
- ☐ The independent variable (the selection of metals to test) is in the left column.

2 Place the test tubes in the test tube rack.

3 Use a 10 mL measuring cylinder to measure 10 mL of 1 M hydrochloric acid into each of four test tubes.

4 Use the sandpaper to clean the surface of each metal sample.

5 Place the aluminium sample into the first test tube. Observe how the metal reacts with the hydrochloric acid. Record your observations in your results table.

6 Repeat step 5 with each of the other metal samples.

7 Consider the position of each metal in the periodic table. Think about this position and how each metal reacted with the acid.

Questions

1 Using the data you have collected and organised in your table, list the metals in order from the most reactive to the least reactive.

2 Do your results match the general trend in the periodic table in relation to the reactivity of metals? Explain your answer.

3 Write a word equation for the chemical reaction that may occur when magnesium is combined with hydrochloric acid.

Conclusion

Copy and complete:

'The results show that: *(respond to the aim)*.'

Investigation 3.3

Part A Part B

Student-designed investigation: properties of common substances

Inquiry skill: Evaluating

Inquiry skill focus: Evaluating the validity and reproducibility of methods

A good investigation needs to collect valid data. Studies are only considered valid if they are based on high-quality data. Data needs to demonstrate a consistent trend across a large sample size and be reproducible. It needs to be collected using specific equipment, using digital tools where necessary, so that the results are precise and increase the accuracy of the investigation. When all of these factors are present, the method of the investigation is considered reproducible.

You will work in groups of two or three to produce a plan to investigate the properties of common substances. You might choose to investigate many properties of one common substance, or one property of many common substances.

Your student-designed investigation must:

- have a clear aim. What are you trying to find out by conducting the investigation?
- be safe. Identify hazards and describe ways to reduce risk and have a clear aim.
- be valid and reproducible. The variables must be clearly identified and controlled.

Hint 1: Consider the sample size and precision of your results.

Hint 2: How could you test solubility? Could the dissolved substances also be tested for things like electrical conductivity?

Hint 3: If you get stuck, refer to 'Evaluating' in the Science how-to section on pages 260–61.

Aim

To produce a plan to investigate the properties of common substances

AN OPEN FLAME IS A HAZARD. BE CAREFUL. IF YOU BURN YOURSELF, TELL YOUR TEACHER IMMEDIATELY AND RUN COLD WATER OVER THE AFFECTED AREA FOR 20 MINUTES.

DISPOSE OF WASTES APPROPRIATELY.

Materials

(Materials will vary for each student group.)

- safety glasses
- laboratory coat
- assorted metallic, ionic and covalent substances:
  - metallic (e.g. copper wire, aluminium foil, magnesium strip, zinc strip, tin foil)
  - ionic (e.g. copper sulfate, sodium chloride, sodium carbonate, calcium carbonate)
  - covalent (e.g. water, sulfur, polystyrene, hard plastic (pen cap vs PVC), sugar, graphite, hydrogen gas (evolves when 0.1 M HCl and magnesium strip are combined), carbon dioxide (evolves when sodium carbonate and vinegar are combined)
- 0.1 M HCl (acid)
- 0.1 M NaOH (base)
- white vinegar
- power supply, electrodes, light bulb, alligator clips
- tap water
- test tube rack
- assorted sizes of beakers, flasks, test tubes and stoppers
- stopper with hole for collection tube
- assorted sizes of measuring cylinders
- glass stirring rod, laboratory spatula, tongs
- thermometer
- watch glass
- electronic balance
- Bunsen burner and matches
- tripod, gauze mat, heat mat
- universal indicator or blue/red litmus paper
- any other materials approved by your teacher

Method

Part A: Producing a plan for the investigation

1 Explore the materials provided by your teacher. Select the substance(s) and property/properties you would like to explore. Get approval from your teacher. Make adjustments as needed.

continues ▶

2 Write a specific aim for your group's investigation.
3 Identify the independent variable, dependent variable and controlled variables. (More information about variables is available in the Science how-to section on pages 235–37.)
4 List the materials and equipment required to address your investigation aim.
5 Identify the hazards associated with the required materials. Describe ways to reduce the risks for each of the hazards you have identified. Include disposal of waste; for example, copper sulfate cannot go down the sink.
6 Write a step-by-step procedure to address the investigation aim that is detailed enough for a classmate to follow.
7 Construct a results table, following the model in Table INV3.3c. Be sure to include units where needed, and indicate if multiple trials are required.
8 Get final approval from your teacher.

Table INV3.3a: Investigation variables

Independent variable (What will you change?)	Dependent variable(s) (What will you measure, and how?)	Controlled variables (What will you keep the same, and how?)

Table INV3.3b: Risk assessment

Hazardous substances and equipment	Describe the hazard	Describe ways to reduce risk	Describe what to do if an accident occurs

Table INV3.3c: Results

	DV (unit)	DV (unit)	DV (unit)
IV			
IV			

Part B: Carrying out the investigation

9 Review your procedure and risk assessment before collecting your materials.
10 Follow your procedure, including safety measures, and record the data in your results table.
11 With support from your teacher, make adjustments to the procedure as needed.
12 Share your results with other groups in the class.
13 Complete the questions below.

Questions

1 a Based on the properties observed, classify the substances that were investigated as metallic, ionic or covalent.
b Conduct a search to determine the true classification of the substances tested. Were there any that did not match your predictions? Propose reasons for any differences.
2 Some tests are not easy to carry out in a school setting. List additional tests that might support the investigation of the materials that were explored.
3 How well were you able to control the controlled variables? Discuss this in relation to the validity of your results.
4 How well were you able to communicate the details of your method steps? Discuss this in relation to the reproducibility of your results. Would a classmate be able to follow your method and generate the same results?
5 For each of the substances you tested, identify links between their properties and how they are commonly used in daily life.

Conclusion

Copy and complete:
'The results show that: *(respond to the aim)*.'

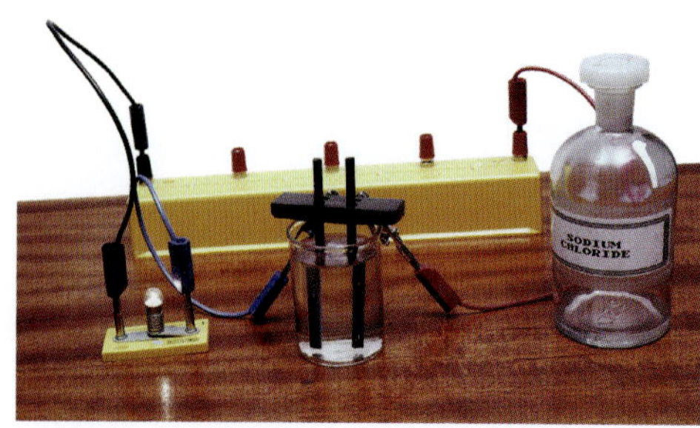

Figure INV3.3: ▶ An electrical circuit can be used to test the conductivity of common substances when they are dissolved in water.

Modelling atomic structure alongside the periodic table

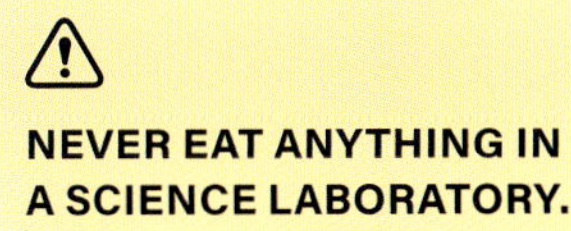

Inquiry skill: Evaluating

Inquiry skill focus: Constructing evidence-based arguments

The results of an investigation can be used as evidence to make claims and construct arguments about scientific concept and skills. How might the results of this investigation be used to make a claim about the relationship between atomic structure and the periodic table?

Hint 1: *How does the arrangement of subatomic particles relate to the organisation of the periodic table?*

Hint 2: *Refer to 'Writing evidence-based essays' in the Science how-to on pages 280–81.*

Aim

To use the periodic table to produce atomic models of the atoms of the first 20 elements of the table and to explore how the periodic table is organised

Materials

- a copy of the periodic table
- 1 large paper plate
- 40 small lollies (20 each of two colours) (or pieces of playdough or plasticine)
- camera

Method

1. Copy the results table (Table INV3.5a on the next page) into your notebook, adding a title and 20 blank rows. Number the rows from 1 to 20 in the 'Atomic number' column.
2. Write the names of the first 20 elements of the periodic table. (Consult the periodic table.)
3. Answer the following questions for each element in the results table. (Consult the periodic table.) Write your answers in the table in your notebook.
   a. Identify the element's symbol.
   b. Identify the number of protons.
   c. Identify the number of electrons.
   d. In the periodic table, which period is the element in?
   e. In the periodic table, which group is the element in?

Note: The number of neutrons in an atom's nucleus cannot be determined from the periodic table, so we will ignore neutrons for the purpose of this activity.

4. On the paper plate, draw five circles (see Figure INV3.5 on the next page). The paper plate represents an atom. The innermost circle represents the nucleus. The second, third, fourth and fifth circles represent electron shells 1, 2, 3 and 4.
5. The lollies represent protons (one colour) and electrons (the other colour). Start with hydrogen. Using your paper plate 'atom' and your lolly 'subatomic particles', create an atomic model of a hydrogen atom. Take a photo. Repeat this process for elements 2 to 20 of the periodic table.

Note: When you are placing electrons in the electron shells, fill the shell closest to the nucleus first (shell 1), then shells 2, 3 and 4. Remember that electrons are distributed across electron shells in a 2, 8, 8, 2 pattern (see page 64).

6. After completing each model, in the results table, record the number of electron shells that have electrons, and the number of electrons in the valence shell.

   For example, a neutral magnesium atom has 12 protons and 12 electrons. Place 12 lollies in the nucleus to represent the protons. Place two electrons in electron shell 1; place eight electrons in electron shell 2; then place the remaining two electrons in electron shell 3. This means that a magnesium atom has three electron shells and two electrons in its valence shell (the shell furthest from the nucleus).

continues ▶

Table INV3.5a: Results

Atomic number	Element name	Symbol	Number of protons in an atom	Number of electrons in an atom	Period of the periodic table	Group of the periodic table	Number of electron shells that have electrons	Number of valence electrons
1								
2								
3								

Questions

1 a Identify any trends or patterns in your results table.

b Link your observations to the organisation of the periodic table.

2 Propose which part of atomic structure is responsible for the properties of an element. Provide a reason for your response.

3 Trends and patterns can be used to make predictions about elements after calcium, because their electron configurations become more complicated.

a Copy Table INV3.5b into your notebook. Without drawing the atoms, use the periodic table to complete the table.

b Provide a general reason for your responses.

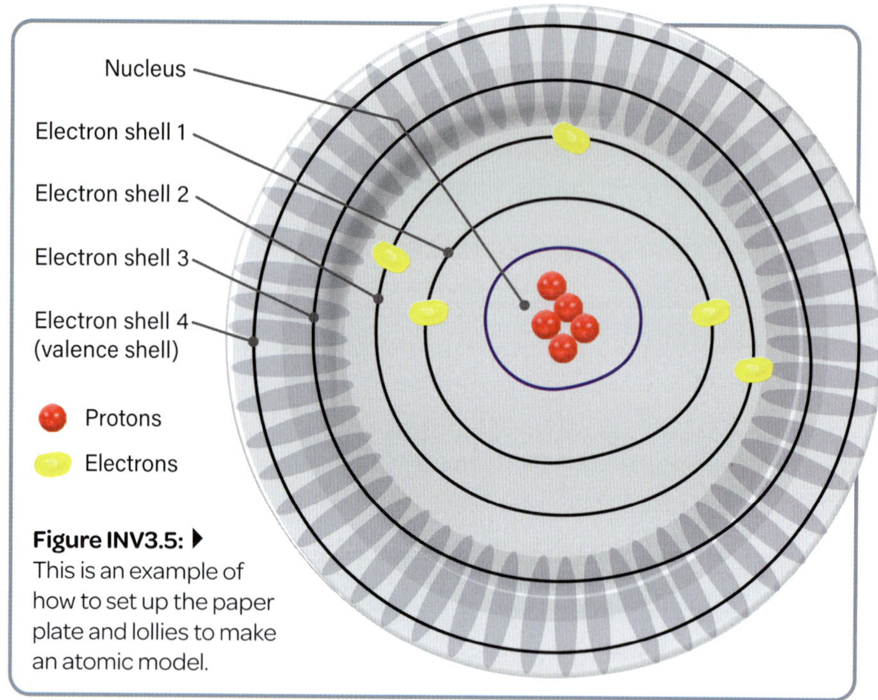

Figure INV3.5: ▶ This is an example of how to set up the paper plate and lollies to make an atomic model.

Table INV3.5b: Question 3a answers

Electron	Predict the number of valence electrons	Predict the number of electron shells
Iodine		
Krypton		
Tin		
Barium		
Potassium		

4 a Arrange the atomic model photos you took of the first 20 elements of the periodic table into a digital table structured like the periodic table.

b Label the groups and periods.

c Highlight the valence electrons and shell patterns that you observe.

Conclusion

Copy and complete:
'The results show that: *(respond to the aim)*.'

Investigation 4.2A

Synthesis of iron oxide

WEAR SAFETY GLASSES AND GLOVES WHILE DOING THIS INVESTIGATION TO PREVENT RUSTED METAL FROM IRRITATING YOUR SKIN OR SERIOUSLY DAMAGING YOUR EYES.

Observe test tubes every second day

Inquiry skill: Planning and conducting

Inquiry skill focus: Describing ways to minimise risk

When conducting any first-hand investigation, it is important to use safe practices to minimise the risk of harm. Before you begin, potential hazards should be identified and strategies put in place to minimise the chance of accidents or harm in the laboratory.

Hint 1: What materials in this investigation could cause harm if used incorrectly?

Hint 2: What strategies minimise the chance of harm occurring?

Hint 3: More information about assessing risks in first-hand investigations is available in the Science how-to section on page 240.

Aim

To investigate the conditions required for the rusting of iron

Materials

- safety glasses
- nitrile gloves
- 5 test tubes
- test tube rack
- masking tape
- marker
- laboratory spatula
- drying agent such as calcium hydroxide
- 2 × 10 mL measuring cylinders
- tap water
- salt water
- boiled water
- 5 ungalvanised iron nails
- electronic balance
- oil
- 5 rubber stoppers
- paper towel

Method

1 Copy the results table from the next page into your notebook, adding a title.

2 Read all the instructions before you begin this investigation.

3 Set five test tubes in a test tube rack.

4 Label the first test tube 'Air only'.

5 Label the second test tube 'Drying agent'. Add one spatula of calcium hydroxide.

6 Label the third test tube 'Water and air'. Add 10 mL of tap water (or enough to cover a nail).

7 Label the fourth test tube 'Salt water'. Add 10 mL of salt water (or enough to cover a nail).

8 Label the fifth test tube 'Oil and boiled water'. Add 10 mL of boiled water (or enough to cover a nail).

9 Weigh the five nails separately. Note each mass and add a nail to each test tube, ensuring that each is covered by water or the drying agent.

10 Add oil to cover the boiled water in test tube 5.

11 Stopper all test tubes with the rubber stoppers.

12 Observe the test tubes every second day for a week.

13 Remove the nails, clean and dry them, and weigh them.

14 Record all your results in the results table.

Questions

1 With reference to data, rank the nails from the least rusted to the most rusted.

2 Predict whether there should have been a gain or a loss in the mass of the nails. Explain.

3 a Identify which environment causes the most rust.

b Research why this environment causes the most rusting of iron.

4 Propose why it was important to measure the mass of the iron nail before and after the investigation.

5 Explain how a nail rusting is an example of a synthesis reaction. Provide a word equation or chemical equation to support your response.

continues ▶

Conclusion

Copy and complete:

'The results show that: *(respond to the aim)*.'

Table INV4.2A: Results

	Test tube				
	1 Air only	2 Drying agent	3 Water and air	4 Salt water	5 Oil and boiled water
Initial mass of nail (g)					
Final mass of nail (g)					
Change in mass (g)					
Observations					

Investigation 4.2B

Decomposition of copper(II) carbonate

Inquiry skill: Planning and conducting

Inquiry skill focus: Generating and recording data

A good investigation involves collecting accurate and precise data. The data collection process must ensure consistency with carefully selected equipment used to take measurements. Precision also comes from correct use of equipment and data tables.

Hint 1: Consider the equipment for this investigation and how you will use it to make accurate measurements.

Hint 2: Ensure the method will result in the collection and recording of precise data.

Hint 3: Refer to the Science how-to section on pages 240–41.

Aim

To observe the decomposition of a metal carbonate

Materials

- retort stand
- bosshead and test tube clamp
- Bunsen burner
- heat mat
- 3 g copper(II) carbonate
- electronic scales
- 2 large test tubes
- test tube rack
- stopper with delivery tube
- 25 mL measuring cylinder
- 15 mL limewater (calcium hydroxide)
- black paper or card
- matches

⚠

AN OPEN FLAME IS A HAZARD. BE CAREFUL. IF YOU BURN YOURSELF, TELL YOUR TEACHER IMMEDIATELY AND RUN COLD WATER OVER THE AFFECTED AREA FOR 20 MINUTES.

DISPOSE OF WASTES APPROPRIATELY.

Method

1. Copy the results table from the next page into your workbook, adding a title and rows as needed.
2. Set up the apparatus as shown in Figure INV4.2B.
3. Weigh about 3 g of copper(II) carbonate into test tube A. Record the mass of the test tube and copper(II) carbonate.
4. Return test tube A to the retort stand, adjusting the clamp so the test tube is at an angle. Seal test tube A with a stopper that has an attached delivery tube.
5. Measure about 15 mL of limewater into test tube B and place it in the test tube rack. Place the black paper behind test tube B. (This will help you to observe any changes during the reaction.)
6. Submerge the end of the long tube in the limewater, so it is close to the bottom of test tube B.
7. Light the Bunsen burner and position the blue flame under the copper carbonate. Record your observations.

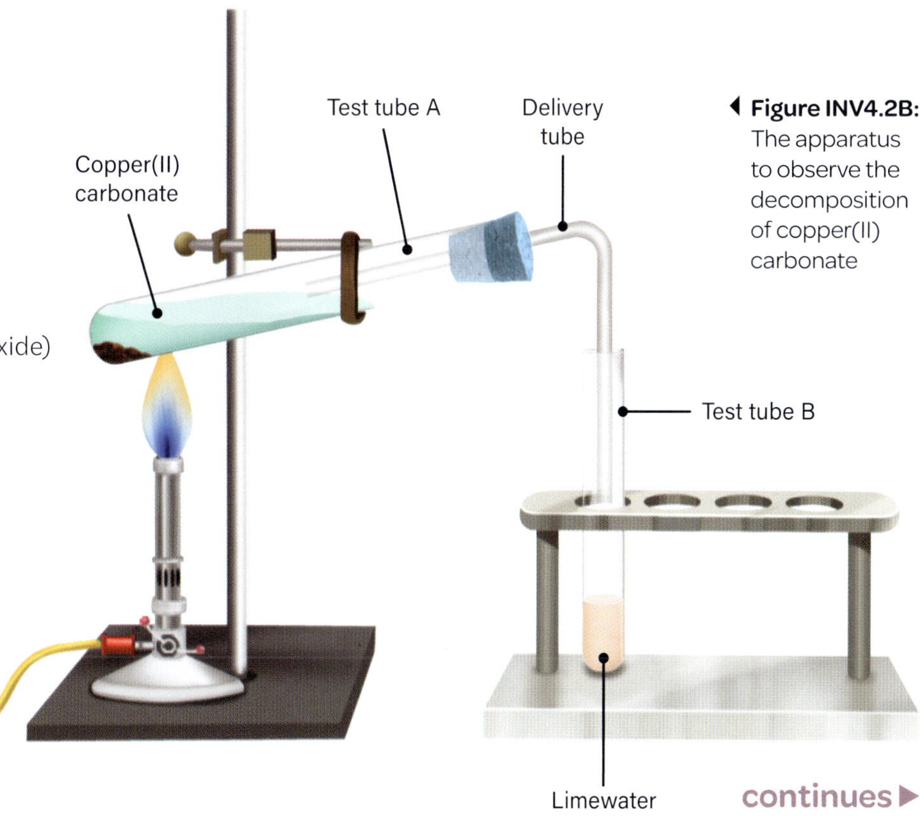

Figure INV4.2B: The apparatus to observe the decomposition of copper(II) carbonate

continues ▶

8 After 1–2 minutes, or when the reaction appears to be complete, turn off the Bunsen burner and remove the tube from the limewater.

9 Once cooled, reweigh test tube A, including its contents.

10 Share your results and observations with other groups in the class.

Questions

1 State the observations that indicate a chemical reaction is taking place.

2 Summarise the observations that indicate this is a decomposition reaction.

3 Describe the purpose of bubbling the gaseous product through the limewater.

4 a Propose the purpose of the flame in this investigation.

b Predict whether the reaction will occur without the flame. Provide a reason.

5 The limewater turns white because calcium carbonate (an insoluble salt) and water form when carbon dioxide reacts with calcium hydroxide. Write a word equation to represent this reaction.

6 **(Extension)**

a Write a skeleton chemical equation for this reaction.

b Write a complete and balanced chemical equation to represent this reaction. (*Hint:* You may have to refresh your skills from previous sections.)

Conclusion

Copy and complete:
'The results show that: *(respond to the aim)*.'

Table INV4.2B: Results

Groups	Mass of test tube A and copper carbonate (g)		Additional (dot-point) observations (e.g. colour change, limewater, change in mass)
	Before	After	
Group 1			

Investigation 4.3A

Observation

Growing metallic crystals: metal displacement reactions

Inquiry skill: Questioning and predicting

Inquiry skill focus: Making predictions

A key component of first-hand investigations is making a prediction of the outcome of the investigation. You may not know much about the concepts under investigation yet, so sometimes it is okay to make a prediction based on your own experiences and prior observations. Answer Question 1 before you start this investigation.

Hint 1: Think about how the activity series can be used to inform your prediction.

Hint 2: More information about making predictions is available in the Science how-to section on page 234.

WEAR SAFETY GLASSES, NITRILE GLOVES AND A LABORATORY COAT.

$AgNO_3$ IRRITATES THE SKIN AND EYES, STAINS SKIN AND CLOTHING AND IS AN ENVIRONMENTAL HAZARD.

$SnCL_2$ IRRITATES THE SKIN AND EYES AND IS AN ENVIRONMENTAL HAZARD.

$CuSO_4$ IRRITATES THE SKIN AND EYES AND IS AN ENVIRONMENTAL HAZARD.

DISPOSE OF ALL WASTES APPROPRIATELY. (YOUR TEACHER WILL PROVIDE WASTE DISPOSAL INSTRUCTIONS.)

Background information

Some metals are more reactive than others. When a piece of solid metal is placed in a solution of less reactive metallic ions, the solid metal reacts to form dissolved ions and the ions react to form solid metallic crystals. The following is an example of a single-displacement reaction.

Magnesium (s) + aluminium chloride (aq) → magnesium chloride (aq) + aluminium (s)

Aim

To investigate single-displacement reactions by growing metallic crystals from ionic compounds

Materials

- safety glasses
- nitrile gloves
- laboratory coat
- 10 mL measuring cylinder
- 10 mL of your allocated 0.1 M ionic solution of silver nitrate, tin chloride or copper sulfate
- 30 mL fresh, warm agar solution (*Tech note:* 1.6 g agar/100 mL water.)
- petri dish and lid
- 1 cm × 4 cm strip of clean zinc, copper, magnesium or aluminium

Method

1. Copy the results table from the next page into your notebook, adding a title and rows as needed.
2. Your group will be allocated one ionic solution and one strip of metal. Write these in your results table. Predict whether a reaction will occur.
3. Measure 10 mL of your allocated ionic solution into a 10 mL measuring cylinder.
4. Measure 30 mL of warm agar solution into a petri dish.
5. Immediately pour 10 mL of the ionic solution gently into the petri dish.
6. Quickly place a clean strip of metal into the centre of the petri dish, submerging it below the surface of the agar solution.
7. Place the lid on the petri dish. **Do not move the petri dish until the agar sets.**
8. Observe the formation of metal crystals over the next 1–2 days. Record whether a reaction occurred.
9. Share your results with the class and record their results.

continues ▶

Questions

1 In one sentence, describe each chemical reaction taking place in the petri dishes.

2 Propose why the crystals were grown in agar rather than in the ionic solution.

3 Explain any colour changes observed.

4 a Construct a word equation for each displacement reaction that occurred.

b **(Extension)** Construct a chemical equation for each of the word equations written in part a.

5 Evaluate your prediction from step 2 of the method.

6 Identify any unexpected results when compared to the activity series of metals. Propose reasons for why your results may not align with the expected outcome.

Conclusion

Copy and complete:
'The results show that: *(respond to the aim)*.'

Table INV4.3A: Results

Metal strip	Ionic solution in agar	Reaction/ No reaction

Investigation 4.3B

Precipitation reactions

Inquiry skill: Questioning and predicting

Inquiry skill focus: Developing a hypothesis

Part of a scientific investigation includes developing a hypothesis, which is a proposed prediction of the outcome of an investigation intended to address a problem or answer a question. To develop a hypothesis, you must first identify the independent, dependent and controlled variables. The independent variable is the one thing that you purposefully change in the investigation, the dependent variable is what you measure, and the controlled variables are all the things you keep the same throughout the investigation. Develop a hypothesis for this investigation, using the format below.

If ..., then ..., because ...

Hint 1: More information about developing a hypothesis is available in the Science how-to section on page 236.

Hint 2: More information about variables is available in the Science how-to section on pages 235–37.

Hint 3: In your conclusion, state whether your hypothesis was supported (correct) or rejected (incorrect).

⚠

WEAR SAFETY GLASSES, NITRILE GLOVES AND A LABORATORY COAT.

KI IRRITATES THE SKIN AND EYES AND IS AN ENVIRONMENTAL HAZARD.

$AgNO_3$ IRRITATES THE SKIN AND EYES, STAINS SKIN AND CLOTHING AND IS AN ENVIRONMENTAL HAZARD.

$BaCl_2$ IRRITATES THE SKIN AND EYES AND IS AN ENVIRONMENTAL HAZARD.

$CuSO_4$ IRRITATES THE SKIN AND EYES AND IS AN ENVIRONMENTAL HAZARD.

DISPOSE OF ALL WASTES APPROPRIATELY (YOUR TEACHER WILL PROVIDE WASTE DISPOSAL INSTRUCTIONS).

Aim

To find out which reactions form a precipitate

Materials

- safety glasses
- nitrile gloves
- laboratory coat
- spotting tile
- piece of blank paper
- 0.1 M solutions in dropper bottles of sodium sulfate, sodium carbonate, sodium chloride, potassium iodide, silver nitrate, barium chloride, copper sulfate and sodium hydroxide
- paper towel

Method

1. Copy the results table from the next page into your notebook, adding a title.
2. Put the spotting tile on the piece of blank paper. Write the names of the solutions being used down the left-hand side and across the top of the spotting tile, in the same order as your results table. Depending on the size of your spotting tile, you may have to wash it and reuse it to get through all the solutions.

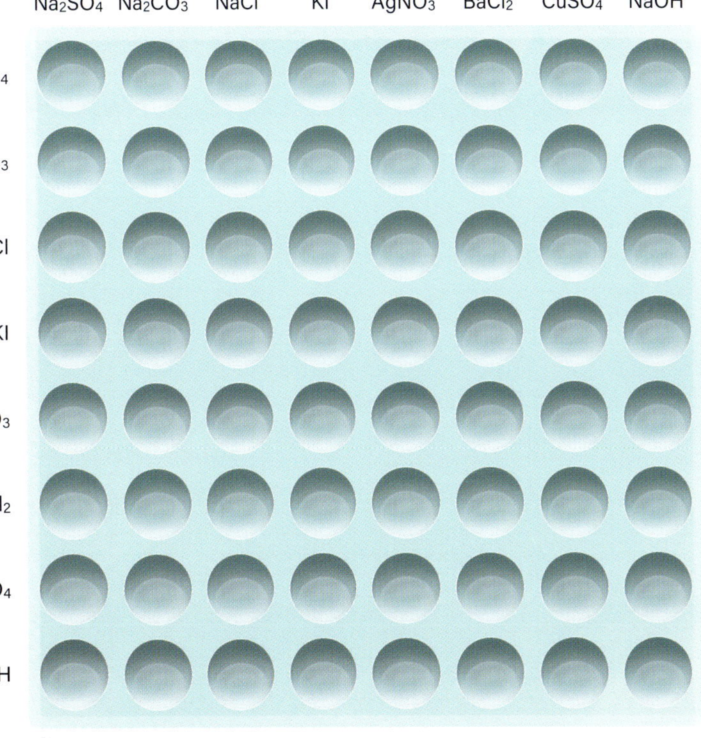

Figure INV4.3B: Spotting tile

continues ▶

3 Into each cavity on the spotting tile, add one drop of the solution written at the top of the column and one drop of the solution written to the left of the row. Observe whether a precipitate forms. Record your results in the table, by using 's' for soluble and 'p' for precipitate. Be careful not to cross-contaminate the solutions.

Questions

1 Propose why you used:
   a dropper bottles of solution instead of solutions from a beaker.
   b spotting tiles and not test tubes to carry out these reactions.

2 Construct a scientific question that would have been suitable for this investigation.

3 a Use your results to predict whether the following salts will be soluble: barium sulfate, silver hydroxide, copper carbonate, silver chloride, lead iodide.
   b Evaluate your predictions using the solubility rules.

4 Explain how precipitation reactions occur.

5 **(Extension)**
   a Construct the net ionic equation for the formation of the silver chloride precipitate.
   b Construct net ionic equations for three more reactions of your choice that formed precipitates.

Conclusion

Copy and complete:
'The results show that: *(respond to the aim)*.'

Table INV4.3B: Results

	Na_2SO_4	Na_2CO_3	NaCl	KI	$AgNO_3$	$BaCl_2$	$CuSO_4$	NaOH
Na_2SO_4								
Na_2CO_3								
NaCl								
KI								
$AgNO_3$								
$BaCl_2$								
$CuSO_4$								
NaOH								

Investigation 4.4A

Neutralising hydrochloric acid

Inquiry skill: Planning and conducting

Inquiry skill focus: Generating and recording data

A good investigation involves collecting accurate and precise data. The data collection process must ensure consistency with carefully selected equipment used to take measurements. Precision also comes from correct use of equipment and data tables.

Hint 1: Consider the equipment for this investigation and how you will use it to make accurate and precise measurements.

Hint 2: Ensure the method will result in the collection and recording of precise data.

Hint 3: Refer to the Science how-to section on page 240.

Aim

To neutralise hydrochloric acid using sodium hydroxide (base) and phenolphthalein indicator

Materials

- safety glasses
- laboratory coat
- 50 mL of 0.1 M HCl
- 10 mL measuring cylinder
- 3 test tubes
- test tube rack
- disposable pipette dropper
- 3 drops of phenolphthalein indicator
- 50 mL of 0.1 M NaOH
- 50–100 mL beaker

Method

WEAR A LABORATORY COAT AND SAFETY GLASSES TO PROTECT YOUR EYES FROM CHEMICALS AND GLASSWARE.

1. Measure 10 mL of 0.1 M HCl into test tube 1. Add 1 drop of phenolphthalein indicator and swirl it. Record your observations in Table INV4.4Aa and set the test tube aside (do not discard it).
2. Measure 10 mL of 0.1 M NaOH into test tube 2. Add 1 drop of phenolphthalein indicator and swirl it. Record your observations in Table INV4.4Aa and set the test tube aside (do not discard it).
3. Carefully add 0.1 M HCl into test tube 3, counting 1 drop at a time until it is about one-quarter full. Aim for each drop to be consistent in size. Record the number of drops added in Table INV4.4Ab.
4. Add 1–2 drops of phenolphthalein indicator to test tube 3 and swirl it.
5. Carefully add 0.1 M NaOH to test tube 3, counting 1 drop at a time, swirling the test tube in between each drop. You should start to see a pink colour persist when you get close to neutralisation.
6. When the entire solution turns light pink for at least 30 seconds before returning to colourless, this is the point of neutralisation (see Figure INV4.4A).
7. Record in Table INV4.4Ab the number of drops of NaOH required to neutralise HCl. Share your results with two other groups in the class.
8. Combine the contents of the test tubes from steps 1 and 2 into a small beaker and swirl. Record your observations in Table INV4.4Aa.

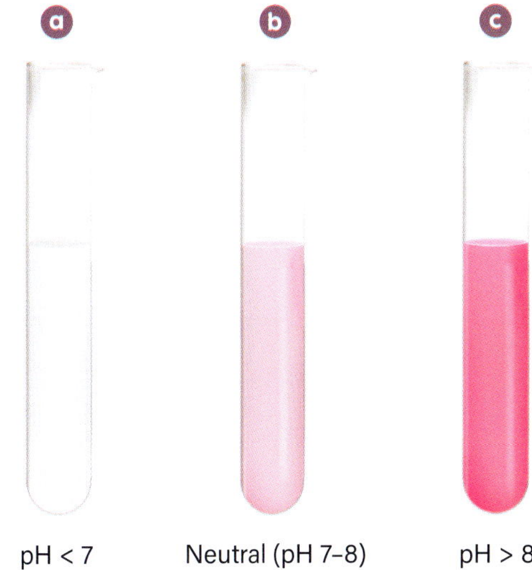

Figure INV4.4A: Phenolphthalein indicator is (a) colourless when acidic and (c) bright pink when basic. (b) A neutral solution will have a very faint pink colour.

continues ▶

Questions

1 a Phenolphthalein changes from colourless to pink at about pH 8. Propose why we still assume the solution is neutralised (pH 7) when the colour is faint pink. (*Hint:* When HCl is neutralised, the pH rapidly changes from 4 to 10 with just 1 drop of NaOH.)
   b When no colour is observed in solution, propose why it is better to call it 'colourless' rather than 'clear'.

2 a Write the general word equation to represent when an acid is neutralised by a base.
   b Represent hydrochloric acid being neutralised by sodium hydroxide using a:
      i word equation.
      ii **(Extension)** chemical equation.

3 a Identify how well you were able to neutralise the HCl by the dropwise method, with reference to results (INV4.4Ab, 'Final colour of solution').
   b Describe how you could improve the accuracy and precision of the results.

4 Did you expect your observations for step 8? Explain with reference to:
   a your data from steps 3 and 7.
   b the accuracy and precision of measurements in steps 1 and 2.

5 Given that the concentrations of the HCl and NaOH are the same for this investigation, comment on the relative strength of the HCl and NaOH. Justify your response with reference to your results.

Conclusion

Copy and complete:
'The results show that: *(respond to the aim).*'

Table INV4.4Aa: Results

Method	Colour of solution	Acidity (acidic, neutral, basic)
Step 1 (HCl only)		
Step 2 (NaOH only)		
Step 8 (HCl and NaOH combined)		

Table INV4.4Ab: Results

	Number of drops of HCl	Number of drops of NaOH	Final colour of solution	Acidity of final solution (acidic, neutral or basic)
Your group's results				
Group 1 results				
Group 2 results				

Investigation 4.4B

The effects of indicators on acids and bases

Inquiry skill: Planning and conducting

Inquiry skill focus: Selecting scientific equipment

Before starting a first-hand investigation, it is important to identify and select equipment that will allow the investigation to be conducted safely and effectively. It is also important that the equipment will allow the precise and accurate generation and collection of data that addresses the investigation aim.

Hint 1: *Is there equipment in this investigation that could be replaced with something safer or more precise?*

Hint 2: *For help, refer to 'Planning and conducting' in the Science how-to section on page 239.*

Aim

To observe the effects of indicators on acids and bases

Materials

- safety glasses
- nitrile gloves
- 4 × 10 mL measuring cylinders
- 4 test tubes
- test tube rack
- 25 mL vinegar
- 25 mL of 0.1 M hydrochloric acid
- 25 mL of 1 M ammonia
- 25 mL of 0.1 M sodium hydroxide
- dropper bottle of universal indicator solution

WEAR SAFETY GLASSES TO PROTECT YOUR EYES FROM CHEMICALS. WEAR GLOVES TO PROTECT YOUR SKIN.

- universal indicator chart
- bromothymol blue
- methyl orange
- phenolphthalein
- litmus paper
- 3 × plastic pipettes

Method

1. Copy the results table from the next page into your notebook, adding a title.
2. Add 5 mL of vinegar, hydrochloric acid, ammonia and sodium hydroxide to each of four test tubes.
3. Add 3 drops of universal indicator to each solution. Use the universal indicator chart to identify each solution as an acid or a base.
4. Remove solutions from test tubes and rinse with water. Refill test tubes with each of the four solutions.
5. Repeat the experiment with the rest of the indicator solutions – bromothymol blue, methyl orange, phenolphthalein, litmus paper (add a small piece of litmus paper to each test tube) – and observe their colours in acidic and in basic solutions.
6. Compare your results with those of other groups.

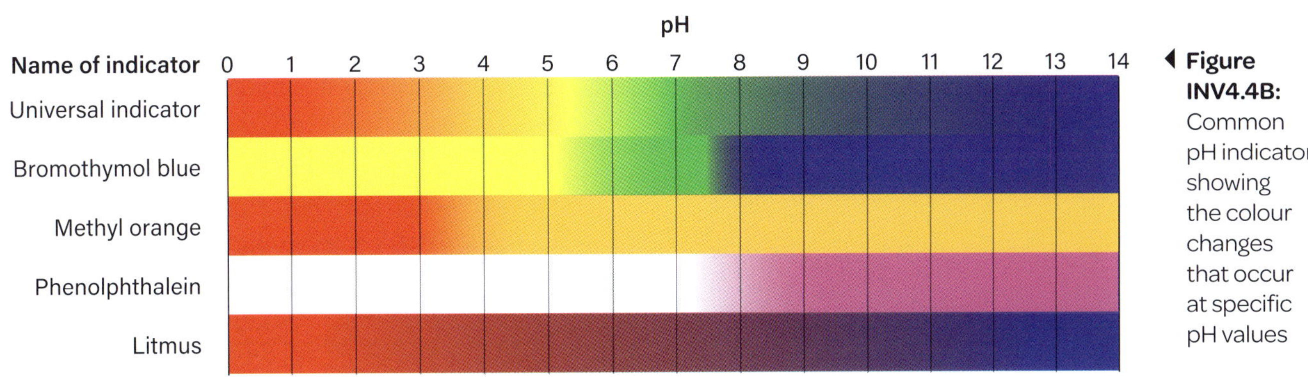

Figure INV4.4B: Common pH indicators showing the colour changes that occur at specific pH values

continues ▶

Questions

1 Identify the substances that are acidic and those that are basic.

2 If lemonade is acidic, propose what colour it would change bromothymol blue to. Provide a reason.

3 Propose the characteristics of indicators that make them useful.

4 Propose whether we would be able to use bromothymol blue to identify whether two unknown substances were acids or bases. Explain your answer.

5 Describe the limitations of using litmus paper as an indicator.

Conclusion

Copy and complete:
'The results show that: *(respond to the aim)*.'

Table INV4.4B: Results

Substance	Acid or base according to indicator				
	Universal indicator	Bromothymol blue	Methyl orange	Phenolphthalein	Litmus paper
Vinegar					
Hydrochloric acid					
Ammonia					
Sodium hydroxide					

Investigation 4.4C

Reactions of acids with metals

Inquiry skill: Planning and conducting

Inquiry skill focus: Developing a risk assessment

In this investigation, it is critical to collect and record your data and observations accurately. Before you start the investigation, consider the measuring increments of the thermometer and practise interpreting the measurement of the room temperature. Consider how you will determine the change in temperature.

Hint 1: Consider how many decimals can be read and recorded from the thermometer you are using.

Hint 2: More information about conducting investigations is available in the Science how-to section on page 240.

ACIDS ARE CORROSIVE. IF YOU BURN YOUR SKIN, RINSE THE AFFECTED AREA UNDER COLD RUNNING WATER FOR 20 MINUTES.

WEAR SAFETY GLASSES TO PROTECT YOUR EYES FROM ACID AND METALS.

WEAR GLOVES.

DISPOSE OF WASTES APPROPRIATELY.

Aim

To investigate the rate of reaction of hydrochloric acid with a range of metals

Materials

- safety glasses
- nitrile gloves
- similar-sized pieces of metals: aluminium, copper, iron, magnesium and zinc
- sandpaper
- 50 mL of 2 M hydrochloric acid
- 10 mL measuring cylinder
- 5 test tubes
- test tube rack
- thermometer
- 5 rubber stoppers
- matches

Method

1 Copy and complete the risk assessment table.

2 Copy the results table from the next page into your notebook, adding a title.

3 Clean all the metals with sandpaper.

4 Add 5 mL of hydrochloric acid to each of the five test tubes.

5 Record the initial temperature of the acid in each test tube.

6 Add aluminium to one test tube and stopper it with a rubber stopper until the reaction stops.

7 Light a match, remove the rubber stopper and hold the match inside the test tube. If a pop sound occurs, it means that the gas produced was hydrogen.

8 Measure the final temperature of the contents of the test tube.

9 Repeat steps 6–8 with the rest of the metals.

Table INV4.4Ca: Risk assessment

Hazardous substances and equipment	Describe the hazard	Describe ways to reduce risk	Describe what to do if an accident occurs	Risk (low, medium, high)

continues ▶

Table INV4.4Cb: Results

Metals	Initial temperature (°C)	Final temperature (°C)	Change in temperature (°C)	Observations
Aluminium				
Copper				
Iron				
Magnesium				
Zinc				

Questions

1 List the controlled variables, independent variable and dependent variable.

2 What can you conclude about the bubbles produced?

3 Comment on the final temperature readings obtained.

4 a Rank the metals from the most reactive to the least reactive.

  b Explain why you ranked the metals in that way.

5 Did all the reactions give a positive pop test? Explain.

Conclusion

Copy and complete:
'The results show that: *(respond to the aim)*.'

Reactions of acids with carbonates

Inquiry skill: Planning and conducting

Inquiry skill focus: Obtaining replicable data

It is important to obtain replicable data to ensure that your results are valid. When completing an investigation, some equipment or methods will be easier to obtain replicable results from than others. Selecting appropriate equipment, using it carefully, and undertaking multiple trials will all help to improve the repeatability of results.

Hint 1: Consider the sources of error in your investigation. If they are making it hard to replicate results, think about how to improve your method.

Hint 2: Try some test measurements before you start collecting data to check the repeatability.

Hint 3: Multiple trials help to reduce random errors. Consider how many trials you need to carry out to ensure your results are valid.

Aim

To observe the reaction between hydrochloric acid and sodium carbonate

Materials

- safety glasses
- 2 test tubes connected by a delivery tube with rubber stoppers
- retort stand
- bosshead and clamp
- 10 mL limewater
- sodium carbonate
- laboratory spatula
- 2 M hydrochloric acid
- 2 × 10 mL measuring cylinders

⚠

HYDROCHLORIC ACID IS CORROSIVE TO EYES AND SKIN AND IS AN IRRITANT. IT MAY CAUSE BURNS. IF SPILLED ON SKIN, FLUSH THE AREA WITH COLD TAP WATER. WEAR SAFETY GLASSES.

DISPOSE OF WASTES APPROPRIATELY.

Method

1 Copy the results table into your notebook, adding a title.

2 Set up the equipment as shown in Figure INV4.4D. Note that the delivery tube must go into the limewater.

3 Add 2 spatulas full of sodium carbonate to the clamped test tube (test tube 1) and have someone hold the test tube containing 10 mL of limewater (test tube 2).

4 Add 5 mL of hydrochloric acid to the test tube containing sodium carbonate and stopper both test tubes. Make sure they are connected by the delivery tube, as shown in the diagram.

5 Observe what happens in both test tubes and record your results.

Table INV4.4D: Results

Test tube	Observations	Colour of limewater	Equation
1			Sodium carbonate + hydrochloric acid → ____________
2			

continues ▶

Figure INV4.4D: Set-up of the investigation

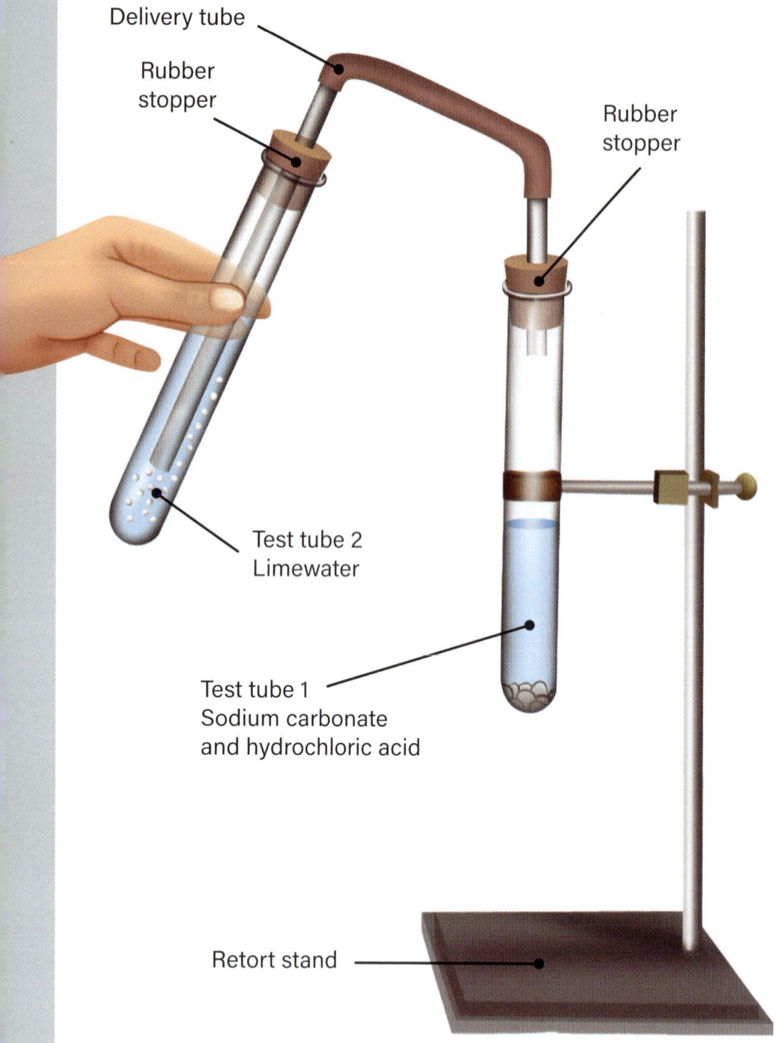

Delivery tube
Rubber stopper
Rubber stopper
Test tube 2 Limewater
Test tube 1 Sodium carbonate and hydrochloric acid
Retort stand

Questions

1 Describe what happened to the test tube containing limewater.

2 An endothermic reaction absorbs heat from the surroundings and the reaction container becomes cold to touch. An exothermic reaction releases heat to the surroundings and the reaction container becomes hot. Do you think this reaction was endothermic or exothermic? Provide a reason to support your response.

3 Based on the reactants, predict the gas that was released during this reaction: oxygen, hydrogen, nitrogen or carbon dioxide? Provide a reason.

4 Complete the sentence:

When ______________ gas is bubbled through limewater, the limewater turns ____________.

5 Complete the general word equation for the reaction observed in this investigation:

Acid + carbonate → salt + _________ + water

6 Write a word equation and a chemical equation to represent the reaction observed in this investigation.

Conclusion

Copy and complete:
'The results show that: *(respond to the aim)*.'

Investigation 4.5A

The effect of concentration on the rate of reaction

Inquiry skill: Questioning and predicting

Inquiry skill focus: Constructing scientific questions

Scientific investigations are based on scientific questions constructed by scientists wanting to learn more about the world. These questions aim to identify a relationship or connection between an independent variable (the factor being changed) and a dependent variable (the factor being measured).

Hint 1: What should a good question look like?

Hint 2: More information about constructing scientific questions is available in the Science how-to section on page 235.

Aim

To investigate how concentration affects the rate of chemical reactions

Materials

- safety glasses
- 5 × 100 mL beakers
- white paper
- thick marker
- 2 × 50 mL measuring cylinders
- distilled water
- 0.15 M sodium thiosulfate
- 2 M hydrochloric acid
- 5 × 10 mL measuring cylinders
- stopwatch

⚠

SODIUM THIOSULFATE IS HARMFUL IF INGESTED. DO NOT INGEST IT. WASH YOUR HANDS AFTER USING IT.

HYDROCHLORIC ACID IS CORROSIVE TO EYES AND SKIN AND IS AN IRRITANT. IT MAY CAUSE BURNS. IF SPILLED ON SKIN, FLUSH THE AREA WITH COLD TAP WATER. WEAR SAFETY GLASSES.

SULFUR DIOXIDE (GENERATED DURING THE INVESTIGATION) IRRITATES THE RESPIRATORY SYSTEM. MAKE SURE THE ROOM IS VENTILATED.

Method

1. Use the aim of the investigation to help you construct a scientific question for this investigation.
2. Copy the results table into your notebook, adding a title.
3. Label the five beakers 1–5.
4. Draw an ‘X’ on the paper about the size of the base of the beakers, as shown in Figure INV4.5A.
5. Use the two 50 mL measuring cylinders to add sodium thiosulfate and distilled water to the beakers in the volumes indicated in the results table.
6. Add 5 mL of hydrochloric acid solution to each 10 mL measuring cylinder and place them next to each beaker.
7. Carefully add the hydrochloric acid to the sodium thiosulfate solution in beaker 1 and immediately start timing. Carefully, yet quickly, swirl the solution and place the beaker over the ‘X’ on the white paper.
8. Stop timing when the X is no longer visible.

Table INV4.5A: Results

Beaker	Volume of sodium thiosulfate (mL)	Volume of water (mL)	Concentration of sodium thiosulfate (M)	Time (s)	$\frac{1}{\text{time}}$ (s^{-1})
1	50	0	0.150		
2	40	10	0.120		
3	30	20	0.090		
4	20	30	0.060		
5	10	40	0.030		

continues ▶

9 Record the reaction time in seconds in the results table.

10 Repeat steps 7–9 with the remaining beakers.

11 Calculate $\frac{1}{\text{reaction time}}$ for each trial.

Questions

1 List the controlled variables, independent variable and dependent variable.

2 With reference to data, answer the scientific question you constructed before you began the investigation.

3 Use collision theory to justify the finding of the investigation.

4 Justify why the total volume of each beaker should be the same.

5 a Plot concentration vs time, and concentration vs $\frac{1}{\text{time}}$, on separate graphs.

b Compare and contrast each graph, indicating which one 'best' represents the findings. Justify your answer with reference to rate of reaction being measured in seconds or 'per second' (s^{-1}).

Conclusion

Copy and complete:

'The results show that: *(respond to the aim).*'

Figure INV4.5A: Time the reaction until the 'X' drawn on the paper beneath the beaker is no longer visible.

Investigation 4.5B

The effect of temperature on the rate of reaction

Inquiry skill: Questioning and predicting

Inquiry skill focus: Developing a hypothesis

Part of a scientific investigation includes developing a hypothesis, which is a proposed prediction of the outcome of an investigation intended to address a problem or answer a question. To develop a hypothesis, you must first identify the independent, dependent and controlled variables. The independent variable is the one thing that you purposefully change in the investigation, the dependent variable is what you measure, and the controlled variables are all the things you keep the same throughout the investigation. Develop a hypothesis for this investigation, using the format below.

If ..., then ..., because ...

Hint 1: In your conclusion, state whether your hypothesis was supported (correct) or rejected (incorrect).

Hint 2: Refer to 'Graphing' in the Science how-to section on page 294.

Aim

To investigate the effect of temperature on the rate of a reaction

Materials

- safety glasses
- distilled water at three different temperatures: cold, room temperature and warm (40 °C)
- 3 × 250 mL beakers
- marker
- thermometer
- 3 Alka-Seltzer tablets
- stopwatch

WEAR SAFETY GLASSES TO PREVENT ANY MATERIAL FROM ENTERING YOUR EYES.

Method

1 Copy the results table into your notebook, adding a title.

2 Measure 200 mL each of warm water, room-temperature water and cold water into three separate labelled beakers.

3 Measure the water temperature in each beaker and record it in the results table.

4 Drop an Alka-Seltzer tablet into each beaker. Use the stopwatch to immediately start timing how long it takes for each tablet to dissolve.

5 Stop the timer when the final tablet has dissolved completely.

6 Record the reaction times (in minutes and seconds) in the results table.

Questions

1 Construct a graph to show the reaction time (seconds) on the *y*-axis against the water temperature (°C) on the *x*-axis.

2 List the controlled variables, independent variable and dependent variable.

3 How does reaction time change with temperature? Do the results support your hypothesis? Why or why not?

4 Use collision theory to justify the finding of the investigation.

5 a If some tablets used are whole and some are broken, propose whether the temperature experiment would still be valid. Explain your answer.

b If some students stirred the water while their tablets were dissolving, propose whether the temperature experiment would still be valid. Explain your answer.

6 If everyone in the class used a different water temperature, propose whether you could average the results. Explain your answer.

Conclusion

Copy and complete:

'The results show that: *(respond to the aim)*.'

Table INV4.5B: Results

Type of water	Temperature (°C)	Time to dissolve (s)
Cold		
Room temperature		
Warm		

Investigation 4.6A

The effect of surface area on the rate of reaction

Inquiry skill: Planning and conducting

Inquiry skill focus: Developing a risk assessment

The level of risk should always be assessed before conducting investigations. Completing a risk assessment allows you to identify hazards, assess their level of risk and establish safe practices to minimise them. A hazard is anything that could harm the person conducting or observing the investigation. The level of risk can be low, medium or high, depending on the likelihood that an accident will occur and how severe the consequences would be. For each hazard, strategies should be put in place to minimise the risks.

Hint 1: What are the hazards for this investigation?

Hint 2: How will you minimise the likelihood that hazards will cause harm or their severity? What will you do if an accident occurs?

Hint 3: Refer to step 2 of 'Planning and conducting' in the Science how-to on page 240.

Hint 4: Refer to 'Graphing' in the Science how-to section on page 294.

⚠ **WEAR SAFETY GLASSES WHILE DOING THIS INVESTIGATION TO PREVENT ANY MATERIALS FROM ENTERING YOUR EYES.**

Aim

To investigate the effect of increasing the surface area on the rate of a reaction

Materials

- safety glasses
- 50 mL measuring cylinder
- 4 Alka-Seltzer tablets
- mortar and pestle
- tap water
- 4 × 100 mL beakers
- disposable knife
- stopwatch

Method

1 Copy and complete the risk assessment table.

2 Copy the results table from the next page into your notebook, adding a title.

3 Add 50 mL of water to each of the four beakers.

4 Of the four Alka-Seltzer tablets:
   - leave the first one whole
   - cut the second one in half
   - cut the third one into quarters
   - crush the fourth one with the mortar and pestle.

5 Add the whole tablet to the first beaker of water and time how long it takes to dissolve.

6 Add the pieces of the second tablet to the second beaker of water and time how long they take to dissolve.

7 Add the pieces of the third tablet to the third beaker of water and time how long they take to dissolve.

8 Add the crushed tablet to the fourth beaker of water and time how long it takes to dissolve.

9 Make your observations and record your results.

Table INV4.6Aa: Risk assessment

Hazardous substances and equipment	Describe the hazard	Describe ways to reduce risk	Describe what to do if an accident occurs	Risk (low, medium, high)

Questions

1 List the controlled variables, independent variable and dependent variable.

2 Do you think your experiment was reproducible? Explain your answer.

3 Construct a column graph to show your results.

4 Explain what you think the rate of reaction depended on. In your response, refer to collision theory.

5 Explain how you changed the surface area of the tablets.

Table INV4.6Ab: Results

Size of Alka-Seltzer tablet	Time taken for the tablet to dissolve (s)
Whole	
Halved	
Quartered	
Crushed	

Conclusion

Copy and complete:
'The results show that: *(respond to the aim)*.'

Investigation 4.6B

The effect of a catalyst on the rate of reaction

Inquiry skill: Planning and conducting

Inquiry skill focus: Obtaining replicable data

It is important to obtain replicable data to ensure that your results are valid. When completing an investigation, some equipment or methods will be easier to obtain replicable results from than others. Selecting appropriate equipment, using it carefully, and undertaking multiple trials will all help to improve the replicability of results.

Hint 1: *Consider the sources of error in your investigation. If they are making it hard to replicate results, think about how to improve your method.*

Hint 2: *Try some test measurements before you start collecting data, to check the repeatability.*

Hint 3: *Multiple trials help to reduce random errors. Consider how many trials you need to conduct to ensure your results are valid.*

HYDROGEN PEROXIDE CAN IRRITATE THE SKIN. WEAR GLOVES AND WASH YOUR HANDS IMMEDIATELY IF HYDROGEN PEROXIDE GETS ONTO THEM.

MANGANESE DIOXIDE CAN IRRITATE IF INHALED, SO AVOID SPILLING IT.

DISPOSE OF WASTES APPROPRIATELY.

Aim

To investigate the effect of a catalyst on a chemical reaction

Materials

- safety glasses
- nitrile gloves
- retort stand, bosshead and clamp
- trough of water
- conical flask with a delivery tube attached with a rubber stopper
- 50 mL and 100 mL measuring cylinders
- 50 mL hydrogen peroxide (6% strength)
- tap water
- electronic balance
- laboratory spatula
- 0.5 g manganese dioxide
- powder funnel
- stopwatch

Method

1. Copy the results table from the next page into your notebook, adding a title.
2. Set up the apparatus as shown in Figure INV4.6B.
3. Using a 50 mL measuring cylinder, add 50 mL of hydrogen peroxide solution to a conical flask.
4. Fill the 100 mL measuring cylinder with water, seal the top with your fingers, and invert it into the water trough, as shown in Figure INV4.6B.
5. Put the rubber stopper loosely into the conical flask and make sure the delivery tube connects to the inverted measuring cylinder.
6. Measure 0.5 g of the catalyst manganese dioxide.
7. Add the catalyst to the conical flask, put the stopper back into the flask and start the stopwatch.
8. Record the volume of gas given off every 10 seconds. Continue timing until no more oxygen appears to be given off. Record all your observations.

Questions

1. Identify the catalyst in this reaction.
2. Describe how the catalyst affects the rate of decomposition of hydrogen peroxide, making reference to the results.
3. Construct a graph to display your results, with time on the *x*-axis and volume of oxygen on the *y*-axis. Be sure to include units.
4. **a** Identify the trends you see on the graph.
   b Describe the relationship between the independent and dependent variables.
5. Explain the role of the catalyst in this investigation.

Conclusion

Copy and complete:

'The results show that: *(respond to the aim)*.'

Table INV4.6B: Results

Time (s)	Volume of $O_2(g)$ with catalyst (mL)	Volume of $O_2(g)$ without catalyst (mL)
10		
20		
30		
40		
50		

Figure INV4.6B: Set-up of the investigation

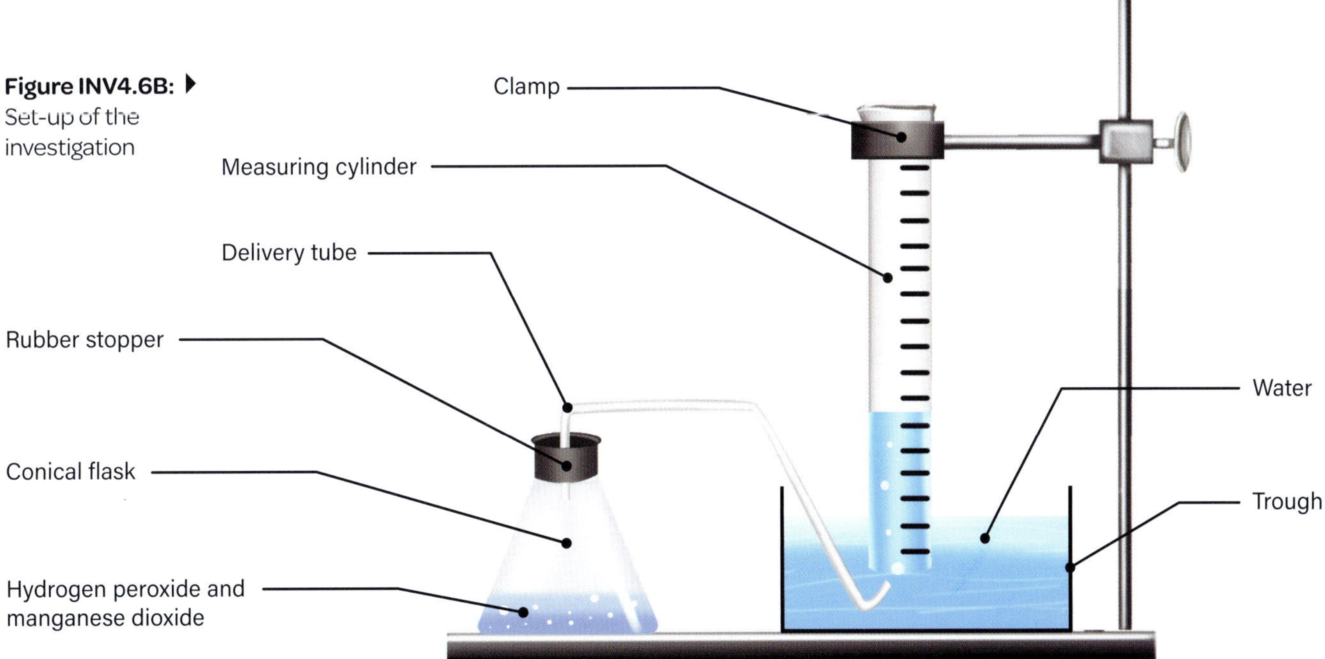

Investigation 4.6C

Student-designed investigation: investigating the rate of chemical reactions

Inquiry skill: Planning and conducting

Inquiry skill focus: Designing scientific investigations

You will work in groups of two or three to produce a plan to investigate how factors affect rates of chemical reaction.

Your student-designed investigation must:

- address a scientific question. What are you trying to find out by conducting the investigation?
- be safe. Identify hazards and describe ways to reduce risk.
- be valid. Variables must be clearly identified and controlled.

Hint 1: *Select a chemical reaction to focus your investigation. It might be one that you have already investigated as part of another investigation, or it could be a new reaction, with teacher approval.*

Hint 2: *Select **one** factor to explore that affects rate of reactions. All other factors should be controlled.*

Hint 3: *Select a dependent variable (e.g. time, mass, volume) and determine a suitable method to measure the dependent variable.*

Hint 4: *Refer to 'Planning and conducting' in the Science how-to section on pages 241–42.*

Aim

To produce a plan to investigate a factor that affects the rate of chemical reactions

Materials

(Will vary for each student group)

- safety glasses
- laboratory coat
- assorted substances used in recent investigations:
  - acids and bases: HCl, H_2SO_4, vinegar, NaOH, ammonia
  - metals: copper, magnesium, zinc, aluminium, tin, steel wool
  - carbonates: copper carbonate, sodium carbonate, calcium carbonate

AN OPEN FLAME IS A HAZARD. BE CAREFUL. IF YOU BURN YOURSELF, TELL YOUR TEACHER IMMEDIATELY AND RUN COLD WATER OVER THE AFFECTED AREA FOR 20 MINUTES. DISPOSE OF WASTES APPROPRIATELY.

  - catalysts: light source, MnO_2, natural enzymes (e.g. liver, potato)
  - miscellaneous: limewater, sodium thiosulfate, Alka-Seltzer tablets, hydrogen peroxide
- assorted equipment used in recent investigations: measuring cylinder, beaker, test tube and test tube rack, flask, pipette dropper, trough of water, retort stand, bosshead and clamps, stopper with delivery tube, stopwatch, electronic balance, stirring rod, tongs, thermometer, Bunsen burner and heat mat

Method

Part A: Producing a plan for the student-designed investigation

1. Explore the materials provided by your teacher.
   a. Select a chemical reaction to focus your investigation. It might be one that you have already investigated as part of another investigation, or it could be a new reaction.
   b. Select one factor to explore that affects the rate of reaction: concentration, temperature, catalysts, surface area, stirring. All other factors should be controlled.
   c. Select a dependent variable (e.g. mass, volume, time), and determine a suitable method to measure the independent variable. The method might be similar to one you are already familiar with from previous investigations.
2. Propose your idea to your teacher to get preliminary approval. Make adjustments as needed.
3. In your group, write a scientific question for your investigation. (More information about how to write a scientific question is available in the Science how-to section on page 235.)

Part A

Part B

4 Identify the independent variable, dependent variable and controlled variables.

5 List the materials and equipment required to address your investigation aim.

6 Complete a risk assessment for your investigation: Identify the hazards associated with the required materials. Describe ways to reduce the risks for each of the hazards you have identified. Include disposal of waste (e.g. sodium thiosulfate cannot go down the sink).

7 Write a step-by-step procedure to address the investigation aim, detailed enough for a classmate to follow.

8 Construct a results table. Remember that the independent variable (IV) goes in the left column and the dependent variable (DV) headings go across the top row. Be sure to include units where needed and indicate if multiple trials are required.

9 Get final approval from your teacher.

10 Construct a hypothesis based on scientific knowledge.

Table INV4.6Ca: Investigation reaction and variables

Word equation for the reaction under investigation:	**Reactants → Products** →	
Independent variable (what you will change)	**Dependent variable(s) (what you will measure and how)**	**Controlled variables (what you will keep the same and how)**

Table INV4.6Cb: Risk assessment

Hazardous substances and equipment	**Describe the hazard**	**Describe ways to reduce risk**	**Describe what to do if an accident occurs**

Table INV4.6Cc: Results template

	DV (unit)	**DV (unit)**	**DV (unit)**
IV			
IV			

Part B: Carry out the investigation

11 Review your procedure and risk assessment before collecting your materials.

12 Follow your procedure, including safety measures, and record the data in your results table.

13 With support from your teacher, adjust the procedure as needed.

Questions

1 Determine whether the data collected supports or refutes your hypothesis.

2 Justify your findings with reference to collision theory.

3 How well were you able to control the controlled variables? Discuss this in relation to the validity of your results.

4 Describe an extension investigation that could further support your findings.

Conclusion

Copy and complete:
'The results show that: *(respond to the aim)*.'

Investigation 4.7

Predicting energy changes for exothermic and endothermic reactions

Inquiry skill: Questioning and predicting

Inquiry skill focus: Developing a hypothesis

Part of a scientific investigation includes developing a hypothesis, which is a proposed prediction of the outcome of an investigation intended to address a problem or answer a question. To develop a hypothesis, you must first identify the independent, dependent and controlled variables. The independent variable is the one thing that you purposefully change in the investigation, the dependent variable is what you measure, and the controlled variables are all the things you keep the same throughout the investigation. Develop a hypothesis for this investigation, using the format below.

If ..., then ..., because ...

Hint 1: *You will be changing the type of reaction being measured (exothermic vs endothermic). What will you be measuring?*

Hint 2: *Refer to step 3 of 'Questioning and predicting' in the Science how-to section on page 236.*

HYDROCHLORIC ACID AND SODIUM THIOSULFATE ARE HAZARDS. BE CAREFUL.

WEAR SAFETY GLASSES AND GLOVES. ENSURE ADEQUATE VENTILATION.

DISPOSE OF WASTE AS INSTRUCTED BY YOUR TEACHER. (SOLIDS, COPPER AND SODIUM THIOSULFATE MUST NOT GO DOWN THE SINK.)

Aim

To explore energy changes of known exothermic and endothermic reactions

Figure INV4.7: Predict whether the temperature of your test tube will increase or decrease for each of your reactions.

Materials

For each group:

- 1 x test tube rack
- 6 large test tubes
- markers (to label test tubes)
- 10 mL measuring cylinder
- 10 mL of 1 M hydrochloric acid (HCl)
- thermometer
- laboratory spatula
- stopwatch
- half a spatula (about 2 g) of magnesium turnings
- 4 × 10 mL of deionised or distilled water
- half a spatula (about 2 g) of potassium nitrate
- 10 mL of 1 M copper(II) sulfate solution
- half a spatula (about 2 g) of zinc powder
- half a spatula (about 2 g) of ammonium chloride
- half a spatula (about 2 g) of sodium thiosulphate
- half a spatula (about 2 g) of anhydrous copper(II) sulfate

Method

1. Copy the results table from the next page into your notebook, adding a title.
2. Label the six test tubes A–F.
3. Predict whether the temperature will increase or decrease for each test tube (A–F) and record your prediction in the results table.
4. Measure 10 mL of 1 M HCl into test tube A.
5. Measure and record the initial temperature.

6 Add half a spatula (about 2 grams) of magnesium turnings to test tube A. Start the timer.

7 After 1 minute, record the final temperature.

8 Repeat steps 4–7 for the remaining reactions, according to the reactants given in the results table (B–F) in their respective test tubes. Be sure to rinse or wipe your equipment between reactions to avoid cross-contamination.

Questions

1 a Describe the theory you used to make your predictions.

b Identify whether your predictions were correct.

2 Describe how you implemented safe practices while conducting this investigation.

3 Explain whether using a digital thermometer would impact the accuracy of your results.

4 Draw a series of energy profile diagrams to represent each of the six chemical changes, relative to each other in terms of energy. *Hint:* The change in temperature indicates how much energy was absorbed or lost overall.

5 a Predict the change temperature value for the following combinations of reactants:

i 10 mL of 1 M HCl and a whole spatula of magnesium turnings

ii 5 mL of $Cu(SO_4)$ and a whole spatula of zinc powder

iii 5 mL water and half a spatula of sodium thiosulfate

iv 5 mL water and a quarter of a spatula of ammonium chloride

b Justify your reasoning for your answers to part a.

Conclusion

Copy and complete:
'The results show that: *(respond to the aim).*'

Table INV4.7: Results

Reaction type (exothermic or endothermic)	**Test tube**	**Reactant 1** (10 mL of solution)	**Reactant 2** (half a spatula of solid)	**Predicted change in temp.** (increase or decrease)	**Initial temp. (°C)**	**Final temp. (°C)**	**Change in temp. (°C) = T(final) – T(initial)**
Exo	A	HCl	magnesium turnings				
Endo	B	water	potassium nitrate				
Exo	C	1 M $Cu(SO_4)$	zinc powder				
Endo	D	water	ammonium chloride				
Endo	E	water	sodium thiosulfate				
Exo	F	water	anhydrous copper(II) sulfate				

Investigation 5.2

The albedo effect

Inquiry skill: Evaluating

Inquiry skill focus: Describing the impact of assumptions and errors

When assumptions and errors in an investigation are not considered or controlled, they can impact on the data that is gathered. For example, they can cause values to be higher or lower than expected. Being able to describe how an assumption or error impacts the data can assist in improving the method and in evaluating the validity of the results.

Hint 1: *Identify assumptions and/or errors in the investigation.*

Hint 2: *Consider how assumptions/errors impact the results. For example: 'The error of … increases/decreases the value of …'*

Hint 3: *Read through the Science how-to section on pages 256–58.*

Aim

To determine how the albedo of the surface affects how it heats and cools

A HEAT LAMP IS A HAZARD. BE CAREFUL. IF YOU BURN YOURSELF, TELL YOUR TEACHER IMMEDIATELY AND RUN COLD WATER OVER THE AFFECTED AREA FOR 20 MINUTES.

Materials

- Blu Tack or plasticine
- 2 × thermometers
- aluminium can, painted black
- aluminium can, painted white
- heat lamp
- ruler
- stopwatch

Method

1. Copy the results table into your notebook.
2. Use the Blu Tack to secure a thermometer in each of the cans so that it does not touch the bottom or sides and no air can escape from the hole at the top.
3. Set up the heat lamp.
4. Use the ruler to place the cans 20 cm from the heat lamp.
5. Record the starting temperature for each of the cans.
6. Turn on the heat lamp and start the stopwatch. Record the temperatures every minute for 10 minutes.
7. At 10 minutes, turn off the heat lamp and remove from the workspace. Continue to record the temperature every minute for a further 10 minutes.
8. Construct an appropriate graph of your results.

Table INV5.2: Results

Time (minutes)		**0**	**1**	**2**	**3**	**4**	**5**	**6**	**7**	**8**	**9**	**10**
Temperature (°C)	Black can											
	White can											
Time (minutes)		**11**	**12**	**13**	**14**	**15**	**16**	**17**	**18**	**19**	**20**	
Temperature (°C)	Black can											
	White can											

Questions

1 Calculate the rate of heating (temperature change/time) for the:
   a black can.
   b white can.

2 Calculate the rate of cooling (temperature change/time) for the:
   a black can.
   b white can.

3 Use the data in your graph and your responses in Questions 1 and 2 to compare the heating and cooling rates of the two cans.

4 Use your data to explain the types of surfaces that you would expect to:
   a absorb more solar radiation.
   b reflect more solar radiation.

5 Relate the results of this investigation to the role the albedo effect plays in Earth's energy budget.

6 Propose what could happen to Earth if light surfaces were replaced by darker surfaces.

7 a Identify any errors or assumptions in the method of this investigation.
   b Suggest how your response to part a would impact the data.
   c Describe how you would modify the method to improve the quality of the data.

Conclusion

Copy and complete:
'The results show that: *(respond to the aim)*.'

Figure INV5.2: How does the albedo of a surface affect how it heats and cools?

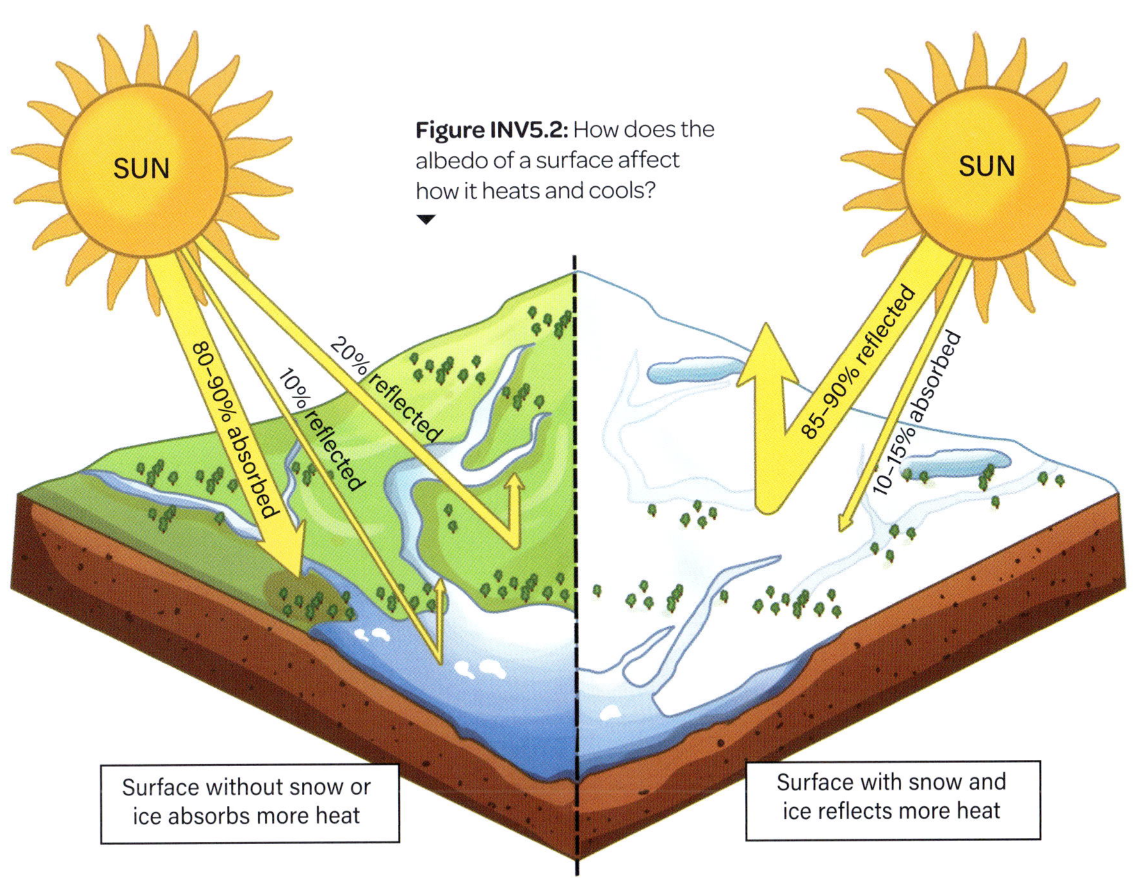

Investigation 5.3

Observing the weather

Inquiry skill: Communicating

Inquiry skill focus: Using digital technologies to organise and present findings

Digital technologies allow us to present scientific findings in many different ways. They can be used to construct tables and graphs, as well as to record images and videos. Think about how you could use digital technologies to organise and present your findings for this investigation.

Hint 1: *Think about the observations you will be making and the results you will be gathering.*

Hint 2: *Think about how digital technologies could help you to display your observations and results. For example, you could construct a digital results table on a computer.*

Aim

To make observations over a period of one week and relate these to the weather observed

Materials

For entire group:

- rain gauge
- ruler
- thermometer
- hygrometer
- barometer
- anemometer

Method

1 Determine where you will take your readings over the next week.

2 The day before you begin, make sure the rain gauge is set up according to the instructions.

3 Copy the results table into your notebook, including space for the weather measurements you will make.

4 Temperature: Use the ruler to measure 1 m above the ground. Hold the thermometer at this point until the reading settles. Record the temperature in the results table.

5 Humidity: Set up the hygrometer according to the instructions. Hold it, or set it up, so it hangs 1 m above the ground. Record the humidity in the results table.

6 Air pressure: Set up the barometer according to the instructions. Record the air pressure in the results table.

7 Wind speed and direction: Set up the anemometer according to the instructions. Record the wind speed and direction in the results table.

8 Rain: Measure the volume of water in the rain gauge, then dispose of any rain collected to reset the gauge. Record the volume in the results table.

9 Cloud cover: Describe the amount of cloud cover (perhaps as a percentage of the sky covered).

10 Other observations: Record other observations you can make about the weather.

11 Consider the best way to represent your raw data. (More information about drawing graphs is available in the Science how-to sections on pages 294–98.)

Questions

1 Use the Bureau of Meteorology website to find the weather observations for your location over the same period. Compare these with your observations.

2 Evaluate the data you gathered with reference to its validity and reproducibility.

3 Propose how you could improve the data you gathered with reference to the validity of conclusions and claims.

Conclusion

Copy and complete:

'The results show that: *(respond to the aim)*.'

Table INV5.3: Results

Date	Time	Temperature (°C)	Humidity (%)	Air pressure (hPa)	Wind speed (km h^{-1})	Wind direction	Rain in the last 24 hours (mm)	Observations	
								Cloud cover	Other

The enhanced greenhouse effect

A HEAT LAMP IS A HAZARD. BE CAREFUL. IF YOU BURN YOURSELF, TELL YOUR TEACHER IMMEDIATELY AND RUN COLD WATER OVER THE AFFECTED AREA FOR 20 MINUTES.

Inquiry skill: Evaluating

Inquiry skill focus: Proposing ways to improve the quality of data

Poor-quality data can impact the validity of the investigation. Evaluating the quality of the data you have collected allows improvements to be made in the future.

Validity is based on the spread of data; does it show a consistent trend? Does it appear to be random, or is the data too limited to be able to tell?

Hint 1: Refer to the Science how-to section on pages 258–60.

Hint 2: Does the investigation allow for multiple sets of data to be collected?

Hint 3: Is the data accurate?

Aim

To investigate how the presence of carbon dioxide affects air temperature

Materials

- 2 × 400 ml conical flasks
- 2 × thermometers
- 100 mL water
- marker
- stopwatch
- 2 × stoppers with holes for the thermometers
- 2 × 50 mL measuring cylinders
- 100 mL vinegar
- 10 g bicarbonate soda
- heat lamp

Method

1. Carefully push the thermometers through the stoppers so that the bulb of the thermometer will sit in the middle of the conical flask when the stopper is in place.
2. Use the measuring cylinders to add 50 mL of water and 50 mL of vinegar to each conical flask.
3. Place the stopper and thermometer into the first conical flask.
4. Label the first flask 'No CO_2' and the second flask 'CO_2'.
5. In two quick movements, add the 10 g of bicarbonate soda to the second conical flask and put in the stopper. (This reaction will produce carbon dioxide.)
6. Record the temperatures of each flask and start the stopwatch.
7. Place both flasks under the heat lamp or in a sunny spot.
8. Record the temperature of each flask every 5 minutes for half an hour.
9. Switch off the heat lamp or move the flasks into the shade.
10. Record the temperature of each flask every 5 minutes for half an hour.

Questions

1. Identify the dependent, independent and controlled variables in this investigation.
2. Construct an appropriate graph of your data.
3. Describe the trend illustrated by the graph.
4. Compare the two flasks with regards to:
   a. the time spent heating.
   b. the time spent cooling.
5. Explain what the results allow you to infer about the relationship between the presence of carbon dioxide and air temperature.
6. Evaluate the method in relation to its accuracy, validity and repeatability.
7. Propose how you could modify the method to improve the quality of the data gathered.

Conclusion

Copy and complete:

'The results show that: *(respond to the aim).*'

Table INV5.4: Results

Flask		Time (minutes)						
		0	5	10	15	20	25	30
Temperature (°C)	No CO_2							
	CO_2							

Modelling thermal expansion

Inquiry skill: Evaluating

Inquiry skill focus: Describing types of errors

Investigations are never perfect. There are often errors in the method or equipment that impact the results and must be identified. Acknowledging and describing various types of errors helps to evaluate the reliability of your results and allows improvements to be made in future investigations. (See step 2 of 'Evaluating' in the Science how-to on page 256.)

Hint 1: Are there systematic errors in the equipment or method steps that consistently affect the results?

Hint 2: Are there random errors, such as variations in measurements or environmental factors, that make the results less consistent?

Hint 3: Could personal errors, such as reading instruments incorrectly or inconsistencies in technique, have affected the investigation?

A HEAT LAMP IS A HAZARD. BE CAREFUL. IF YOU BURN YOURSELF, TELL YOUR TEACHER IMMEDIATELY AND RUN COLD WATER OVER THE AFFECTED AREA FOR 20 MINUTES.

Aim

To investigate how water expands when heated

Materials

- 600 mL conical flask or bottle
- paper towel
- food colouring (optional)
- tap water
- thermometer
- glass tubing or transparent straw
- 30 cm ruler
- rubber stopper with two holes, or plasticine
- marker
- heat lamp
- stopwatch

Method

1. Copy the results table from the next page into your notebook, adding a title and rows as needed.
2. Place the flask or bottle on the paper towel. Place a few drops of food colouring in the flask before filling it with water to the top.
3. Carefully place the thermometer and tubing/straw in the flask so that at least 5 cm sits within the flask and 5 cm sits above. Use the stopper or the plasticine to help you keep it in place. Some of the water will overflow.
4. Ensure there are no gaps and that no water can leak out.
5. Use the marker to mark the water level in the straw. This will be your zero mark.
6. Measure the starting temperature of the water and record it in the results table.

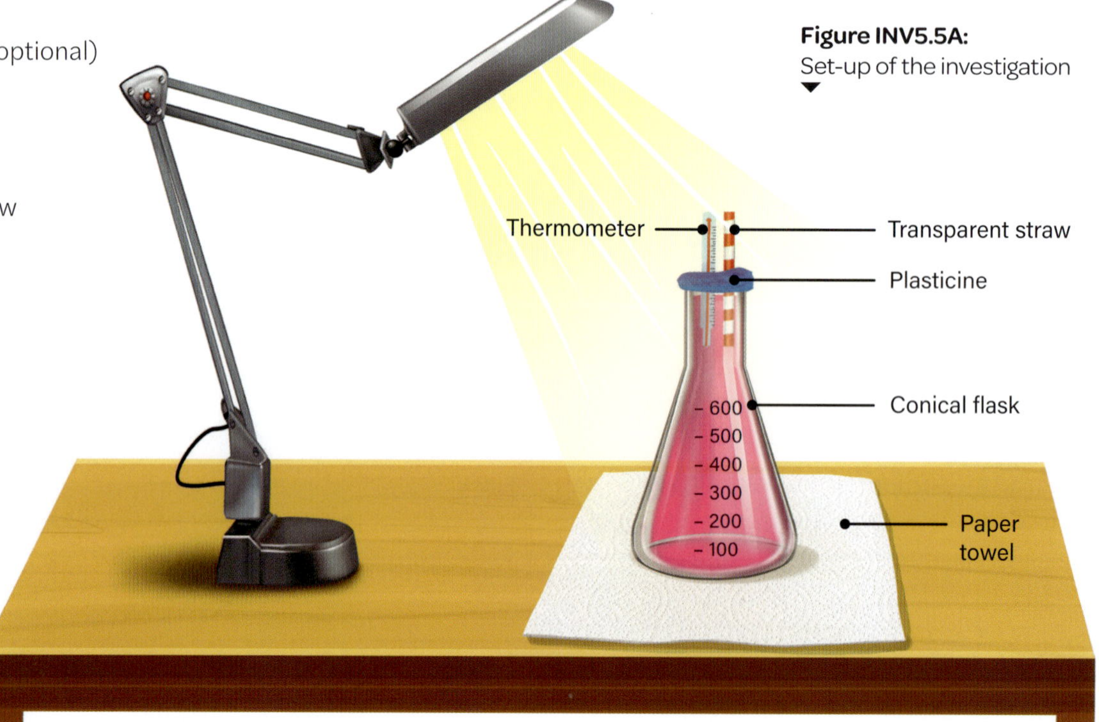

Figure INV5.5A: Set-up of the investigation

7 Place the bottle underneath the heat lamp and turn it on.
8 Measure and record the water level in the straw and the temperature of the water every 5 minutes for 30 minutes.
9 Construct a suitable graph for your data.

Table INV5.5A: Results

Time (minutes)	Water level (mm)	Temperature (°C)

Questions

1 Identify the dependent, independent and controlled variables for this investigation. (More information about variables is available in the Science how-to section on pages 235–37.)
2 Describe the trends in your data.
3 Compare your results with those of your classmates. Discuss the reliability and validity of your data.
4 Use your findings from this investigation to explain how warmer ocean temperatures also result in sea-level rise.
5 Propose how you could modify the method of this investigation to improve the validity of the results.

Conclusion

Copy and complete:
'The results show that: *(respond to the aim)*.'

Investigation 5.5B

Melting ice and sea-level rise

Inquiry skill: Evaluating

Inquiry skill focus: Constructing evidence-based arguments

The results of an investigation can be used as evidence to make claims and construct arguments about scientific concepts and skills. How might the results of this investigation be used to support claims about the impact of melting glaciers and sea ice on sea-level rise?

Hint 1: How does this model represent the real world?

Hint 2: Refer to 'Writing evidence-based essays' in the Science how-to on page 280.

A HEAT LAMP IS A HAZARD. BE CAREFUL. IF YOU BURN YOURSELF, TELL YOUR TEACHER IMMEDIATELY AND RUN COLD WATER OVER THE AFFECTED AREA FOR 20 MINUTES.

Aim

To model how melting sea ice and melting glaciers contribute to sea-level rise

Materials

- 2 × 500 mL beakers or containers
- tap water
- heat lamp
- clay or plasticine
- 8 ice cubes
- marker
- 30 cm ruler

Method

1. Copy the results table into your notebook, adding a title.
2. Place a lump of clay or plasticine in one beaker; this will model a continent. Model it so there is a platform for the ice cubes to sit on. The top of the 'continent' should sit just above the 200 mL mark.
3. Place four ice cubes (the glacier) on the 'continent' and add four ice cubes to the second beaker (to model sea ice).
4. Add water to both beakers up to the 200 mL mark or to the same height.
5. Use the marker to mark the initial water level in both beakers.
6. Place both beakers under the heat lamp until the ice melts.
7. Use the marker to record the new water level in both beakers.
8. Use a ruler to measure the initial water level and the final water level. Calculate the change.

Table INV5.5B: Results

Beaker	Initial water level (mm)	Final water level (mm)	Change in water level (mm)
Sea ice			
Glaciers			

Questions

1. Describe your results. Compare your results with those of other groups.
2. Explain any differences in the results between the beaker that represented sea ice and the beaker that represented glaciers.
3. What does this model tell you about how melting glaciers and sea ice contributes to sea-level rise?
4. Consider the headline 'Melting Arctic sea ice causes sea-level rise'. Use your understanding to suggest an improvement to the headline.

Conclusion

Copy and complete:

'The results show that: *(respond to the aim)*.'

▼ **Figure INV5.5B:** Set-up of the investigation

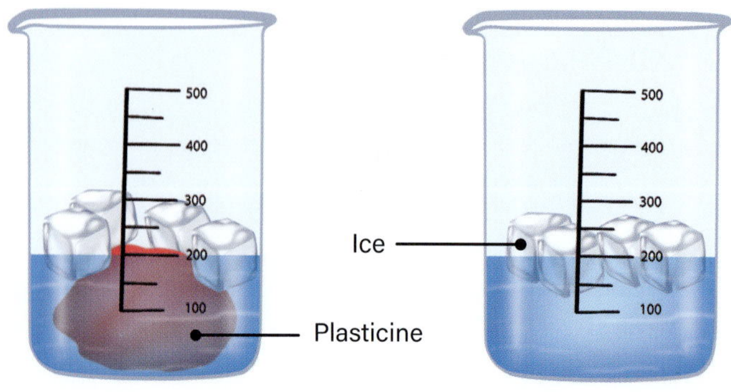

Investigation 5.6

Modelling ocean acidification

30 min
Initial set-up

5 days
Leave shells in beakers for 3 days. Remove and dry for 2 days

30 min
Complete mass calculations

Inquiry skill: Evaluating

Inquiry skill focus: Evaluating the validity and reproducibility of methods

A good investigation needs to collect valid data. Studies are only considered valid if they are based on high-quality data. Data needs to demonstrate a consistent trend, across a large sample size, and be reproducible. It needs to be collected using specific equipment, using digital tools where necessary, so that the results are precise and increase the accuracy of the investigation. When all of these factors are present, the method of the investigation is considered reproducible. What factors might impact the reproducibility of this investigation?

Hint 1: How does using percentage mass change, instead of mass change, impact the consistency of your results?

Hint 2: Refer to step 5 of 'Evaluating' in the Science how-to on page 260.

Aim

To model the effect of ocean acidification on organisms with calcium carbonate shells

Materials

- 6 × 100 mL beakers
- marker
- 1 × 50 mL measuring cylinder
- distilled water
- 50 mL of solutions of 5%, 10%, 15%, 20% and 25% acetic acid
- electronic balance
- 6 seashells of identical type and shape *or* eggshells (you may need to use pieces of broken shell)
- paper towel

Method

1 Copy the results table into your notebook, adding a title and rows as needed.

2 Label the beakers 0%, 5%, 10%, 15%, 20% and 25%.

3 Use a measuring cylinder to measure 50 mL of distilled water. Place it in the 0% beaker.

4 Use the measuring cylinder to measure 50 mL of each of the acetic acid solutions into each beaker. Rinse the cylinder between each use.

5 Identify which beaker each shell will go into. Use the electronic balance to weigh each shell. Record this in the results table, taking care not to mix up the shells.

6 Place the shells in their corresponding beakers.

7 Leave the shells in the beakers for 3 days.

8 Remove the shells from the beakers and place them on labelled paper towel to dry for 2 days.

9 When the shells are dry, measure their mass on the electronic balance. Record this in the results table.

10 Calculate the mass change:
initial mass – final mass = mass change

11 Calculate the percentage mass change:
$\frac{\text{mass change}}{\text{initial mass}} \times 100$

12 Construct a graph of your results.

Questions

1 Identify the dependent, independent and controlled variables in this investigation.

2 Describe the trends in your results.

3 Explain how this investigation models the process of ocean acidification.

Table INV5.6: Results

Concentration of acid (%)	Initial mass of shell (g)	Final mass of shell (g)	Mass change (g)	Per cent mass change $\frac{\text{mass change}}{\text{initial mass}} \times 100$

continues ▶

4 Discuss the importance of calculating the percentage mass change to ensure your data is reliable.

5 Evaluate the method. Did it allow you to collect valid and reliable data? Redesign the method to improve the validity and reproducibility of your data. (More information about validity and reproducibility is available in the Science how-to on pages 260–61.)

Conclusion

Copy and complete:
'The results show that: *(respond to the aim)*.'

Investigation 6.1

Modelling the expanding universe

Inquiry skill: Communicating

Inquiry skill focus: Using digital technologies to organise and present findings

Digital technologies allow us to present scientific findings in many different ways. They can be used to construct tables and graphs, as well as to record images and videos. Think about how you could use digital technologies to organise and present your findings for this investigation.

Hint 1: Think about the observations you will be making and the results you will be gathering.

Hint 2: Think about how digital technologies could help you to display your observations and results. For example, you could construct a digital results table on a computer.

Aim

To determine how a galaxy's distance from a reference point affects its speed

Materials

- 1 round balloon
- 5 different coloured stick-on dots
- piece of string approximately 50 cm
- 30 cm ruler
- stopwatch

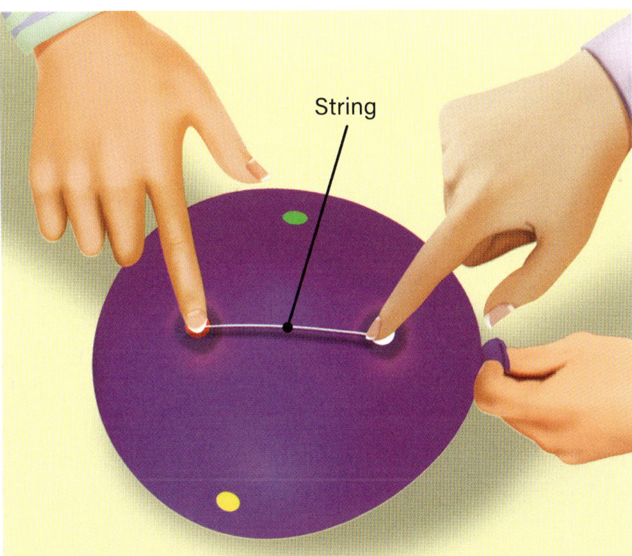

Figure INV6.1: Using a piece of string, you can measure the distance on the outside of the balloon.

Method

1. Copy the results table from the next page into your notebook, adding a title.
2. Blow up the balloon to about 150 mm and hold the nozzle closed (do not tie it up).
3. Stick the 5 dots (galaxies) onto the balloon (universe). Try to spread them evenly around the balloon.
4. Select one of the dots to be your home galaxy. Use the string to measure the distance between your home galaxy and the other galaxies, as shown in the diagram. Measure the string with the ruler to determine this distance in centimetres.
5. Record these distances in the results table. (*Note:* The distance from your home galaxy will be 0 cm.)
6. Fully inflate the balloon and use the stopwatch to time how long it takes. Tie the balloon and record the time taken in the results table.
7. Use the string and ruler to again measure the distance from your home galaxy to the other galaxies. Record these distances in the results table.
8. Calculate the change in distance by subtracting the first measurements from the second measurements.
9. Calculate the speed of expansion of each galaxy by dividing the change in distance (cm) by the time it took the balloon to inflate (s). (*Note:* The speed of expansion of your home galaxy will be 0 cm s^{-1}.)
10. Plot a line graph with speed (cm s^{-1}) on the *y*-axis and distance (cm) on the *x*-axis.

continues ▶

Questions

1 a Identify whether all galaxies move at the same speed.
  b If not, describe the relationship between speed and the distance from the home galaxy.

2 Propose what would happen to the results if you had used a different home galaxy.

3 Explain how the results of this investigation simulate the way the universe is expanding.

4 Use your observations from this investigation to explain why we observe light from distant galaxies more redshifted than those closer to us.

Conclusion

Copy and complete:
'The results show that: *(respond to the aim)*.'

Table INV6.1: Results

Colour of dot	Distance from home galaxy (cm)	Distance from home galaxy after inflating (cm)	Change in distance (cm)	Speed = $\frac{\text{distance}}{\text{time}}$ (cm s^{-1})
Red				
Green				
Blue				
White				
Yellow				
Time to fully inflate balloon(s)				

Investigation 6.2

Investigating orbits

Inquiry skill: Communicating

Inquiry skill focus: Preparing representations to communicate findings

Being able to represent findings in effective and appropriate ways is important for communicating science to others clearly and precisely. After conducting this investigation, you should be able to prepare a representation of your results to communicate your findings.

Hint: Refer to step 3 of 'Communicating' in the Science how-to on page 265.

Aim

To determine the relationship between orbit radius and orbital period

Materials

- string
- ruler
- scissors
- rubber stopper
- glass or plastic tubing 10–20 cm long
- 50 g mass
- mass carrier (50 g)
- marker pen
- stopwatch

Method

1. Copy the results table into your notebook, adding a title.
2. Measure and cut 110 cm of string.
3. Tie one end of the string to the rubber stopper.
4. Thread the string through the tubing.
5. Tie the 50 g mass carrier to the other end of the string. Make sure there is 100 cm between the mass and the stopper.
6. Use the ruler and the marker to mark 10 cm intervals between the mass and the stopper.
7. With the stopper at a 50 cm distance from the tubing, hold the tube above your head and swing the stopper around until it is spinning steadily. The 50 g mass should hang in the same position. If it does not, you will need to increase or decrease the speed of spinning.
8. Use the stopwatch to time how long it takes for the stopper to complete 10 orbits of the tubing.
9. Record your results in the table and calculate the time taken for one orbit.
10. Repeat steps 7–9 for the other distances.
11. Draw a line graph of your results.

Table INV6.2a: Results

Orbit radius (cm)	Time to complete 10 orbits (s)	Time to complete 1 orbit (s)
50		
40		
30		
20		
10		

continues ▶

Table INV6.2b:
Planet orbit distances and time to orbit

Planet	Distance from Sun (AU)	Time to orbit (Earth years)
Mercury	0.4	0.2
Venus	0.7	0.6
Earth	1.0	1.0
Mars	1.5	1.9
Jupiter	5.2	11.9
Saturn	9.5	29.5
Uranus	19.2	84.0
Neptune	30.2	164.8

Questions

1 Identify what represents the Sun and the planets in this model.

2 Describe the relationship between the distance from the tubing and the time it took for the stopper to orbit.

3 Identify what you can infer from this model about the movement of the inner and outer planets.

4 Table INV6.2b contains data for the distance from the Sun and orbital times for each planet in our solar system. Create a line graph of this data.

5 Compare the trendline of your results with that of this second graph. Describe how they are similar and how they are different.

Conclusion

Copy and complete:
'The results show that: *(respond to the aim)*.'

Investigation 6.3

Bottle rockets

Inquiry skill: Questioning and predicting

Inquiry skill focus: Developing a hypothesis

Part of a scientific investigation includes developing a hypothesis, which is a proposed prediction of the outcome of an investigation intended to address a problem or answer a question. To develop a hypothesis, you must first identify the independent, dependent and controlled variables. The independent variable is the one thing that you purposefully change in the investigation, the dependent variable is what you measure, and the controlled variables are all the things you keep the same throughout the investigation. Develop a hypothesis for this investigation, using the format *If ..., then ..., because ...*

Hint: For help, refer to the Science how-to section on page 236.

Aim

To determine how the ratio of water to air in a bottle rocket impacts its launch height

Materials

- large outdoor space
- bottle rocket kit (may include stopper, tubing and launch stand)
- scrap plastic and cardboard
- large soft drink bottle (empty)
- scissors
- tape
- bike pump
- ruler
- recording device
- 500 mL measuring cylinder
- water

Method

1. Answer Question 1 in your notebook.
2. Copy the results table into your notebook, providing enough space for each of your trials.
3. Use the scrap plastic and cardboard to construct a nose cone and fins for your bottle rocket. Stick them to your bottle with the tape.
4. Set up the launch stand and bike pump in your outdoor space. *Note:* Your teacher will give you a safety briefing and set up a safe space for your class to launch the rockets.
5. Set up the recording device so that you can record the rocket launch and the height of launch.
6. Determine how you will measure launch height. Are there buildings, goal posts or trees you can use as a reference?
7. Determine the different volumes of water you will be testing. You might like to conduct a test flight first to help you with this.
8. Use the measuring cylinder to add the first volume of water into your bottle rocket.
9. Carefully place the stopper in the rocket, connect it to the bike pump and the launch stand. Start the recording device.
10. Pump the bike pump until the rocket launches. Record the launch height.
11. Repeat twice more for each volume of water.
12. Construct an appropriate graph of your results.

Questions

1. a Identify the dependent, independent and controlled variables for this investigation.
   b Use these variables to formulate a scientific question that could be answered by this investigation.
   c Use an 'if ..., then ...' statement to construct a hypothesis that addresses the relationship between the dependent and independent variables.
2. Describe the trends shown in your data.
3. Identify whether your results allowed you to identify an optimal amount of water to get the best launch height. If yes, what was it? If not, why do you think it did not?
4. Discuss how you could modify this investigation to improve the validity and reliability of your results.

Conclusion

Copy and complete:
'The results show that: *(respond to the aim)*.'

Table INV6.1: Results

Volume of water in rocket (mL)	Launch height (m)		
	Trial 1	Trial 2	Trial 3

Investigation 6.4A

Making a telescope

Inquiry skill: Questioning and predicting

Inquiry skill focus: Constructing scientific questions

Scientific investigations are based on scientific questions constructed by scientists wanting to learn more about the world. These questions aim to identify a relationship or connection between an independent variable (the factor being changed) and a dependent variable (the factor being measured).

Hint 1: What should a good question look like?

Hint 2: More information about constructing scientific questions is available in the Science how-to section on page 235.

Aim

To use lenses to make a telescope

Materials

- two convex lenses of different sizes
- book, or sheet of paper with writing on it
- ruler
- cardboard tube
- box cutter
- sticky tape
- Blu Tack

Method

1. Hold the larger lens between you and the book or sheet of paper. The writing should look blurry when looking through the lens. This will be the primary or objective lens.
2. Hold the smaller lens between you and the larger lens. This will be the eyepiece lens. Move the lenses so that when you view the writing through the smaller lens it is in focus.
3. Record the distance between the first and second lenses.
4. Cut a slot in the cardboard tube about 2 cm from one end to hold the larger lens.
5. Cut a second slot for the smaller lens, the same distance away from the larger lens that you recorded in step 3.
6. Cut away any excess tubing, leaving about 2 cm.
7. Place the two lenses in the slots, holding them in place with tape or Blu Tack.

Questions

1. Draw and label a scale diagram of your telescope. Include the objective lens, eyepiece lens and the measurements.
2. Describe the purpose of the objective lens.
3. Describe the purpose of the eyepiece lens.
4. Explain what you would expect to happen with the detail you can see if you increase the size of these lenses.
5. Explain what you would expect to happen with the length of the telescope if you increase the size of these lenses.

Conclusion

Copy and complete:

'The results show that: *(respond to the aim)*.'

Lens diameter and resolution

Inquiry skill: Questioning and predicting

Inquiry skill focus: Making predictions

A key component of first-hand investigations is making a prediction of the outcome of the investigation. You may not know much about the concepts under investigation yet, so sometimes it is okay to make a prediction based on your own experiences and prior observations.

Hint 1: Think about how close to a poster you need to be before you can read it.

Hint 2: More information about making predictions is available in the Science how-to section on page 234.

Aim

To investigate how lens diameter affects resolution

Materials

- selection of binoculars and/or optical telescopes
- ruler
- A4 piece of paper
- marker pen
- sticky tape
- trundle wheel or long tape measure

Table INV6.4B: Results

Optical instrument	Diameter of lens (mm)	Distance from 'stars' where they are first observed as two separate objects (m)
Human eye	10	
Binoculars		
Telescope		

Method

1. Copy the results table into your notebook, adding a title.
2. Measure and record the diameter of the lens of the binoculars and optical telescope and record it in the results table.
3. On the piece of paper, draw two thick lines 2 cm long and 2 cm apart. These will represent stars.
4. Use the sticky tape to tape the paper to a wall.
5. Position an observer at a point away from the wall where they observe the two lines as one.
6. The observer should walk towards the wall, stopping when they first observe the two lines as being distinct from one another.
7. Use the tape measure or trundle wheel to measure the distance between the observer and the wall. Record this in your results table.
8. Repeat steps 5–7 using the binoculars and telescopes that you have.
9. Construct a graph of your results.

Questions

1. Describe the relationship between the diameter of the lens and the distance where the 'stars' are first observed as two separate objects.
2. 'Resolution' is the ability to distinguish two separate objects or to see more detail. Identify which optical instruments had the best and worst resolutions.
3. Alpha Centauri is the brightest star in the constellation of the Pointers and the third brightest star in the night sky. To the naked eye, Alpha Centauri appears as one star but it is a triple star system. Explain how resolution of an instrument is important for making these discoveries.
4. Compare the resolution of a telescope with the resolution of a microscope in terms of its ability to distinguish between objects that are close together.

Conclusion

Copy and complete:

'The results show that: *(respond to the aim)*.'

Investigation 6.6

Stargazing

Inquiry skill: Questioning and predicting

Inquiry skill focus: Constructing scientific questions

Scientific investigations are based on scientific questions constructed by scientists wanting to learn more about the world. These questions aim to identify a relationship or connection between an independent variable (the factor being changed) and a dependent variable (the factor being measured).

Hint 1: *What should a good question look like?*

Hint 2: *More information about constructing scientific questions is available in the Science how-to section* *on page 235.*

Aim

To investigate the night sky

Materials

- star map for the month and year you are viewing, or a stargazing app
- red torch, or torch with a red cellophane cover
- compass
- blank A4 paper
- pencil and highlighter
- binoculars or telescope (optional)

Method

Part 1: Exploring the night sky

1. On a clear night, find a dark area where you will have a good view of the night sky.
2. Use the red torch to help you see your papers.
3. Use the compass to identify north, south, east and west in your location.
4. Orientate your star map.
5. Use your star map to identify:
   a. the Southern Cross and Pointers
   b. at least two other constellations
   c. a planet (if visible at the time you are observing).
6. Use the highlighter to highlight the objects on your sky map that you locate.
7. Compare what you observe with the naked eye to what you can see with a telescope or binoculars.

Part 2: Locate south using the Southern Cross and Pointers

1. Draw an imaginary line with your finger from the left side of the cross to the right side and extending beyond it, as shown on Figure INV6.6a.
2. Draw an imaginary line joining the two Pointers. Midway along this line, draw another line at right angles to it.
3. Where lines 1 and 2 meet is the South Celestial Pole. This is the point in the sky around which all the stars seen from the southern hemisphere rotate.
4. Locate south by dropping a vertical line from the South Celestial Pole to the horizon.

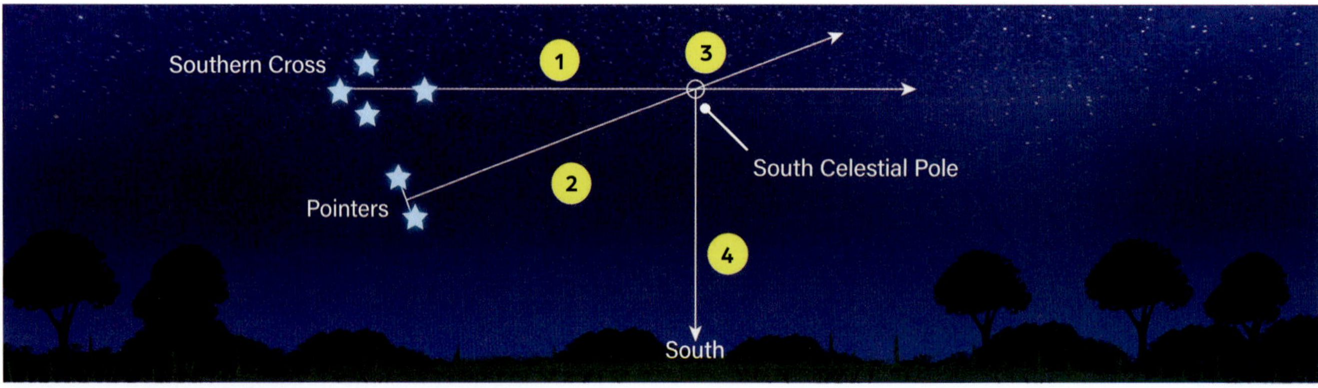

▲ **Figure INV6.6a:** Locating the South Celestial Pole

45 min

Figure INV6.6b: Diagram for marking the motion of stars

Part 3: Observing the motion of the stars

1 Identify a bright star that will be easy for you to identify again. It should be close to an object such as a roof, tree or chimney.
2 Take note of where you are standing so that you can return to it later.
3 On your piece of paper, draw a silhouette of the object that you are comparing your bright star to, as shown in Figure INV6.6b.
4 Mark the location of the bright star on your paper. Make a note of the time you make your observation.
5 For the period of your observations, return to the same spot every 30 minutes and note the location of the star.

Questions

Part 1

1 If you used binoculars or a telescope, describe the differences between the observations you made with them and those you made with the naked eye.

Part 2

2 Compare the technique using the Southern Cross to locate south with using the compass.

Part 3

3 Describe the movement of the star as time passed.
4 Identify the point that the movement is occurring around. Add this to your diagram if appropriate.
5 Propose what causes this movement.

Conclusion

Copy and complete:
'The results show that: *(respond to the aim).*'

Cars and pedestrians

Inquiry skill: Processing, modelling and analysing

Inquiry skill focus: Processing data

Data can be processed in different ways. One way is by collecting raw measurements and using mathematical relationships to calculate a new value. This investigation requires you to measure the time taken for cars and pedestrians (walking and running) to travel a set distance, and to then calculate their speed from the data collected.

Hint 1: Refresh your understanding of how to calculate speed, and the units for each measurement you will take.

Hint 2: Consider how to improve the accuracy of your data collection to ensure accurate results.

WORKING ON THE SIDE OF THE ROAD CAN BE DANGEROUS. DO NOT CARRY OUT ANY ACTIVITIES TO DISTRACT DRIVERS, AND STAY AS FAR BACK FROM THE EDGE OF THE ROAD AS POSSIBLE.

Aim

To investigate the speed of cars and pedestrians

Materials

- trundle wheel
- coloured markers or cones
- stopwatch

Method

Risk assessment

Conduct a risk assessment for this activity. Consider your chosen location when undertaking your risk assessment – the risk will depend on the road and the conditions you are working under.

Part A: Recording the speed of vehicles

1. Copy the results table for Part A on the next page into your notebook, adding a title and rows as needed.
2. Use a trundle wheel to measure 50 m along the footpath beside the road. Place cones or markers at each end of the 50 m distance.
3. Position a student at each end of the 50 m distance.
4. Both students use a stopwatch to time how long it takes a car to travel from the starting cone or marker to the finishing cone or marker. Record this time in the results table.
5. Repeat step 4 so that you have results for 10 cars. If any pedestrians or cyclists travel the length of the 50 m, record their time to compare it later to the cars' speeds.

Part B: Recording the speed of pedestrians

1. Draw up a results table for Part B in your notebook. It will be similar to the one for Part A.
2. Use the trundle wheel to measure 10 m along the footpath beside the road. Place cones or markers at each end of the 10 m distance.
3. Position a student at each end of the 10 m distance.
4. Both students use a stopwatch to time how long it takes for another student to walk the 10 m. Record this time in your results table.
5. Repeat step 4 several times, but record the speed for walking backwards, hopping, running, skateboarding, cycling or an activity of the student's choosing.

Figure INV7.2a: Use a trundle wheel to measure the distance between your cones.

Questions

1 Outline the formula used to calculate speed, including units for each value.

2 For Part A, compare your results to the speed limit of the road.

3 Which student activity was fastest and which was slowest? How do you know?

4 Identify some of the causes of errors in this investigation.

5 Propose some ways to improve the method and increase the accuracy of the results. Think about how technology could help.

Conclusion

Copy and complete:
'The results show that: *(respond to the aim)*.'

Table INV7.2: Results for Part A

Distance: 50 m Speed limit of road: ______	Time (s)			Speed	
	Time 1	Time 2	Average	(m s^{-1})	(km h^{-1})
Car 1					
Car 2					
Truck 1					

▲ **Figure INV7.2b:** Set up your cones 50 m apart.

Investigation 7.3

Ticker timers

Inquiry skill: Processing, modelling and analysing

Inquiry skill focus: Constructing graphs

Graphing data shows the visual relationship between the independent (changed) variable and the dependent (measured) variable.

Hint 1: *Identify the type of graph to use for the data gathered in this investigation.*

Hint 2: *Ensure your data goes onto the correct graph axis. See page 294 of the Science how-to section for more help with graphing.*

Hint 3: *Consider what the gradient of the graph represents in each graph you draw.*

Aim

To use a ticker timer to analyse and calculate the relationship between speed, distance and acceleration

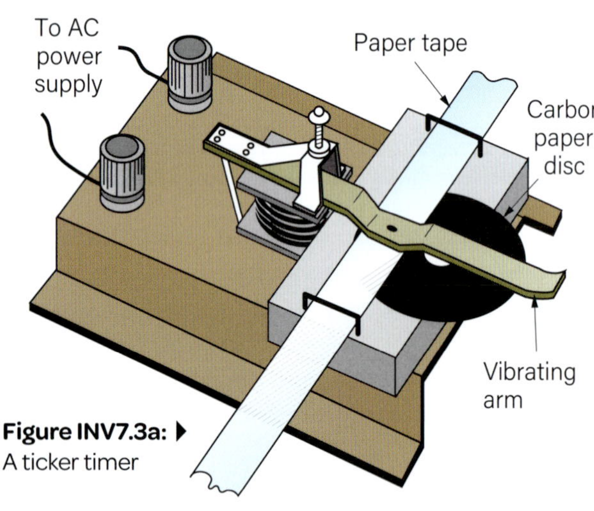

Figure INV7.3a: A ticker timer

Figure INV7.3b: A ticker timer tape showing motion

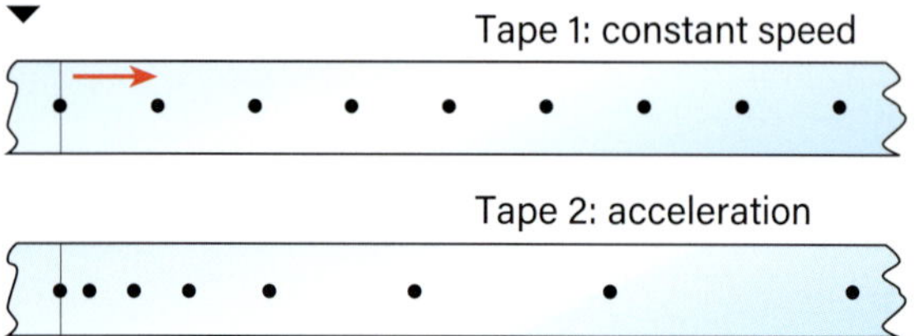

Introduction

A ticker timer is a device that has a vibrating arm that hits a ring of carbon paper (Figure INV7.3a). As a strip of paper (ticker tape) is pulled between the arm and carbon paper by a moving object, the ticker timer leaves dots on the ticker tape. The dots on the tape represent motion of the object, as shown in Figure INV7.3b.

Materials

- ticker timer
- G-clamp
- ticker tape
- dynamics trolley
- sticky tape
- scissors
- power-pack
- 2 electrical leads

Method

Part A: Constant speed

1. Copy the results table from the next page into your notebook, adding a title and rows as needed.
2. Set up the ticker timer and attach it to the edge of the bench with a G-clamp.
3. Cut off about 1 m of ticker tape and thread it through the ticker timer. Attach one end of the tape to the trolley.
4. Start the ticker timer and pull the trolley and tape through the ticker timer at a constant speed. Label this tape and the direction of motion.
5. Repeat with a new piece of tape at a:
   a. higher constant speed.
   b. lower constant speed.

   Note: You will need a separate results table for each piece of tape.
6. To analyse your tape, draw a line through the first clear dot, and then every fifth dot after that. Every five dots represents a time interval of 0.1 s.

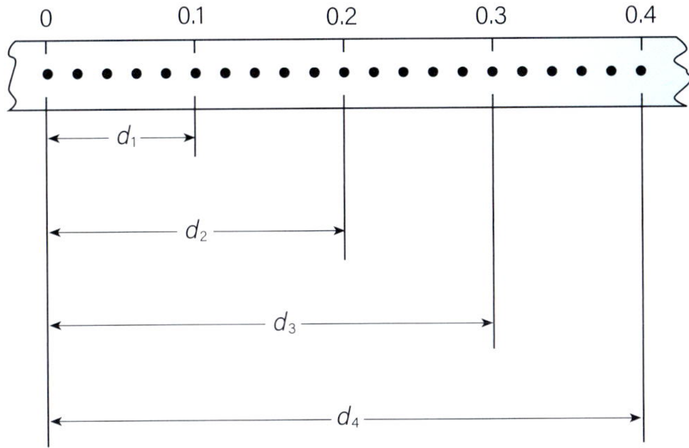

Figure INV7.3c: A ticker timer tape analysis

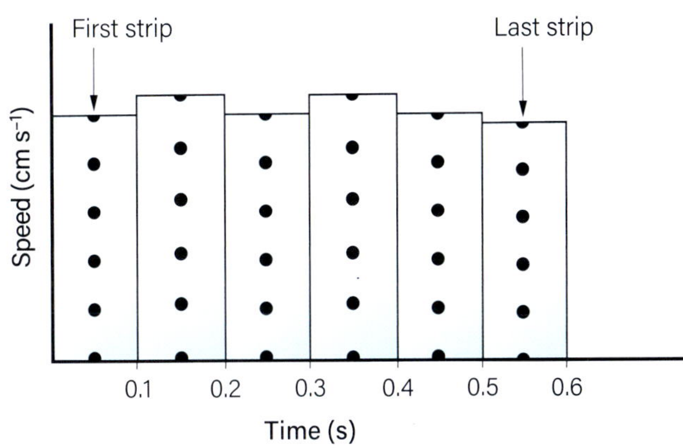

Figure INV7.3d: A ticker tape speed–time graph

7 Measure the distance to each marked dot (or interval) as shown in Figure INV7.3c. Record the distance travelled and total time to each interval in your results table.

8 Draw a distance–time graph of your results. Make sure your line of best fit is straight.

9 Calculate the slope or gradient to find the average speed (your answer will be in cm s^{-1}).

10 Repeat your analysis of the tape for the lower and higher speeds.

Part B: Acceleration

1 Repeat the steps in Part A, this time pulling the trolley so that the speed steadily increases. If it helps, tilt the table by placing one end on a block, or place the trolley on a ramp, so that the trolley rolls downhill and steadily accelerates.

2 Analyse the tape and draw a distance–time graph.

Questions

1 Compare the gradient of each graph in Part A for constant speed. What do you notice about the relationship between the slope and the speed?

2 Outline the formula used to calculate acceleration, including units for each value.

3 a Explain the shape of the distance–time graph for acceleration in Part B.

b Describe how this is different from the graphs from Part A of constant speed.

4 Calculate the average acceleration for Part B from your graph.

5 Select one piece of tape for constant speed, and one for acceleration. Cut them up at the 0.1 s marks and number each piece to keep them in order. Stick them on a piece of paper in order, vertically, to create a speed–time graph, as shown in Figure INV7.3d. Compare the graphs you have just created for constant speed and acceleration. Explain what you observe.

Conclusion

Copy and complete:
'The results show that: *(respond to the aim).*'

Table INV7.3: Results

Tape speed: ______________	
Distance (cm)	**Total time (s)**
1	
2	
3	
4	

Investigation 7.4A

Car crashes and inertia

Inquiry skill: Planning and conducting

Inquiry skill focus: Generating and recording data

A good investigation involves collecting accurate and precise data. The data collection process must ensure consistency with carefully selected equipment used to take measurements. Precision also comes from correct use of equipment and data tables.

Hint 1: Consider the equipment for this investigation and how you will use it to make accurate measurements.

Hint 2: Ensure the method will result in the collection and recording of precise data.

Hint 3: Refer to the Science how-to section on page 240.

Aim

To investigate and demonstrate Newton's laws by modelling a car crash with crash-test dummies

Materials

- plasticine
- dynamics trolley
- talcum powder
- wooden plank
- textbooks to prop up plank
- brick
- metre ruler

Method

1 Copy the results table from the next page into your notebook, adding a title.

2 Make three crash-test dummy models from the plasticine – small, medium and large. These will represent different-sized people. Make sure they are very different sizes (masses), to get better results.

3 Lightly dust the top of the dynamics trolley with talcum powder and sit the smallest dummy on top. The powder is to make sure the crash-test dummy is not stuck down.

4 Set up the ramp and the brick as shown in Figure INV7.4A. Place the metre ruler next to the brick so that the 0 cm mark is at the near edge of the brick.

5 Place the trolley with the dummy at the top of the ramp and then release it. The trolley should travel down the ramp and strike the brick at the bottom.

6 Record how far the crash-test dummy travels after striking the brick.

7 Repeat steps 5 and 6 twice to obtain two more results. Calculate the average distance from the three trials.

8 Repeat steps 5–7 for the other two larger crash-test dummy models.

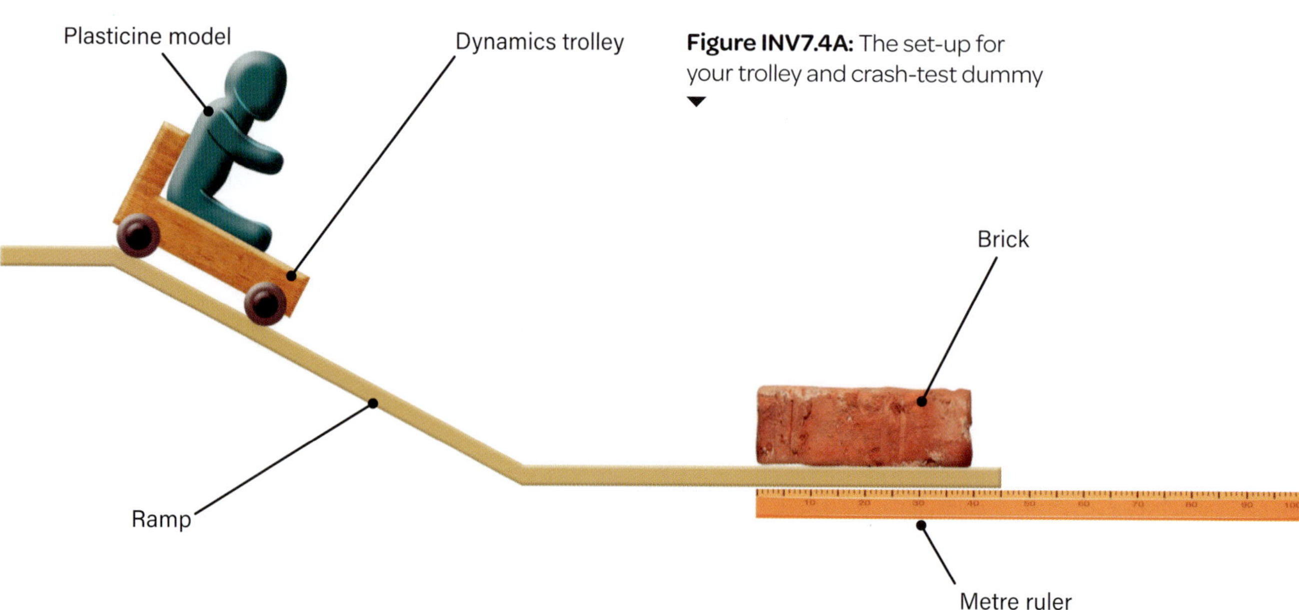

Figure INV7.4A: The set-up for your trolley and crash-test dummy

Questions

1 Using your data, explain which crash-test dummy travelled the furthest and the least after striking the brick. Propose reasons for the differences in distance travelled.

2 Explain the term 'inertia' and how it relates to this investigation (in particular, the crash-test dummies).

3 Explain how this investigation demonstrates:

a Newton's first law.

b Newton's second law.

c Newton's third law.

4 Using evidence from this investigation, propose reasons for why wearing a seatbelt in a car is important.

Conclusion

Copy and complete:
'The results show that: *(respond to the aim)*.'

Table INV7.4A: Results

Crash-test dummy	Distance (cm)			
	Trial 1	Trial 2	Trial 3	Average
Small				
Medium				
Large				

Investigation 7.4B

Balloon rockets

Inquiry skill: Planning and conducting

Inquiry skill focus: Describing ways to minimise risk

When conducting a first-hand investigation, it is important to minimise the risk of harm. Before you begin, potential hazards should be identified and strategies should be put in place to minimise the chance of accidents or harm in the laboratory.

Hint 1: *Identify any materials or methods in this investigation that could cause harm.*

Hint 2: *Develop strategies to minimise the chance of harm occurring.*

Hint 3: *More information about assessing risks is available in the Science how-to section on page 240.*

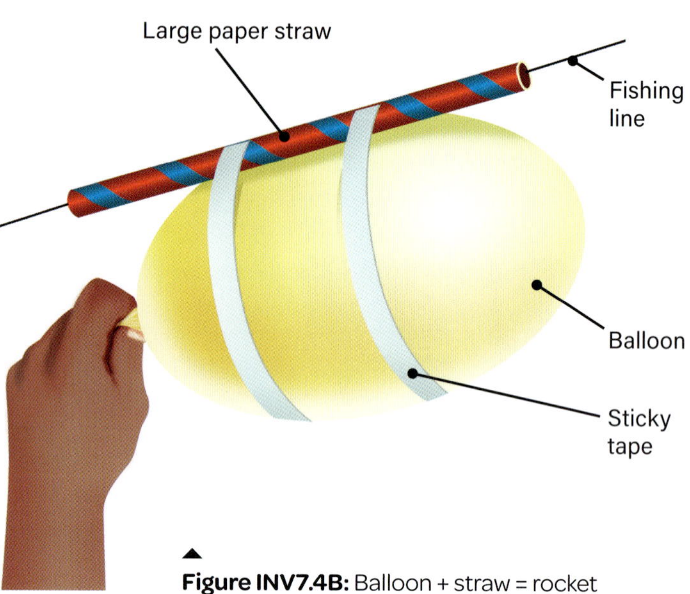

Figure INV7.4B: Balloon + straw = rocket

Aim

To demonstrate and verify Newton's third law, using balloon rockets

Materials

- balloon
- large paper straw
- fishing line
- sticky tape

Method

1. Inflate the balloon to about the size of a basketball. Hold the mouth of the balloon closed, but do not tie it off.
2. Attach the straw to the balloon as shown in Figure INV7.4B. Thread the fishing line through the straw.
3. Ask another student to hold one end of the fishing line while you hold the other, keeping it tight.
4. Make sure the balloon is near one end of the fishing line so that most of the line is out the front of the balloon rocket.
5. Let go of the mouth of the balloon. Record your observations.

Questions

1. Describe the motion of the balloon and the direction of the air escaping from it.
2. Draw a diagram to show the action and reaction force pairs.
3. Explain how the motion of the balloon demonstrates Newton's third law of motion.
4. Propose how you could modify the set-up to allow the balloon to travel further and faster.
5. If you inflate a balloon and then let it go when it is not attached to a straw or the line, it flies all over the place in an unpredictable path. Explain why this occurs in relation to the force of escaping air.

Conclusion

Copy and complete:

'The results show that: *(respond to the aim).*'

Investigation 7.5

Acceleration and mass

Inquiry skill: Planning and conducting

Inquiry skill focus: Obtaining replicable data

It is important to obtain replicable data to ensure that your results are valid. When completing an investigation, some equipment or methods will be easier to obtain replicable results from than others. Selecting appropriate equipment, using it carefully, and undertaking multiple trials to minimise random errors will all help to improve repeatability of results.

Hint 1: Consider the sources of error in your investigation. If they are making it hard to replicate results, think about how to improve your method.

Hint 2: Try some test measurements before you start collecting data to check the repeatability.

Hint 3: Multiple trials help to reduce random errors. Consider how many trials you need to ensure your results are valid. Adjust the results table if required.

Aim

To investigate the effect of increasing mass on an object's acceleration

Materials

- masking tape
- fishing line
- 100 g mass carrier
- 5 × 100 g masses
- bench-mounted pulley
- dynamics trolley
- stopwatch

(Optional: This investigation can be expanded to use ticker timers or motion sensors to capture the acceleration.)

Method

1 Copy the results table from the next page into your notebook, adding a title.

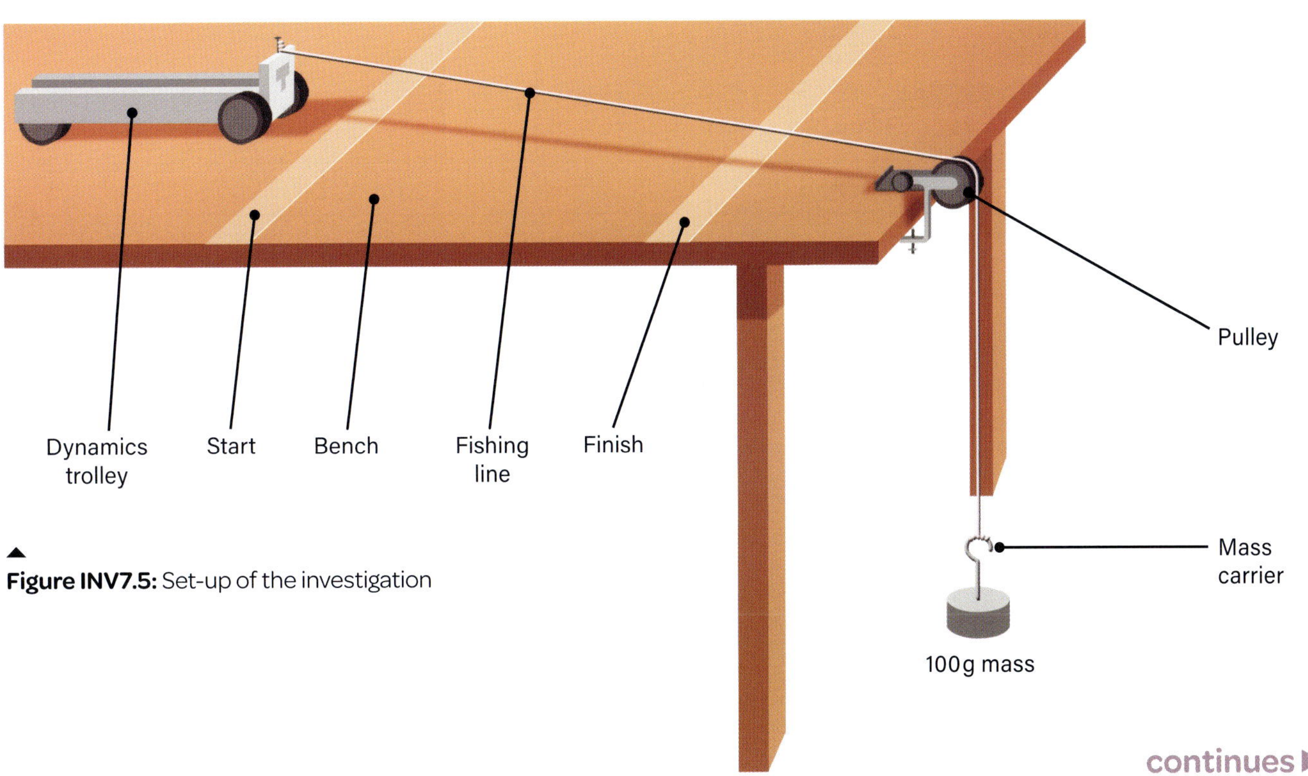

Figure INV7.5: Set-up of the investigation

continues ▶

2 On a desk or benchtop, place two strips of masking tape 1 m apart. These strips represent the start and finish lines. Make sure there is at least a dynamics trolley's length of benchtop before and after each strip of tape.
3 Attach a 100 g mass to one end of the fishing line. Thread the other end over the pulley and attach it to the trolley, as shown in Figure INV7.5.
4 Release the mass so that it drops straight down to the ground. Use the stopwatch to time how long the trolley takes to travel from the start line to the finish line. Make sure you catch it before it falls off the table. Perform three trials and record the times in your results table.
5 Add a 100 g mass to the trolley and repeat step 4.
6 Repeat step 4 with masses of 300 g, 400 g and 500 g on the trolley. Record all your data.
7 Calculate the average acceleration for each mass in your results table.

Questions

1 What can you infer about the time taken to travel 1 m and the acceleration of the trolley?
2 Describe any pattern you notice in your results.
3 Explain how your results confirm Newton's second law.
4 a Which part of the apparatus provided the force that accelerated the trolley?
   b Calculate the force pulling the trolley.
5 Identify any other forces acting on the trolley and describe the effect they had on it.
6 Identify any sources of error in this investigation and propose improvements to the method that could reduce these.

Conclusion

Copy and complete:
'The results show that: *(respond to the aim)*.'

Table INV7.5: Results

Mass (g)	Time (s) to travel 1 m			
	Trial 1	Trial 2	Trial 3	Average
100				
200				
300				
400				
500				

Investigation 7.8

Car crashes and momentum

Inquiry skill: Processing, modelling and analysing

Inquiry skill focus: Analysing scientific models

Scientists can use modelling and simulations to simplify and explain ideas. Models make things easier to understand and visualise, which helps us to make predictions and understand what is happening during an investigation. However, there can be limitations when using models. This is why we analyse scientific models: to evaluate their effectiveness in simulating what is happening in the real world.

Hint 1: Review the concept of momentum and think about how this investigation models this concept.

Hint 2: As you complete the investigation and explore momentum and collisions using this model, think about whether the model is accurate. (Does it demonstrate momentum and what really happens in a collision?) Consider how the model may be improved.

Aim

To investigate and demonstrate momentum by modelling a car crash with crash-test dummies

Materials

- plasticine
- 2 × dynamics trolleys
- talcum powder
- wooden plank
- textbooks to prop up plank
- metre ruler
- weights
- masking tape
- toilet rolls, paper, pieces of sponge or other similar material

Method

Part A: Momentum

1 Copy the results table from the next page into your notebook, adding a title.

2 Make two crash-test dummy models of different sizes from the plasticine.

3 Lightly dust the top of the dynamics trolley with talcum powder and sit the smallest dummy on top. The powder is to make sure the crash-test dummy is not stuck down.

4 Set up the ramp and trolleys as shown in Figure INV7.8 with the dummies at the front of the trolleys. Place the metre ruler on the ramp so that the 0 cm mark is near the bottom of the ramp.

5 Place trolley B about 40 cm in front of the ramp. Trolley A should be about halfway up the ramp. Mark your release point for trolley A to use the same release point in future steps.

TEST A: **Trolley A: low velocity, low mass**
Trolley B: low mass

6 Release trolley A from about halfway up the ramp so that it collides with trolley B.

7 Repeat the collision three times.

8 Record your observations, including the motion (velocity) of trolley A and trolley B, and of the passengers in each case.

TEST B: **Trolley A: high velocity, low mass**
Trolley B: low mass

9 Repeat the collision three times, this time releasing the trolley from the top of the ramp. Alternatively, you could make the ramp steeper to increase the speed.

10 Record your observations.

TEST C: **Trolley A: low velocity, high mass**
Trolley B: low mass

11 Add weights to trolley A to increase the mass. The weights must be attached to the trolley so they cannot move during the collision. (Think about inertia!) Masking tape may be useful.

12 Repeat the collision three times, releasing trolley A from halfway up the ramp.

13 Record your observations.

continues ▶

TEST D: Trolley A: high velocity, high mass
Trolley B: low mass

14 Repeat the investigation three times, again releasing trolley A from the top of the ramp.

15 Record your observations.

TEST E: Trolley A: low velocity, low mass
Trolley B: high mass

16 Now swap trolleys A and B so the trolley with more mass is trolley B.

17 Repeat the collision three times, releasing trolley A from halfway up the ramp.

18 Record your observations.

TEST F: Trolley A: high velocity, low mass
Trolley B: high mass

19 Repeat the collision three times, releasing trolley A from the top of the ramp.

20 Record your observations.

Part B: Crumple zones

1 Redesign your investigation by adding a crumple zone to one trolley. This could be done with a paper cylinder attached to the front, a ball of paper, a sponge or other material.

2 Repeat one or more of the crash-test scenarios and record your observations, noting whether there is any change to the collision.

Questions

1 a Describe what happens to the passengers in the moving car during a collision.

b Describe what happens to the passengers in the stationary car during a collision.

2 Identify from your results the relationship between the:

a momentum and the velocity of a trolley.

b momentum and the mass of a trolley.

3 Identify which test had the highest momentum and which had the lowest. Justify your answer.

4 Describe the effect of higher momentum in an accident.

5 Describe how adding a crumple zone changed the collision, including the impact on passengers.

6 Explain how a crumple zone works in terms of momentum and energy.

7 a Explain whether you think this investigation clearly modelled the concept of momentum.

b Propose how this model could be improved.

Conclusion

Copy and complete:
'The results show that: *(respond to the aim).*'

Table INV7.8: Results: Part A

Test	Variables in test			Observations
	Speed of trolley A (high/low)	**Mass of trolley A (high/low)**	**Mass of trolley B (high/low)**	**Ensure you include observations of:** ▪ Trolley A ▪ Trolley B ▪ Passenger A ▪ Passenger B
A				
B				
C				
D				
E				
F				

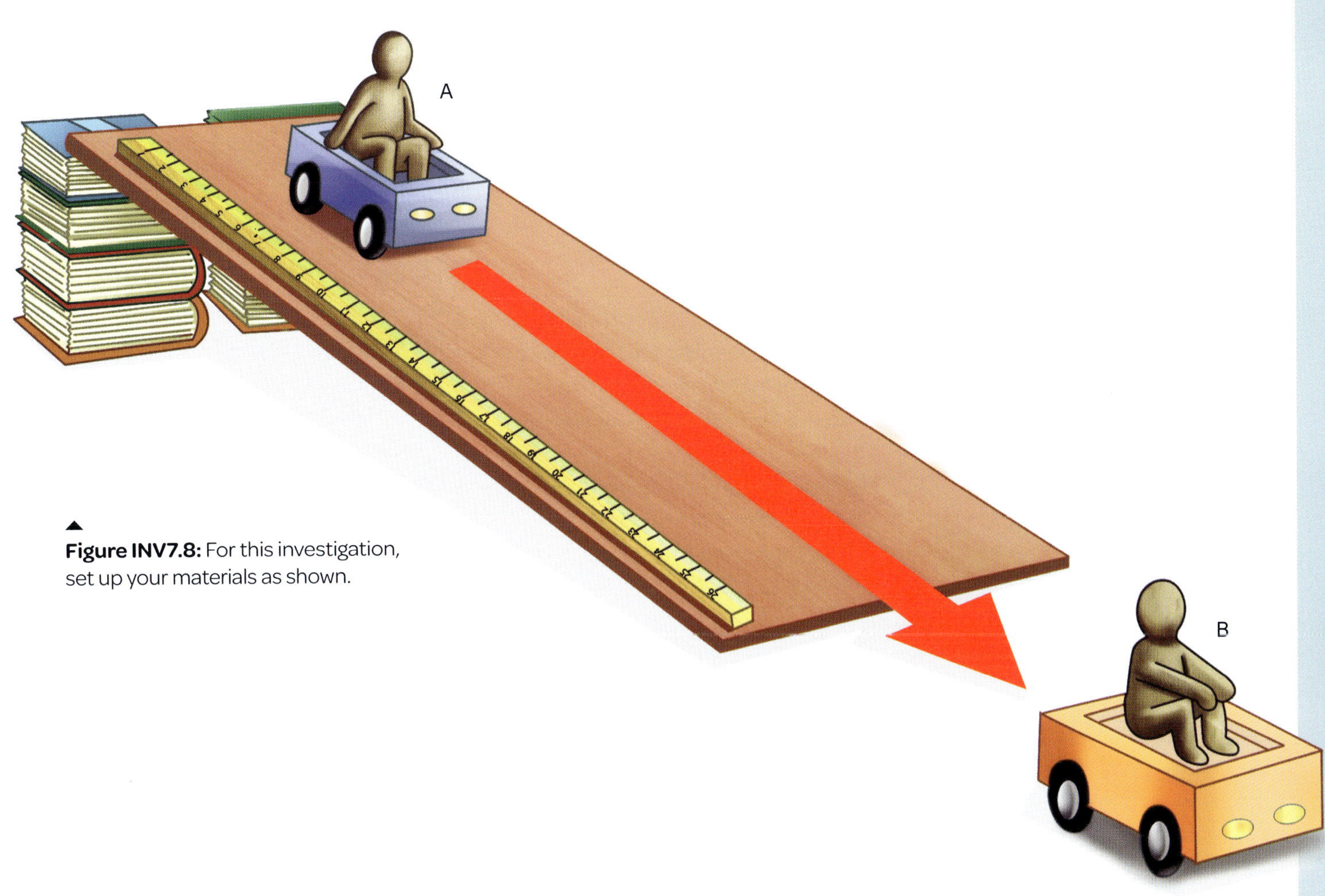

Figure INV7.8: For this investigation, set up your materials as shown.

Glossary

A

abiotic: non-living

absolute dating: determining the age range of an object, such as a fossil, in numbers of years

acceleration: a change in speed over time

accuracy: how close a measured value is to the true, exact value; how closely a recorded value matches the expected outcome of an investigation

acid: a substance with a pH of less than 7

activation energy (E_a): the energy required for successful collisions and a chemical reaction to occur

albedo: the amount of light reflected from a surface

alkali: a base that is dissolved in water

allele: a variation of a gene

alpine: areas of high elevation

amino acid: an organic molecule that is the building block of a protein

ancestral species: an older species from which one or more recent species have evolved

anion: negative ion

anomalies: data points or findings that do not follow the normal trend or expected findings

antimatter: particles that have properties opposite to that of normal matter

assumption: something accepted as true that may not be tested, which impacts the results of an investigation

astronomer: a scientist who studies space, stars and celestial objects

astronomical unit (AU): the average distance between Earth and the Sun (about 150 million kilometres)

astrophysicist: a scientist who studies the physics of the universe

atmosphere: the layer of gas that surrounds Earth

atomic mass: the average mass of the atoms in a sample of an element

atomic number: the number of protons in an atom

average speed: the speed averaged for an entire journey

B

balanced forces: forces acting on an object are equal and cancel out, so that motion remains the same

base: a substance with a pH of more than 7

big bang theory: the current accepted model of the beginning of the universe, where the universe rapidly expanded from a single dense, extremely hot region

biodiversity: the variety of life on Earth; the variety of organisms in an ecosystem

biogeography: the study of the past and present distribution of living organisms

biosphere: all the living things on Earth

black hole: a body with such a strong gravitational field that it attracts all nearby matter

Bohr diagram: a diagram showing the electron configuration of an atom

braking distance: the distance travelled before a car stops after a driver pushes the brake

braking time: the time taken for a car to stop after a driver pushes the brake

C

carbon cycle: the natural cycling of carbon through ecosystems

carbon farming: where carbon storage in soil and vegetation is maximised

carbon footprint: a measure of the amount of carbon dioxide emitted as a result of the activities and choices of an individual, organisation or community

carbon sequestration: the process of capturing and storing atmospheric carbon dioxide

carbonate: a substance containing the elements carbon and oxygen

catalyst: a substance that increases the rate of a chemical reaction without being used up

cation: positive ion

caustic: able to burn or corrode organic tissue through chemical action

cell cycle: the process a cell goes through each time it divides

chemical bond: an electrostatic force that connects atoms to one another

chemical equation: a chemical reaction represented using chemical formulas of reactants and products

chemical formula: an expression of the elements that make up a chemical compound, usually presented as a ratio using letters and numbers – for example, H_2O

chemical property: a property of a substance that is observed during a chemical reaction

chemical reaction: a process in which one or more substances are changed to form new substances; a rearrangement of the way atoms are joined

chemical symbol: a symbol of one or two letters used to represent an element

chromosomal mutation: a mutation caused by a change to a large section of a chromosome

chromosome: a tightly coiled strand of DNA

citation: a way of giving credit to a source, usually in the same text where the information appears

climate: trends in weather over a period of at least 30 years

climate zone: a region with a specific climate

collision theory: a theory which states that for a chemical reaction to occur, particles must collide with the correct orientation and with enough energy; can be used to predict the rate of reactions

comet: chunks of rock and ice moving through space

common ancestor: a previous version of a species that has evolved into two or more recent species

comparative anatomy: the study of similar anatomical structures in different species in order to understand their evolution

comparative embryology: the study of the similarity of the embryos of different species as they develop

compound: a combination of two or more different elements, joined together in a fixed ratio

concave: curved or rounded inwards

concentration: the amount of a substance in a volume of solution

contact force: a force that is applied by touching

controlled variable: variable in an investigation that must be kept the same for all trials. Only the independent variable should change, otherwise it is not a fair test

convection: the transfer of energy by movement of a liquid or a gas

conversion factor: a number used to change one unit of measurement to another

cool burning: a traditional land management practice that involves the controlled use of low-intensity fire

coral atoll: a ring-shaped island formed by a coral reef

coral bleaching: a phenomenon where corals lose their colour due to the absence of symbiotic algae

correlation: a relationship between two factors or variables that shows them changing together, in either the same or opposite way

corrosion: the degradation of a metal due to its reaction with its environment

corrosive: highly reactive and damaging or destructive to another substance

cosmic rays: high-energy particles that move through space at the speed of light

cosmologist: a scientist who studies the origins and structure of the universe

covalent: made up of non-metals

crossing over: the exchange of genes by homologous chromosomes

crystalline solid: a solid with a highly ordered arrangement of particles at the microscopic and macroscopic levels

cytokinesis: the last stage of mitosis, where a cell divides into two identical daughter cells

D

Dalton's atomic theory: the theory that all matter is made up of tiny particles

dark energy: a theoretical form of energy causing the expansion of the universe

dark matter: a theoretical form of matter contributing to the mass of the universe

data: facts and information collected for reference or analysis

data point (datum): a single identified element in a dataset

dataset: a collection of data, often from numerous trials related to a single factor

deceleration: a decrease in speed over time

decomposition reaction: a reaction in which one reactant breaks down to form multiple products

degradation: deterioration of physical properties of a material

density: how heavy something is for its size; mass divided by volume

deoxyribonucleic acid (DNA): the carrier of genetic information; located in the nucleus of a cell

dependent variable: the thing that is measured in a first-hand investigation

diploid cell: normal body cells that contain a complete set of chromosomes ($2n$)

displacement: the distance an object moves from its starting position

displacement reaction: a reaction in which an element replaces (displaces) another element from a compound

dissociate: to split apart into ions in water; to dissolve ionic compounds

distance: the total length an object travels

dominant allele: an allele that only requires one copy to be seen in a phenotype

double helix: the structure of a DNA molecule; a double-stranded spiral

double-displacement reaction: a reaction in which parts of two compounds replace each other to form two new compounds

Dreaming stories: complex First Nations stories that hold significant knowledge about history, creation, law, lore, and many other knowledges; have been used by First Nations Peoples for tens of thousands of years to care for Country and to live well

E

Earth's energy budget: the amount of energy coming into and leaving Earth's climate system

electromagnetic spectrum: all the different electromagnetic waves

electron configuration: the number of electrons in each shell of an atom, starting with the smallest – for example, 2, 8, 8, 2

electron shell: the space around an atom's nucleus, in which electrons circulate at different energy levels

element: a pure substance made of only one type of atom

endothermic: a reaction that absorbs energy in the form of heat

energy efficiency: how much usable energy is produced compared to how much energy has been supplied

energy profile diagram: a diagram that shows energy changes as a reaction progresses from the reactants to products, including activation energy and whether energy is released or absorbed

enhanced greenhouse effect: an increase in the greenhouse effect due to human greenhouse gas emissions

environmental factor: something in the environment that affects gene expression

enzyme: a biological catalyst that increases the rate of reactions in cells; a protein that increases the rate of a specific chemical reaction in the body

equation: a mathematical statement which shows that two things are equal; for example, $2x + 6 = 14$ is an equation that needs to be solved so that $2x + 6$ does actually equal 14

error bar: used on graphs to show uncertainty in repeated measurements

ethical: in science, minimising harm to those involved, and ensuring investigations are conducted honestly and data is collected and recorded accurately

evaluate: to judge value based on scientific evidence

evolution: the way in which organisms change over generations as a result of adaptations that suit their environment

exoplanet: a planet outside our solar system

exothermic: a reaction that releases energy in the form of heat or light

exponent: the superscript value that says how many times to use a number in a multiplication; for example, when we write 10^3, '3' is the exponent. It means we need to multiply 10 by itself 3 times

expression: the process of converting the instructions in DNA into a trait, such as a protein

F

fair test: a test where all variables are kept the same, except for the independent variable and the dependent variable

force: a push or pull between objects that changes the motion speed and/or direction of their motion or their shape

force diagram: a simplified diagram showing the direction and size of forces acting on an object

force multiplier: a machine that increases the force applied to an object

fossil: the geologically altered remains of a previously living organism

fossil record: a record of all fossils that are on Earth

friction: a force that resists motion when two surfaces rub on each other

frost line: a boundary just inside Jupiter's orbit

G

galaxy: a system of millions or billions of stars

gamete: a sex cell – an ovum or a sperm

gas giants: large planets with low density that are primarily made up of hydrogen and helium gas

gene: a segment of DNA, the basic functional unit of heredity

genome: an organism's entire sequence of DNA

genotype: the genetic code for an organism

giant network solid: a continuous network of atoms joined together through covalent bonds (not discrete); also called 'covalent network solid'

glacier: a slowly moving mass of ice formed by the accumulation of snow

Gondwanaland: the supercontinent that broke from Pangaea approximately 250 million years ago and was made up of the modern-day continents of Africa, South America, Australia and Antarctica

gravity: a force of nature where two objects with mass attract each other

greenhouse effect: the trapping of the Sun's heat by Earth's atmosphere

greenhouse gas: a gas that traps the Sun's heat energy in Earth's atmosphere

greenhouse gas emission: the production of a greenhouse gas

groundwater: water that flows underground in spaces between rocks and within soils

group: a column in the periodic table

H

half-range: shows how much a set of experimental values might vary

haploid cell: sex cells that contain half a complete set of chromosomes (n)

hazard: something that can harm living things, objects or the environment

heredity: the passing on of traits from parents to their offspring

heterozygous: a genotype of two different alleles

homologous: chromosomes that have the same genes in the same order

homologous structures: features in different organisms that are similar in structure, but different in function

homozygous: a genotype of two of the same allele

hydrosphere: all the water on Earth

hypothesis: a suggested explanation or prediction of a scientific problem that can be tested with an investigation

I

impact force: the force applied by an object when it impacts another object

independent variable: the thing that is deliberately changed in a first-hand investigation

indicator: a substance used to determine the acidity of a solution

Industrial Revolution: a period in the late 1700s when manufacturing transformed to large-scale factories that were powered by the burning of fossil fuels

inertia: a property of matter that causes it to resist change in speed or direction (to remain at rest or in a state of uniform motion)

infrared light: an electromagnetic wave with longer wavelength than red light

instantaneous speed: the speed at a particular time in a journey

interstellar: the areas of space between stars

inversely proportional: as one quantity increases, the other decreases

ion: an atom that has lost or gained an electron to become charged (positive or negative)

ionic: generally made up of metals and non-metals

ionic salt: a compound made up of a cation and an anion

ionise: to remove an electron from an atom

isolation: when a population is cut off from others

isotopes: atoms of the same element with the same number of protons but a different number of neutrons

K

karyotype: a picture of an organism's full set of chromosomes

kinetic energy: energy of motion

L

land justice: recognising First Nations Peoples' rights to land

landfill: disposal of waste by burying it

lever: a simple machine that increases the force applied

light-year (ly): the distance that light travels in one Earth year

lithosphere: Earth's rigid outer zone (crust and upper mantle), made up of tectonic plates

Local Group: the group of galaxies that the Milky Way is part of

lore: traditions and knowledge about a subject

luminosity: a measure of the brightness of an object

M

magnetic field: a region surrounding Earth where magnetic forces are exerted

magnitude: the size of a measurement

main sequence star: a star that is fusing hydrogen in its core

mass: the amount of matter in an object, measured in grams (g) or kilograms (kg)

mathematical formula: a rule or principle that helps you to find the answer to a question or to understand the relationship between variables

mean: a measure of centre (an average) calculated by adding all the numbers together and dividing by how many numbers there are

median: the middle number in a set of numbers when they are arranged in order

megapode: a class of flightless bird that had large legs and feet and built nesting mounds in which to lay its eggs

meiosis: complex cell division that produces unique haploid gamete cells

meniscus: the curve seen at the top of a liquid in its container

metallic: made up of metals

meteorologist: a scientist who studies the atmosphere and its effects on Earth, including weather patterns

microgravity: a condition where gravity appears to be almost absent, making objects and people seem weightless

mitosis: simple cell division that produces identical cells

mode: the number that appears most frequently in a set of numbers

momentum: a property of motion that is dependent on the mass and velocity of an object

monogenic: influenced by a single gene

mosaic burning: cool burns conducted in patterns to protect habitat and biodiversity

motion: a change in position of an object over time

mutation: an error in DNA that changes the DNA sequence, which can be caused by DNA replication, or cell division

N

natural selection: the process in which organisms that are better suited to their environment tend to survive and reproduce, passing their traits on to further generations

nebula: a vast region of gas and dust

net force: the sum of all forces acting on an object

net ionic equation: a chemical equation that only includes the ions and precipitate involved in the reaction, not spectator ions

net zero emissions: greenhouse gas emissions that are produced are balanced out by those that are absorbed by other processes

neutralisation reaction: a reaction involving an acid and a base to produce water and a salt

neutralise: to make something chemically neutral; neither acidic nor basic

neutron star: the small, extremely dense core left over from the collapse of a star, when matter is compressed into a small space

nitrogenous: containing nitrogen

noble gas configuration: the valence electron configuration that occurs through chemical bonding

non-contact force: a force that is applied without touching

non-uniform motion: a motion where there is a change in speed or direction

normal force: the force of the ground pushing up in opposition to gravity

nuclein: the name given to DNA when it was first discovered

nucleotide: the building block of DNA, consisting of a sugar, a phosphate and a nitrogenous base (adenine, guanine, thymine or cytosine)

O

ocean acidification: a decrease in the pH of the oceans due to the absorption of more carbon dioxide

ocean circulation: the movement of water in the oceans due to major currents

octet rule: the tendency of atoms to achieve eight electrons in their valence (outer) shell to be stable

orbit: the curved path of an object as it travels around a planet or moon

P

parallax error: the apparent shift in something's position when it is viewed from different angles

pattern (data): when data repeats in a predictable way

payload: the products carried into orbit by the space shuttle

pedigree chart: a chart that shows the inheritance of a trait through generations of a family

percentage error: measures how close the experimental value of an investigation is to the theoretical or true value

period: a row in the periodic table

periodic table: a table of all known elements and their chemical symbols

pH: a figure expressing the acidity or alkalinity of a solution

phenotype: how the genotype is physically expressed

physical property: a property of a substance that can be observed and measured, such as colour, texture, melting and boiling points, density and hardness

plagiarise: to copy someone else's work and present it as your own

plasmid: a small circular DNA molecule in bacteria and some other microscopic organisms

plausible: could be reasonably accepted based on available evidence

point mutation: a mutation caused by a change to a single nucleotide base

polygenic: influenced by two or more genes

population: a group of organisms of the same species that live and interact within the same area

power: the result of multiplying a quantity by itself one or more times; for example, 2 to the *power* of 3 is 2 × 2 × 2, which is 8, so the *power* of 2^3 is 8

precipitate: an insoluble product that forms a solid in solution

precipitation: liquid or solid water that forms in the atmosphere and falls to Earth's surface

precipitation reaction: a type of double-displacement reaction that forms a precipitate when two solutions are combined

precision: how close measured values are to each other within a dataset

predation: when an organism preys on another organism within an ecosystem

proportional: as one quantity increases, so does the other

proteins: long polymer chains made up of amino acid monomers

protostar: a contracting mass of gas that is in the early stages of forming a star

Punnett square: a diagram that shows the possible genotypes of offspring

R

range: in a set of numbers, a measure of spread between the highest number and the lowest number

ratio: a way of comparing like quantities without units

raw data: data that is collected directly from the investigation

reaction distance: the distance travelled during the reaction time

reaction time: the time taken to react

reactive: undergoes chemical reactions easily

recessive allele: an allele that requires two copies to be seen in a phenotype

red giant: a star that has stopped fusing hydrogen in its core

redshift: a change in the wavelength of light towards the red end of the visible spectrum

relationship: a link between two factors

relative dating: arranging geological objects and rocks in a sequence from oldest to youngest

repeatable: similar results can be obtained using the same method and equipment

replicable data: data that can be reproduced using the same method and equipment

reproducible: similar results can be obtained by someone else using the same method and equipment

resolution: the ability of a telescope to identify the light of two objects that are close together as separate objects

ribonucleic acid (RNA): a nucleic acid similar to DNA that helps to produce proteins from the genes of DNA

rusting: the corrosion of iron to form iron oxides

S

scale factor: the ratio between corresponding measurements of an object and a copy of that object

scientific notation: a way to write very large or very small numbers in a simple form

selection pressure: a factor that can influence which traits are beneficial to the survival of organisms

significant figures (or sig figs): digits in a number that indicate the precision of a measurement

single-displacement reaction: a reaction in which a more reactive element replaces a less reactive element from a compound

singularity: an infinitely dense point of matter that existed before the big bang

solar mass: a unit of measurement based on the mass of the Sun

solar system: a system of planets that orbits around one or more stars, held together by gravity

spearthrower: a device used to increase the velocity of a spear (known as a woomera in some parts of Australia)

spectator ion: an ion that does not take part in the reaction

speed: the distance an object travels divided by the time taken to travel that distance

spiral galaxy: a spiral-shaped galaxy with 'arms' that extend out of its dense, rotating centre

star: a burning ball of gas, mostly hydrogen and helium, held together by its own gravity

stopping distance: the sum of the reaction distance and the braking distance

subscript: a letter or number written slightly below and to one side of another – for example, '2' in H_2O

supernova: the explosion of a massive star at the end of its life

surface area: the area of the outermost layer of an object

surface tension: where the molecules at the surface of a liquid are more attracted to each other than to the air above the liquid

symbiotic relationship: an ecological relationship where both parties cannot survive without the other

synthesis reaction: a reaction in which two or more reactants combine to form a more complex product

T

telescope: a device that uses mirrors and lenses to magnify the size of objects

theory of evolution by natural selection: a theory that says species have evolved adaptations over time that helped them to survive in their environment

thermal expansion: the increase in volume of a substance due to an increase in temperature

thermohaline circulation: the movement of ocean currents due to differences in temperature and salinity in different regions of water

transcription: the process in which a cell makes an RNA copy of a piece of DNA

trend (data): when data moves in a general direction, usually up or down

true value: the actual, exact value of a measurement, error-free, which would be obtained if a perfect measurement were made

U

ultraviolet light: an electromagnetic wave with shorter wavelength than violet light

unbalanced forces: forces acting on an object are different in size and do not cancel out but cause a change in motion

uncertainty: a measure of the extent to which data is known (i.e. its variability)

uniform motion: a motion with constant speed and direction

V

valence electron: an electron in the outermost electron shell

valence shell: outermost electron shell (furthest from the nucleus)

validity: when an investigation meets its intended purpose

variable: a factor in an investigation or a model that can be changed, measured or controlled

variation: the difference between individuals in a species, in both genetics and phenotypes

vector: a quantity that has direction and magnitude

velocity: a measure of how quickly displacement changes

vestigial structure: a feature of an organism that has lost some or all of its function through evolution

visible spectrum: the part of the electromagnetic spectrum that humans can see

W

water cycle: the cycle of processes by which water circulates between Earth's oceans, atmosphere, land and biosphere

weather: what is happening in the atmosphere at a specific place and time

weight: a force acting downwards on a mass due to gravity

white dwarf: a small, very dense star formed at the end of a small star's lifetime

word equation: a representation of a chemical reaction using the names of the reactants and products

Z

zooxanthellae: algae that live in coral tissues

Index

D

E

H

I

J

K

L

M

N

S

T

U

V

W

X

Z

Acknowledgements

The authors and publisher are grateful to the following for permission to reproduce material:

Cover: Fabrice Coffrini/Getty Images

Images: APP Image /Alastair Grant, **137** /JEAN-CHRISTOPHE BOTT/EPA, **2–3**; Adobe Stock /Aaron, **342** (top) /Alano Design, **268** /Aldona, **26** (top), **52** /Alessandro, **99** (top) /Alexander Kaludov, **26** (bottom) /AlexMas, **72 & 81** (diamond) /Ali, **10** /amnaj, **294** /Andrey Popov, **267** /Andrey Solovev, **245** (right) /Around Ball, **43** (left) /Baillie Photography, **55** /blacksalmon, **222** /bsd studio, **266** /Cavan, **32** /Chalabala, **117** (left) /chones, **235** /Chris Jiacheng Sun, **138** (bottom), **226–27** /closeupimages, **285** /comicsans, **245** (centre) /concept w, **63** (centre), **66–67** /DC Studio, **301** /dimlight, **231** /Douglas, **198** (top) /Eric Isselee, **vii** (left & right) /filin174, **345** /Frank, **375** /freshidea, **239** (top) /Gabrielle, **135** (top) /geptays, **18** (right) /havepino, **117** (right) /hodagmedia, **36** (bottom) /illustrissima, **72 & 81** (carbon dioxide) /Inna, **370** /Jessada Boonlao, **268** /jkraft5, **58** /joy666, **326** /kateina, **305** (ruler) /KKT Madhusanka, **38** (right) /konradbak, **215** (right) /ktsdesign, **225** /Leah-Anne Thompson, **120**, **145** /lial88, **212** /Liaurinko, **289** /librakv, **72 & 81** (silver) /Lifeking, **4** /Lightning Strike Pro, **360** /Luis, **51** (bottom) /Luke, **281** /Marek R. Swadzba, **40, 56** (bottom) /Mediteraneo, **104** /moho_cp, **276** /myphotobank.com.au, **139**, **151** (top), **224** /onairjiw, **330** /p a w e l, **116** /P.j.Hickox, **48** (top) /Paitoon, **332** /Paul Reverve, **85** /Pavlo Vakhrushev, **105** /peefay, **13** /petrroudny, **87** (top), **107, 110** (right) /PH-HY, **20, 31** (top right) /Phoenix Icons, **268** /Photo Sesaon, **197** (top) /PhotoSG, **72 & 81** (glucose) /pikovit, **11** /pressmaster, **265** /radiorio, **366** /sablinstanislav, **140** (top right) /Sanhanat, **254** (right) /Sasha, **140** (bottom) /Sensvector, **245** (left) /skumer, **190** (top) /south west images, **126** (left) /spinetta, **297** /sudok1, **72 & 81** (oxygen) /Supernova, **156** /surassawadee, **72 & 81** (marble) /Swapan, **350** /Syda Productions, **190** (bottom) /Tasha Vector, **23** (centre), **24**, **31** (top left & centre) /tendo23, **72 & 81** (silica) /Tim, **234** /Tony Skerl, **357** /Tunatara, **146–47** /Tunatara, **v** /VectorMine, **106**, **122** (bottom), **123**, **125** /Vencav, **38** (left) /vesvocrea, **264** /virojt, **229** /vitalyzorkin, **305** (pencil) /vladim_ka, **98** (left) /Warning Signs, **99** (bottom) /Winai Tepsuttinun, **72 & 81** (sulfur) /Winston Link, **306** (left) /yanik88, **202** (top) /Yuliya, **16** /Zoran Zeremski, **230**; Alamy /Andrew Brookes, **33** /Andrew Michael, **220–21** /blickwinkel, **12** (top), **19** (bottom) /Corbin17, **53** (top) /Donna Ikenberry/Art Directors, **34** /dpa picture alliance, **83** (top right) /E.R. Degginger, **97** (bottom left) /Excitations, **247** (bottom) /Geopix, **152** /Geopix/NASA/Matthew Dominick, **177**, **185** (bottom left) /George Blonsky, **22** /GRANGER Historical Picture Archive, **166** (right), **173** (left), **184** (centre right) /H. Armstrong Roberts/ClassicStock, **167** (top left), **184** (top left) /Historic Images, **75** (bottom right) /Howard Chew, **147** (centre left) /ImageDJ, **244** /ImagineThat, **279** /Inaki Relanzon/Nature Picture Library, **51** (top) /Jean-Paul Ferrero/AUSCAPE, **135** (bottom) /KamilSD, **60–61** /Kateryna Kon/Science Photo LIbrary, **21** (bottom) /Lee Hudson, **209** (bottom) /Library Book Collection, **54** /Maurice Savage, **261** /Mike Veitch, **147** (centre) /Monica Schroeder, **14** /Olga Yastremska, **336** /PCN Black/PCN Photography, **193** (top) /Penny Tweedie, **48** (bottom) /Penta Springs Limited, **211** (top) /Pictorial Press, **164** (left), **167** (top right) /Pictures from History, **166** (top left) /Planet Observer, **146** (top) /Purple Marbles Garden, **314** /Rob Jones, **165** /S.E.A. Photo, **124** (top) /Schwab Lukas/Prisma by Dukas Presseagentur GmbH, **39, 56** (centre) /Science History Images, **29, 31** (bottom right), **44**, **57** (centre right), **75** (top left) /Science Photo Library, **9** (top), **75** (bottom left) /Science Picture Co, **277** /sciencephotos, **70** (bottom) /Scott Camazine, **8–9** (bottom) /seanscott/RooM the Agency, **36** (top) /Seumas Christie-Johnston, **35** /Stefen Binke, **148** (top right) /Stephanie Jackson, **144** /Steve Gschmeissner/Science Photo Library, **8–9** (top) /Terra Images, LLC, **18** (left) /The Print Collector, **317** (bottom) /Trent Townsend, **148** (top left) /Trevor Clifford Photography/SPL, **69** (top), **80** (bottom right) /Universal Images Group North America LLC, **167** (bottom right) /Wavebreak Media, **269** /World History Archive, **41** /zerocreatives/Westend61, **204–5**; © The Trustees of the British Museum, **166** (bottom left), **209** (top); CSIRO, photo by Alex Cherney, **172** (top); Dreamstime /Pavle Matic, **76** (top); Designer inventor: Duncan Foster Fitzsimons, Photo credit: Colin Ross, **83** (bottom right); Event Horizon Telescope Collaboration, **181** (bottom); ESA /ATG medialab, **163** (top) /C. Letelier/ESO, **171** /NASA, ESA, J. Dalcanton (University of Washington, USA), B. F. Williams (University of Washington, USA), L. C. Johnson (University of Washington, USA), the PHAT team, and R. Gendler, **164** (top); ESO, **178**; Fairfax /Craig Abraham, **132** (bottom); Getty Images /Amy Toensing, **47** /Auscape, **48** (centre) /Jason LeVeris, **83** (left) /Monica Bertolazzi, **160** (top) /Tim Wright, **189**; iStock, **364** /antpkr, **317** (blunt pliers) /arisotoo, **124** (bottom), **150** (top) /Ayakochun, **306** (right) /Betka82, **317** (crucible tongs) /blueingmedia, **214** (bottom), **217** (bottom right) /BookyBuggy, **317** (kitchen tongs) /burroblando, **253** /Children of Dune, **302** (top & left) /Chris Gordon, **136** /Dimitry_Chulov, **208** /DoraDalton, **210**, **217** (centre) /Douglas Cliff, **138** (top) /fotoVoyager, **71** /Ifor Suka, **140** (top left) /jeridu, **317** (forceps) /Kanawa_Studio, **100** (top) /Ken Griffiths, **252** /Lemonan, **100** (bottom) /Lo-So-Ma, **86** /miljko, **219** (bottom) /NicolasMcComber, **196** /niunlu, **87** (bottom) /P_Wei, **215** (left) /Peopleimages, **241** /Pongasn68, **369** (top) /recep-bg, **371** /shapecharge, **274** /simonkr, **84** /Sue Thatcher, **214** (top) /supsktiypumpy, **247** (centre top) /Tenedos, **153** /Vitus72, **88** /vm, **219** (top); Mark Godfrey /The Nature Conservancy, **142**; Narelle MacDonald, **iv**; NASA /Astronaut Jasmin Moghbeli services microbe samples by NASA Johnson, CC BY-NC-ND 2.0, **168** (top); ESA/Hubble, M. Kornmesser, **162** (bottom); NSA/ESA, **175** (top); NASA/ESO, **179** (top); NASA, **122** (top globe), **157** (top), **175** (bottom), **176** (top & bottom), **180** (bottom), **184** (bottom), **185** (top), **282**; NASA / WMAP Science Team, **157** (bottom); NASA Earth Observatory, **149**; NASA, ESA, and S. Beckwith (STScI) and the HUDF Team, **187**; NASA, ESA, and The Hubble Heritage Team (STScI/AURA)-ESA/Hubble Collaboration, **362**; NASA, ESA, and the Hubble SM4 ERO Team, **161** (bottom), **182** (top), **185** (bottom right); NASA, ESA, CSA, STScI, **167** (bottom left), **172** (bottom), **173** (right), **184** (top right); NASA, ESA, CSA, STScI, Webb ERO Production Team, **154–55**; NASA, ESA, G. Dubner (IAFE, CONICET-University of Buenos Aires) et al.; A. Loll et al.; T. Temim et al.; F. Seward et al.; VLA/NRAO/AUI/NSF; Chandra/CXC; Spitzer/JPL-Caltech; XMM-Newton/ESA; and Hubble/STScI, **183** (left); NASA/ Goddard Space Flight Center Scientific Visualization Studio, The Blue Marble data is courtesy of Reto Stockli (NASA/GSFC), **132** (top & centre), **151** (bottom right); NASA/Jacques Descloitres, MODIS Rapid Response Team, NASA/GSFC, **118–19**; NASA/JPL-Caltech/MSSS, **169**; NASA/JPL-Caltech/SSC, **180** (top); NASA/SDO, **161** (top); NASA/Timothy Kopra, **174**; NASA/ JPL, **168** (bottom); NASA/National Aeronautics and Space Administration, Science Mission Directorate. (2010). Introduction to the Electromagnetic Spectrum. Retrieved [insert date – e.g. August 10, 2016], from NASA Science website: http:// science.nasa.gov/ems/01_intro, **170**; NASA/SpaceX, **206** (top); Nature Picture Library /Tony Heald, **143** / Tui De Roy, **59**; Science Photo Library, **76** (bottom), **113**, **121** (left), **239** (bottom) /A. Barrington Brown © Gonville & Caius College/Coloured by Science Photo Library, **28** (bottom) /Andrew Lambert Photography, **91** (top right) /Beauty of Science, **62** /Charles D. Winters/Science Source, **108** (top) /Claus Lunau, **162** (top) /David Nunuk, **163** (bottom) /Eckhard Slawik, **183** (right) /European Space Agency, **158** (right) /GIPHOTOSTOCK, **89** (top), **90** (bottom), **93** (top), **96** (bottom), **98** (top left), **102**, **110** (left) /Jim West, **53** (bottom) /Jurgen Freund/Nature Picture Library, **147** (centre right) /Martyn F. Chillmaid, **68**, **69** (bottom), **95** (right), **97** (top & bottom right), **98** (bottom left), **199**, **240**, **302** (bottom right), **338**, **342** (bottom) /Maurico Anton/Science Photo Library, **46** /Petr Jan Juracka, **63** (top) /Science History Institute, **77** /Tim Brown, **163** (centre), **179** (bottom) /Trevor Clifford Photography, **236** /Turtle Rock Scientific LLC, **243**; Science Source /Spencer Sutton, **43** (right), **57** (top); Shutterstock /Darkydoors, **114–15**; Wellcome Collection (CCBY 4.0), **74** (top), **75** (top right).

Other material: The Victorian Curriculum F-10 content elements are © VCAA, reproduced by permission. The VCAA does not endorse or make any warranties regarding this resource. The Victorian curriculum G-10 and related content can be accessed directly from the VCAA website, https://f10.vcaa.vic.edu.au/, **vi–1.**

While every care has been taken to trace and acknowledge copyright, the publisher tenders their apologies for any accidental infringement where copyright has proved untraceable. They would be pleased to come to a suitable arrangement with the rightful owner in each case.